Irrigation and Soil Nutrition

Irrigation and Soil Nutrition

Edited by Donald Cronin

SYRAWOOD
PUBLISHING HOUSE

New York

Published by Syrawood Publishing House,
750 Third Avenue, 9th Floor,
New York, NY 10017, USA
www.syrawoodpublishinghouse.com

Irrigation and Soil Nutrition
Edited by Donald Cronin

© 2019 Syrawood Publishing House

International Standard Book Number: 978-1-68286-669-6 (Hardback)

Cataloging-in-Publication Data

Irrigation and soil nutrition / edited by Donald Cronin.
   p. cm.
Includes bibliographical references and index.
ISBN 978-1-68286-669-6
1. Irrigation. 2. Soils, Irrigated. 3. Soils and nutrition. I. Cronin, Donald.
S613 .I77 2019
631.587--dc21

TABLE OF CONTENTS

Permissions

List of Contributors

Index

PREFACE

Irrigation refers to the controlled application of water to agricultural crops and plants to ensure healthy growth and production. A soil is considered ideal for agriculture if it can sustain and yield crops of high quality. To ensure this, the soil must be sufficiently irrigated and provided the essential nutrients required for plant growth. Ensuring suitable soil pH, internal drainage, thriving population of a range of microbes that support growth are some other considerations of a good agricultural practice. These can be achieved by strategic use of nutrients in the form of fertilizers, organic matter and the judicious use of water. This book presents the recent developments in this domain with an emphasis on irrigation water management and good agricultural practices to improve crop health and soil fertility. From theories to research to practical applications, studies related to all contemporary topics of relevance to this field have been included in this book. It will prove to be immensely beneficial to professionals and students in this field.

The researches compiled throughout the book are authentic and of high quality, combining several disciplines and from very diverse regions from around the world. Drawing on the contributions of many researchers from diverse countries, the book's objective is to provide the readers with the latest achievements in the area of research. This book will surely be a source of knowledge to all interested and researching the field.

In the end, I would like to express my deep sense of gratitude to all the authors for meeting the set deadlines in completing and submitting their research chapters. I would also like to thank the publisher for the support offered to us throughout the course of the book. Finally, I extend my sincere thanks to my family for being a constant source of inspiration and encouragement.

Editor

Simulation Study of Yield and Soil Water Balance Responses of a Maize Crop to Farmers' Irrigation Scheduling Practices in Tanzania

HE Igbadun[1]* and BA Salim[2]

[1]*Department of Agricultural Engineering, Ahmadu Bello University, P.M.B. 1044, Zaria, Kaduna State, Nigeria*
[2]*Department of Agricultural Engineering and Land Planning, Sokoine University of Agriculture, P.O. Box 3003, Chuo Kikuu, Morogoro, Tanzania*

Abstract

Maize (*Zea mays. L*) farmers in the traditional irrigation schemes in middle Mkoji sub-catchment, Tanzania observes three irrigation scheduling practices. This paper presents a simulation study of the impacts of these scheduling practices on yield and soil water balance of the maize crop. The three scheduling practices include irrigating at 5 days and 7 days intervals throughout the crop growing season, respectively; and irrigating at 7 days interval from planting to vegetative growth stage and 5 days interval from flowering to crop maturity stages. ISIAMOD, a crop growth cum irrigation simulation model, was used to simulate grain yield, soil water balance components and crop water productivity responses for the three scheduling practices over a range of water application depths. Simulated grain yield varied from 1338 to 4023 kg/ha. Seasonal water applied and deep percolation varied from 425 to 1800 mm, and 50 to 113 mm, respectively. The crop water productivity in terms of water applied varied from 0.22 to 0.57 kg/m3. These values closely agree with field measured values reported by some researchers for the study area. Irrigating maize fields at 5 days interval throughout the crop growing season or at flowering to crop maturity gave higher water productivity output only when application depths per irrigation did not exceed 30 mm. Water application beyond this depth only led to very high deep percolation losses without appreciable difference in crop yield compared to irrigating at 7 days interval throughout the crop growing season. Moreover, the productivity of water applied dropped by about 30 and 50 %. This implies that farmers who irrigate at 5 days interval because of they have access to water do not have any advantage (in terms of yield and water productivity) over those who irrigate at 7 days interval except they minimize water applied to their fields. Water application depth for higher productivity under the 7 days irrigation interval for the maize crop in the study area was 40 to 45 mm depth. Beyond this depth, there was no appreciable increase in grain yield but a fall in productivity of applied water and a buildup of deep percolation. To avoid over irrigation and the consequences associated with it, maize farmers at any sector of the irrigation scheme in the study area are advised to observe 7-day irrigation interval and keep water application depth within 40-45 mm per irrigation.

Keywords: Irrigation practice; Maize crop; Simulation model; Crop water productivity

Introduction

Traditional and smallholders irrigation schemes are very common in Tanzania. These schemes usually cover relatively small areas of 5 to 50 ha, and are managed by the groups of farmers in the communities via Water Users Association. These schemes are common around the mountainous eastern regions of Tanga, Kilimanjaro, and Arusha and also in south western regions of Morogoro, Iringa, and Mbeya. There source of water are the perennial/semi-perennial streams and rivers that originate from the mountains and hill tops around the schemes. Crop cultivated in these schemes include paddy, maize and vegetables. The numbers of these schemes are on the increase in those regions and are contributing meaningful to the rapid growth of irrigated agriculture in the country largely because of government, foreign and local organizations are assisting many of the schemes to develop the intake structures to abstract water from the streams and rivers.

One of the very active traditional irrigation schemes in the South-western region of the country is the Igurusi ya Zamani Irrigation Scheme (IZIS) located in Igurusi village of Mbeya Region. IZIS is community-managed, and water abstraction from the networks of perennial streams flowing through the scheme follows a 7-day rotational arrangement so that farmers at the middle- and down-stream of the scheme can have access to water. However, some farmers at the upstream irrigate their maize fields every five days as against weekly. Some others observe weekly irrigation until the maize begins to tassel, and thereafter turn to 5 days interval until the crop matures. They take advantage of the fact that their fields are located at the upstream of the water source to irrigation more frequently. However, their activities sometimes deprive those downstream ample accesses to water. A

field survey on water application regimes by farmers in the study area showed that farmers upstream apply water as much as 60 to 70 mm per irrigation while those with limited access downstream irrigate as low as 25 mm, as they minimize water application depth to ensure that the available water spread over their command area. A knowledge gap has remained on the implications of these different irrigation schedules practiced by the farmers on yield, soil water balance components and crop water productivity in the study area.

The effect of irrigation regimes on crop yield, soil water balance and water use efficiency varies with crops, soils and water application depths. Crop yield and water use response to irrigation is known to be climate specific; hence the continuous study of the subject by different researchers. Singh et al. [1], Al-Jamal et al. [2], Imtiyaz et al. [3], Camposeo and Robino [4], Mermoud et al.[5], Sun et al.[6], Nazeer [7], Ayana [8], Quanqi et al. [9] are few examples of such studies. According to Kang et al. [10], the responses of grain yields and water use efficiency to irrigation varied considerably due to differences in soil water content and irrigation schedule. Mermoud et al. [5] has also reported that the

*Corresponding author: HE Igbadun, Department of Agricultural Engineering, Ahmadu Bello University, P.M.B. 1044, Zaria, Kaduna State, Nigeria
E-mail: igbadun20@yahoo.com

impact of irrigation regimes on yield and water use vary with climate. They reported that in Camboiné, Burkina Faso, West Africa, irrigating onion twice a week instead of once, leads to increase in root zone water storage, a better crop water availability throughout the whole root zone and higher yield. They also argued that changing irrigation frequency have a strong influence on the components of the water balance. A decrease in the irrigation frequency causes a decrease in the water stored in the root zone, an augmentation of the crop transpiration, and a decrease of the water content in the immediate vicinity of the soil surface leading to reduced evaporation. Jin et al. [11] also reported that excessive irrigation leads to decrease in crop water use efficiency (WUE) and that effective deficit irrigation may result in higher production and WUE. Sun et al. [6] supported this view when he noted that excessive irrigation might not produce greater yield or optimal economic benefit, and advised that suitable irrigation be established for crops, soil types and specific climates.

The general objective of this study was to provide quantitative information on yield and soil water balance responses to the different irrigation scheduling practiced by farmers in the IZIS, Tanzania. The specific objectives were to use a computer-based simulation model to simulate grain yield, soil water balance (seasonal water applied, evapotranspiration and deep percolation) and crop water productivity of a maize crop cultivated based on the aforementioned irrigation scheduling practiced by farmers in the area, and to analyze the effect of the scheduling protocols on the simulated variables. The study is aimed at providing insight to the effect of the irrigation scheduling practiced by the farmers and to see if there is a window of opportunity for farmers to regulate water utilization at field level, and release excess of what they actually need to other water users in the downstream sector of the scheme.

Materials and Methods

The study location

The Igurusi ya Zamani irrigation scheme (IZIS) in the middle Mkoji sub-catchment (MSC) of the Great Ruaha River Basin, Mbeya Region of Tanzania. The River Basin itself is also a sub-catchment of the Rufiji River Basin which empties its water into the India Ocean. Figure 1a shows the map of the Mkoji sub-Catchment and Igurusi village located. Inserted is the map of Tanzania, showing the location of MSC. The IZIS lies on latitude 8.33°S, and longitude 33.53°E, and an altitude of 1100–1120 m above mean sea level. The Mkoji sub-catchment covers an area of about 3400 km² [12]. The study area has

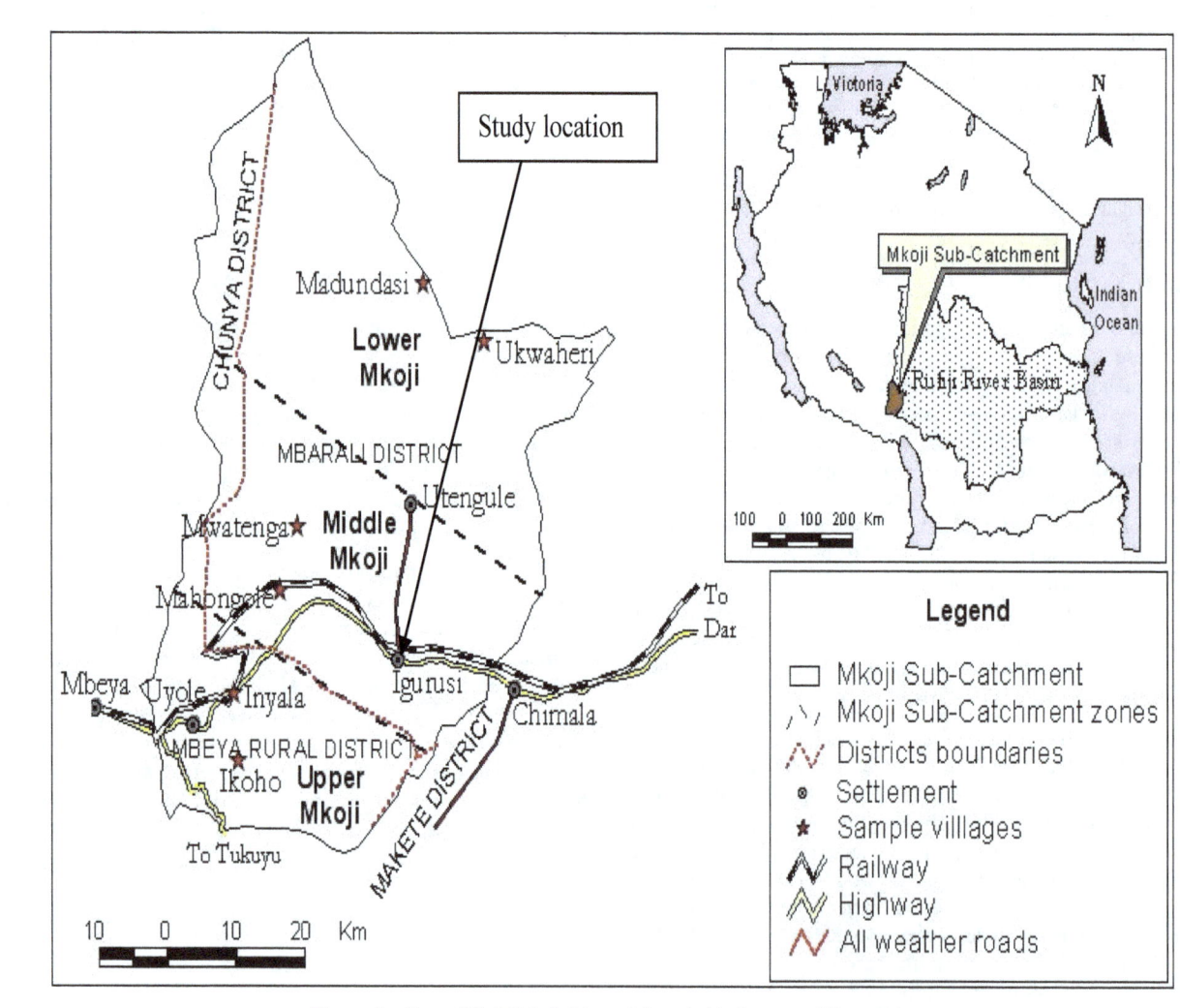

Figure 1a: Map of Mkoji Sub-Catchment (Inserted is the map of Tanzania).

a unimodal rainfall pattern which occurs between October and April with mean annual rainfall of 800 mm. The months of May to October are usually dry, but the weather favors the cultivation of arable crops like maize, cowpea, vegetables and fruits under irrigation. The mean daily maximum temperatures range from 28°C to 32°C, while the mean daily minimum temperatures range from 9.5°C to 19.5°C, respectively. The highest temperature values are recorded in October and November while the lowest values are experienced in June and July. The mean daily net solar radiation varies from 7.5 MJ/m²/day to 12.3 MJ/m²/day. The average annual open pan evaporation is about 2430 mm and the total open pan evaporation from June to October when dry season irrigation takes place is about 1080 mm [13]. The soils of the scheme vary from sandy clay to clay loam. The water holding capacity of a typical soil profile is about 97 mm/m with an average bulk density of 1.42 g/cm³.

The computer simulation model used for the study

The computer simulation model used for this study was the Irrigation Scheduling Impact Assessment Model [14, 15]. ISIAMod consist of eleven modules which were integrated in hierarchical manner to simulate crop growth process, soil water balance of a cropped field, and water management response indices (WMRI) which are used to explain the impact of an irrigation scheduling decision. The input data required in the model include weather, soil, crop, rainfall, and irrigation scheduling decisions. The minimum weather data required are daily maximum and minimum ambient temperatures for the duration of crop growth. Other weather parameters may include wind speed, maximum and minimum relative humidity, sunshine hour or solar radiation. The model uses the weather data to simulate reference evapotranspiration either by Penman-Monteith or Hargreaves Methods [6], depending on available data. The soil input data include volumetric soil moisture content at field capacity and at wilting point, initial soil moisture contents, bulk density, and the percentage of sand in the soil texture. The crop input data include maximum rooting depth, maximum leaf area index, potential (non-water limited) harvest index, radiation use efficiency (RUE), radiation extinction coefficient, and peak crop water use coefficient (K_c). Others include crop base and optimum temperatures; leaf area index shape factors; water-limited harvest index adjustment factors; crop planting, emergence, and physiological maturity dates; days from planting for the start of each of the four crop growth stages, and fraction of the crop growth duration at which leaf area index started to decline. The four crop growth stages to be used in the model are crop establishment, vegetative, flowering and maturity (which include seed formation through to maturity). A unique feature of the model which makes it an improvement on existing model is the WMRI modules which generate the waters accounting indices, crop productivity indices and the seasonal relative deficit/losses indices used to define the level of impact of an irrigation scheduling decision on the crop and the environment.

ISIAMOD runs on daily time step from planting to maturity dates which are entered as part of the crop input data. The output simulated by the model include crop growth response like leaf area index, crop rooting depth, crop biomass, final harvest index and grain yield; soil water balance components such as daily soil moisture content, evaporation, transpiration, runoff, deep percolation, and rainfall interception. The crop yields and water balance components outputs are further processed by the model to generate the water management

response indices. The detailed of model development, calibration and validation for a maize crop in the study area has been reported [14, 15].

Simulation procedure

The three irrigation scheduling practices that were simulated include 5-day irrigation interval labeled E5V5F5G5 (E stands for establishment, V for vegetative, F for flowering, and G for grain filling growth stages, and the numbers stands for the irrigation intervals); 7-day irrigation interval labeled (E7V7F7G7), and 7-day intervals at establishment and vegetative growth stage and a switch to 5 days interval at flowering and grain filling labeled E7V7F5G5. The weather data inputted into the model were daily maximum and minimum air temperature, relative humidity, wind speed and sunshine hour obtained from Igurusi Weather Station located about 4 km away from the irrigation scheme. The average weather data for 10 years (1985-1994) covering the crop growing season is presented as Table 1. The soil input data (Table 2) were those obtained from the research fields of the

Month of the year	Max. Temp (°C)	Min. Temp (°C)	Max. Rel. humidity (%)	Wind speed (m/sec)	Sun Shine (hr)
Jun	27.6	12.1	68.5	1.0	9.4
Jul	26.8	10.7	61.1	1.1	9.3
Aug	28.3	12.2	60.0	1.3	9.0
Sep	30.0	13.3	59.6	1.4	8.5
Oct	31.2	15.6	57.6	1.4	7.1

Table 1: Weather data from Igurusi Weather Station used to run the model (average of 10-year: 1985-1994).

Soil profile depth (mm)	MC @ field capacity (m³/m³)	MC @ wilting point (m³/m³)	Bulk density (dry) (g/cm³)	Clay %	Silt %	Sand %	Soil Textural Class
0-150	0.262	0.127	1.44	19	18	64	Sand loam
150-400	0.295	0.163	1.39	31	17	52	Sand clay loam
400-700	0.305	0.226	1.45	33	22	45	Sand clay loam
700-1000	0.278	0.212	1.38	36	19	45	Sandy clay

Table 2: Soil physical properties of the research fields of MATI.

Parameters	Value
Maximum rooting depth	1.2 m
Maximum harvest index	0.34
Harvest index adjustment factor for the flowering stage	0.45
Harvest index adjustment factor for the maturity stage	0.5
Radiation extinction coefficient	0.55
Maximum leaf area index	0.35m²/m²
RUE (establishment and vegetative stages)	0.25 g/MJ
RUE (flowering and maturity stages)	0.23 g/MJ
Base temperature	8°C
Optimal temperature	24°C
Fraction of the growth duration at which leaf area index starts to decline	0.75
Days after planting at which establishment growth stage starts	0
Days after planting at which vegetative growth stage starts	23
Days after planting at which flowering growth stage starts	64
Days after planting at which maturity growth stage starts	93
Peak crop water use (kc) coefficient	1.2
Soil dependent transpiration constant	0.018 mm/day
Evaporation coefficient for bare soil	1.05

Source: [14]

Table 3: Crop and other input parameters used for running the model.

Ministry of Agriculture Igurusi (MATI). The crop input data (Table 3) were those used to calibrate the ISIAMOD [16].

The three irrigation schedules were simulated over water application depths ranging from 25 to 70 mm at an increment of 5 mm; thus making a total of 10 simulation runs per schedule and 30 simulation runs per cropping season. The simulation was done for 10 cropping seasons covering the period of available weather data (1985-1994). The total simulation runs in the study were 3 scheduling practices x 10 water application depths x 10 seasons = 300. It was assumed in each run that soil moisture content at planting was at field capacity. Planting was in the third week of June and crop attained physiological maturity was at 120 days after planting. The maize variety was TMV1-ST.

Data analysis

The model output variables that were analyzed in this study were grain yield, seasonal water applied, evapotranspiration, deep percolation and crop water productivity. Two-way Analysis of Variance tests were carried out for each of the output variables to study the effect of the water application depths and the scheduling practices on the output variables. The ten years simulation runs were regarded as replications in the analyses. The pooled data (averages of the 10 years simulation runs) for each of the output variables are presented graphically in the result and discussion section.

Results and Discussion

Simulated grain yield

Figure 1b shows the averages of the simulated grain yields for the three irrigation scheduling practices for the range of water application depths (WAD). The simulated grain yields ranged from 1338.08 to 4023.46 kg/ha. The lowest yield was obtained from the E7V7F7G7 schedule irrigated with 25 mm depth of water per irrigation event, while the highest yield was obtained when the maize crop was irrigated with 45 mm depth of water and above per irrigation event at 5 days interval throughout the crop growing season (E5V5F5G5). The simulated grain yields are found to relatively compared with the ranges of yield reported for irrigated maize by the National Irrigation Master Plan [17] and the Soil Water management Research Group (SWMRG-FAO, 2003) for

the study area, which were 1800-2000 kg/ha and 1778 to 3703 kg/ha, respectively. More so, in a series of field experiments conducted in the study area, Igbadun et al. [18] reported grain yields ranging from 1580 to 3780 kg/ha for seasonal water application depths ranging from 400 to 700 mm.

Analysis of variance test indicated that there was a significant difference (P < 0.05) among the mean grain yields of the three scheduling practices. Moreover, while there was also significant difference (P< 0.05) among the mean grain yields of the water application depths, there was no significant difference in the interaction of water application depths and scheduling practices. This implies that the grain yields are largely affected either by water application or scheduling practices. Further analyses revealed that the differences among the mean grain yields happened when water application depths were below 40 mm. It may be observed from Figure 1 that there is no visible difference between the grain yields for E7V7F7G7 and the other two irrigation schedules (E5V5F5G5 and E7V7F5G5) except when the water application depths per irrigation were below 40 mm. When water was applied at 25 to 35 mm depth, the grain yield for E5V5F5G5 and E7V7F5G5 were higher than E7V7F7G7 by between 22 and 60 %. There was also no appreciable difference between the grain yield of E5V5F5G5 and E7V7F5G5 except at WAD of 25 mm where the grain yield for E5V5F5G5 was found to be higher than the E7V7F5G5 by about 13 %. The reason for these trends of result may not be far fetch. If the readily available moisture (RAM) of the effective root zone depth of the maize crop has been consumed by the crop before the next irrigation, water application depths below 40 mm may not raise soil moisture content to field capacity. Therefore, the crop will suffer moisture deficit and the resultant consequent is low yield. But when the interval of irrigation is at 5 days or it is shifted to 5 days at flowering and grain filling stages of the crop growth, the impact of moisture deficit on yield will be reduced and the grain yields will improve. This agrees with Yazar et al. [19] and Pandey et al. [20] who reported that maize grain yield increased significantly with irrigation, and Hsiao [21] and Jamieson et al. [22] who reported that the reduction in crop yield and its degree depend upon the timing, severity and duration of water stress.

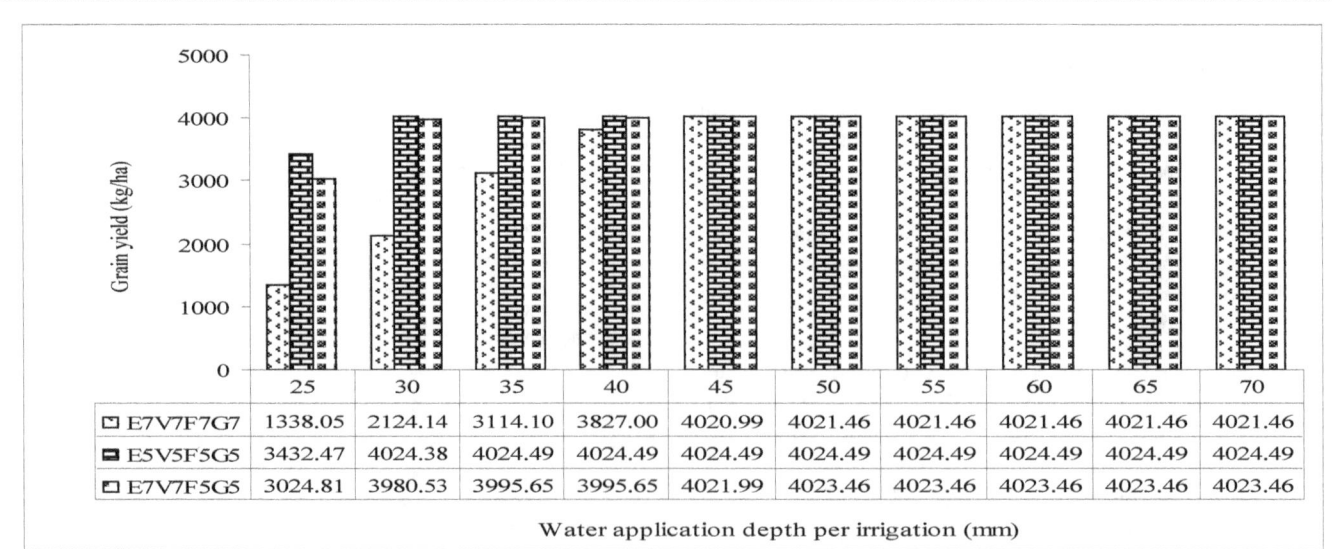

	25	30	35	40	45	50	55	60	65	70
▢ E7V7F7G7	1338.05	2124.14	3114.10	3827.00	4020.99	4021.46	4021.46	4021.46	4021.46	4021.46
▬ E5V5F5G5	3432.47	4024.38	4024.49	4024.49	4024.49	4024.49	4024.49	4024.49	4024.49	4024.49
▢ E7V7F5G5	3024.81	3980.53	3995.65	3995.65	4021.99	4023.46	4023.46	4023.46	4023.46	4023.46

Water application depth per irrigation (mm)

Figure 1b: Simulated grain yield for the E7V7F7G7, E5V5F5G5 and E7V7F5G5 irrigation schedules.

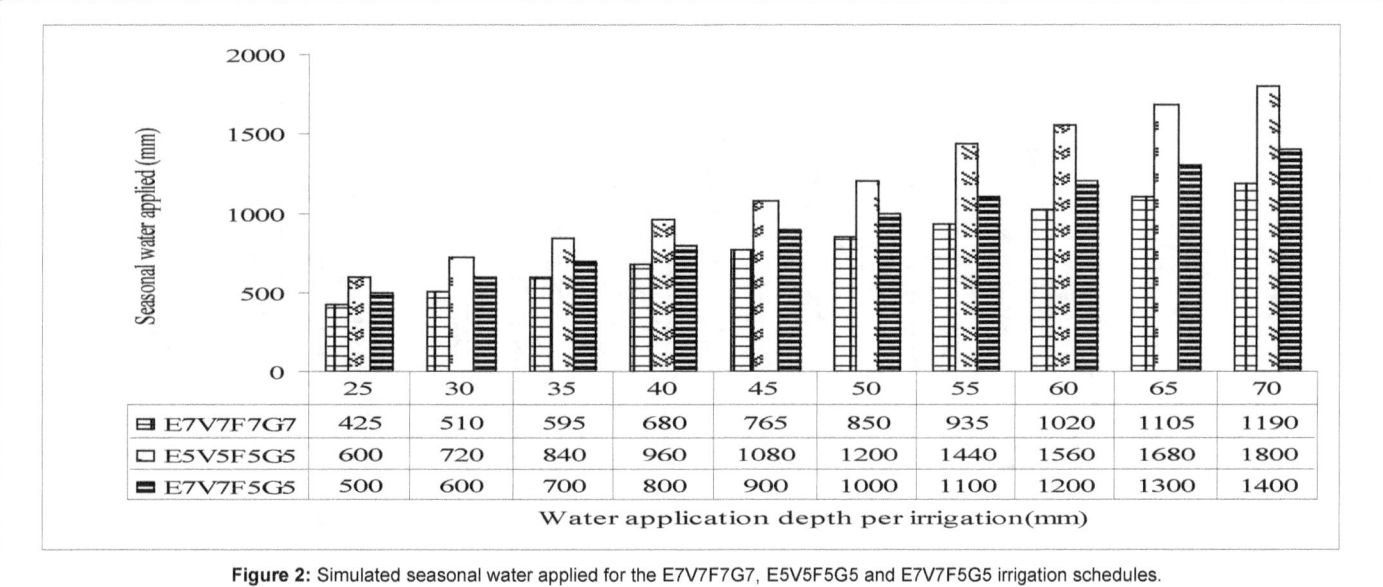

Figure 2: Simulated seasonal water applied for the E7V7F7G7, E5V5F5G5 and E7V7F5G5 irrigation schedules.

	25	30	35	40	45	50	55	60	65	70
E7V7F7G7	425	510	595	680	765	850	935	1020	1105	1190
E5V5F5G5	600	720	840	960	1080	1200	1440	1560	1680	1800
E7V7F5G5	500	600	700	800	900	1000	1100	1200	1300	1400

Water application depth per irrigation(mm)

Simulated soil water balance components

The soil water balance components considered in this study were seasonal water applied, evapotranspiration and deep percolation only. Figure 2 show the simulated seasonal water applied for the different water application depths per irrigation event. The seasonal water applied for E7V7F7G7 varied from 425 to 1190 mm, depending on water application depth. The seasonal water applied for E5V5F5G5 varied from 600 to 1800 mm, while the seasonal water applied for E7V7F5G5 varied from 500 to 1400 mm. These values are without the pre-planting irrigation since the model assumed that the soil moisture was at field capacity at planting. The highly significant differences (P <0.01) in seasonal water applied among the three schedules were as a result of the frequencies of irrigation. The E5V5F5G5 schedule which was the most frequently irrigated recorded the highest seasonal water applied. Its range of water application far exceeds the crop water requirement of the maize crop given as 500 to 800 mm [23]. But the fact that some farmers used such amount of water in their fields is supported by Rajabu et al. [24] who reported that farmers in the study area abstracts water ranging from 825 to 1628 mm depth to irrigate their crops. Farmers who observe a weekly irrigation interval and apply water at 25 mm depth (E7V7F7G7) are irrigating fell below the seasonal crop water requirement, and this explains the reason for the poor grain yield recorded in that schedule.

Figure 3 show the simulated seasonal evapotranspiration (SET) for the different water application depths per irrigation event. The SET ranged from 412 to 563 mm for the three irrigation schedules. Analysis of Variance test showed the mean SET values of the three scheduling practices were significant at P<0.01. The mean SET values associated with the water application depths were significantly different at P<0.05, but the interaction were not significantly also different, which implies that the variation was not as a result of the combine effect of the scheduling and water application depth. Further analyses indicated that only the E5V5F5G5 schedule that was significantly different from the other two. The mean SET values of the E7V7F5G5 and the E7V7F7G7 were not significantly different. The highly significant difference associated with the E5V5F5G5 scheduling may be due to high evaporation rate since the field is frequently irrigated. However, the simulated SET values for the three scheduling practices were found to be within the range of seasonal

consumptive use of maize crop reported by Howell et al.[25] and Farré and Faci [26] being 465-802 mm and 234-578 mm, respectively. Figure 4 shows that simulated deep percolation losses for the three schedules for the ranges of WAD. It may be observed from the figure that deep percolation increased with increase in WAD per irrigation, and this was expected. The deep percolation losses for E7V7F7G7 varied between 50 and 675 mm, while that of E5V5F5G5 varied between 86 and 1133 mm. The E7V7F5G5 also recorded deep percolation of 50 to 874 mm depth of water. It may be noticed from the figure that applying water above 40 mm depth for any of the three irrigation schedules lead to high loss of water to deep percolation. The amount loss to deep percolation when water was applying water above 45 mm depth either at 5 days interval from flowering stage or throughout the crop growing season was found to be about 30 to 100 % of water loss to evapotranspiration.

The trend of results in this study imply that only at WAD of between 25 and 30 mm is the 5-day irrigation interval either throughout the crop growing season or at flowering and grain filling growth stages of any advantage over the 7-day irrigation interval. Water application beyond this 35 mm will only lead to high water losses without appreciable increase in grain yield. The consequences of high deep percolation losses as outlined in Michael [27] include rapid buildup of water table, soil salinity, water logging which leads to poor yield due to low soil temperatures and poor aeration of plant roots. It is not unlikely that in a practical field setup, the high percolation losses under the 5-day irrigation interval especially at WAD exceeding 50mm could lead to poor crop yield. However, this is not reflected by the model as the current version does not have the module to capture the effect of water logging on crop yield performance.

Crop water productivity

Table 4 shows the simulated crop water productivity expressed in term of yield per seasonal water applied (CWP$_{(wa)}$) and yield per seasonal evapotranspiration (CWP$_{(ET)}$) for the three irrigation schedules under study. The values of the CWP$_{(wa)}$ ranged from 0.31 to 0.53, 0.22 to 0.57 and 0.29 to 0.66 kg/m3 for E7V7F7G7, E5V5F5G5, and E7V7F5G5 schedules, respectively. The CWP$_{(ET)}$ were also found to range from 0.31 to 0.74, 0.63 to 0.71 and 0.60 to 0.74 kg/m³, for E7V7F7G7, E5V5F5G5, and E7V7F5G5 schedules, respectively. These ranges of water

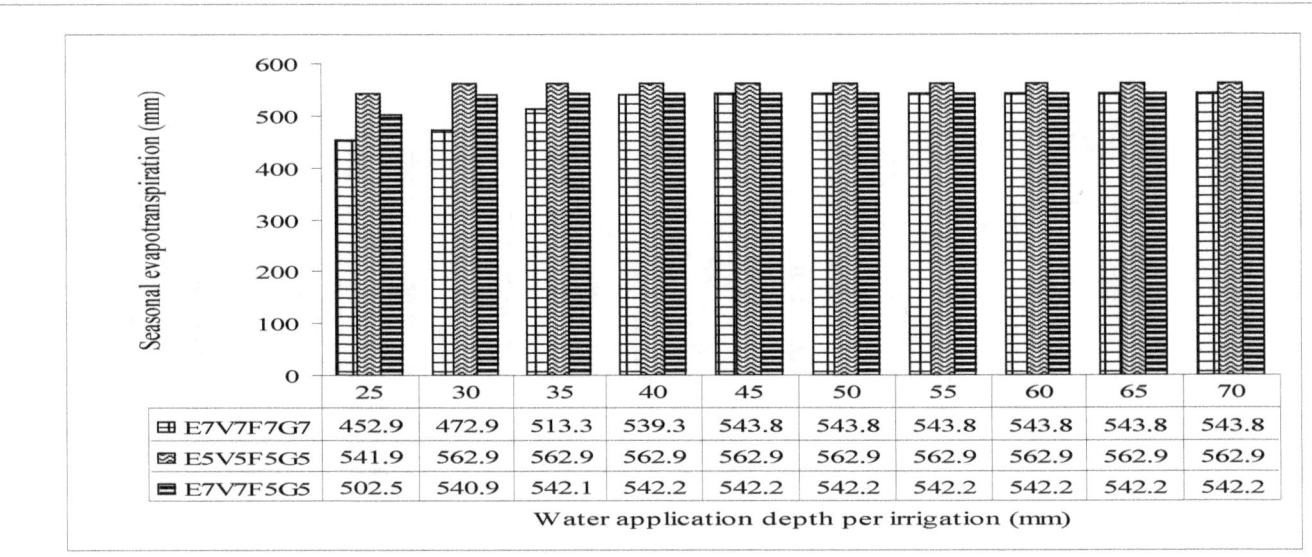

Figure 3: Simulated seasonal evapotranspiration for the E7V7F7G7, E5V5F5G5 and E7V7F5G5 irrigation schedules.

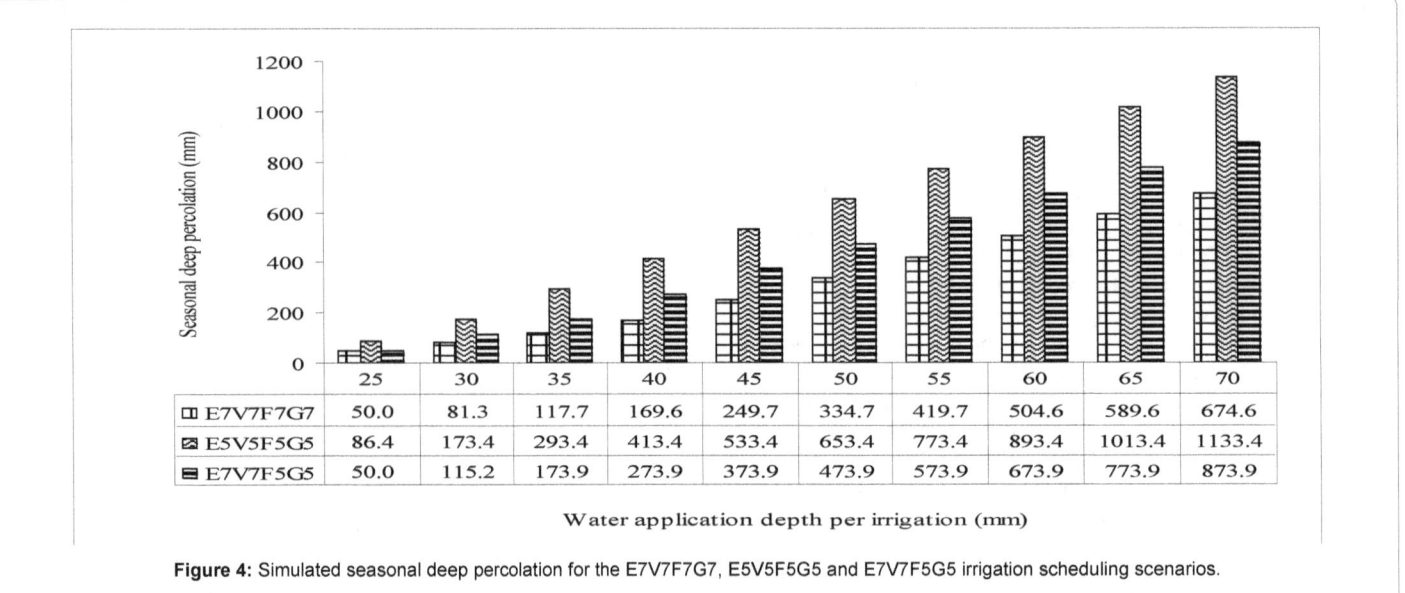

Figure 4: Simulated seasonal deep percolation for the E7V7F7G7, E5V5F5G5 and E7V7F5G5 irrigation scheduling scenarios.

Treatment	Water application depth									
	25	30	35	40	45	50	55	60	65	70
	Crop water productivity (water supply) CWP$_{(wa)}$									
E7V7F7G7	0.31	0.42	0.52	0.56	0.53	0.47	0.43	0.39	0.36	0.34
E5V5F5G5	0.57	0.56	0.48	0.42	0.37	0.34	0.28	0.26	0.24	0.22
E7V7F5G5	0.60	0.66	0.57	0.50	0.45	0.40	0.37	0.34	0.31	0.29
	Crop water productivity (evapotranspiration) CWP$_{(ET)}$									
E7V7F7G7	0.31	0.45	0.61	0.71	0.74	0.74	0.74	0.74	0.74	0.74
E5V5F5G5	0.63	0.71	0.71	0.71	0.71	0.71	0.71	0.71	0.71	0.71
E7V7F5G5	0.60	0.74	0.74	0.74	0.74	0.74	0.74	0.74	0.74	0.74

Table 4: Crop water productivity in terms of water applied (CWP$_{(wa)}$) and evapotranspiration (CWP$_{(ETa)}$) for the irrigation scheduling intervals.

productivity were found to be within 0.25 to 1.80 kg/m³ range obtained for irrigated maize in the Mediterranean environment by Farré and Faci [26]. Pandey et al. [20] also reported water productivity range for various deficit irrigation and nitrogen regimes in Niger Republic of 4.91 to 6.46 kg/ha-mm (0.49 to 0.65 kg/m³).

It may be noticed from the Table that the productivity of water applied of E5V5F5G5 schedule was highest (0.57 kg/m³) only at 25 mm WAD, and declined steadily with increase in water application depth. The productivity of water applied for E7V7F5G5 schedule peaked at 30 mm WAD, while that of E7V7F7G7 schedule peaked at 40 mm WAD and declined thereafter. The productivity of water in terms of seasonal evapotranspiration (CWP_{ET}) of E5V5F5G5 and E7V7F5G5 schedules were noticed to reach their peaked at 30 mm WAD while that of E7V7F7G7 was at peak at 45 mm WAD.

These results clearly show the water application regimes at which best water productivity level can be achieved. Among the three irrigation schedules, the best productivity of water may be obtained if water is applied 30 mm depth per irrigation event if the E7V7F5G5 schedule is followed. The $CWP_{(wa)}$ for E5V5F5G5 and E7V7F5G5 were found to be higher than E7V7F7G7 by between 25 and 48 % only at WAD of 25 and 30 mm. In other words, grain yield production per cubic meter of water applied will be 25 to 48 % higher if water is applied at 25-30 mm depth at 5 days irrigation interval throughout the crop growing season or at flowering to maturity stage. If water is applied at higher depths, water productivity of such schedules will drop lower that of weekly irrigation interval by about 30 to 50 %.

Conclusion

The implications of the irrigation schedules practiced by farmers in Igurusi ya Zamani irrigation scheme (IZIS) in the middle Mkoji sub-catchment of the Great Ruaha River Basin was studied using a simulation model. Irrigating maize fields at 5 days interval throughout the crop growing season or at flowering to crop maturity gave better water productivity output only when water application depths did not exceed 30 mm depth. Water application beyond this depth leads to very high deep percolation losses without significant difference in crop yield when compared to irrigating at 7 days interval throughout the crop growing season. Moreover, the productivity of applied irrigation water will drop by about 30 and 50%. The water application depth for best productivity under the 7 days irrigation interval for the maize crop in the study area was 40 to 45 mm depth. Beyond this depth, there was no appreciable increase in grain yield but a fall in productivity of applied water and a buildup of deep percolation. It is advised that farmers in the upstream of the irrigation scheme should maintain water application regime at 40 to 45 mm depth per irrigation while observing a weekly irrigation scheduling throughout the crop growing season. If they must observe a more frequent irrigation, the water application depths should not exceed 30 mm so as not to harm the soil and waste water and labor.

References

1. Singh PK, Mishra AK, Imtiyaz ZM (1991) Moisture stress and the water use efficiency of mustard. Agr Water Manage 20: 245-253.

2. Al-Jamal MS, Sammis TW, Ball S, Smeal D (1999) Yield-based, irrigated onion crop coefficient. Appl Eng Agric 15: 659-668.

3. Imtiyaz ZM, Mgadla NP, Chepete B, Manase SK (2000) Response of six vegetable crop to irrigation scheduling. Agr Water Manage 45: 331-342.

4. Camposeo S, Rubino P (2003) Effect of irrigation frequency on root water uptake in sugar beet. Plant Soil 253: 301-309.

5. Mermoud A, Tanini TD, Yacouba Y (2005) Impact of different irrigation schedules in water balance components of an onion crop in a semi-arid zone. Agr Water Manage 77: 282-295.

6. Sun HY, Liu CM, Zhang XY, Shen YJ, Zhang YQ (2006) Effect of irrigation on water balance, yield and water use efficiency on winter wheat in the North China Plains. Agr Water Manage 85: 211-218.

7. Nazeer M (2009) Simulation of maize crop under irrigated and rainfed conditions with CROPWAT model. ARPN Journal of Agricultural and Biological Science 4: 68-73.

8. Ayana M (2011) Deficit irrigation practices as alternative means of improving water use efficiencies in irrigated agriculture: case study of maize crop at Arba Minch, Ethiopia. African Journal of Agricultural Research 6: 226-235.

9. Quanqi L, Xunbo Z, Yahai C, Songlie Y (2012) Water consumption characteristics of winter wheat grown under different planting patterns and deficit irrigation regime. Agr Water Manage 105: 8-12.

10. Kang S, Zhang L, Liang Y, Hu X, Cai H, et al. (2002) Effects of limited irrigation on yield and water use efficiency of winter wheat in the loess Plateau of China. Agr Water Manage 55: 203-216.

11. Jin MG, Zhang RQ, Gao YF (1999) Temporal and spatial soil water management: A case study in the Heiloonggang Region, PR China. Agricultural Water Management 42: 173-187.

12. SMUWC (2001) Irrigation water management and efficiency. Directorate of Water Resource, Ministry of Water and Livestock, Government of Tanzania, Dar es Salaam, Tanzania.

13. FAO (2003) Comprehensive Assessment of Water Resources of Mkoji sub catchment, its Current Uses and Productivity. FAO-Netherlands Partnership Programme, Water and Food Security.

14. Igbadun HE (2006) Evaluation of Irrigation Scheduling Strategies for Improving Water Productivity: Computer-Based Simulation Model Approach. Sokoine University of Agriculture, Morogoro, Tanzania.

15. Igbadun HE (2012) Irrigation scheduling impact assessment model (ISIAMOD): A decision tool for irrigation scheduling. Indian Journal of Science and Technology 5: 3090-3099.

16. Allen RG, Pereira LS, Raes D, Smith M (1998) Crop Evapotranspiration: Guideline for Computing Crop Water Requirements. FAO Irrigation and Drainage.

17. JICA/MAFS (2002) The study on the National Irrigation Master Plan in the United Republic of Tanzania. Ministry of Agriculture and Food Security 1: 240.

18. Igbadun HE, Mahoo HF, Tarimo AKPR, Salim BA (2006) Crop water productivity of an irrigated maize crop in Mkoji sub-catchment of the Great Ruaha River Basin, Tanzania. Agricultural Water Management 85: 141-150.

19. Yazar A, Howell TA, Dusek DA, Copeland KS (1999) Evaluation of crop water stress index for LEAP irrigated corn. Irrigation Science 18: 171-180.

20. Pandey RK, Maranville JW, Admou A (2000) Deficit irrigation and nitrogen effects on maize in a sahelian environment I. Grain yield and yield components. Agr Water Manage 46: 1-13.

21. Hsaio TC (1990) Measurement of plant water status, Annual Review. Plant Physiology 24: 519-570.

22. Jamieson PD, Martin RJ, Francis GS, Wilson DR (1995) Drought effects on biomass yield production and radiation use efficiency in barley. Field Crop Research 43: 77-86.

23. Doorenbos J, Kassam AH (1979) Yield Response to Water. FAO Irrigation and Drainage, Rome, Italy.

24. Rajabu KRM, Mahoo HF, Sally H, Mashauri DA (2005) Water abstraction and use patterns and their implications on downstream river flows: A case study of Mkoji Sub-catchment in Tanzania. Proceedings of the East Africa Integrated River Basin Management Conference, Morogoro, Tanzania.

25. Howell TA, Tolk JA, Arland DS, Evertt R (1998) Evapotranspiration, yield and water use efficiency of corn hybrid differing in maturity. Agronomy Journal 90: 3-9.

26. Farre I, Faci JM (2006) Comparative response of maize (Zea mays L.) and sorghum (Sorghum bicolor L. Moench) to deficit irrigation in a Mediterranean environment. Agr Water Manage 83: 135-143.

27. Michael AM (1999) Irrigation Theory and Practice. Vikas Publishing House Pvt Ltd, New Delhi.

2

Environmental Factors Influencing Fish Species Distribution in Irrigation Channels around Ariake Sea, Kyushu, Japan

Greshishchev V*, Onikura N and Iyooka H

Fishery Research Laboratory, Kyushu University, Tsuyazaki, Fukutsu, Fukuoka, Japan

Abstract

Freshwater ecosystems have suffered a long history of anthropogenic disturbances that are responsible for declining fish populations. Fish populations in irrigation channels, in particular, still remain unprotected and relatively unstudied. This study looked to investigate what environmental variables and anthropogenic disturbances had the greatest influence on fish species distribution in irrigation channels. Fish sampling and environmental data collection was conducted throughout six municipalities, spanning 43 sites, in southern Fukuoka, Kyushu, Japan. A total of 37 fish species and 17 environmental variables were statistically analyzed using detrended correspondence analysis (DCA) and receiver operating characteristic (ROC) curve analysis. DCA results showed that channel width, channel substrate and submerged plants had a statistical correlation with fish species distribution. We believe proper management of irrigation channel substrate conditions and macrophyte cover will go a long way in supporting fish species biodiversity and reducing pressure from invasive species. ROC curve analysis revealed a disparity in the environmental preferences of endangered and invasive species. This result suggests that the distribution of endangered species in irrigation channels may be heavily influenced by invasive species. Further research must be done before drafting an effective management solution.

Keywords: Paddy field; Threatened fish species; Water quality; Freshwater fish; Irrigation ditch

Introduction

The start of Japanese wet-rice cultivation corresponds to the beginning of the Yayoi period 2300-2800 years ago [1]. Archaeological evidence suggests that some of the oldest rice paddies in the Japanese archipelago are those that have been found in northern Kyushu [2]. During this period land was plentiful and cultivation took place along natural flood plains. There was little need for large scale human induced irrigation of paddy fields. Naturally, as the population increased the demand for more land and cultivation increased along with it. Wet-rice cultivation would need to take place away from the natural flood plains in order to meet the demands of the rising population. It was not until the Heian period (794-1159) that natural flood plains were dug out as permanent channels and the rerouting of streams and rivers for the irrigation of paddy fields took place [3]. It was of little concern at that time how the rerouting and construction of irrigation channels would impact the local freshwater fauna.

Japanese rivers differ from most continental rivers because they are short (max. length: 367 km), steep (average slope: 0.44%), and exhibit very flashy flow regimes [4]. Japanese rivers and streams are host to a rich freshwater fish fauna. There are a total of 321 sub-species, 18 orders, 53 families, and 145 genera [5]. Of these species 88 are endemic to Japan and only 11% of the fish are introduced species [6]. In Kyushu the average number of species per major catchment is 43 [7]. However, the number of those species found in irrigation channels is much lower due to the harsher environmental conditions and lentic nature of the water.

Freshwater ecosystems have suffered a long history of anthropogenic disturbances that are responsible for declining fish populations [8]. Some of the prevailing anthropogenic pressures that threaten biodiversity include habitat destruction and fragmentation, eutrophication and chemical pollution, invasive alien species, over-exploitation and climate change [9]. Up until the 1970's environmental issues were mostly ignored in Japan. Most people used their rivers for waste disposal due to their steep nature. Fortunately, during the past decade water quality has significantly improved, fostered by

the enforcement of the Environmental Pollution Prevention Act in 1970 [7]. However, irrigation channels have not encountered the same pressure for rehabilitation. Irrigation channels are fed into by industrial wastewater, agricultural runoff, surface runoff and human sewage that often undergo limited to no treatment. These compounded anthropogenic pressures have drastic effects on the water quality of the irrigation channels.

Due to previous studies largely focusing on fish populations in riverine environments, irrigation channel fish populations have remained relatively unstudied. To increase our understanding of how natural and anthropogenic disturbances shape aquatic communities and individual performance, it is crucial to identify the most vulnerable species and the cause of decline [10]. Scientific identification of the factors threatening biodiversity and understanding of the relative importance of such factors are prerequisite to drafting effective plans [11]. In this study we looked to identify what environmental variables and anthropogenic disturbances had the greatest influence on fish species distribution in irrigation channels and assessed what species may be at risk.

Materials and Methods

Study area

The study was conducted in the irrigation channels of southwestern Fukuoka Prefecture, Kyushu, Japan (Figure 1). The water in these channels originates from the mountains in the east. As water flows downstream it is artificially split into multiple channels and is

---

*Corresponding author: Vladimir Greshishcher, Fishery Research Laboratory, Kyushu University, Tsuyazaki, Fukutsu, Fukuoka, Japan
E-mail: vgreshis@connect.carleton.ca

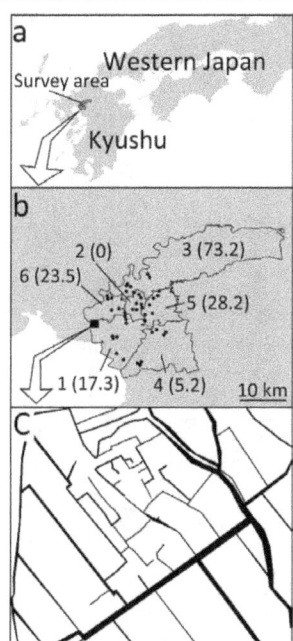

Figure 1: Map of survery area located in northern Kyushu Island in western Japan (a), a total of 43 sampling sites across six municipalities (1: Yanagawa City; 2: Oki City; 3: Kurume City; 4: Miyama City; 5: Chikugo City; 6: Okawa City) with the sewage treatment percentage of each municipality writen in parenthesis (b). Aerial view of Irrigation channel network in select area (c).

dispersed throughout the city. These channels serve to transport water and to irrigate paddy fields. Water is pumped up from the channel and is used to irrigate the nearby fields. The unused water is then returned to the channel, subsequently altering the chemical composition of the water. As this process is repeated several times traveling downstream, the chemical composition of the water downstream becomes drastically different from that of upstream. The irrigation channels are also fed into by industrial wastewater, agricultural runoff, surface runoff and human sewage that undergo varying stages of treatment depending on the municipality. A total of 43 sites spanning six cities were selected. The Study areas were chosen due to the varying levels of sewage treatment that takes place across the six municipalities (Figure 1). The cities do not have access to large sewage treatment plants and must rely on on-site treatment units. Roughly seven survey sites per city were chosen. Potential survey sites were selected based on size, connectivity, location and estimated species abundance (high and low). Survey site and sewage treatment maps (Figure 1) were created using QGIS ver.2.4.0 [12].

Environmental data collection

On site measurement of pH, electric current (EC), current velocity and dissolved oxygen (DO) was conducted throughout multiple days in the months of June and November, 2014. The measurement of pH was done using a pHTestr 30 (Eutech/Oakton instruments, Vernon Hills, IL USA). EC was measured using an ECTestr (Eutech/Oakton instruments, Vernon Hills, IL USA). Current velocity was measured using a VR-301 propeller-type current meter (KENEK Corporation, Tokyo, Japan). DO was measured using a DO-5509 dissolved oxygen meter (Lutron Electronic Enterprise, Taipei, Taiwan).

Channel depth and width were measured on-site using a 2 m measuring rod. Each site was divided into ten evenly spaced segments where the depth and width of each segment was measured and recorded.

The average depth and width across the ten segments was then used to represent the site as a whole. Concrete revetment and aquatic plant coverage was calculated in a similar fashion. Each site was once again divided into ten evenly spaced segments. In each segment the presence of concrete in the bed and bank was noted. Through the summation of all segments, we were able to calculate the total concrete revetment for that specific site. For aquatic plant coverage, the presence of emergent, floating and submerged plants was recorded for each segment. Again, by totalling each category we were able to calculate the percentage of aquatic plant coverage for each site (Figure 2).

Water and soil samples were collected at each site and stored for future analysis. The water samples were tested for suspended solids (SS), chemical oxygen demand (COD), total nitrogen (TN) and total phosphate (TP) in accordance to the manuals of the Japan Society for Analytical Chemistry [13]. The soil samples were analyzed through loss on ignition testing in accordance to the standards of the Japanese Geotechnical Society [14].

Fish sampling and data analysis

Fish sampling was done concurrently with environmental data collection. Sampling was carried out by a team of four researchers equipped with casting nets and hand nets. A minimum of 30-60 minutes was allotted for sample collection and species identification at each site. As irrigation channels differed greatly in size, the allotted time for each site was properly adjusted. Fish fauna identification was accomplished in accordance to Nakabo [15]. After identification, a picture of the fauna was taken and presence/absence data was recorded. All fish captured were released back into the channel.

We used detrended correspondence analysis (DCA) to analyze the

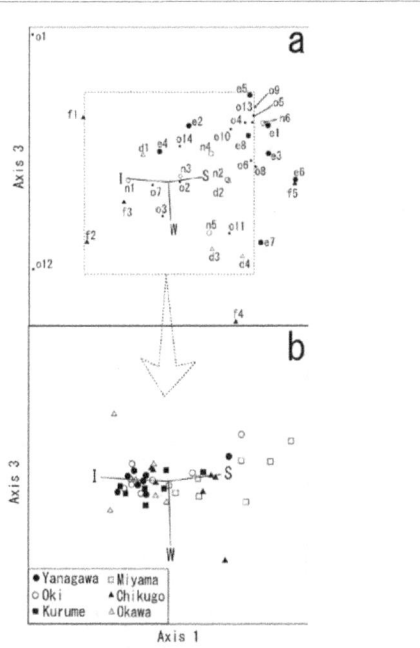

Figure 2: Detrended correspondence analysis ordination between axes 1 and 3. (a) Plot of species distributions where solid and open circles represent endangered and near threatened species respectively, solid and open triangles represent foreign and domestic alien species, small solid squares indicate other species. Species IDs are shown near each plot and are referenced in Table 2. (b) Survey site distribution. Environmental variables (I: ignition loss; S: submerged plants; W: width) correlated strongly with axis 1 and/or axis 3 and are indicated as straight lines.

relationship between 17 environmental variables and species presence/absence distribution of 37 species across 43 sites. The Box-Cox power transformation [16] was applied to improve distribution normality prior to DCA. Statistical analysis software PC-ORD ver. 5.31 (MjM software, Gleneden Beach, Oregon, U.S.A.) was used to perform DCA. A receiver operating characteristic (ROC) curve was used to further analyze the relationship between prominent environmental variables and the presence/absence of each species by using EXCEL add-in software (EXCEL-TOUKEI 2012, SSRI, Shinjuku, Tokyo, Japan). The presence/absence was set to 1/0 in ROC analysis for confirming positive correlation with each environmental variable and 0/1 for confirming negative correlation. When an area under the ROC curve (AUC) was >0.7 it was evaluated as a high or moderate correlation [17]. In addition, ROC curves were also used to determine the optimal cutoff point (COP) for each environmental variable [17].

Results

Environmental variables

The average, standard deviation, minimum and maximum values of all measured environmental variables across 43 sites were calculated and recorded in Table 1. Nearly all environmental variables, aside from pH, had high variability. This is evident from the large standard deviation values which were often greater or equal to the average. Although pH only had a standard deviation of 0.78, the logarithmic nature of the measurement exacerbates this result. However, the min-max range of pH only varied by a factor of 3 and never exceeded the critical level for fish survival. The wide range between min and max of the variables shows that environmental conditions varied drastically among channels.

Fish species

The average, min and max species richness across the 43 sites was recorded in Table 1. The highest species richness was 17, observed in Miyama City. The lowest species richness was 3, observed in Chikugo City. The average number of species observed per site was 9.67. A total of 37 different species were observed throughout the 43 sites (Table 2). Of the 37 species observed eight species were listed on the local red data book as endangered species [18], four species were domestic

Variable	Avg ± SD	Min–Max
pH	7.77 ± 0.78	6.58-9.60
EC (µS/cm)	561.21 ± 236.68	3.07-1254
Velocity	5.38 ± 13.55	0-61
Oxygen	6.92 ± 3.02	2.7-22.9
SS (mg/L)	31.52 ± 32.85	1.29-157.00
COD-Mn (mg/l)	6.90 ± 2.46	2.08-10.93
TN (mg/)	2.66 ± 0.95	1.03-6.45
TP (mg-P/l)	0.31 ± 0.30-	0.03-1.58
Ignition Loss	9.41 ± 4.31	2.23-20.18
Depth (cm)	90.40 ± 42.98	18.5-250
Width (cm)	443.50 ± 298.96	26.70-1325
Emergent (%)	23.00 ± 29.01	0-100
Floating (%)	11.52 ± 19.93	0-80
Submerged (%)	7.91 ± 17.12	0-70
Conc. Bank (%)	80.42 ± 26.92	0-100
Conc. Bed (%)	21.63 ± 30	0-100
Sp. Richness	9.67 ± 2.95	3-17

Table 1: The average (Avg), standard deviation (SD), minimum (Min) and maximum (Max) of all measured environmental variables.

Species ID	Scientific Name	Num. of Sightings	Total sighting rank
Critically Endangered			
e1	*Acheilognathus tabira* ssp.	4	22
e2	*Hemigrammocypris rasborella*	2	29
Endangered			
e3	*Cobitis kaibarai*	2	24
e4	*Rhodeus ocellatus kurumeus*	37	4
e5	*Rhodeus atremius atremius*	1	34
Vulnerable			
e6	*Misgurnus anguillicaudatus*	1	30
e7	*Sarcocheilichthys variegatus variegatus*	2	25
e8	*Tanakia lanceolata*	7	16
Near Threatened			
n1	*Abbottina rivularis*	25	5
n2	*Acheilognathus rhombeus*	1	35
n3	*Biwia zezera*	14	13
n4	*Nipponocypris sieboldii*	10	14
n5	*Oryzias latipes*	20	9
n6	*Tanakia limbata*	6	17
Domestic Alien			
d1	*Carassius cuvieri*	38	3
d2	*Gnathopogon elongatus*	1	36
d3	*Opsariichthys uncirostris*	17	11
d4	*Squalidus chankaensis tsuchigae*	6	20
Foreign Invader			
f1	*Channa argus*	3	23
f2	*Gambusia affinis*	21	8
f3	*Lepomis macrochirus*	24	6
f4	*Micropterus salmoides*	1	37
f5	*Paramisgurnus dabryanus*	1	31
Other (Native Species)			
o1	*Carassius auratus beurgeri*	1	33
o2	*Carassius auratus langsdorfi*	41	1
o3	*Cyprinus carpio*	18	10
o4	*Hemibarbus barbus*	2	27
o5	*Odontobutis obscura*	6	18
o6	*Pseudogobio esocinus*	6	19
o7	*Pseudorasbora parva*	41	2
o8	*Pungtungia herzi*	2	26
o9	*Silurus asotus*	5	21
o10	*Squalidus gracilis gracilis*	9	15
o11	*Zacco platypus*	15	12
o12	*Mugil cephalus cephalus*	2	28
o13	*Tridentiger brevispinis*	1	32
o14	*Rhinogobius* sp.	22	7

Table 2: Total number of sightings and sighting rankings for all observed species across 43 sites.

aliens and four were foreign invaders. The eight endangered and six near threatened freshwater fish species were classified as follows: Critically Endangered: *Acheilognathus tabira* ssp. 2 (Species ID: e1), *Hemigrammocypris rasborella* (e2); Endangered: *Cobitis kaibarai* (e3), *Rhodeus ocellatus kurumeus* (e4), *Rhodeus atremius atremius* (e5); Vulnerable: *Misgurnus anguillicaudatus* (e6), *Sarcocheilichthys variegatus variegatus* (e7), *Tanakia lanceolata* (e8); and Near Threatened: *Abbottina rivularis* (n1), *Acheilognathus rhombeus* (n2), *Biwia zezera* (n3), *Nipponocypris sieboldii* (n4), *Oryzias latipes* (n5), *Tanakia limbata* (n6). The domestic alien species were *Carassius cuvieri* (d1), *Gnathopogon elongatus* (d2), *Opsariichthys uncirostris* (d3) and *Squalidus chankaensis tsuchigae* (d4). The foreign invader species were *Channa argus* (f1), *Gambusiaaffinis* (f2), *Lepomis macrochirus* (f3), *Micropterus salmoides* (f4) and *Paramisgurnus dabryanus* (f5). The remaining species were designated as other and were not listed on the red list. These species include freshwater fish (o1: *Carassius auratus beurgeri*, o2: *C. a. langsdorfi*, o3: *Cyprinus carpio*, o4: *Hemibarbus barbus*, o5: *Odontobutis obscura*, o6: *Pseudogobio esocinus*, o7: *Pseudorasboraparva*, o8: *Pungtungia herzi*, o9: *Silurus asotus*, o10: *Squalidus gracilis gracilis*, o11: *Zacco platypus*), brackish water fish o12: *Mugil cephalus cephalus*, o13: *Tridentiger brevispinis* and diadromous fish o14: *Rhinogobius* sp.

The total number of sightings for each individual species is recorded in Table 2. The two most frequently observed species were *C. a. langsdorfi* and *P. parva*. Both were found in 41 out of 43 sites. There were eight species that were observed only once throughout the 43 sites. These species were *A. rhombeus*, *C. a. beurgeri*, *G. elongates*, *M. anguillicaudatus*, *M. salmoides*, *P. dabryanus*, *R. a. attremius* and *T. brevispinis*.

Relationship between fish fauna and environmental variables

The DCA created 3 axes that explained approximately 60% of the ordination of fish fauna in our survey area. The first axis correlated positively with submerged plants and negatively with ignition loss while the third axis correlated negatively with width of channel. Most threatened and near threatened species were plotted on right side, suggesting that these species prefer sites with high submerged plants and low ignition loss. Whilst three foreign invaders were plotted on the left side, suggesting that they are able to adapt to poor sediment quality. Although most sites were plotted around the center of this ordination, all sites of Miyama City were plotted on the right side.

Relationship between fish distribution and correlated environmental variables

Of the 37 species analyzed, ten had a positive and four had a negative correlation with submerged plants. Six species had positive while another six had negative correlations with width of channel. Fourteen species had a positive while twelve had a negative correlation with ignition loss (Table 3). Whilst six endangered species had a positive correlation with submerged plants, three domestic and foreign invader species had a negative correlation with submerged plants. Three endangered species had a negative correlation with channel width where as two invader species had a positive correlation. Five endangered species had a negative correlation with ignition loss while four invader species were positively correlated with ignition loss.

Discussion

Water and soil quality in Japanese irrigation channels

Japanese rivers and lakes are subject to strict water quality guidelines that outline the acceptable levels for both fish and human use [19]. However, Japanese irrigation channels have not faced the same regulation pressures and to date have not received any strict guidelines regarding acceptable water quality levels for fish populations [19]. There are, however, lesser unenforced guidelines for water quality that outline potential goals. The only form of regulation in regards to water quality in irrigation channels deals with the maintenance of non-toxic levels to human health. Results of our environmental variable analysis (Table 1) show that there is a very large range of variability in nearly all measured variables. This means that the water and soil quality of an

Species ID	Submerged			Width			Ignition Loss		
	P/N	AUC	COP	P/N	AUC	COP	P/N	AUC	COP
e1	P	0.79	10		0.36	N/A	N	0.92	6.08
e2	P	0.93	20		0.48	N/A		0.43	N/A
e3	P	0.97	40	N	0.79	204.5	N	0.85	6.08
e4		0.34	N/A	N	0.71	569		0.68	N/A
e5	P	0.80	10		0.33	N/A		0.31	N/A
e6	P	0.92	40	N	0.81	148	N	0.76	6.08
e7	P	0.95	40		0.55	N/A	N	0.91	4.67
e8		0.69	N/A		0.34	N/A	N	0.74	8.62
n1		0.46	N/A		0.33	N/A		0.56	N/A
n2		0.38	N/A	P	0.90	753	N	0.90	3.25
n3		0.42	N/A	N	0.70	468		0.35	N/A
n4		0.62	N/A		0.51	N/A		0.32	N/A
n5		0.61	N/A		0.53	N/A		0.46	N/A
n6	P	0.84	10		0.33	N/A	N	0.90	8.62
d1	N	0.73	20		0.39	N/A	P	0.79	7.52
d2		0.38	N/A	P	0.90	753	N	0.90	3.25
d3		0.60	N/A		0.51	N/A	N	0.76	9.77
d4		0.62	N/A		0.63	N/A	N	0.84	8.62
f1		0.58	N/A		0.58	N/A	P	0.87	11.61
f2	N	0.73	0		0.45	N/A	P	0.65	N/A
f3	N	0.72	0		0.45	N/A	P	0.57	N/A
f4		0.38	N/A	P	1	1325		0	N/A
f5	P	0.92	40	N	0.81	148	N	0.76	6.08
o1		0.38	N/A		0.62	N/A	P	0.90	13.81
o2		0.62	N/A	P	0.87	244		0.49	N/A
o3		0.58	N/A		0.50	N/A		0.65	N/A
o4		0.59	N/A	P	0.71	434.5	N	0.94	3.25
o5		0.64	N/A		0.33	N/A	N	0.73	8.62
o6		0.66	N/A		0.51	N/A	N	0.96	4.67
o7	N	0.95	30		0.51	N/A		0.88	7.52
o8	P	0.87	10		0.40	N/A	N	0.90	4.67
o9		0.58	N/A	N	0.82	365		0.36	N/A
o10		0.64	N/A		0.39	N/A	N	0.72	8.62
o11		0.63	N/A		0.54	N/A	N	0.91	8.62
o12		0.38	N/A	P	0.93	715		0.61	N/A
o13	P	0.80	10		0.50	N/A	N	0.95	2.46
o14		0.49	N/A		0.58	N/A		0.48	N/A

Table 3: Receiver operating characteristic (ROC) curve analysis results comparing the three prominent environmental variables with species presence/absence and absence/presence. Positive and negative correlations are noted as P and N respectively. Area under the curve (AUC) values below 0.7 were determined as having a weak correlation and consequently have had their cut-off point (COP) values omitted. Species ID and scientific names are shown in Table 2.

irrigation channel differs drastically depending on the location of the channel itself and where along the channel the measurement was taken.

Loss on ignition is a method used to estimate the organic matter composition of soil as well as the substrate of marine and freshwater ecosystems [20]. In this study loss on ignition was used to assess channel substrate and indirectly the quality of water. The lowest recorded pH out of all the sites was 6.58 recorded in Chikugo City. Uncoincidentally, this site was also recorded as having the highest loss on ignition recorded at 20.18%. However, despite the poor water quality and substrate conditions, this site was recorded as having average species richness [9]. This observation hints that there are more factors at play that govern the distribution of fish species in irrigation channels than simply the overall quality of water and substrate. It appears that sewage treatment coverage had little to do with the overall species richness of a municipality. In fact, the top three cities with the greatest sewage treatment coverage had lower species richness than the bottom three treated cities. Some factors that were more important in determining the overall species richness include survey site location and complexity of the irrigation network. Unfortunately, it is difficult to truly gauge the impacts of anthropogenic pressures on fish species as they might be masked or modified by natural factors that govern the performance of individuals and the structure of aquatic communities [21,22].

The quality of channel substrate was directly correlated to current velocity. Sites with a higher current velocity yielded a much lower loss on ignition. It is also important to note that sites with the highest species richness were all lotic environments. This observation can be explained by the proximity of the irrigation channel to the nearest river or stream. Areas of the irrigation channel that are directly fed into by a river or stream possess a current. These areas act as transitional zones between a river and the agricultural channel. Transitional zones have a higher species richness because fish encompassing all lifestyle preferences (both lentic and lotic) must pass through in order to traverse the channel [23]. Another explanation for this observation is that when water in an irrigation channel branches and travels further downstream, the effects of sedimentation become stronger. Areas upstream that possess a current have a mechanism for cycling nutrients and waste disposal while lotic areas further downstream act as deposition sites and subsequently experience lower water and soil quality.

Relationships among fish fauna/each species and environmental variables

Results of the detrended correspondence analysis produced three statistically correlated variables. These variables were channel width, channel substrate (Loss on Ignition) and submerged plants. Although most sites were plotted around the center of this ordination, all sites of Miyama City were plotted on the right side. This is an interesting observation because Miyama City was also noted as having the highest species richness. The reason for this has to do with the complexity of the irrigation channel network and the quality of water. Sites of cities plotted around the center of the ordination have heavily branched networks. Although these cities also possess sites with moving water, they are heavily dominated by areas of still water. The abundance of lentic waters reduces the overall water and soil quality and consequently species richness of these cities. Miyama City on the other hand possesses a rather simple irrigation channel network. This simplicity means far less branches on the network which leads to a stronger overall current velocity present throughout the channel. This characteristic has allowed Miyama City to maintain higher water and soil quality levels as well as overall higher species richness.

Previous studies [24,25] have documented that species diversity and stream fish assemblages are affected by wetted width. In this study, channel width had a statistical correlation with species distribution. However, this does not mean that increasing the width of a channel will

directly increase fish species diversity. This result supports the findings of Yan et al. [24] and suggests that channel width has an influence on fish species assemblages at the species level.

The quality of channel substrate plays a crucial role in the selection of spawning sites for many fish species found in the channel. Five species observed within the channel have been previously found to directly rely on the substrate as a life history trait and for spawning purposes. These species are *Z. platypus*, *P. esocinus*, *S. v. variegatus*, *N. sieboldii* and *A. rivularis* [26]. Six species of bitterling found in the channel have been found to indirectly rely on the quality of substrate. These species are *R. o. kurumeus*, *T. limbata*, *A. tabira* ssp. 2, *T. lanceolata*, *R. a. atremius* and *A. rhombeus* [26]. Bitterling are parasitoids that lay their eggs into live bivalves where their embryos develop inside the gills of the bivalve [27]. For this reason, bitterling species are highly dependent on the presence of bivalves in their environment. As substrate quality decreases the presence of bivalves decreases as well. Naturally, as bivalve presence declines, the population size of bitterling species in that environment will decline as well [28].

ROC curve analysis results from this study support the findings of Nakajima et al. and Onikura et al. [26-28]. In this study 7 out of the 11 species previously found to directly and indirectly rely on channel substrate had a very strong negative correlation ($AUC>0.90$) with Loss on Ignition (Table 3). The remaining four species (*N. sieboldii*, *A. rivularis*, *R. o. kurumeus* and *R. a. atremius*) did not have a significant statistical correlation. However, the lack of correlation from these species can be justified. The lack of correlation in *R. a. atremius* may be explained by its very small sample size. Although *R. o. kurumeus* did have a significant sample size it was only marginally off ($AUC=0.68$) from having a significant correlation. On the other hand, both *N. sieboldii* and *A. rivularis* had meaningful sample sizes but showed no statistical correlation. The reason for this is that both species rely on the substratum for spawning purposes. However, sample collection was conducted during feeding and spawning season which is outside of spawning season. During this time they do not rely on the substratum. If sample collection were to be done during spawning season there would likely be a stronger measurable correlation.

The presence or absence of macrophytes can drastically alter fish species diversity and community dynamics of freshwater environments. Macrophytes in lakes and rivers provide cover for fish and macro invertebrates, alter substrate composition, produce oxygen and act as food [29]. In this study the presence of submerged macrophytes was one of the three environmental variables that showed a statistical correlation with fish species distribution. Submerged macrophytes improve the structural heterogeneity of microhabitats in aquatic ecosystems, often providing an important habitat for zooplankton [30]. Submerged macrophytes are utilized by fish for the nursing of juveniles, feeding, and predator avoidance [31]. However, excessive development of free-floating macrophytes on the water surface can reduce the biomass of submerged macrophytes and results in a relatively simple habitat structure [30]. Poor water quality arising from floating weed mats is considered to be the main determinant of reduced fish abundance and diversity [32].

In this study, Of the 37 fish species surveyed, ten had a positive correlation and four had a negative correlation with the presence of submerged macrophytes (Table 3). An important observation to note is that three of the four negatively correlated species were domestic and foreign aliens. This shows that irrigation channels are still foreign environments for these species and that they generally prefer environmental similar in conditions to that of their native range of rivers and lakes. The other negatively correlated species was *P. parva*. This species has been noted as having both high life history plasticity and wide-scale habitat use [27]. It was noted by Onikura and Nakajima [33] that structural revetment conditions and macrophyte cover were unimportant for this species. However, the results of this study suggests that when presented with multiple environmental conditions, *P. parva* has a strong aversion (AUC=0.95) to channels with submerged macrophyte cover. The possible reason for this discrepancy may have to do with the scale of the survey area where said research was conducted. The survey sites chosen by Onikura and Nakajima [33] were much smaller in size and all possessed heavy macrophyte cover. In this scenario *P. parva* was found to have no preference as its options were limited and it was forced to adapt to environments with heavy macrophyte cover. However, the survey sites chosen in this study encompassed sites with varying degrees of macrophyte cover. It was in this type of scenario that *P. parva* was found to have a strong aversion to submerged plants due to having the option of preference.

Relationship between native and invasive species

The most commonly observed domestic alien species was *C. cuvieri*. It was the third most observed species overall and was observed at 38 out of 43 sites (Table 2). *C. cuvieri* originates from Lake Biwa and has been widely distributed throughout Japan as a game fish. *C. cuvieri* has been shown to be fairly hardy and adept at avoiding predation at the juvenile stage [34]. DCA results show that *C. cuvieri* is not particularly averted to environments with poor soil and water quality conditions. ROC curve analysis (Table 3) shows that *C. cuvieri* may in fact have a statistical preference (AUC=0.79) to areas with poor soil conditions. This is a unique characteristic shared only by three other invasive species (Table 2). *Carassius auratus beurgeri* was also noted as having a strong positive correlation, however, the small sample size puts the validity of the result into question. We hypothesize that the natural hardiness of these invasive species has allowed them to succeed and even show an affinity to areas with poor substrate conditions. Native species are naturally averted to such conditions and leave the invasive species with little competition.

The most commonly observed foreign invasive species was *L. Macrochirus*. It was the sixth most observed species overall and was present at 24 out of 43 sites (Table 2). *L. Macrochirus*, commonly known as bluegill, is one of the most commonly introduced fish species in the world [35]. It is characterized by an opportunistic behaviour that can lead to aggressive dominance [36]. Studies have shown that the introduction of blue gill into Japanese waters induces a range of changes at multiple trophic levels [37]. In irrigation channels *L. Macrochirus* has been shown to have an allopatric distribution with *O. latipes*. Although no direct correlation between these two species was found in this study. ROC curve analysis suggests that *L. Macrochirus* has a negative correlation (AUC=0.72) with submerged plants. Two other invasive species (d1, f2) were also found to have a negative correlation with submerged plants. On the other hand, six out of eight endangered species and one threatened species had positive correlations to submerged plants. This observation shows that there is a large disparity between the environmental preferences of endangered and invasive species in terms of submerged plants. We believe a possible explanation for this is that the ability of invasive species to outcompete the endangered species in areas with little macrophyte cover has caused the endangered species to have a strong preference or simply an illusion of preference for environments with heavy macrophyte cover.

Endangered species showed a preference for irrigation channels with a narrow width (Table 3). Whereas invasive species tended to prefer

wider channels. We believe this can be attributed to invasive species a) showing a natural preference to environmental conditions similar to their native range of rivers and lakes b) outcompeting endangered species in wider channels causing a decline in their numbers in those environments.

Conclusions

Fish species distribution was found to be statistically correlated with channel width, channel substrate and submerged plants. Nearly all domestic and endagered species had a positive correlation with submerged plants. Although these macrophytes, in the eyes of a layperson, may seem like unwanted debris clogging up the channel, they are infact a vital part of a healthy ecosystem. In order to support fish species biodiversity we advise against the total clearing of macrophyte cover in irrigation channels. All domestic and endagered species had a negative correlation with substrate quality while nearly all domestic and foreign invaders had a positive correlation. This result sugests that domestic and endagered species strongly rely on good substrate conditions while invasive species are more adaptive. We believe the management of irrigation channel substrate conditions will go a long way in supporting fish species biodiversity and reducing pressure from invasive species. However, further research must be done before drafting an effective management solution.

Acknowledgement

We wish to thank all of the members of the Aquatic Field Science lab at Kyushu University for helping with data collection. We would like to thank Jasso, the Rotary Yoneyama Foundation and the Ogori Rotary Club for providing financial support that made this research possible. A special thanks to the anonymous reviewers for their valuable comments and contribution.

References

1. Imamura K (1996) Prehistoric Japan: new perspectives on insular East Asia. UCL Press, London.

2. Ishikawa H (1992) Zukai Nihonno Jinruiiseki. Tokyo-daigaku Shyuppan, Tokyo.

3. Environment Agency of Japan (1981) The 2nd National Survey on Natural Environment: Rivers. Marketing Intelligence Corporation, Tokyo

4. Kawanabe H, Mizuno N (1989). Freshwater Fishes of Japan. Yama-kei Publishers, Tokyo.

5. Taniguchi Y, Inoue M, Kawaguchi Y (2001) Stream fish habitat science and management in Japan. Aquatic Ecosystem Health and Management 4: 357-365.

6. Yoshimura C, Omura T, Furumai H, Tockner K (2005) Present state of rivers and streams in Japan. River Res Appl 21: 93-112.

7. Dudgeon D, Arthington AH, Gessner MO, Kawabata Z, Knowler DJ, et al. (2006) Freshwater biodiversity: importance, threats, status and conservation challenges. Biological Reviews 81: 163-182

8. Japan Biodiversity Outlook Science Committee (JBOSC) (2010) Report of comprehensive Assessment of Biodiversity in Japan.

9. Maceda AV, Green AJ, De Sostoa A (2014) Scaled body-mass index shows how habitat quality influences the condition of four fish taxa in north-eastern Spain and provides a novel indicator of ecosystem health. Freshwater Biology 59: 1145-1160.

10. Yoshioka A, Miyazaki Y, Sekizaki Y, Suda S, Kadoya T, et al. (2013) A "lost biodiversity" approach to revealing major anthropogenic threats to regional freshwater ecosystems. *Ecological Indicators* 36: 348-355.

11. QGIS Development Team (2015) QGIS Geographic Information System. Open Source Geospatial Foundation Project.

12. Hokkaido Branch of the Japan Society for Analytical Chemistry (2005) Water analysis (5thedn.) KAGAKUDOJIN, Shimogyo, Kyoto, Japan.

13. The Japanese Geotechnical Society (2008) Loss on ignition test method meeting the criteria of soi.l

14. Nakabo T (2002) Fishes of Japan with pictorial keys to the species, English Edition. Tokai University Press. Shibuya-ku, Tokyo, Japan.

15. Box GEP, Cox DR (1964) An analysis of transformations. Journal of the Royal Statistical Society 26: 211-252.

16. Akobeng AK (2007) Understanding diagnostic tests 3: receiver operating characteristic curves. Acta Paediatrica 96: 644-647.

17. Fukuoka Prefecture (2014) Fukuoka red data book 2014. Fukuoka Pref, Hakataku, Fukuoka, Japan.

18. Urase T (2011) Clear explanations of water quality and environmental science. Pleiades Publishing House, Azumino, Nagano, Japan.

19. Frangipane G, Pistolato M, Molinaroli E, Guerzoni S, Tagliapietra D (2008) Comparison of loss on ignition and thermal analysis stepwise methods for determination of sedimentary organic matter. Aquatic Conservation 19: 24-33.

20. Gasith A, Resh V (1999) Streams in Mediterranean climate regions: abiotic influences and biotic responses to predictable seasonal events. Annual Review of Ecological Systems 30: 51-81.

21. Ferreira T, Caiola N, Casals F, Cortes R, Economou A, et al. (2007) Ecological traits of fish assemblages from Mediterranean Europe and their responses to human disturbances. Fisheries Management and Ecology 14: 473-481.

22. Terra BF, dos Santos ABI, Aurajo FG (2010) Fish assemblage in a dammed tropical river: an analysis along the longitudinal and temporal gradients from river to reservoir. Neotrop Ichthyol 8: 599-606.

23. Yan YZ, Xiang XY, Chu L, Zhan YJ, Fu CZ (2011) Influences of local habitat and stream spatial position on fish assemblages in a dammed watershed, the Qingyi Stream, China. Ecol Freshwater Fish 20: 199-208.

24. Yan YZ, Wang H, Zhu R, Chu L, Chen YF (2012) Influences of low-head dams on the fish assemblages in the headwater streams of the Qingyi watershed, China. Environ Biol Fish 96: 495-506.

25. Hossain MM, Perhar G, Arhonditsis GB, Matsuishi T, Goto, et al. (2013) Examination of the effects of largemouth bass (Micropterus salmoides) and bluegill (Lepomis macrochirus) on the ecosystem attributes of lake Kawahara-oike, Nagasaki, Japan. Ecological Informatics 18: 149-161.

26. Onikura N, Nakajima J, Eguchi K, Miyake T, Kawamura K, et al. (2008) Present Distributions of Exotic Species in Creeks Around Sea of Ariake and Yatsushiro, Northwestern Kyushu, Japan. Journal of Japan Society on Water Environment 31: 395-401.

27. Nakajima J, Shimatani Y, Itsukushima R, Onikura N (2010) Assessing riverine environments for biological integrity on the basis of ecological features of fish. Advances in River Engineering 16: 449-454.

28. Onikura N, Nakajima J, Miyake T, Kawamura K, Fukuda S (2011) Predicting distributions of seven bitterling fishes in northern Kyushu, Japan. Icthyol Res 59: 124-133.

29. Fleming JP and Dibble ED (2014) Ecological mechanisms of invasion success in aquatic macrophytes. Hydrobiologia 746: 23-27.

30. JY Choi, Jeong KS, La GH, Joo KH (2014a) Effect of removal of free-floating macrophytes on zooplankton habitat in shallow wetland. Knowledge and Management of Aquatic Ecosystems 414: 1-11.

31. Vritilek M, Reichard M (2012) An indirect effect of biological invasions: the effect of zebra mussel fouling on parastisation of unionid mussels by bitterling fish. Hydrobiologia 696: 205-214

32. Onikura N, Nakajima J (2012) Age, growth and habitat use of the topmouth gudgeon, Pseudorasbora parva in irrigation ditches on northwestern Kyushu Island, Japan. Journal of Applied Ichthyology 29: 186-192.

33. Choi JY, Jeong KS, Kim SK, La GH, Chang KH, et al. (2014b) Role of macrophytes as microhabitats for zooplankton community in lentic freshwater ecosystems of South Korea. Ecological Informatics 24: 177-185.

34. Perna CN, Cappo M, Pusey BJ, Burrows DW, Pearson RG (2012) Removal of aquatic weeds greatly enhances fish community richness and diversity: an example from the Burdekin River floodplain, tropical Australia. River Research and Applications 28: 1093-1104.

35. Taguchi T, Miura Y, Krueger D, Sugiura S (2014) Dispersal, migration, and predation of juvenile crucian carps after release from rice fields. Environ Biol Fish 98: 679-690.

36. Lowe S, Browne M, Boudjelas S (2000) 100 of the World's Worst Invasive Alien Species: A Selection from the Global Invasive Species Database. Invasive Species Specialist Group, Auckland New Zealand.

37. Marchetti MP, Moyle PB, Levine R (2004) Invasive species profiling? Exploring the characteristics of non-native fishes across invasion stages in California. Freshwater Biology 49: 646-661.

3

Irrigation with Treated Wastewater: Quantification of Changes in Soil Physical and Chemical Properties

Pradip Adhikari[1], Manoj K. Shukla[2]*, John G. Mexal[2] and David Daniel[3]

[1]Postdoctoral Fellow, Dept. of Soil Environment and Atmospheric Sciences, Univ. of Missouri, Columbia, Missouri, USA
[2]Dept. of Plant and Environmental Sciences, New Mexico State University, New Mexico 88003-8003, USA
[3]Dept. of Economics and International Business, New Mexico State University, Las Cruces, New Mexico 88003-8003, USA

Abstract

Land application of treated wastewater is increasing particularly in areas where water stress is a major concern. The primary objective of this study was to quantify the effect of irrigation with aerated lagoon treated wastewater on soil properties. Core and bulk soil samples were collected from areas under the canopies of mesquite and creosote and intercanopy areas from each of the three plots. Irrigation water quality from 2006 to 2008 showed that average sodium adsorption ratio (SAR), electrical conductivity (EC) and pH of irrigation water were 37.16, 5.32 dS m^{-1} and 9.7, respectively. The sprinkler uniformity coefficients of irrigated plot-I was 49.34 ± 2.23 % and irrigated plot-II was 61.57 ± 2.11 %. Within irrigated and between irrigated and un-irrigated plots, most soil physical properties remained similar except saturated hydraulic conductivity (K_s) which was significantly higher under mesquite canopies than in the intercanopy areas. Chloride (Cl$^-$) concentrations below 60 cm depth were higher under creosote than mesquite canopies in irrigated plots indicating deeper leaching of Cl$^-$. Nitrate (NO$_3^-$) concentrations below 20 cm depth under canopy and intercanopy areas were low indicating no leaching of NO$_3^-$. The average SAR to 100 cm depth under shrub canopies was 18.46 ± 2.56 in irrigated plots compared to 2.94 ± 0.79 in the un-irrigated plot. The Na$^+$ content of creosote was eleven times higher un-irrigated than irrigated plot and Na$^+$ content of herbaceous vegetation was three times higher in the irrigated than unirrigated. Thus irrigation with high sodium wastewater has exacerbated the soil sodicity and plant Na$^+$ content. Since the majority of mesquite roots are found within 100 cm, and creosote and herbaceous vegetation roots are found within 25 cm from soil surface, a further increase in sodicity may threaten the survival of woody and perennial herbaceous vegetation of the study site.

Keywords: Wastewater; Chemical properties; Herbaceous vegetation; Irrigation

Introduction

Southern New Mexico is characterized as semi-arid region where wastewater reclamation and reuse for irrigation has become important part of water resources planning. This has occurred as a result of the increasing fresh water scarcity, high nutrients in wastewater, and the high cost of advanced treatment required for other wastewater uses. United Nations Millennium Development Goal also targets the use of wastewater as irrigation to reduce the water deficit [1]. Certain quality criteria should be met prior to using wastewater for irrigation. Some of the parameters requiring close attention are electrical conductivity (EC), total dissolved solids (TDS), sodium adsorption ratio (SAR), suspended heavy metals and organic matter (OM). Without proper management, wastewater application can pose serious risks to human health and the environment [1]. Treatment of urban and industrial wastewater is complex, expensive, and requires energy and technology. The safe disposal of the treated wastewater is also a challenge because the effect of wastewater on the soil and plant environment is complex and depends upon the amount of various elements present in the wastewater. Reuse of effluent could be beneficial especially in areas where water stress is a major concern primarily due to limited water resources, higher water demands and limited economic resources. Wastewater can add nutrients to the soil system stimulating plant growth, increasing plant NO$_3^-$ uptake, and the turnover of soil NO$_3^-$ and denitrification. A major objective of land application systems is to allow the physical, chemical, and biological properties of the soil-plant environment to assimilate wastewater constituents without adversely affecting beneficial soil properties [2]. However, when wastewater is irrigated beyond the assimilation capacity of the soil-plant system, it can provide a source of readily leachable nutrient or contaminant [3]. Waste water can also affect soil physical properties, including bulk density (BD), drainable porosity (*d*), soil moisture retention and

hydraulic conductivity (K_s). Recent study on the same location reported lower Ks and macroporosity in the wastewater irrigated areas than in the unirrigated areas [4]. The levels of dissolved OM and suspended solids in effluent depend on the quality of the raw sewage water and the degree of treatment [5,6] Suspended solids present in effluents accumulate in soil voids and physically block water-conducting pores leading to a sharp decline in soil hydraulic properties [4,5]. The reduction in K_s could be due to the retention of OM during infiltration and the change of pore size distribution as a result of expansion or dispersion of soil particles. Application of wastewater with sodium (Na$^+$) content to soil increases sodicity, causes clay dispersion, changes pore geometry, and reduces K_s [7,8]. In contrast, [9] found no adverse impact on the hydraulic parameters while applying standard domestic effluents to soil in Israel. Soils in the arid region are generally calcareous with high pH in the upper soil horizons favoring the precipitation of most heavy metals and reduce the risk of groundwater pollution [10]. The primary goal of land application of wastewater is to maximize vegetative cover to increase the capacity of the soil to serve as a sink for wastewater contaminants, minimize salt accumulation in the root zone, and avoid NO$_3^-$ leaching to the groundwater [11,12]. In this context application of treated wastewater on common arid and semiarid shrubs could be more

*Corresponding author: Manoj K. Shukla, Dept. of Plant and Environmental Sciences, New Mexico State University, New Mexico 88003-8003, USA
E-mail: shuklamk@nmsu.edu

economical and environmentally beneficial. Soil chemical properties are one of the most researched aspects of wastewater irrigation. Changes due to irrigation vary greatly and are largely dependent on the quality of the irrigation water. However, little work has been conducted on the impact of wastewater irrigation in Chihuahuan desert ecosystem on the native vegetation. An earlier study conducted on part of the West Mesa irrigated site reported that the sprinkler distribution uniformity was low (53.7%) and could have caused the variability in soil chemical and physical properties between canopies and intercanopy areas [11]. In spite of the variability of application, the previous study did not report statistical differences in chemical and physical properties between vegetation canopies and intercanopy areas likely due to low sample size. Similarly, NO_3^- and OM content of wastewater listed by the Environmental Protection Agency (EPA) as a method of recycling nutrients and OM were not addressed in that study. The present study overcomes these limitations of the earlier study and provides a detailed account of the impact of wastewater on physical and chemical properties under different vegetation canopies and intercanopy areas within the entire irrigated site. The objectives of this study were to: (1) determine the influence of lagoon treated wastewater interception by shrub canopies on physical and chemical properties of canopy soil, and (2) compare physical and chemical properties among the canopy and intercanopy areas.

Materials and Methods

Experimental site

The West Mesa industrial and municipal wastewater land application facility (West Mesa) is located near Las Cruces, NM(longitude W 106° 54.408' latitude N 32° 15.99', altitude 1298 m). This includes a wastewater treatment plant and a land application system. The untreated industrial and municipal wastewater generated from dairy processing and metal wire fabrication industries is treated in a 1,500 m^3d^{-1} capacity treatment plant, which can discharge 200 m^3d^{-1} of wastewater to the 36-ha study site. Additional details about the study locations could also found in [13,14]. Aerated lagoon effluent application on this site began on February 5, 2002 to the Chihuahuan Desert upland adjacent to the wastewater treatment plant by 1,243 fixed-head sprinklers operated by automated pumps [15]. The treated plots received variable amounts of effluent due to the temporal fluctuations in tenant-generated wastewater and the high evaporation losses from the wastewater lagoons through the peak summer months. During the late summer the application onto the treated site increased usually due to the decreased evaporation and increased tenant's wastewater discharge. From 2006 to 2008, the entire 36-ha received an average of 57.66 cm of water of which 34.68 cm came from the effluent application (Table 1). Total average non stressed ET for mesquite and creosote shrubs was 154.06 cm during 2006-08 and the ratio of total water applied to ET was about 0.37 ± 0.03. Overall, vegetation in the experimental site was water stressed because little or no wastewater was available for application during the summer months when ET demands were high. This area is dominated by woody perennials such as creosote (*Larreatridentata,* (DC) Cov.) and honey mesquite *(Prosopisglandulosa Torr. varglandulosa)* whose percent

groundcover in 2002 were approximately 8.7 and 14.4%, respectively (Babcock et al. [11]). The visual observation during the spring and early summer months of 2008 revealed that approximately 80% of the irrigated area was covered with perennial vegetation including, desert daisy (*BebbiajunceaBenth.),snakeweed (*GutierreizaLag.), pigweed (*AmaranthusL.), spiderling (*BoerhaviaL.), sagebrush (*Artemisia* L.), and chinchweed (*PectisL.). Coppice dunes occur under mesquite canopies and occasionally under creosote canopies over most of the experimental site. Before the development of coppice dunes the area was level and surface horizons consisted of coarse textured materials [16] that provides better condition for infiltration and leaching of Na+ and other soluble salts. Soil texture of the coppice dunes and the intercanopy areas varies from sand to light sandy loam with little or no gravel. Soil series identified in and around the West Mesastudy site are Onite (coarse-loamy, mixed, superactive, thermic TypicCalciargids), Pintura (Mixed, thermic Typic Torripsamments), Bucklebar (TypicHaplargid), Pajarito (Coarse-loamy, mixed, superactive, thermic TypicHaplocambids), and Bluepoint (Mixed, thermic TypicTorripsamments) [16].

Soil sampling and analysis

Three plots were identified for soil sampling: an unirrigated plot, irrigated plot-I, and irrigated plot- II. The soils in unirrigated and irrigated plot-I were classified as Blue point loamy sand whereas in irrigated-II, it was Onite-Pajarito association. Amount of wastewater received was approximately 10 % higher in the Irrigated plot-I than the Irrigated plot-II due to the head differences from the wastewater holding point. Three mesquite and three creosote shrubs were selected randomly in each plot. Shrubs within the irrigated plot-I and II were located on the periphery of the sprinkler uniformity test site. Four sampling points were selected in the center of each canopy (four cardinal directions within the canopy) and three on each intercanopy area. Intact soil cores were taken by a core sampler (19 cm length and 5.5 cm diameter) from each sampling point at 0-20 cm and 20-40 cm depths. Similarly, bulk soil samples were taken by a metal auger (3 cm diameter) from each sampling point at 0-20, 20-40, 40-60, 60-80, 80-100 and 100-150 cm depths. Thus, a total of 162 core and 486 bulk soil samples were collected from all three plots. Visual observations were made to detect the signs of stress and leaf burn caused by wastewater application. Particle size analysis (PSA) was performed by hydrometer method using air-dried sample < 2 mm [17]. Soil cores were trimmed and the BD was determined by soil core method [18]. Cores were saturated with tap water by slowly raising water level in the trough and K_s was determined by the constant head method [19]. Volumetric moisture content (θ) of each core was determined at 0, 0.003, 0.006 MPa suctions using tension table and 0.03, 0.1, 0.3, 1, 1.5 MPa using pressure plate apparatus [20].The difference in θ at 0 MPa and 0.006 MPa was calculated to estimate drainable porosity (θ_d) or soil macroporosity, the difference in θ at 0.03MPa (field capacity; FC) and 1.5 MPa (wilting point; WP) was used to estimate plant available water content (AWC). The van Genuchten (1980) model was fitted to the measured soil moisture retention [$h(\theta)$] curves to obtain the air entry value (1/α),the pore size distribution parameter(λ), and empirical

Year	Wastewater	Precipitation	Total water applied	Creosote ET	Mesquite ET	Average crop ET	Deficit
				--cm---			
2006	17.62	33.93	51.55	170.30	179.83	175.18	123.63
2007	36.79	20.45	57.24	177.66	143.63	158.53	101.29
2008	49.65	14.55	64.20	135.73	121.23	128.48	64.27
Ave.	34.68	22.97	57.66	161.23	148.23	154.06	96.39

Table 1: Amounts of wastewater, precipitation, and evapotranspiration (ET) during 2006-2008 at West Mesa land application site.

Year	----------TDS (mg L^{-1})-----------		---------Chloride (mg L^{-1})---------	
	Influent	Effluent	Influent	Effluent
2006	1866.66 ± 450.41	3160.00 ± 900.68	320.66 ± 43.07	528.00 ± 169.00
2007	810.00 ± 28.86	3455.00 ± 293.72	247.00 ± 28.61	633.25 ± 67.71
2008	982.50 ± 221.74	3607.50 ± 455.60	252.50 ± 73.86	855.00 ± 127.19
Average	1219.72 ± 223.67	34075.50 ± 550.00	273.39 ± 145.54	672.08 ± 121.30
	--------------Nitrate (mg L^{-1})----------		------------Sodium (mg L^{-1})---------	
2006	1.47 ± 0.47	13.49 ± 13.04	332.00 ± 83.57	1175.33 ± 149.69
2007	1.61 ± 0.97	1.19 ± 0.32	215.00 ± 47.61	1094.00 ± 17.87
2008	0.66 ±0.22	0.36 ± 0.10	184.75 ± 48.58	1097.75±94.63
Average	1.24 ± 0.55	5.01 ± 4.48	389.40 ± 64.76	1122.36 ± 87.39
	-----------EC (dS m^{-1})--------------		--------------------SAR------------------	
2006	2.91 ± 1.21	4.93 ± 1.40	7.55 ± 1.91	41.47 ± 4.33
2007	1.26 ± 0.04	5.39 ± 0.45	5.46 ± 1.08	36.89 ± 1.60
2008	1.54 ± 0.34	5.64 ± 0.72	4.13 ± 0.89	33.14 ± 1.52
Average	1.90 ± 0.53	5.32 ± 2.57	5.71 ± 1.29	37.16 ± 2.48

Source-City of Las Cruces, Water Quality Lab

Table 2: Influent and effluent chemical values means and standard errors from 2006-2008.

parameters (n and m) using the retention curve (RETC) program of van Genuchten et al., (1991).

$$S_e = \frac{\theta - \theta_r}{\theta_s - \theta_r} = \left[1 + (\alpha h)^n\right]^{-m} \ldots h < 0 \qquad (1)$$

$$= \theta_s \ldots\ldots\ldots\ldots\ldots\ldots\ldots\ldots h \geq 0$$

Where S_e is the degree of saturation $0 \leq S_e \leq 1$, θ_s and θ_r are saturated and residual water contents. The RETC uses a non-linear least-square optimization approach to estimate the unknown model parameters and empirical constants affecting the shape of the retention curve. Chemical properties, like EC and pH were determined on 1:2 ratio of soil: water. NO_3^- concentration was measured using auto analyzer [21]. For NO_3^- concentration, 2.5 g of sieved soil sample was mixed with 25 ml of 2N sodium chloride (KCL) solution in 125 ml Erlenmeyer flask and shaken for one hour using mechanical shaker. The solution was filtered through Whatman no. 2 filter paper before analysis. The extract was used to analyze the amount of nitrate-nitrogen (NO_3^--N) through the Technicon auto analyzer [22]. The amount of NO_3^- was calculated from NO_3^--N. For Cl$^-$ analysis, about 5 g of soil and 25 ml of DI water was mixed in a centrifuge tube, shaken for an hour in a mechanical shaker, and centrifuged for15 minutes at 2000 rpm speed. A mixture consisting of 5-ml of final soil solution, 35 ml of DI water and 2 ml of nitric acid was titrated with the 0.1 N silver nitrate by 798 MPT Titrinotitrator. Only one sample was analyzed for OM, SAR, ESP and Na$^+$ from unirrigated plot because no wastewater was applied to this plot and an earlier study showed no significant differences in soil chemical properties between 2002 and 2006 for the unirrigated plot. In addition, 126 composite soil samples were analyzed for pH, EC, Cl$^-$, NO_3^-, OM, ESP and SAR (Harris Lab, Columbus, Nebraska). Plant samples of mesquite, creosote and perennial weeds from intercanopy areas were collected from both irrigated and unirrigated plots. Each sample was washed, oven dried at 60°C, ground and analyzed for Na$^+$ and NO_3^- (Harris Lab, Columbus, Nebraska). Chemical properties including heavy metal concentrations of wastewater influent and effluent from 2006-2008 were provided by the City of Las Cruces, Water Quality Lab. All the wastewater analysis was conducted in the Continental Analytical Service Inc., Salina Kansas, following the United States Environmental Protection Agency (USEPA) guidelines. Sprinkler uniformity tests were conducted to determine the effectiveness of sprinklers to discharge the wastewater uniformly. The

sprinklers in irrigated I were installed on a trapezoidal grid rather than on a square grid. The spacing of sprinklers was 11 m by 12.7 m and 11.5 m by 14.2 m in irrigated I and 11.9 m by 12.6 m and 12.0 m by 11.4 m in the irrigated II. Uniformity of wastewater application with sprinkler irrigation system was calculated by Christiansen's coefficient (Cu) (Christiansen, 1942) using the American Society of Agricultural Engineers standard #3301 (ASAE Standards, 1993)

$$Cu = 100\left(1.00 - \sum |dv| / nX\right) \qquad (2)$$

Where Dv = deviations of volume of water collected in the catch funnel from the mean catch volume; n= number of catch funnels; X =mean volume collected in catch funnel.

Statistical analysis

To assess differences in soil chemical and physical properties among plots, one-way analysis of variance (ANOVA) with contrasts was performed. Similarly, ANOVA was also performed to assess differences in soil chemical and physical properties between the canopies within the plots. The SAS General Linear Model Procedure (Proc GLM) was used to assess plot, vegetation and plot x vegetation interaction due to the application of wastewater for soil physical and chemical properties at 0-20 and 20-40 cm depths. All statistical analyses were performed using SAS® software version 9.1.3 (SAS Institute Inc., 2002-2003). All statistical analyses were performed for a significance level of $P \leq 0.05$.

Results and Discussion

Wastewater quality and application

Evaporation losses at the experimental site ranged from 50 to 90% similar to the typical values reported for arid regions, which can result in 2 to 20 fold increases in soluble salt concentrations [23]. Water quality for the irrigation water was based on the SAR, total salinity, EC, and specific ion concentrations. Analysis of the wastewater showed higher amounts of TDS, Cl$^-$, Na$^+$, EC, and SAR in the effluent than influent primarily due to high rate of evaporation in the holding ponds (Table 2). Wastewater generated from meat and dairy processing industry is reported to contain elevated concentrations of Na$^+$, with SAR ranging between 4 and 50 [24]. The average SAR and Na$^+$ concentration of applied wastewater was 37.16 ± 2.48 and 1122.36 ± 87.39 mg L^{-1}, respectively. Irrigation with water having high Na$^+$concentrations is reported to cause an accumulation of exchangeable Na$^+$ on soil

Properties	-----Unirrigated----		------Irrigated-I--------		-------Irrigated-II----------	
	0-20	20-40	0-20	20-40	0-20	20-40
				-P-values-			
Sand	(>.08) [1,2,3]	(>.08) [1,2,3]	(<.05)[1*](>.08) [2,3]	(>.12) [1,2,3]	(>.07)[1,2,3]	(<.01)[1*](>.06)[2,3]
Silt	(>.55) [1,2,3]	(>.18) [1,2,3]	(>.09)[1,2,3]	(>.07) [1,2,3]	(<.05)[1,3*] (>.48)[2]	(>.10)[1,2,3]
Clay	(>.17) [1,2,3]	(>.05) [1,2,3]	(>.69) [1,2,3]	(>.41) [1,2,3]	(>.18)[1,2,3]	(>.05)[1,2,3]
BD	(<.005)[2,3], (>.08)[1]	(>.66) [1,2,3]	(>.05) [1,2,3]	(>.06) [1,2,3]	(>.14) [1,2,3]	(>.07)[1,2,3]
K_s	(<.001)[2,3*] (<.08) [1]	(0.09) [1,2,3]	(<.001) [2*](>.09) [1,3]	(>.08) [1,2,3]	(<.005)[1,2,3*]	(>.48)[1,2,3]
AWC	(>.07) [1,2,3]	(<.005)[1*] (>.38) [2,3]	(>.61) [1,2,3]	(>.44) [1,2,3]	(>.52) [1,2,3]	(>.38)[1,2,3]
FC	(>.17) [1,2,3]	(<.05)[1*] (>.07) [2,3]	(>.21) [1,2,3]	(>.21) [1,2,3]	(>.05) [1,2,3]	(>.07)[1,2,3]
θd	(>.33) [1,2,3]	(>.27) [1,2,3]	(>.37) [1,2,3]	(>.39) [1,2,3]	(>.42) [1,2,3]	(>.25)[1,2,3]

[1]= one-way ANOVA contrast between mesquite vs. creosote, [2]= mesquite vs. intercanopy, [3]= creosote vs. intercanopy. Numbers inside the parenthesis indicate the P-values
* Indicates significant differences at $P < 0.05$

Table 3: One-way ANOVA contrasts between vegetation canopies and intercanopy areas for particle size, bulk density (BD), hydraulic conductivity (K_s) available water content (AWC), field capacity (FC) and drainable porosity (θd) at 0-20 and 20-40 cm depth during 2007.

Source	DF	F value	Pr>F	F Value	Pr>F
			0-20 cm		20-40 cm
			----Sand----		
Plot	2	1.07	0.365	2.20	0.227
Vegetation	2	2.45	0.114	2.14	0.233
Plot x vegetation	4	0.16	0.956	1.35	0.288
			----Silt----		
Plot	2	1.49	0.328	2.06	0.156
Vegetation	2	0.56	0.611	4.62	0.024
Plot x vegetation	4	0.99	0.438	0.23	0.911
			----Clay----		
Plot	2	3.73	0.121	2.53	0.194
Vegetation	2	4.62	0.091	2.40	0.206
Plot x vegetation	4	0.86	0.503	0.63	0.647
			----K_s----		
Plot	2	5.28	<.05*	129.43	<.0005*
Vegetation	2	29.04	<0.0001*	22.83	<.005*
Plot x vegetation	4	2.64	<.05*	0.05	0.994
			----BD----		
Plot	2	1.97	0.253	1.47	0.331
Vegetation	2	4.65	0.090	2.07	0.155
Plot x vegetation	4	1.89	0.156	1.00	0.434
			----AWC----		
Plot	2	4.95	0.082	0.29	0.760
Vegetation	2	3.35	0.139	0.60	0.593
Plot x vegetation	4	0.76	0.564	5.34	0.005*
			----FC----		
Plot	2	20.19	<.005*	57.03	<.001*
Vegetation	2	6.66	0.053	2.66	0.069
Plot x vegetation	4	0.78	0.555	0.27	0.894
			----θd----		
Plot	2	3.34	0.140	8.87	<.05*
Vegetation	2	1.21	0.065	7.28	0.05
Plot x vegetation	4	0.36	0.832	1.03	0.418

* Indicates significant differences at $P < 0.05$

Table 4: Values of F statistic and the probability (Pr) from analysis of variance (n=27) for sand, silt, clay, K_s, BD, available water content (AWC), field capacity (FC), and drainable porosity(θd) at 0-20 and 20-40 cm depth during 2007

colloids and affect the survival of vegetation in the long run [25]. Visual observations during field visits also indicated sign of stress e.g., leaf burn in creosote and wilting in the mesquite possibly due to the application of sodic wastewater. The EC tolerance limit for mesquite is 9.36 dS m^{-1} [26] and for creosote is 7.51 dS m^{-1} [27]. The highest measured wastewater EC form 2006 -2008 was 5.64 dS m^{-1}. Thus, with regard to EC of wastewater, there is no immediate danger for the sustainability of native shrubs in the area. However, shallow rooted annual and perennial weed mustard may be threatened due to higher SAR irrigation water (37.16 ± 2.48).

Water transport and retention parameters

There are several attributes of wastewater, such as SAR, EC and OM content that can affect the soil hydraulic properties. Soil porosity can change due to the blockage of the inter-soil spaces by suspended materials [6] and can also impact soil hydraulic conductivity [28,29]. A one -way ANOVA contrast detected significant difference for K_s and d between irrigated -I and unirrigated plots at 0-20 cm depth (Table 3).The plot and vegetation interactions were significant for Ks at both depths and plot x vegetation interaction at 0-20 cm depth (Table 4). The average K_s of canopies and intercanopy areas at 0-20 cm depth in the unirrgated plot was 15.18 ± 1.50 cm h^{-1}, irrigated plot -I was 11.16 ± 1.42cm h^{-1} and in irrigated plot-II was 12.33 ± 0.80cm h^{-1} (Table 4). The K_s was higher under mesquite canopies (18.20 ± 1.29 cm h^{-1}) followed by creosote (14.20 ± 0.78 cm h^{-1}) and intercanopy areas (4.80 ± 0.34 cm h^{-1}) in all three plots (Table 3). Higher K_s under mesquite canopies than intercanopy areas and creosote canopies were likely due to higher sand content and higher amounts of macrospores associated with coppice dunes. In addition, differences in K_s between vegetation canopies might be due to the differences in morphological structure of the vegetation and differences in the interception of wastewater by vegetation canopies. A white coating on the soil surface was observed only in the intercanopy areas, which was due to the repreciptation of salt due to evaporation and could have caused reductions in the K_s at the intercanopy areas. The water content at FC and AWC are reported to increase due to the application of wastewater [30]. In this study, significant differences for water content at FC were detected between irrigated and unirrigated plots at both depths; some differences were observed among vegetations but were not significant (Table 4). No significant plot, vegetation or plot x vegetation interactions were detected for AWC and θd at 0-20 cm depth (Table 5) but θd was significantly different among plots and vegetation at 20-40 cm depth (Table 5). The θd was higher under the mesquite than creosote canopy and was in accord with high macroporosity of coppice dunes.

Soil moisture content variations under vegetation and intercanopy areas in different plots expressed as standard errors were generally lower at most suctions for vegetation canopies as well as for intercanopy areas in unirrigated than in irrigated-I and irrigated-II plots at 0-20 cm depth (Figure 1). The coefficient of determination (R^2) between measured and [31] model fitted $h(\theta)$ ranged from 0.96 to 0.99 (Table 6). The bubbling pressure, which is the inverse of α, was higher under vegetation canopies and intercanopy areas of unirrigated plot than both the irrigated plots. The irrigated plots have higher SAR and EC and lower bubbling pressure, which could be due to the higher osmotic potential than the unirrigated plot.

Soil nitrate and chloride concentration

Significant plot, vegetation and plot x vegetation interaction effects were obtained for NO$_3^-$ at 0-20 and 20-40 cm depths (Table 7). One-way ANOVA contrasts also detected differences for NO$_3^-$ between creosote canopies and intercanopy areas at 0-20 cm depth, between mesquite and creosote canopie at 60-80 and 100-150 cm in the irrigated plot-I (Table 8). NO$_3^-$ concentration was higher under mesquite canopies in both irrigated and unirrigated plots than under creosote canopies and intercanopy areas at 0-20 cm depth (Figure 2A). Mesquite is N fixing tree and that may be the reason for higher NO$_3^-$ under mesquite canopies because nitrate concentration of effluent water was low. It is reported that mesquite can store soil nitrogen 3 to 7 times greater beneath its canopies than in the interspaces between species [32]. Higher NO$_3^-$ at upper depths than deeper depths indicated no leaching of NO$_3^-$. Significant plot, vegetation and plot x vegetation interaction effects were observed for Cl$^-$ at 0-20 cm and only plot interaction was significant at 20-40 cm depth (Tables 7). Chloride concentration was higher under creosote canopies than mesquite and intercanopy areas in irrigated plot-I at all depths (Figure 2B; Table 8). The Cl$^-$ concentration almost linearly increased with depth under creosote and intercanopy areas. Higher Cl$^-$ concentration under creosote canopies than intercanopy areas and mesquite canopies could be due to the higher wastewater interception by creosote canopies. Soil Cl$^-$ accumulation was observed between 60 and 150 cm depth under creosote and intercanopy areas (Figure 3). However, a lower level of Cl$^-$ under mesquite might be the effect of higher K_s that resulted in the deeper leaching of Cl$^-$ below the sampling depths. This is also supported by larger errors in the Cl$^-$ balance (total applied-available at 0-150 cm depth) under mesquite

Vegetation	Sand (%)	Silt (%)	Clay (%)	BD (Mg m^{-3})	K_s (cm h^{-1})	AWC (cm^3 cm^{-3})	FC (cm^3 cm^{-3})	d (cm^3 cm^{-3})
				Unirrigated				
Mesquite	89.77 ± 0.31	3.61 ± 0.24	6.62 ± 0.37	1.52 ± 0.00	22.20 ± 2.82	1.85 ± 0.13	0.11 ± 0.00	0.14 ± 0.01
Creosote	89.69 ± 0.41	3.83 ± 0.41	6.48 ± 0.72	1.57 ± 0.01	12.35 ± 0.30	2.02 ± 0.15	0.13 ± 0.00	0.11 ± 0.14
Intercanopy	88.64 ± 1.15	4 .00 ± 0.57	7.36 ± 0.57	1.59 ± 0.03	11.00 ± 1.40	1.27 ± 0.19	0.11 ± 0.00	0.12 ± 0.00
Average	89.37 ± 0.62	3.81 ± 0.40	6.82 ± 0.55	1.56 ± 0.01	15.18 ± 1.50	1.71 ± 0.15	0.12 ± 0.00	0.12 ± 0.05
				Irrigated-I				
Mesquite	89.19 ± 0.06	4.67 ± 0.05	7.14 ± 0.13	1.54 ± 0.01	13.54 ± 1.58	2.06 ± 0.25	0.16 ± 0.01	0.12 ± 0.03
Creosote	88.94 ± 0.16	3.41± 0.22	7.62 ± 0.08	1.49± 0.00	11.65± 1.97	2.90 ± 0.19	0.21 ± 0.01	0.11 ± 0.01
Intercanopy	87.98 ± 0.57	4.2 ± 0.33	7.84 ± 0.09	1.57± 0.01	8.20 ± 0.72	2.21 ± 0.29	0.17 ± 0.01	0.10 ± 0.01
Average	88.70 ± 0.26	3.76 ± 0.20	7.53 ± 0.10	1.53 ± 0.01	11.16 ± 1.42	2.39 ± 0.24	0.18 ± 0.01	0.11 ± 0.01
				Irrigated-II				
Mesquite	89.35 ±0.66	3.67 ± 0.72	6.98 ± 0.21	1.51 ± 0.01	18.20 ± 1.29	2.08 ± 0.21	0.17 ± 0.01	0.13 ± 0.00
Creosote	88.98 ± 0.43	3.90 ± 0.36	7.12 ± 0.16	1.50 ± 0.03	14.00 ± 0.78	2.37 ± 0.14	0.20 ± 0.01	0.10 ± 0.00
Intercanopy	89.12 ± 1.33	2.83 ± 1.20	8.05 ± 0.33	1.55 ± 0.01	4.80 ± 0.34	2.07 ± 0.53	0.16 ± 0.00	0.10 ± 0.02
Average	89.15 ± 0.80	3.47 ± 0.76	7.38 ± 0.23	1.52 ± 0.01	12.33 ± 0.80	2.17 ± 0.29	0.17 ± 0.00	0.11 ± 0.02
	One way ANOVA Contrast							
Irri-I vs. Uni	0.055	0.315	0.201	0.074	<0.001*	0.823	0.047*	0.029*
Irri-II vs. Uni	0.093	0.106	0.319	0.285	0.496	0.446	0.005*	0.094
Irri-I vs. Irri-II	0.085	0.523	0.057	0.603	0.459	0.62	0.39	0.29

*Indicates significant differences at $P< 0.05$

Table 5: Mean, standard errors and one-way ANOVA contrasts between plots for particle size, bulk density (BD) and hydraulic conductivity (K_s) available water content (AWC), field capacity (FC), and drainable porosity (θd) at 0-20 cm depth during 2007.

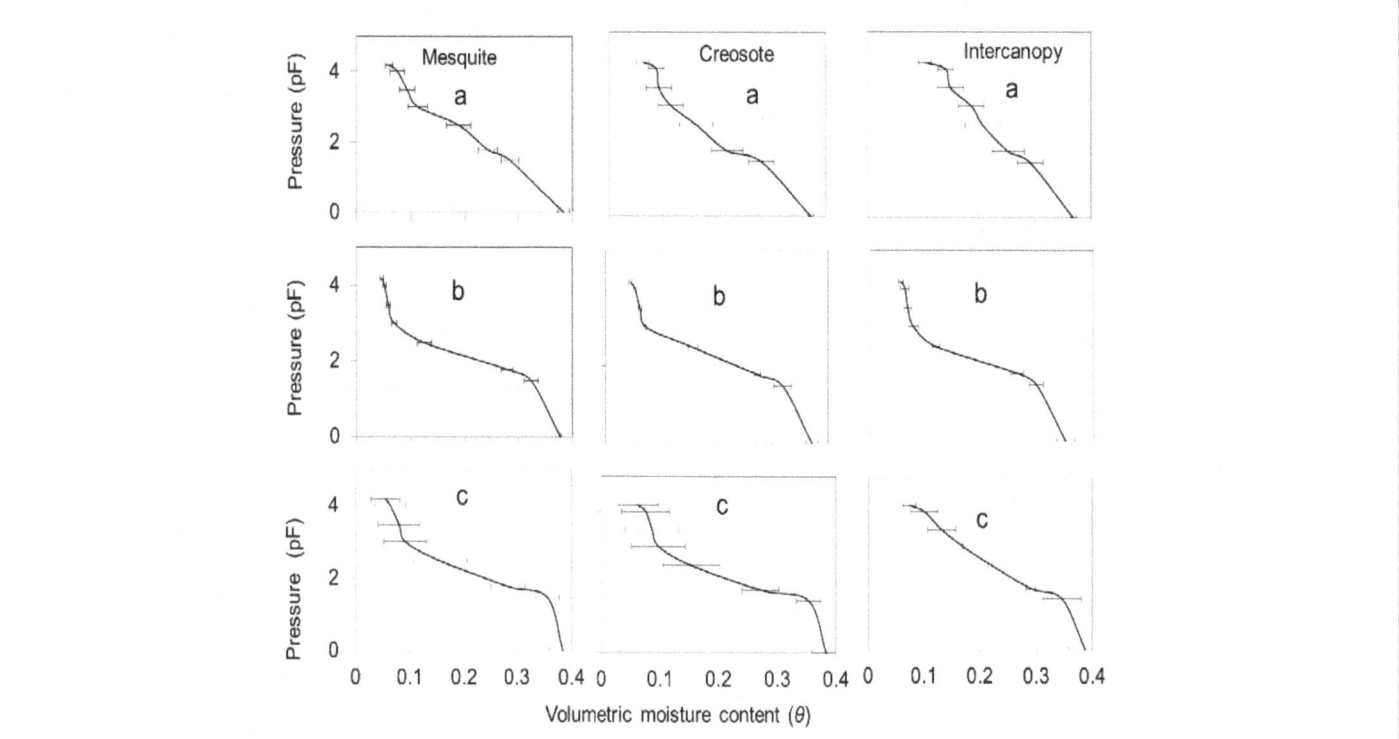

Figure 1: Soil moisture release curves of mesquite, creosote, and intercanopy areas at 0-20 cm depth by plot where pF is log of pressure in centimeters (a) irrigated plot-I (b) unirrigated (c) irrigated plot-II during 2007 [1-5].

Plots	Vegetation	θ_r	θ_s	α	η	R^2	α^{-1}cm
Irrigated-I	Mesquite	0.03 ± 0.02	0.38 ± 0.00	0.65 ± 0.15	1.35 ± 0.03	0.98	1.54
	Creosote	<0.001	0.36 ± 0.01	0.94 ± 0.47	2.10 ± 0.89	0.98	1.06
	Intercanopy	<0.001	0.35 ± 0.05	0.83 ± 0.47	1.13 ± 0.00	0.99	1.22
Unirrigated	Mesquite	0.04 ± 0.00	0.37 ± 0.00	0.17± 0.05	1.93 ± 0.13	0.99	5.88
	Creosote	0.03 ± 0.01	0.36 ± 0.00	0.17 ± 0.05	1.77 ± 0.19	0.98	5.56
	Intercanopy	0.04 ± 0.00	0.36 ± 0.00	0.18 ± 0.00	1.79 ± 0.05	0.99	5.56
Irrigated-II	Mesquite	0.05 ± 0.01	0.37 ± 0.00	0.38 ± 0.04	1.35 ± 0.04	0.98	2.63
	Creosote	0.09 ± 0.02	0.39 ± 0.00	0.44 ± 0.04	1.39 ± 0.08	0.96	2.27
	Intercanopy	0.01± 0.01	0.37 ± 0.02	0.50 ± 0.04	1.21 + 0.02	0.99	2.00

Where θ_r is residual soil moisture, θ_s is saturation soil moisture, α and η are equation parameters, R^2 is coefficient of determination

Table 6: Mean and standard errors for the van Genuchten (1980) parameters at 0-20 cm depth in both irrigated and unirrigated plots during 2007.

canopy than creosote or intercanopy area. An earlier study conducted on the same site reported high Cl⁻ concentration in the upper profile (0-15cm) of intercanopy areas due to wastewater ponding that could not be supported by this study and the white precipitate observed in the intercanopy areas were primarily due to Na⁺. The NO_3 and Cl⁻ are weakly held anions and can leach to greater depths with percolating water; however, most of the applied NO_3^- was accounted for within 0-150 cm depth. This study demonstrated that Cl⁻ but not the NO_3^- was leached below the sampling depths of 150 cm.

Soil electrical conductivity and pH

Significant interactions in EC were obtained only among plots (Table 7). The EC was higher under creosote than mesquite canopies at 0-20 cm depth of the irrigated plot-I (Figure 2C). Higher EC under creosote canopies was also in accord with the higher wastewater interception by the canopies. The EC was similar under vegetation canopies at all sampled depths in the unirrigated plot. Similar to Cl⁻, EC in irrigated-I increased by depth under both vegetation canopies and intercanopy areas.

Increased irrigation with salty water generally tended to increase soil EC with soil depth except at shallow (2.5-5 cm) depths because of the evaporation at the soil surface [33]. Similar patterns of increases in EC were observed except under mesquite canopies in irrigated plot-II. These values were lower in 2007 than those reported in 2005 [11]. This might be due to the time of the sampling, amount of wastewater application and precipitation. Samples were collected during July 2007 after some rainfall events and no application of wastewater was made during March 2007 to July 2007. Whereas in 2005 samples were collected during December and wastewater was continuously applied from September onwards with no precipitation recorded during the past three months.

Soil pH was similar (9.20 ± 0.01 to 9.80 ± 0.09) under vegetation canopies and intercanopy areas in both irrigated plots until 60 cm depth. Although plot interaction for pH was significant at 0-20 and 20-40 cm depths (Table 7), one-way ANOVA contrasts for pH did not detect differences between vegetation canopies and intercanopy areas in the irrigated plots (Table 8). Irrigation with wastewater with a pH of 9.70 ± 0.10 on soils in irrigated plots raised the soil pH to >9. Although

Source	DF	F value	Pr>F	F Value	Pr>F
			0-20 cm		20-40 cm
			--------NO$_3^-$--------		
Plot	2	16.33	<.0001*	16.24	<0.0001*
Vegetation	2	8.12	<.005*	8.5	<.005*
Plot x vegetation	4	4.7	<.005*	4.3	<.05*
			--------Cl$^-$--------		
Plot	2	24.45	<0.0001*	9.3	<.05*
Vegetation	2	10.67	<.0005*	2.84	0.177
Plot x vegetation	4	4.84	<.005*	2.09	0.124
			--------EC--------		
Plot	2	11.92	<.05*	13.96	<.05*
Vegetation	2	2.08	0.240	3.07	0.155
Plot x vegetation	4	2.14	0.117	1.62	0.213
			--------pH--------		
Plot	2	45.69	<0.0001*	66.82	<.005*
Vegetation	2	9.57	<.05*	1.87	0.267
Plot x vegetation	4	0.25	0.908	1.31	0.303
			--------SAR--------		
Plot	2	7.14	<.001*	3.47	0.133
Vegetation	2	1.66	0.298	0.06	0.946
Plot x vegetation	4	0.68	0.61	2.11	0.141
			--------Na$^+$--------		
Plot	2	19.53	<.005*	18.52	<.005
Vegetation	2	1.5	0.327	1.08	0.421
Plot x vegetation	4	0.51	0.731	0.58	0.684
			--------ESP--------		
Plot	2	9.48	<.005*	5.21	0.076
Vegetation	2	0.93	0.420	0.06	0.946
Plot x vegetation	4	0.64	0.645	2.01	0.157
			--------OM--------		
Plot	2	0.1	0.905	0.31	0.738
Vegetation	2	0.96	0.456	2.91	0.083
Plot x vegetation	4	3.04	0.05*	3.04	0.05

* Indicates significant differences at $P < 0.05$

Table 7: Values of F statistic and the probability (Pr) from analysis of variance (n=27) for nitrate (NO$_3^-$), chloride (Cl$^-$), electrical conductivity (EC), pH, sodium adsorption ratio (SAR), sodium (Na$^+$), exchangeable sodium percentage (ESP), and organic matter (OM) at 0-20 and 20-40 cm depth during 2007

Chemical properties	0-20	20-40	40-60	60-80	80-100	100-150
			---------Irrigated-I---------			
pH	(>.28)[1,2,3]	(>.72)[1,2,3]	(>.321)[1,2,3]	(>.15)[1,2,3]	(>.11)[1,2,3]	(>.25)
EC	(<.05)[1*] (>.08)[2,3]	(>.72)[1,2,3]	(>.32)[1,2,3]	(>.15)[1,2,3]	(>.12)[1,2,3]	(>.12)[1,2,3]
NO$_3^-$	(<.009)[3*] (>.13)[1,2]	(>.23)[1,2,3]	(>.32)[1,2,3]	(<.01)[1*] (>.29)[2,3]	(>.15)[1,2,3]	(<.01)[1*] (>.07)[2,3]
Cl$^-$	(<.05)[1,3*] (>.45)[2]	(<.05)[2*] (>.10)[1,3]	(>.08)[1,2,3]	(>.07)[1,2,3]	(>.18)[1,2,3]	(>.25)[1,2,3]
SAR	(>.24)[1,2,3]	(>.35)	(>.31)[1,2,3]	(>.64)	(>.18)[1,2,3]	(>.25)[1,2,3]
Na$^+$	(>.37)[1,2,3]	(>.35)[1,2,3]	(>.40)[1,2,3]	(>.30)[1,2,3]	(>.12)[1,2,3]	(>.15)[1,2,3]
ESP	(>.26)[1,2,3]	(>.35)[1,2,3]	(>.37)[1,2,3]	(>.11)[1,2,3]	(>.25)[1,2,3]	(>.11)[1,2,3]
OM	(>.12)[1,2,3]	(<.05)[2,3*] (>.97)[1]	(<.05)[1,2,3] (>.45)[1]	(<.05)[1,2*] (>.59)[3]	(<.04)[1*] (>.27)[2,3]	(>.15)[1,2,3]
			---------Irrigated-II---------			
pH	(>.28)[1,2,3]	(>.51)[1,2,3]	(>.54)[1,2,3]	(>.58)[1,2,3]	(>.36)[1,2,3]	(>.56)[1,2,3]
EC	(<.05)[1*] (>.08)[2,3]	(>.06)[1,2,3]	(>.08)[1,2,3]	(>.13)[1,2,3]	(>.15)[1,2,3]	(>.27)[1,2,3]
NO$_3^-$	(>.13)[1,2,3]	(>.34)[1,2,3]	(>.05)[1,2,3]	(>.09)[1,2,3]	(>.37)[1,2,3]	(>.12)[1,2,3]
Cl$^-$	(<.05)[1,3*] (>.45)[2]	(>.10)[1,2,3]	(>.21)[1,2,3]	(>.26)[1,2,3]	(>.55)[1,2,3]	(>.26)[1,2,3]
SAR	(>.29)[1,2,3]	(>.33)[1,2,3]	(>.51)[1,2,3]	(>.38)[1,2,3]	(<.05)[3*] (>.54)[1,2]	(>.54)[1,2,3]
Na$^+$	(<.05)[2*] (>.09)[1,3]	(>.05)[1,2,3]	(>.36)[1,2,3]	(>.31)[1,2,3]	(>.25)[1,2,3]	(>.42)[1,2,3]
ESP	(>.05)[1,2,3]	(>.33)[1,2,3]	(>.53)[1,2,3]	(>.38)[1,2,3]	(<.05)[3*] (>.26)[1,2]	(>.55)[1,2,3]
OM	(>.09)[1,2,3]	(<.05)[2,3*] (>.59)[1]	(<.05)[2,3*] (>.14)[1]	(>.12)[1,2,3]	(<.05)[2,3*] (>.09)	(<.05)[2] (>.06)[1,3]

[1]= mesquite vs. creosote, [2]= mesquite vs. intercanopy, [3]= creosote vs. intercanopy. Numbers inside the parenthesis indicated the contrast P-values
*Indicates significant differences at $P < 0.05$

Table 8: One way ANOVA contrast for chemical properties at different depths between vegetation canopies and intercanopy areas in irrigated-I and irrigated-II plots during 2007.

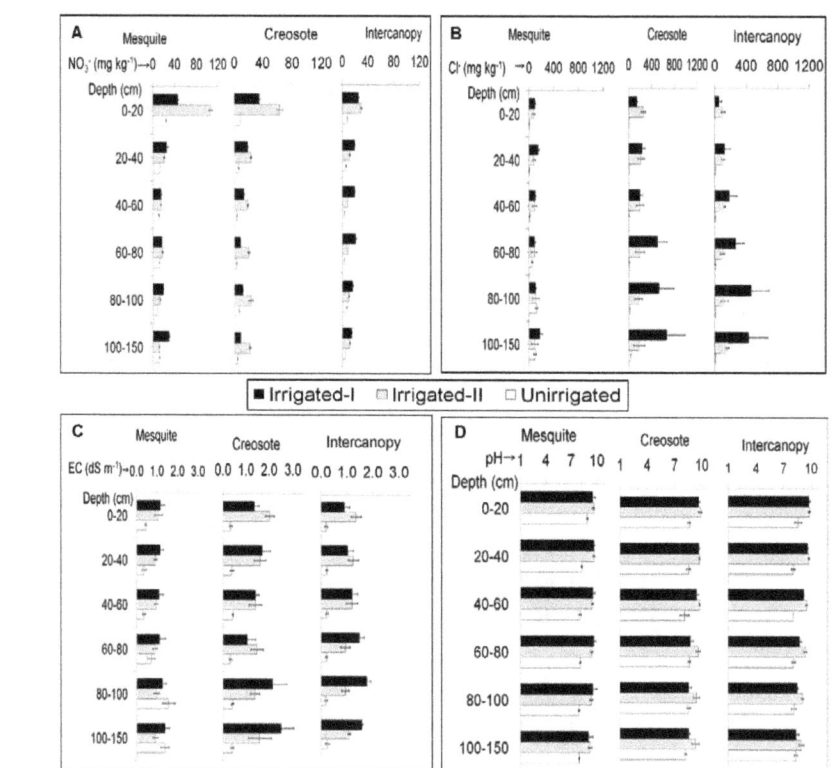

Figure 2: Concentration of (A) nitrate, NO3;(B) chloride, Cl-; (C) electrical conductivity, EC and (D) pH in three plots under the canopies of mesquite, creosote and intercanopy area during 2007 [1-5].

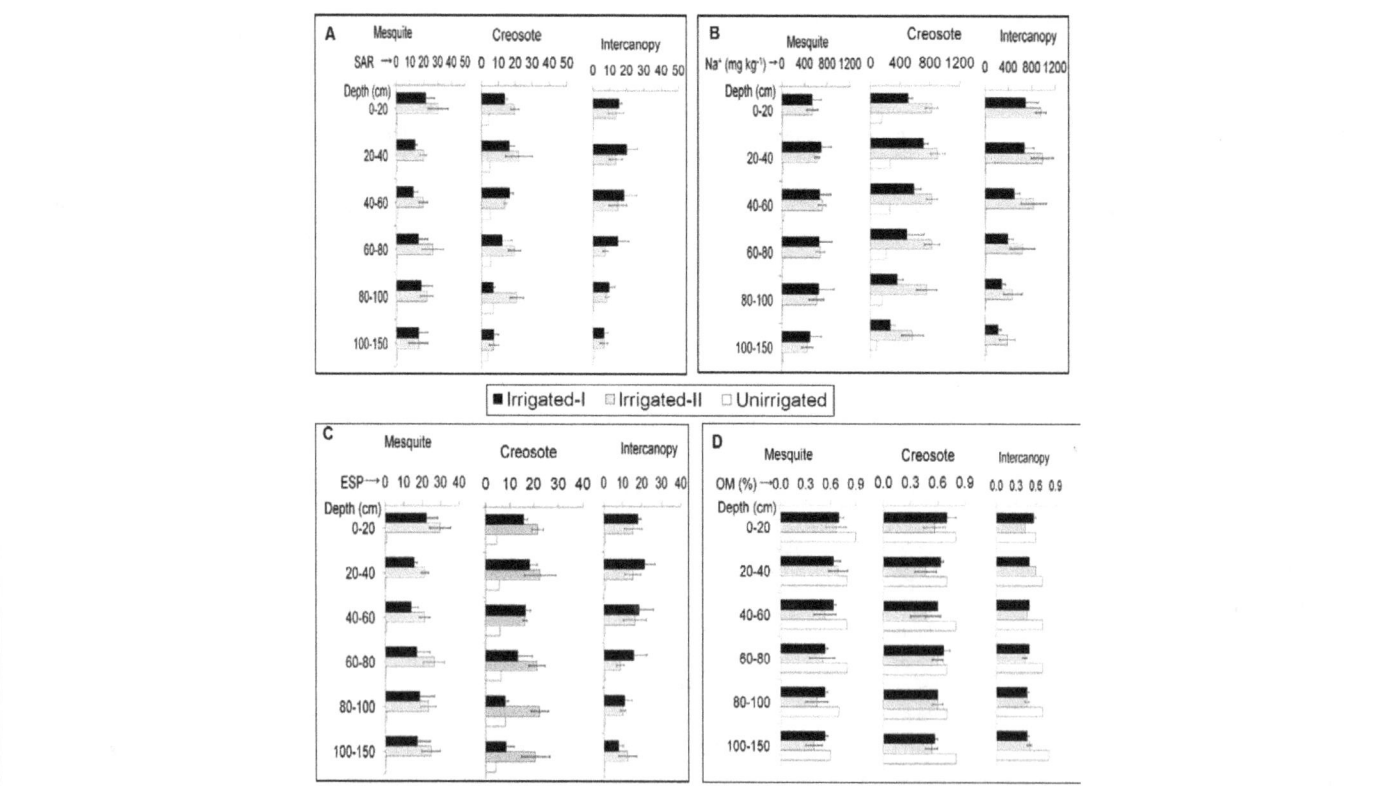

Figure 3: Concentration of (A) sodium adsorption ratio, SAR; (B) sodium, Na+; (C) exchangeable sodium percentage, ESP; and (D) organic matter, OM, in three plots under the canopies of mesquite, creosote and intercanopy area during 2007 [1-5].

mesquite and creosote are deep rooted bushes, it is difficult to assess the exact influence of high surface pH on their survival. However, high pH can certainly have an effect on survival and growth of native perennial and herbal vegetation by reducing the availability of certain micronutrients, particularly iron (Fe) and manganese (Mn).

Sodium adsorption ratio and exchangeable sodium percentage

Application of high SAR wastewater raised soil SAR in both irrigated plots and the SAR was higher in irrigated than unirrigated plots at most depths (Figure 4A). Significant plot interactions for SAR were observed at 0-20 cm depth alone (Table 7). One way ANOVA contrasts for SAR did not detect differences between vegetation canopies and intercanopy areas among the plots (Table 8). The SAR under vegetation canopies and intercanopy areas was >15 and pH> 8.5 within 0-100 cm depth which is characterized by reduced nutrient and micronutrient availability (Brady and Weil, 2000). Mesquites are deep-rooted plants which can survive with less moisture [32]. The rooting depth is about 12 m for mesquite and 3 m for creosote. However, majority of mesquite roots are distributed within 0-100 cm depth [34] and creosote within 0-25 cm depth [35]. Therefore high SAR and Na^+ concentration would affect the survival of mesquite and creosote bushes along with other perennial vegetation. Significant plot interactions were observed for Na^+ concentration at 0-20 and 20-40 cm depths (Table 7). The Na^+ concentration was higher in the intercanopy areas at 0-20 cm depth than under vegetation canopies in both irrigated plots (Figure 3B). Higher Na^+ in upper depths in the intercanopy areas were likely due to lower K_s and θd at the intercanopy areas than under the vegetation canopies which accumulated Na^+ in the upper depths. The ESP showed a similar trend as SAR and only plot interaction was significant (Figure 3C, Table 7). Differences in ESP were also detected between creosote canopies and the intercanopy areas at 0-20 and 80-100 cm depth in the irrigated plot-II (Table 8). However, no significant plots, vegetations and plot x vegetation interactions were observed for ESP at 0-20 and 20-40 cm depth (Table 7).

Soil organic matter

Few differences were detected for OM between mesquite canopies and intercanopy areas, between creosote canopies and intercanopy areas at 20-40, 40-60, 80-100 and 100-150 cm depth of the irrigated plots (Table 7). The EPA has recommended wastewater application as a method of recycling nutrients and organic matter. However, organic matter content was lower in both irrigated plots than in the unirrigated plot. Soil microorganisms and plants prefer a near neutral pH range of 6 to 7 for better performance [36]. Since the pH of irrigated plot is >9 at upper depths, it may have decreased the performance of microorganisms and the decomposition of OM in the irrigated plots. This study did not support that land application of solid organic residuals increases the OM content and soil moisture retention [3].

Vegetation analysis

The analysis of plant samples showed higher amount of Na^+ in the vegetation of irrigated than the unirrigated plots. The Na^+ content of creosote was eleven times higher in irrigated plots (880 mg.kg⁻¹) than the unirrigated plot (80 mg kg⁻¹), mesquite Na^+ content was two times higher in irrigated (1600 mg kg⁻¹) than unirrigated (800 mg.kg⁻¹) and perennial vegetation Na^+ content was three times higher in irrigated (240 mg.kg⁻¹) than unirrigated (80 mg kg⁻¹). Total percentage N in irrigated mesquite was 3.5 %, unirrigated mesquite was 2.9%, irrigated creosote 2.5% and unirrigated creosote 1.9%. The N percentage in the irrigated perennial vegetation was three times higher (4.952) then in the unirrigated weeds. Thus native vegetations were taking up chemical

constituents from the soil added through wastewater. The SAR under vegetation canopies and intercanopy areas was >15 and pH > 8.5 within 0-100 cm depth which is characterized by reduced nutrient and micronutrient availability (Brady and Weil, 2000). As the primary vegetation in the study areas are mesquite and creosote with rooting depths of about 12 m and 3 m, respectively. Majority of mesquite roots are distributed within 0-100 cm depth [34] and creosote within 0-25 cm depth [35]. Therefore high SAR and Na^+ concentration would affect the survival of mesquite and creosote bushes along with other perennial vegetation.

Conclusions

Chemical parameters were higher in the effluent than in the influent primarily due to evaporation in the holding pond. Low sprinkler uniformity in both irrigated plots was observed primarily due to the non uniform sprinkler distances, wind velocities and wastewater interception by vegetation canopies. Application of wastewater containing high EC, SAR, and Na^+ concentration decreased the K_s of the irrigated west mesa soil. NO_3^- did not leach to the deeper depths but Cl^- did leach below the sampling depths. High Na^+ concentration (>693 mg kg⁻¹), SAR (>15) and pH (>8.5) at 0-100 cm depth of the irrigated plots threaten the survival of woody as well as annual and perennial forbs and grass in the study areas as can be seen from high Na^+ content of vegetation of the irrigated area. Necessary steps should be taken to schedule uniform application of wastewater all around the year and measures should be taken to reduce the evaporation in the holding pond. Wastewater application in the site should also take into account the relative differences and importance of intercanopy and under the canopy soils.

Acknowledgement

Authors thank New Mexico State University Agricultural Experiment Station, Water Resources Research Institute, and City of Las Cruces for help and support to this study.

References

1. Hamilton AJ, Stagnitti F, Xiong X, Kredil SL, Benke KK, Maher P (2007) Wastewater irrigation: The State of Play. Vadose Zone J 6: 823-840.

2. Magesan GN (2001) Changes in soil physical properties after irrigation of two forested soils with municipal wastewater. N Z J For Sci 31: 188-195.

3. Magesan GN, Wang H (2003) Application of municipal and industrial residuals in New Zealand forests. Aust J Soil Res 41: 557-569.

4. Adhikari P, Shukla MK, Mexal JG (2012-a) Spatial variability of infiltration rate and soil chemical properties of desert soils: implications for management of irrigation using treated wastewater. Transaction of the ASABE 55: 1711-1721.

5. Mamedov AI, Shainberg I, Levy GJ (2000) Irrigation with effluent water: effects of rainfall energy on soil infiltration. Soil Sci Soc Am J 64: 732-737.

6. Abedi-Koupai, Mostafazadeh JB, Bagheri MR (2006) Effect of treated wastewater on soil chemical and physical properties in an arid regions. Plant Soil Environ 52: 335-344.

7. Halliwell DJ, Barlow MK, Nash MD (2001) A review of the effect of wastewater sodium on soil physical properties and their implications for irrigation systems. Aust J Soil Res 39: 1259-1267.

8. Sparks DL (2003) Environmental soil chemistry.2nd edition, Academic Press, San Diego, California, USA.

9. Agassi M, Tarchitzky J, Keren R, Chen Y, Goldstein D, et al. (2003) Effects of prolonged irrigation with treated municipal effluent on runoff rate. J Environ Qual 32: 1053-1057.

10. Rostango CM, Sosebee RE (2001) Biosolids application in the Chihuahuan desert: Effects on runoff water quality. J Environ Qual 30: 160-170.

11. Babcook M, Shukla MK, Picchioni GA, Mexal JG, Daniel D (2009) Chemical and physical properties of Chihuahuan desert soils irrigated with industrial effluent. Arid Land Research and Management 23: 47-66.

12. Adhikari P, Shukla MK, Mexal JG (2012-b) Treated wastewater application

in Southern New Mexico: Effect on soil chemical properties and surface vegetation. New Mexico Journal of Science 46: 105-120.

13. Adhikari P, Shukla MK, Mexal JG (2011-a) Spatial variability of electrical conductivity of desert soil irrigated with treated wastewater: implications for irrigation management. Applied and Environmental Soil Science 2011: 1-11.

14. Adhikari P, Shukla MK, Mexal JG, Sharma P (2011-b) Assessment of the soil physical and chemical properties of desert soils irrigated with treated wastewater using principal component analysis. Soil Sci 176 : 356-366.

15. Adhikari P, Shukla MK, Mexal JG (2012-c) Spatial variability of soil properties in an arid ecosystem irrigated with treated municipal and industrial wastewater. Soil Sci 177: 458-469.

16. Gile LH, Hawley JW, Grossman RB (1981) Soils and geomorphology in the basin and range area of southern New Mexico-Guide book to the Desert Project. New Mexico Bureau of Mines and Mineral Resources, Memoir 39: 1-222.

17. Gee GW, Bauder JW (1986) Particle size analysis. Methods of soilanalysis Part 1 (2nd edition), ASA and SSSA, Madison, WI.

18. Blake GR, Hartge KH (1986) Bulk density. Methods of soil analysis. Part 1, 2nd edition. ASA and SSSA, Madison, WI.

19. Klute A, Dirksen C (1986) Hydraulic conductivity and diffusivity. Methods of soil analysis Part I (2nd edition): WI, USA.

20. Klute A (1986) Water retention. Methods of soil analysis Part I (2nd edition): WI, USA.

21. Black CA (1965) Methods of soil analysis. Chemical and microbiological properties, ASA Inc, Madison, WI.

22. Maynard DG, Kalra YP (1993) Nitrate and exchangeable ammonium nitrogen. Soil sampling and methods of analysis. CSSS, Lewis Publishers, USA.

23. Sparks DL (2003) Environmental soil chemistry.2nd edition, Academic Press, San Diego, California, USA.

24. Menner JC, McLay CDA, Lee R (2001) Effects of sodium contaminated wastewater on soil permeability of two New Zealand soils. Aust J Soil Res 39: 877-891.

25. Jalali M, Merrikhpour H (2008) Effects of poor quality irrigation waters on the nutrient leaching and groundwater quality from sandy soil. Environ Geology 53: 1289-1298.

26. Felker P, Clark PR, Laag AE, Pratt PF (1981) Salinity tolerance of the tree legumes: mesquite (Prosopisglandulosa var. torreyana, P. velutina and P. articulata), algarrobo (P. chilensis), Kiawe (P. pallida), and Tamarugo (P. tamarugo) grown in sand culture on nitrogen free media. Plant Soil 61: 311-317.

27. Al-Jibury LK (1972) Salt tolerance of some desert shrubs in relation to their distribution in the southern desert of North America. Arizona State University, Tempe AZ, USA.

28. Coppola A, Santini A, Boti P, Vacca S (2003) Urban wastewater effects on water flow and solute transport in soils. J Environ Sci Health 38: 1469-1478.

29. Al-Haddabi M, Ahmed M, Kacimov A, Rahman S, Al-Rawahy S (2004) Impact of treated wastewater from oil extraction process on soil physical properties. Commun Soil Sci Plan 35: 751-758.

30. Ibrahim A, Agunwamba JC, Idike FI (2005) Effects of wastewater effluent re-use on irrigated agricultural soils. J Sust Agri and the Environ 7: 60-80.

31. VanGenuchten MT (1980) A closed-form equation for predicting the hydraulic conductivity of unsaturated soils. Soil Sci Soc Am J 44: 892-898.

32. Ansley RJ, Huddle JA, Kramp BA (1997) Mesquite ecology. Texas Agricultural Experiment Station, Vernon, California, USA.

33. Costa JL, Prunty L, Montgomery BR, Richardson JL, Alessi RS (1991) Water quality effects on soils and alfalfa: II. Soil Physical and Chemical Properties. Soil Sci Soc Am J 55: 203-209.

34. Heitschmidt RK, Ansley RJ, Dowhower SL, Jacoby PW, Price DL (1988) Some observations from the excavation of honey mesquite root systems. Journal of Range Management Archives 41: 227-231.

35. Baynham P (2004) Sonoran originals: The unappreciated smell of rain. Master Garden Journal: 22-24.

36. Sylvia MD, Fuhrmann JJ, Hartel PG, Zuberer DA (2005) Principles and applications of soil microbiology. (2nd edition), Prentice Hall, Upper Saddle River, New Jersey, USA.

4

Contributions of Social Networking to Accessing Resources for Irrigation Farming among Farming Households in North Central Nigeria

Ifabiyi JO* and Komolafe SE

Department of Agricultural Extension and Rural Development, University of Ilorin, Ilorin, Nigeria

Abstract

This study focused on the contributions of social networking in accessing resource for irrigation farming among farming households in north central Nigeria. Systematic random sampling was used to select 194 respondents from Oke-oyi and shonga (Kwara State) and Ejiba (Kogi State). Data were obtained using structured questionnaire. The data were analyzed using descriptive statistics and Pearson Product Moment Correlations. The findings indicated that market information with the mean score of 2.75, was the most important contribution of social networking in accessing resource for irrigation farming while lack of input with mean score of 2.27 was the highest ranked constraint faced by the farmers using irrigation for crop production. The result of correlation analysis revealed that there was no significant relationship between the contributions of social networking in accessing resources for irrigation and the constraints faced by the farmers using irrigation for crop production. Based on this results, it is therefore recommended that there should be creation of awareness on various opportunities that are available for irrigation farmers participating in social networking.

Keywords: Contribution; Social networking; Irrigation; Resources

Introduction

Irrigation is the artificial application of water to the soil for the purpose of supplying moisture for plant growth to supplement insufficient soil moisture especially during the dry season in Nigeria. Irrigation based farming in Nigeria only covers 7 percent of irrigable land [1]. The public irrigation sector in Nigeria accounts for 13% of the irrigated area and an estimated 0.25% of total agricultural area [2]. The place of irrigation in Nigeria's agricultural development cannot be overemphasised. The National Irrigation Policy is based on boosting domestic agricultural production using irrigation because rain-fed production alone cannot meet demand [3].

A social network is a set of individuals or groups who are connected to one another through socially meaningful relationships [4]. A social network can consist of groups and sub-groups of actors. Examples of such socially meaningful relationships include family, friends, or relations based on trust, giving advice, or sharing information [5]. Understanding the individual, groups or entity that enable people to access resources and collaborate to achieve shared goals is an important part of the concept of social capital. Social networkings could be spontaneous, informal, and unregulated exchanges of important information and resources among farmers, as well as efforts at cooperation, coordination, and mutual assistance that help maximize the utilization of available resources. Such networks often clearly stated rules that govern how group members cooperate to achieve common goals. These social networks have the potential to nurture self-help, mutual help, solidarity, and cooperative efforts among farmers.

In Nigeria, there are many irrigation schemes which are constructed and managed by the River Basin and Rural Development Authourities (RBRDAs) for provision of food especially during the dry season. In these irrigation schemes there are many organizations and institutions that are working together to ensure that farmers can access resources for irrigation farming. These organizations and institution are known as social networking. For this study the social networking that are available within the River Basin in Nigeria are water users association, agric extension agency, input suppliers, religious group, cooperative societies, farmers group, community based organization, non-governmental organization, neighbourhood and family. It is therefore necessary to carry out a research in order to determine the contribution of social networking in accessing resources for irrigation farming. In the light of the above, the general objectives of the study is to determine the contributons of social networking in accessing resources for irrigation farming among farming household in North Central Nigeria.

The specific objectives of the study are to:

1. Identify the socio-economic characteristics of farmers using irrigation facilities in the study area.

2. Determine the degree of participation of farmers in social networking.

3. Determine the contributions of social networking in accessing resource for irrigation farming.

4. Determine the constraints faced by farmers using irrigation for crop production.

Statement of hypothesis

The hypothesis tested;

H1: There is no significant relationship between contributions of social networking in accessing resource for irrigation farming and the constraint faced by farmers using irrigation for crop production.

Methodology

This study was carried out in two states of North Central Zone Nigeria (Kwara and Kogi State) under the Lower Niger Irrigation Development Scheme. The respondents for the study were Lower Niger

---

*Corresponding author: Ifabiyi JO, Department of Agricultural Extension and Rural Development, University of Ilorin, Ilorin, Nigeria
E-mail: oluwole4love@yahoo.com

River basin irrigation farmers who used their facilities in Kwara and Kogi States. The villages were Oke-oyi, shonga and Ejiba of which 194 farmers were interviewed (Table 1).

The population of the study comprises of all farmers using the Irrigation scheme of Lower Niger River Basin in Kogi and Kwara State (North Central, Nigeria). Lower Niger River Basin Authority (LNRBA) irrigation farmers (Kogi and Kwara States) were purposively selected for the study. The list of all the farmers at lower Niger River Basin Authority Irrigation sites were provided of which Simple Random Sampling technique was used to select 50% of the farmers in each irrigation sites (Oke-oyi 15 farmers, Shonga 25 farmers and Ejiba 154 farmers). A total of 194 respondents were interviewed for the study.

Results and Discussion

Demographic characteristics of respondents

Age: The age of farmers to a large extent affect their labour productivity and output. It also affects the adoption of innovation in traditional farming. The results showed that majority of respondents were youths and less than 40 years of age. The average age is 35.80 years. The result from Table 2 showed that 46.40% of the respondents' age falls between 31-40 years, 40.20% of them are in the age bracket 20-30 years, 8.25% of the respondents fall between 41-50 years of age while 5.1% are within the age bracket 51-60 years of age. The result therefore indicates that a good proportion of the study respondents are youths based on African Union definition of youths to be every person between 15 – 35 years of age [6]. The fact that most of the respondents are youths could imply that irrigation farming is viable and profitable, this is more so since youths would most likely undertake only paying jobs.

Gender: The findings in Table 2 showed that all the respondents 100% are male. This suggests that irrigation farming is dominated by male farmers in the study area. This may be as a result of fact that operation of the irrigation farming may be cumbersome for females. The result is in line with what Salisu reported that irrigation was a male affair only in northern Nigeria [7].

Marital status: Table 2 shows that 96.4% of the respondents are married while 3.6% are single. The reason for this result may be because irrigation farming requires high labour demand, many farmers relied on their family members to supply the needed labour requirement.

Religion: Findings from Table 2 showed that majority 94.8% of the respondents are Muslim while 5.2% are Christians. This shows that there was no religious barrier to participation in irrigation farming.

Years of Experience: Table 2 shows that 37.19% of the respondent's years of experience ranges between 6 –10 years, 30.9% of them had their years of experience ranges between 11—15 years, 14.9% of the respondents had their years of experience ranges between 1—5 years, 13.4% ranges between 16—20 years while 3.61% between 21---25 years and the mean farmers' irrigation farming experience is 10.87 years. This implies that the farmers have considerable experience in irrigation farming which could be as result of their participation in social networks over the years.

Educational level: The results from Table 2 showed that more than

Demographic Characteristics	Frequency	Percentage	Mean
Age Years			
20-30	78	40.2	35.8
31-40	90	46.4	
41-50	16	8.25	
51-60	10	5.1	
Total	194	100	
Gender			
Male	194	100	
Female	0	0	
Total	194	100	
Marital Status			
Married	187	96.4	
Single	7	3.6	
Total	194	194	
Religion			
Islam	184	94.8	
Christian	10	5.2	
Traditional religion	0	0	
Total	194	100	
Years of Experience (years)			
1-5	29	14.9	10.87
6-10	72	37.19	
11-15	60	30.9	
16-20	26	13.4	
21-25	7	3.61	
Total	194	100	
Educational Level			
No-formal	105	54.1	
Primary	76	39.2	
Secondary	13	6.7	
Total	194	100	
Quantity of land under irrigation(ha)			
≤ 1	142	73.19	1.0832
1.1-1.5	19	9.79	
1.5-2.0	20	10.33	
2.1-2.5	14	7.2	
Irrigation farming income (Naira)			
≤ 100000	125	64.43	1,16,881.44
100000-200000	54	27.84	
210000-300000	12	6.18	
310000-400000	3	1.55	
Source: Field survey, 2013			

Table 2: Demographic characteristics of respondents.

half (54.1%) of the respondents had no–formal education, 39.2% had primary education while 6.7% had secondary education. The above result confirmed the findings of Fakayode et al. who reported that majority of the irrigation farmers have not had any form of formal education [8].

Quantity of land under irrigation: The result showed that 73.19% of the respondents practiced irrigation based farming on a farm size that is less than 1 hectare, 10.33% of them had a farm size that ranges between 1.5 - 2 hectares, 9.79% have farm size ranges between 1.1 - 1.5 hectares while 7.2% have farm size ranges between of 2.0—2.5 hectares. The result shows that majority of the irrigation farmers cultivate on a farm land that is less than 1 hectare. This implies that all the respondents are all small scale irrigation farmers with little or no farm mechanization. The result is in line with what Dixon et al. reported

Irrigation site	Sample frame	Sample size
Oke-oyi	30	15
Shonga	50	25
Ejiba	308	154

Table 1: Sampling population.

that smallholder farmers usually cultivate less than one hectare of land, which may increase up to 10 ha or more in sparsely populated semi-arid areas [9].

Irrigation farming income: The results from Table 2 showed that more than half (64.43%) of the respondents' income from irrigation farming were less than # 100,000 Naira, 27.84% of the respondents had between #100,000 --- #200000 as their annual income from irrigation farming. 6.18% of the respondents had between #210,000---#300,000 as annual irrigation farming income while 1.55% got between #310,000---- #400,000 as their income from irrigation farming. The mean farmers' annual income is # 116,881.44 Naira, this implies that irrigation farmers are better off. Besides, irrigation farming had greatly reduce poverty level and improved the living standard of the farmers in the study area.

Degree of participation in social networks: The data in Table 3 showed that all the respondents (100%) are ordinary members of religious groups, neighbourhood and family, 5.12% are executive members of the water users association while 1.5% are executive members of cooperative societies.

Contribution of social networking in accessing resources for irrigation farming

The results of data analysis in Table 4 showed that market information has the highest mean score of 2.75, this is followed by access to input/technology with the mean score of 2.58, provision of information on labour with the mean score of 2.56, access to information on irrigation (2.53), provision of information on land acquisition (2.26), training of farmers / extension services (1.91) and access to credit (1.76). The result shows that market information is the most important contribution of social networks to the use of irrigation by the respondents. This implies that more farmers would likely show interest in participating in social networking and irrigation farming since the contributions are vast and visible. The result is in agreement

with Dauda et al. which states that the benefit of irrigation in Nigeria is not limited to food supply alone but it also serves as a source of income and employment during the slack period of rain-fed agriculture [10].

Constraints faced by farmers using irrigation for crop production

The result in Table 5 showed that lack of inputs was the most severe constraint and was ranked first with the highest mean score of 2.27, this result is in agreement with what Ifabiyi et al. reported that lack of input and access to credit facilities was the least motivating factors in irrigation farming [11]. This is followed by poor maintenance of irrigation facilities with mean score of 2.14, lack of access road (2.11), lack / inadequate of labourers (1.94), lack of transportation (1.91), lack of access to irrigation facilities (1.63), lack of extension services (1.56), flooding / erosion (1.49) and lack of market (1.37). This implies that farmers' productivity would be reduced due to lack of access to input such as viable seeds, fertilizer, herbicides and insecticides etc.

Testing of hypothesis

H1: There is no significant relationship between contributions of social networking in accessing resource for irrigation farming and the constraint faced by farmers using irrigation for crop production.

The result shows that there is no significant relationship between the contributions of social networking in accessing resources for irrigation farming and constraint faced by farmers using irrigation facilities (r= -0.106; p=0.141). This study therefore accepts the null hypothesis Table 6. This implies that the contribution of social networking in accessing resources for irrigation farming does not have significant influence on the constraints faced by the farmers using irrigation to discourage farmers from engaging in irrigation farming.

Conclusion

The contributions of social networking were discovered to be vast

		Executive member		Ordinary member		No participation	
	Social networks	Freq	%	Freq	%	Freq	%
1	Water user association	10	5.12	184	94.86	-	-
2	Cooperative society	3	1.5	191	98.5	-	-
3	Agriculture extension service	-	-	194	100	-	-
4	Input supplier	-	-	177	91.2	17	18.8
5	Non-governmental agency	-	-	22	11.3	172	88.7
6	Farmers groups	-	-	60	30.9	134	69.1
7	CBO	-	-	23	11.9	171	88.1
8	Religious group	-	-	194	100	-	-
9	Neighbourhood	-	-	194	100	-	-
10	Family	-	-	194	100	-	-
Source: Field survey, 2013							

Table 3: Distribution of respondents based on their degree of participation in social networks.

		High		Undecided		Low		Mean	Rank
S.No	Contribution of social networking	Freq	%	freq	%	Freq	%		
1	Market information	170	87.6	-	-	24	12.4	2.75	1
2	Access input/ technology	133	68.6	41	21.1	20	10.3	2.58	2
3	Provision of information on labour	141	72.7	20	10.3	33	17	2.56	3
4	Access information on irrigation farming	132	68	33	17	29	15	2.53	4
5	Provision of information on land acquisition	110	56.7	24	12.4	60	30.9	2.26	5
6	Training of farmers/extension service	57	29.4	63	32.5	74	38.1	1.91	6
7	Access to credit facilities	57	29.4	34	17.5	103	53.1	1.76	7
Source: Field survey, 2013									

Table 4: Distribution of respondents based on the contribution of social networking in accessing resources for irrigation farming.

S.No	Constraints	Severe freq.	%	Undecided freq.	%	Not severe freq.	%	Mean	Rank
1	Lack of inputs	118	60.8	10	5.2	66	34	2.27	1
2	Poor maintenance	99	51	24	12.4	71	36.6	2.14	2
3	Lack of access road	91	46.9	34	17.5	69	35.6	2.11	3
4	Lack/inadequate labourers	86	44.3	10	5.2	98	50.5	1.94	4
5	Lack of transportation	76	39.2	24	12.4	94	48.5	1.91	5
6	Lack of access to irrigation facilities	61	31.4	-	-	133	68.6	1.63	6
7	Lack of extension services	47	24.2	14	7.2	133	68.6	1.56	7
8	Flooding/erosion	39	20.1	17	8.8	138	71.1	1.49	8
9	Lack of market	32	16.5	7	3.6	155	79.9	1.39	9
Source: Field survey, 2013									

Table 5: Distribution of respondents based on the constraints faced by farmers using irrigation for crop production.

Variable	r value	p value	Decision
Contribution of social networking in accessing resource	-0.106	0.141	Not significant

Table 6: The result of correlation between contribution of social networking in accessing resource for irrigation farming and the constraint faced by farmers using irrigation for crop production.

and rewarding among irrigation farmers in north central Nigeria. The result shows that market information was the most important contribution of social networkings to the use of irrigation while lack of input was the highest ranked constraint faced by farmers using irrigation farming.

Recommendations

The following recommendations were suggested;

1. Government and Non-governmental organization should create more awareness about irrigation farming.

2. Farmers participating in irrigation farming should form functional groups through which they can access resources for irrigation farming.

3. Government should expand the various irrigation facilities in the country so as to attract new farmers.

There should be training of farmers.

References

1. IFAD (2012) Climate-start smallholder agriculture: what's different? International fund for agricultural development. Italy.

2. Federal Ministry of Water Resources (2004) Review of Public irrigation Sector in Nigeria.

3. FGN (2006) National Irrigation Policy and Strategy for Nigeria.

4. Wellman B, Berkowitz SD (1988) Social structures a network approach. Cambridge: Cambridge University Press.

5. Stone W (2001) Measuring social capital. Towards a theoretically informed measurement framework for researching social capital in family and community life. Research paper No.24.

6. African Union (2010) Plan of action; decade for youth development and empowerment third ordinary session of the conference of the African Union Ministers.

7. Salisu SA (2001) Individual pump ownership and associated service providers in Fadama irrigation in Northern Nigeria.

8. Fakayode SB, Ogunlade I, Ayinde O, Olabode P (2010) Factors affecting farmers' ability to pay for irrigation facilities in Nigeria: the case of oshin irrigation scheme in kwara state. Journal of sustainable development in Africa.

9. Dixon J, Tanyeri-Abur A, Wattenbach H (2003) Framework for analysing impacts of globalization on smallholders. Food and Agricultural Organization United Nations.

10. Dauda TO, Asiribo OE, Akinbode SO, Saka JO, Salahu BF (2009) An assessment of the roles of irrigation farming in the millennium development goals. African Journal of Agricultural Research 4: 445-450.

11. Ifabiyi JO, Adesiji GB, Komolafe EO, Ajibola BO (2014) Irrigation farmers motivation for participating in social networking in North Central Nigeria. Journal of environmental studies and management 7: 572-580.

5

Yield and Physiological Response of Tomato to Various Nutrient Managements under Container Grown and Drip Irrigated Conditions

Etissa E[1]*, Dechassa N[2] and Alemayehu Y[1]

[1]*Ethiopian Institute of Agricultural Research, Addis Ababa, Etiopia*
[2]*College of Agriculture and Environmental Sciences, Haramaya University, Etiopia*

Abstract

Two field experiments were conducted to study fruit yield and physiological responses of field grown tomato. The first experiment was conducted with container grown during rainy season by combining three factors namely, two levels of N (0 and 25 kg N ha-1), and two levels of P (0 and 23 kg P ha^{-1}) fertilizers and with six locally available media mix ratios (MR). The experiment was laid down on CRBD in a factorial arrangement and replicated three times. The second experiment was conducted under drip irrigation during hot dry season with three levels of daily irrigation applications: full irrigation, 80% and 60% of daily ETc irrigations. Data on marketable, unmarketable and total fruit yield were recorded and some physiological responses: quantum yield, leaf chlorophyll content and fluorescence, and stomatal conductance were assessed from sample plant leaves using various sensors. The results of container grown experiment indicated that use of combinations of starter N and MR showed a significant effect (P<0.05) on the marketable fruit yield, similarly use of media mixtures had highly significant (P<0.01) influence on the unmarketable fruit yield and finally use of media mix ratio showed a highly significant effect (P<0.01) on the total fruit yield of container grown tomato. MR3 yielded the highest total fruit yield while MR6 gave the lowest total fruit yield. Application of starter N, P or media mix did not bring any combined effect (P<0.01) on the leaf chlorophyll content. However, application of starter N caused a highly significant (P<0.01) effect on leaf quantum yield. The results of drip irrigated experiment indicated that use of various irrigation depth brought a significant (P<0.01) effect on the marketable yield of tomato. The highest fruit yield was recorded in response to full irrigation, while the lowest marketable fruit yield was recorded from 60% of full irrigation. Irrigation depth significantly (P<0.01) affected the tomato leaf chlorophyll content. The highest irrigation level increased leaf chlorophyll content and lowest irrigation depth reduced leaf chlorophyll content. Irrigation depth brought significant (P<0.01) effect on the stomatal conductance as irrigation depth decreased, leaf stomatal conductance was highly reduced. Measuring quantum yield, leaf chlorophyll content, and fluorescence and stomatal conductance of tomato plant there is indicated direct relation with yield performance that would give instant improvement of management practices of the crop. Thus further research is required to fine tune and use of this physiological response with the crop yield.

Keywords: Leaf chlorophyll; Leaf fluorescence; Stomatal conductance; Tomato

Introduction

Tomato (*Lycopersicum esculentum* M.) is grown under various production conditions such as under furrow irrigated and under rainfed in open field, on containers of variable sizes and media mixes in home gardens, under drip irrigated throughout the year in Ethiopia [1]. All smallholder growers in the Central Rift Valley of Ethiopia cultivate tomato using furrow irrigation in open field, where as commercial and semi-commercial growers cultivate in the protected structures [2,3]. The majority of tomato production in the country comes from furrow irrigated open field cultivated almost throughout the year except during the rainy season the production is lower due to diseases pressure.

Based on many literatures, plant physiological parameters such as leaf fluorescence, leaf quantum yield, leaf chlorophyll contents and stomatal conductance are among indexes that would provide indirect estimation of plant growth, development and yield [4-6].

Light energy is absorbed by chlorophyll, carotenoids and other pigment molecules present in the photosynthetic antenna molecules present in the thylakoid membranes of green plants [6-9]. Chlorophyll fluorescence is the light re-emitted by chlorophyll molecules during return from excited to non-excited states during the light absorption and used as indicator of photosynthetic energy conversion in higher plants [4]. Excited chlorophyll dissipates the absorbed light energy by driving photosynthesis (photochemical energy conversion), as heat in non-photochemical quenching or by emission as fluorescence radiation. As these processes are complementary processes analysis of

chlorophyll fluorescence is an important tool in plant research with wide spectra of applications [4].

For light to be active in leaf photosynthesis it must be absorbed; light energy that is absorbed can be dissipated in one of three ways A) lost as heat, B) re-irradiated as red light (fluorescence) or C) used in photosynthesis (photochemistry). These three processes are competitive [4,9]. That is, an increase in the efficiency of one process will decrease the yield of the other two. Even though chlorophyll fluorescence is only about 1 to 2% of the total light energy absorbed; we can use the fluorescence signal of a leaf to estimate rates of photosynthesis (photochemistry) and heat dissipation. Generally, chlorophyll fluorescence production is inversely related to photosynthesis, except when non-photochemical quenching of fluorescence (thermal dissipation) occurs [9]. Under stress or in moderate to high irradiance conditions, plant tissues increase heat production to dissipate excess energy (*Anon.*). This tends to decrease fluorescence emission, at

*Corresponding author: Etissa E, Ethiopian Institute of Agricultural Research, Addis Ababa, Etiopia, E-mail: edossa.etissa@gmail.com

least in the initial and intermediate stages of stress. Therefore, the relative balance between the three major dissipation mechanisms-photosynthesis, heat production, and chlorophyll fluorescence emission-ultimately determines the actual pattern of response observed for fluorescence [9].

Chlorophyll *a* fluorescence from light excited vegetation emanates in specific red and far-red spectral regions, and is produced by photosystems I and II (PSI and PSII) which are pigment– protein complexes involved in the initial stages of photosynthesis [4,9]. The steady-state fluorescence, known to be highly responsive to changes in environmental conditions, is widely used as indicator of plant photosynthetic function, and can be used in the early, pre-visual detection of physiological strain. Early detection may facilitate remedial action before survival, growth and productivity are constrained, and may help to forecast long term resource quality [9].

Leaf quantum yield is light reactions of photosynthesis Taiz and Zeiger [5] quantum yield measures light that has actually been absorbed. The quantum yield for a particular process can range from 0 (if that process does not respond to light) to 1.0 (if every photon absorbed contributes to the process). Photosynthetic rate is often expressed as number of molecules of CO_2 fixed or O_2 evolved per unit leaf area per unit time (for example, μ mol CO_2 m^{-2} s^{-1}), while quantum yield is expressed as number of molecules of CO_2 fixed or O_2 evolved per photon absorbed [10]. Taiz and Zeiger described that in functional chloroplasts kept in dim light, the quantum yield of photochemistry is approximately 0.95, the quantum yield of fluorescence is 0.05 or lower, and the quantum yields of other processes are negligible [5]. The vast majority of excited chlorophyll molecules therefore lead to photochemistry.

The most active photosynthetic tissue in higher plants is the mesophyll of leaves. Mesophyll cells have many chloroplasts, which contain the specialized light-absorbing green pigments, the chlorophylls. In photosynthesis, the plant uses solar energy to oxidize water, thereby releasing oxygen, and to reduce carbon dioxide, thereby forming large carbon compounds, primarily sugars. The absorption spectrum of chlorophyll *a* indicates approximately the portion of the solar output that is utilized by plants. The photosynthetic pigments absorb the light that powers photosynthesis [5].

CO_2 diffuses (stomatal conductance) from the atmosphere into leaves-first through stomata, then through the intercellular air spaces, and ultimately into cells and chloroplasts [5]. In the presence of adequate amounts of light, higher CO_2 concentrations support higher photosynthetic rates. The reverse is also true; that is, low CO_2 concentration can limit the amount of photosynthesis.

The quantum yield is another important parameter of the light reactions of photosynthesis Taiz and Zeiger described as the quantum yield of photosynthesis (ϕ) is defined as follows [5]:

$$\phi = \frac{\text{Number of photochemical products}}{\text{Total number of quanta absorbed}}$$

The quantum yield for a particular process can range from 0 (if that process does not respond to light) to 1.0 (if every photon absorbed contributes to the process). Thus estimation of any one or all of these parameters from growing plants could provide an estimation of the biomass yield of crop plants under consideration. The objectives of these experiments were to assess physiological and yield response of tomato under various nutrient managements under container grown and drip irrigated conditions at Melkassa.

Materials and Methods

Two experiments were conducted at Melkassa Agricultural Research Centre; Central Rift Valley of Ethiopia, the first experiment was with container grown tomato conducted during rainy season and the second experiment was with drip irrigated tomato during the hot-dry season.

Container grown tomato

The first experiment was conducted during the rainy season using container grown tomato. Three factors, namely, media proportions (mixes); and supplementary N and P applications rates were combined factorially. The treatments consisted of two levels of inorganic N (0 and 25 kg N ha^{-1}) fertilizer, and two levels of inorganic P (0 and 23 kg P ha^{-1}) fertilizers and with six media mix ratios. Locally available decayed animal manure and livestock droppings were collected and mixed with sand and topsoil at various proportions. Field soil (top soil), well decomposed manure, sand were mixed forming six treatment combinations (v/v) such as 1) 6:0:0, 2) 5:1:1, 3) 4:1:1, 4) 3:2:1, 5) 2.5:2.5:1 and 6) 1:3:2. These media mix ratios are in the form of Field Soil: Manure: Sand order. The details of the methodology was published in Science, Technology and Arts Research Journal [11].

The inorganic nitrogen fertilizer at two levels: without N (N at 0 kg rate ha^{-1}) and with at 25 kg N rate ha^{-1}. Similarly the inorganic phosphorous fertilizer at two levels: without P (at 0 P kg ha^{-1}) and with P at 23 kg P rate ha^{-1}. All inorganic P fertilizer were applied once through mixing with the media before filling into the pots and N were applied twice.

Two levels of N and P were combined to give four factorially combined N and P treatments, which were again combined with six media mix ratios giving twenty four treatment combinations. The container (pot) size used was 0.17 m^3 volume. The treatments were replicated three times. The experiment was laid out as a completely randomized block design in a factorial arrangement and replicated three times per treatment. The variety used was *Melakshola* tomato variety. This experiment was conducted under rainfed condition during the main season. However, supplementary irrigation was provided for the containerized tomato plants during establishment in the dry spell days and since then left as a rainfed crop.

Drip irrigated tomato

Similarly the second drip irrigated experiment was conducted during the hot and dry season, combining two factors namely, irrigation scheduling at three levels and nutrient management at five levels were combined. Three irrigation treatments 1) Full irrigation, IRI 2) 80% ETc, IRII and 3.60% ETc, IR III were arranged randomly in vertical strips to adjust irrigation depth uniformly along the strips. Integrated nutrient managements 1) rates obtained from field survey (farmers' rate) (N_FP_F) (N185 P60) [INM-I]), 2) N and P rate finding from on station experiment (N75 P50) [INM-II], 3) On station NP rate (N75 P50) +15 tone ha^{-1} (FYM) [INM- III], 4) 15 tone ha^{-1} (Manure only) [INM-IV] and 5) Zero nitrogen, phosphorous and manure [INM-V] were randomly arranged in horizontal strip plots. The treatments were replicated three times and the experiment was laid down on fifteen total number of plots. Similarly the released *Melkasholla* tomato variety was used for this experiment; a multipurpose variety, used for both fresh and processing type with semi-determinate growth habit.

Data collection

Yield data: Yields of tomato (includes both marketable, and unmarketable and total fruit yield) were obtained by summing from continuous harvesting until the last harvest.

Physiological parameters: Data on physiological traits such as leaf chlorophyll content was estimated non-destructively using a portable hand held Chlorophyll Meter (Minolta SPAD-502, Konica Minolta Sensing, Inc. Japan). One leaf per plant and five leaves per plot were measured and averaged. The SPAD readings were measured at 90 DAT on fully expanded leaves. The leaf quantum yield measurements were taken using the same SPAD readings similar to leaf chlorophyll content measurement from 9:00 to 11:00 at 90 DAT from fully expanded leaves of 5 plants per plot and averaged. Leaf chlorophyll fluorescence (Ft) was also taken at same time as quantum yield using a hand-held SPAD readings instrument. Samples of five matured top leaves from five branches were taken from compound leaf. Sample leaves were taken from the third to fourth nodal insertion of the compound leaves and assessed for physiological parameters.

Results and Discussions

The physiological responses of each treatment under each experiment were summarized under the following independent experiments:

Container grown tomato experiment

Tomato yield: Application of starter N, P or media mix did not bring any combined effect ($P<0.01$) on the leaf chlorophyll content. However, application of starter N caused a highly significant ($P<0.01$) effect on leaf quantum yield and similarly use of media mixtures alone showed a highly significant ($P<0.01$) effect on the marketable fruit yield of container grown tomato (Table 1).

Marketable fruit yield: Use of combination of starter N and MR showed a significant effect ($P<0.05$) on the marketable fruit yield of container grown tomato variety *Melkasholla* (Table 1).

The LSD test at $P=0.05$ probability level showed that the highest marketable fruit yield was recorded from MR3, while the lowest marketable fruit yield was recorded from MR6 (Table 2). There yield increase over the grand mean of MR3 has 149.13%, while MR2 has 136.58%, MR6 has 135.89%, MR1 has 85.98%, MR5 has 69.86% and MR6 has 22.53% over the grand mean. Thus MR3 media mixture was found to be the best media mixtures produced highest tomato fruit

Source of variations	df	Mean square values		
		Marketable fruit weight [a]	Unmarketable fruit weight [a]	Total yield
Nitrogen (N)	1	0.021 NS	0.040 NS	1423546.9 NS
Phosphorous (P)	1	0.0002 NS	0.022 NS	2036162.0 NS
Media mix ratio	5	1.261**	0.307**	156180838.2**
N X P	1	0.001 NS	0.007 NS	3595668.1 NS
NX MR	5	0.052**	0.015 NS	3959058.2 NS
P X MR	5	0.010 NS	0.046 NS	4604436.1 NS
N X P X MR	5	0.051 NS	0.007 NS	5105008.3 NS
Error	48	0.0160	0.0165	2873479
Total	71			
R²		0.950	0.749	0.871
CV (%)		2.423	3.741	19.330
Root MSE		0.088	0.128	1695.134

Note NS=Indicates non-significant at P<0.05; *significant at P<0.05 and ** significant at P<0.01 probability levels, respectively, a=Transformed data

Table 1: Mean square values of marketable fruit yield, unmarketable fruit and total yield response of container grown tomato under different media mix ratios, N and P fertilizer applications planted under rainfed condition.

Media mix ratio	Marketable fruit weight [a] (g per plot)	Unmarketable fruit weight [a] (g per plot)	Total yield per plot (g)
MR1	(4935.9) B	(1832.4) C	6768.3 C
MR2	(7840.4) A	(3528.2) AB	11368.6 B
MR3	(8560.4) A	(4389.3) A	12949.7 A
MR4	(7799.5) A	(2987.8) B	10787.2 B
MR5	(4010.3) C	(3559.2) AB	7569.4 C
MR6	(1293.8) D	(1878.4) C	3172.2 D
Mean	5740.021	(3029.201)	8769.22
LSD (0.05)	30.113	0.1057	1560
Nitrogen (kg ha⁻¹)			
0	(5747.2)	(2881.4)	8628.6
25	(5732.8)	(3177.0)	8909.8
Mean			
LSD (0.05)	NS	0.0422	NS
P (kg ha⁻¹)			
0	(5684.3)	(2916.8)	8601.1
23	(5795.7)	(3141.7)	8937.4
Mean			
LSD (0.05)	NS	NS	NS

*= Average of three replications. Means followed in a column with the same letters are not significantly different using LSD at P = 0.05 level of significance respectively; a = Data were transformed, means in brackets are original data

Table 2: Mean response values of marketable fruit weight, unmarketable fruit weight total yield response of tomato planted under different media mixes, N and P application rates under rainfed growing conditions*.

yield. The tomato plants produced the highest marketable fruit yield in the medium that had 4:1:1 (60% soil: 15%: farmyard manure and 15% sand). The sole soil medium treatment (check) in this experiment suffered from moisture stress in the afternoons at the time of dry spells, which was disastrous during the critical growth stages prior, at and after flowering, which caused reduced growth and yield.

Total fruit yield:

Use of media mix ratio showed a highly significant effect ($P<0.01$) on the total fruit yield of container grown tomato (Table 1). The LSD test at $P = 0.05$ probability level showed that MR3 yielded the highest total fruit yield while MR6 gave the lowest total fruit yield (Table 2). It is concluded from this experiment that use of the ratio of 4 field top soil: 1 manure: 1 sand order gave the highest fruit yield for container grown tomato around Melkassa during the rainy season; increasing the ratio of manure beyond this proportion reduced the tomato fruit yield.

Unmarketable fruit yield

Similar to marketable fruit, use of media mixtures had highly significant ($P<0.01$) influence on the unmarketable fruit yield of container grown rainfed tomato (Table 1). The highest unmarketable fruit yield record was obtained from MR3 treatment while the lowest unmarketable fruit yield was obtained from the check and (Table 2).

Physiological responses of tomato

Application of starter N, P or media mix brought variable effects on the physiological responses of container grown tomato.

Leaf quantum yield

Among the physiological responses, leaf quantum yield, application of starter N caused a highly significant ($P<0.01$) effect on leaf quantum yield of container grown tomato (Table 3). The LSD test at $P=0.05$ probability level indicated that the highest leaf quantum yield was obtained from application of starter N and the lowest leaf quantum yield was from the check (Table 3). This would imply the leaves of

Source of Variations	df	Mean square values		
		Leaf chlorophyll fluorescence	Leaf quantum yield	Leaf chlorophyll content
Nitrogen (N)	1	2804.17 NS	0.021**	387.811NS
Phosphorous (P)	1	288.00 NS	0.0008 NS	0.623 NS
MR	5	1296.98 NS	0.005 NS	21.617 NS
N X P	1	1104.50 NS	0.0001 NS	1.201 NS
NX MR	5	381.50 NS	0.005 NS	25.532 NS
P X MR	5	707.422 NS	0.001 NS	7.448 NS
N X P X MR	5	353.011 NS	0.0024 NS	16.746 NS
Error	48	635.595	0.0033	23.297
Total	71			
R²		0.419	0.448	0.419
CV (%)		8.438	10.126	11.812
Root MSE		25.211	0.057	4.826

NS = Indicates non-significant at P < 0.05; * significant at P < 0.05 and ** significant at P < 0.01 probability levels

Table 3: Mean square values of physiological, yield and early blight response of container grown tomato under different media mix ratios, N and P fertilizer applications planted under rainfed condition.

tomato plant fertilized with N nutrient would indicates that every photon absorbed would contributes to the photosynthesis process and vice versa. However under this experiment use of application of starter P and use of media mix ratio did not bring any significant difference on the leaf quantum yield.

Leaf chlorophyll content

Use of starter N, starter P and use of media mix rations either separately or in a combined ways did not affect the tomato leaf chlorophyll content.

Leaf chlorophyll fluorescence

The results container grown tomato indicated that use of media mixtures, starter N and starter P either in a combined form or separately, did not affect tomato leaf fluorescence at 5% probability level (Table 3).

In general, tomatoes grown under different media mix ratios and supplementary N and P fertilizers actually faced two major problems that limited their growth and development; one problem was root confinement and another was the intermittent moisture stress as they were grown under rainfed conditions. Tomato plants grown in field were less affected by low moisture stress during the dry spell as compared to container grown tomato; probably, field grown tomato plant roots explored more volume of soil.

Treatments with FYM were found to be less stressed during dry spells as compared to the check plot, indicating that FYM improves water holding characteristics of the media. In addition, the media containing FYM were less affected by crusting and sealing as this problem was observed in the greenhouse experiment, probably due to association with soil structure, low organic matter content of the topsoil. This is probably because farmyard manure increases organic matter content of the media, which is food for soil biota that enhance the availability of nutrients such as phosphate through increased solubilization. In addition, the organic matter holds moisture like a sponge, avoiding stress during dry days throughout the growth period of tomato plant.

Drip irrigated tomato experiment

Tomato yield

Marketable fruit yield

Use of various irrigation depths brought a significant (P<0.01)

effect on the marketable yield of tomato (Table 4). The mean separation at $P = 0.05$ probability level indicated that the highest marketable fruit yield was recorded in response to full irrigation, while the lowest marketable fruit yield was recorded from 60% of full irrigation (Table 5); similar results were reported by Kirnak et al. where full water supply significantly increased fruit yield of eggplant [12]. However application of integrated nutrient management (INM) did not bring about any significant (P<0.05) difference on marketable fruit yield probably the fertility status of the soil was good (the details of soil analysis was published in African Journal of Agricultural Research.

Unmarketable fruit yield

Application of irrigation did not bring significant difference (P<0.05) on unmarketable fruit yield of tomato (Table 4).

Total fruit yield

Use of irrigation depth had a significant (P<0.05) effect on total fruit yield; the highest fresh fruit yield was obtained from full irrigation and the lowest was obtained from 60% irrigation water with saving of 40% of irrigation water (Table 5). Thus, the fresh fruit yield obtained from fully irrigated tomato plot exceeded the fresh fruit yield obtained from tomato plot irrigated with only 60% of full irrigation water by 62.8%. The results showed that with decrease in the depth of irrigation, there was a decrease in total fruit yield in tomato due to reduced uptake of water (Table 5). The result of this study corroborate that of Muchovej who reported that high quality and yield of vegetable crops are directly associated with proper water management. They also found that the fresh fruit yields of *Melkasholla* variety was reduced under deficit irrigation level. Similar findings were reported by Kirnak et al. where egg plants grown under high water stress had less fruit yield and quality than those in the control treatment [12]. Consistent with the results of this study Kirnak et al. also found that water stress in the container grown eggplants produced a very significant reduction in both dry biomass, they found that eggplant fruit yield was reduced by up to 68% in the water stressed plants compared with unstressed plants [12].

Media mix ratio	Leaf chlorophyll fluorescence	Leaf quantum yield	Leaf chlorophyll content
MR1	308.11	0.57122 AB	39.317
MR2	290.17	0.58801 A	39.600
MR3	289.86	0.53793 B	42.408
MR4	288.03	0.54971 AB	40.525
MR5	310.14	0.57329 AB	40.858
MR6	306.19	0.59222 A	42.450
Mean	298.750	0.568	40.859
LSD (0.05)	NS	0.0473	NS
Nitrogen (kg ha⁻¹)			
0	292.509	0.55160 B	38.539 B
25	304.991	0.58586 A	43.181 A
Mean			
LSD (0.05)	NS	0.0273	2.29
Phosphorous (kg ha⁻¹)			
0	296.750	0.56525	40.953
23	300.750	0.57220	40.767
Mean			
LSD (0.05)	NS	NS	NS

*= Average of three replications. Means followed in a column with the same letters are not significantly different using LSD at P = 0.05 level of significance respectively; a = Data were transformed, means in brackets are original data

Table 4: Mean response values of yield and selected physiological response of tomato planted under different media mixes, N and P application rates under rainfed growing conditions.

They also reported that restricted water supply for tomato can suppress new leaf development, resulting in a shortened yield formation period. Similar findings were reported by Cetin that water stress significantly reduced final yield of field-grown sweet pepper. Similar findings were obtained where increasing irrigation increased total tomato fruit yield.

Irrigation positively influenced tomato productivity; the result was attributed to the increase in the number of berries per plant and the fruit average weight as irrigation increased. The authors concluded that the total yield and marketable tomato yields were decreased significantly as the deficit level was increased. The reduction in total yield of tomato with an increased amount of water stress level of this test was consistent with previous work conducted on tomato and other crops such as cotton as reported.

Physiological responses of tomato to irrigation regimes

Leaf chlorophyll fluorescence

The results of variance analysis indicated that applying various irrigation depths did not bring any significant effect on tomato leaf chlorophyll fluorescence of tomato at all (Table 5).

However, the results of chlorophyll fluorescence measurement indicated that as irrigation depth increased the chlorophyll fluorescence was reduced (Figure 1). Based on the review of Maxwell and Johnson, light energy absorbed by chlorophyll molecules in a leaf can undergo one of three fates: it can be used to drive photosynthesis (photochemistry), excess energy can be dissipated as heat or it can be re-emitted as light-chlorophyll fluorescence [7]. These three processes occur in competition, such that any increase in the efficiency of one will result in a decrease in the yield of the other two. The results from this experiment showed that as irrigation depth increased, the portion of light energy absorbed by chlorophyll molecules in a leaf can undergo to drive photosynthesis (photochemistry) performance would be increased so that yield of the tomato plant increased. On the other hand deficit irrigation increased leaf chlorophyll fluorescence of tomato

probably suggesting much light is not used in the photosynthesis performance.

Leaf chlorophyll content

Applying various irrigation depth significantly ($P <0.01$) affected the tomato leaf chlorophyll content (Table 5). The mean separation indicated that the highest irrigation level increased leaf chlorophyll content and lowest irrigation depth reduced leaf chlorophyll content (Table 5). This would indicate that water availability in the soil would contribute to the N nutrient uptake [13-17].

The regression function analyze indicated that as irrigation depth increased, the leaf chlorophyll content increased linearly. The result further indicated that as irrigation depth increased the leaf chlorophyll content found to be increased in power function R^2 = 82%, further indicating the tomato plant more nutrient N uptake due to moisture availability (Figure 2). Similar findings were reported by Kirnak et al. where water stress resulted in significant decreases in chlorophyll content of egg plants [12].

The larger the irrigation depth the greener tomato plant leaf became; this correspondingly contributed to better growth and development and further yield. Similar results were found by Kirnak et al. who reported that water stress in the container grown eggplants produced a very significant reduction in total chlorophyll content [12].

The regression function analyze indicated that as irrigation depth increased, the leaf chlorophyll content was found to be increased in power function (R^2 = 82%). Similar findings were reported by Kirnak et al. where water stress resulted in significant decreases in chlorophyll content of egg plants [12].

Leaf quantum yield

The results of this experiment shown that irrigation depths did not bring any significant effect on tomato leaf quantum yield (Table 5).

Stomatal conductance

The irrigation depth brought significant ($P < 0.01$) effect on the stomatal conductance of tomato plant (Table 5). As irrigation depth decreased, leaf stomatal conductance was highly reduced [18]. The leaf stomatal conductance was the highest for full irrigation, and the lowest for the lowest irrigation depth as indicated in Table 6. This indicates that under low moisture conditions tomato leaves had low stomatal conductance that contributed to low CO_2 assimilation and further low corresponding fruit yield [19,20]. Well-watered eggplant had high transpiration rate than stressed plants that would contributed to higher fruit yield [12]. They found that transpiration rate gradually decreased

Irrigation regimes	Marketable fruit (t ha⁻¹)	Unmarketable yield (t ha⁻¹)	Total fruit yield (t ha⁻¹)
IR I (100% ETc) (Full irrigation)	63.63 A	18.267	81.902 A
IR II (80% ETc)	33.83 B	22.413	56.250 B
IR III (60 % ETc)	27.82 B	23.062	50.868 C
Mean	41.765	20.813	62.916
LSD (0.05)	9.712	NS	5.689

*= Average of three replications. Means within each column with different letters are significantly different at LSD at P = 0.05 level of probability

Table 5: Mean values of various irrigation regimes on marketable, unmarketable and total fruit yield of tomato grown under drip irrigated condition.

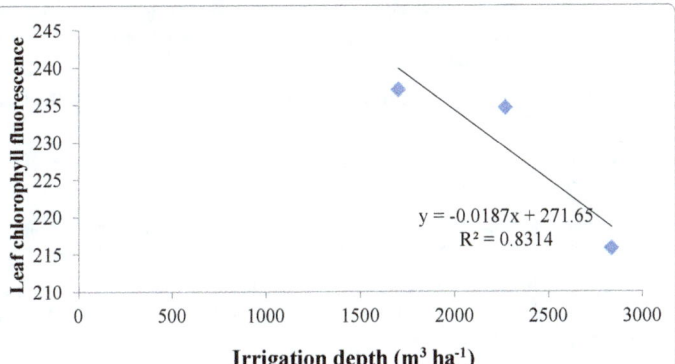

Figure 1: Graphical relationship of leaf chlorophylls fluorescence responses of tomato as a function of irrigation regimes.

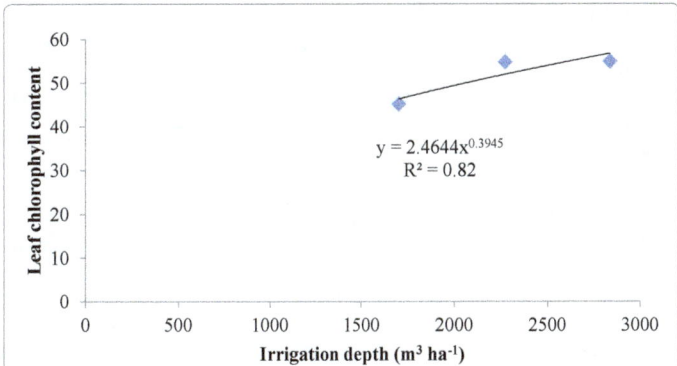

Figure 2: Graphical relationship of leaf chlorophyll content responses of tomato as a function of irrigation regimes.

with increasing the incidence of water stress. Delfine et al. also found that water stress rapidly affected stomatal conductance of field-grown sweet pepper [13].

Figure 3 indicates that at higher irrigation regimes, there would be higher stomatal conductance with regression function linearly at R^2 = 90%, relationship. Low stomatal conductance indicates significant stomatal closure associated with reduced transpiration Taiz and Zeiger and low dry matter production [5]. Low stomatal conductance is related to low water supply to the stomata, which implies relatively dried conditions in the root zone Table 7. Mild water stress does usually affect both leaf photosynthesis and stomatal conductance [5].

Summery and Conclusions

Field experiments were conducted at Melkassa Agricultural Research centre to study tomato fruit yield such as marketable, unmarketable and total fruit yield and some physiological response such as quantum yield, leaf chlorophyll content, leaf fluorescence and stomatal conductance of tomato. The first experiment was conducted with container grown tomato during rainy season while the second field

Source of variations	df	Mean square values			
		Leaf fluorescence	Leaf quantum yield	Leaf chlorophyll content	Stomatal conductance
Irrigation	2	2025.41 NS	0.00153 NS	466.172**	22349.2**
Error	4	1367.90	0.00783	3.216	403.40
CV		16.13	16.57	3.47	15.12

Note NS = Indicates non-significant at P < 0.05; * significant at P < 0.05 and ** significant at P < 0.01 probability levels

Table 6: Mean square values of physiological and yield response of tomato as influenced by integrated nutrient managements and application of various moisture regimes.

Irrigation regimes	Leaf quantum yield	Leaf chlorophyll content	Stomatal conductance	Leaf chlorophyll fluorescence
IR I (100% ETc, Full irrigation)	0.52267	55.02 A	176.74 A	215.89
IR II (80% ETc)	0.53733	54.88 A	117.29 B	234.73
IR III (60 % ETc)	0.54200	45.30 B	104.36 B	237.09
Mean	0.534	51.737	132.80	229.23
LSD (0.05)	NS	1.818	20.362	NS

*= Average of three replications. Means within each column with different letters are significantly different at LSD at *P* = 0.05 level of probability

Table 7: Mean values of yield and physiological response of tomato grown under various irrigation regimes and integrated nutrient management practices.

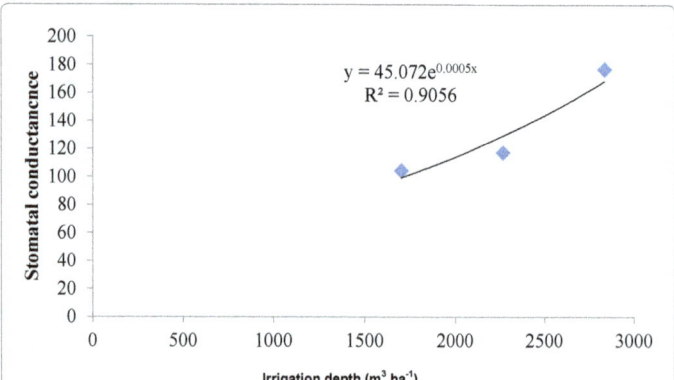

Figure 3: Graphical relationship of leaf stomatal conductance responses of tomato as a function of irrigation regimes.

experiment was conducted under drip irrigation during hot dry season; the treatments consisted of two levels of N (0 and 25 kg N ha^{-1}), and two levels of P (0 and 23 kg P ha^{-1}) fertilizers and with six media mix ratios. For the second drip irrigated experiment, factorial combinations of five levels of nutrient management and of three levels of daily irrigation treatments such as full irrigation, 80% and 60% of daily ETc irrigation regimes on the strip plot design was used irrigation as vertical and nutrient management laid as horizontal strip. The results of container grown experiment indicated that use of combination of starter N and MR showed a significant effect on the marketable fruit yield, similarly use of media mixtures had highly significant influence on the unmarketable fruit yield and finally use of media mix ratio showed a highly significant effect on the total fruit yield of container grown tomato. Media mix ratio 3 (4:1:1, in the form of field soil: manure: sand order) yielded the highest total fruit yield while MR6 (1:3:2) gave the lowest total fruit yield. Application of starter N, P or media mix did not bring any combined effect on the leaf chlorophyll content. However, application of starter N caused a highly significant effect on leaf quantum yield. The results of drip irrigated experiment indicated that use of various irrigation depths brought a significant effect on the marketable yield of tomato; highest fruit yield was recorded in response to full irrigation, while the lowest marketable fruit yield was recorded from least irrigation depth. Irrigation depth significantly affected the tomato leaf chlorophyll content and the stomatal conductance of tomato plant. As irrigation depth decreased, leaf stomatal conductance was highly reduced. The leaf stomatal conductance was the highest for full irrigation, and the lowest for the lowest irrigation depth. Thus use of these sensors should be further fine-tuned for the field management applications.

References

1. Etissa E (2014) Irrigation and Nutrient Management Studies on Vegetable Crops with Particular Reference to Tomato (Lycopersicum esculentum M.) in the Central Rift Valley of Ethiopia. Electronic Thesis and Dissertation, Haramaya University. p: 310.

2. Etissa E, Dechassa N, Alamirew T, Alemayehu Y, Dessalegne L (2014) Irrigation Water Management Practices in Small Scale Household Vegetable Crops Production System: The Case of the Central Rift Valley Area of Ethiopia. Sci Technol Arts Res J 3: 74-83.

3. Etissa E, Dechassa N, Alamirew T, Alemayehu Y, Desalegne L (2013) Household Fertilizers Use and Soil Fertility Management Practices in Vegetable Crops Production: The Case of Central Rift Valley of Ethiopia. Sci Technol Arts Res J 2: 47-55.

4. Misra AN, Misra M, Singh R (2012) Chlorophyll Fluorescence in Plant Biology. In: Misra AN (ed.) Biophysics. pp: 171-192.

5. Taiz L, Eduardo Z (2003) Plant Physiology.

6. Strasser RJ, Srivastava A, Tsimilli-Michael M (2000) The Fluorescence Transient as a Tool to Characterize and Screen Photosynthetic Samples. In: M Yunus, U Pathre, P Mohanty (eds.) Probing Photosynthesis: Mechanism, Regulation and Adaptation. pp: 445-483.

7. Maxwell K, Johnson GN (2000) Chlorophyll Fluorescence-A Practical Guide. J Exp Bot 51: 659-668.

8. Falkowski PG, Raven JA (2007) Aquatic Photosynthesis (2nd edn.). Princeton University Press, Princeton. p: 484.

9. Roberto P, Moya I, Goulas Y, Jacquemoud S (2008) Chlorophyll Fluorescence Emission Spectrum inside a Leaf. Photochem Photobiol Sci 7: 498-502.

10. Vu Joseph CV, Allen LH, Gallo-Meagher M (2001) Crop Plant Responses to Rising CO$_2$ and Climate Change. In: Pessarakli M (ed.) Handbook of Plant and Crop Physiology. CRC Press.

11. Etissa E, Dechassa N, Alemayhu Y (2015) Effect of Growth Media, Starter N and P Fertilizers on Growth and Yield of Tomato under Rainfed Condition in the Central Rift Valley of Ethiopia. Sci Technol Arts Res J 4: 23-31.

12. Kirnak H, Cengiz K, Ismail T, David H (2001) The Influence of Water Deficit on Vegetative Growth, Physiology, Fruit Yield and Quality in Eggplants. Bulg J Plant Physiol 27: 34-46.

13. Delfine S, Alvino A, Loreto F, Centritto M, Santarelli G (2000) Effects of Water Stress on the Yield and Photosynthesis of Field-grown Sweet Pepper (Capsicum annuum L.). Crops Acta Hort.

14. Strasser RJ, Tsimilli-Michael M, Srivastava A (2004) Analysis of the Chlorophyll Fluorescence Transient. Advances in Photosynthesis and Respiration 41: 321-362.

15. Govindjee (2004) Chlorophyll a Fluorescence: A Bit of Basics and History. Advances in Photosynthesis and Respiration 19: 1-41.

16. Allen RG, Pereira LS, Raes D, Smith M (1998) Crop Evapotranspiration (Guidelines for Computing Crop Water Requirements). FAO. Irrigation and Drainage Paper No 56, Italy.

17. Etissa E, Dechassa N, Alamirew T, Alemayehu Y, Dessalegne L (2014c) Growth and Physiological Response of Tomato to Various Irrigation Regimes and Integrated Nutrient Management Practices. African Journal of Agricultural Research 9: 1484-1489.

18. Etissa E, Dechassa N, Alamirew T, Alemayehu Y, Dessalegne L (2013b) Growth and Yield Components of Tomato as Influenced by Nitrogen and Phosphorus Fertilizer Application in Different Growing Seasons. Ethiopian Journal of Agricultural Science 25: 57-77.

19. Etissa E, Dechassa N, Alamirew T, Alemayehu Y, Dessalegne L (2014) Growth and Physiological Response of Tomato to Various Irrigation Regimes and Integrated Nutrient Management Practices. African Journal of Agricultural Research 9: 1484-1489.

20. Jones JB (2008) Tomato Plant Culture: In the Field, Greenhouse and Home Garden. CRC Press.

6

One Dimensional Numerical Simulation of Bed Changes in Irrigation Channels using Finite Volume Method

Xin Liu*, Abdolmajid Mohammadian and Julio Angel Infante Sedano

Dept. of Civil Engineering, University Of Ottawa, Ottawa, Canada

Abstract

A one-dimensional Saint Venant model is developed using the Einstein sediment transport formula for simulating hydraulics and bed changes in irrigation channels. The governing equations are discretized using Finite Volume Method. The Central-Upwind method is used in this study to calculate the numerical flux. The model gives a stable and good prediction of sub, super and transcritical flows and captures harp changes such as hydraulic jump without leading to numerical oscillation or diffusion. An experimental test case is adopted to examine the performance of this mode. The simulated results show a good agreement with experimental data in the upstream section, and can predict an average value of measured profile in the downstream section of a steep channel. The knick point movement is successfully predicted by the model. The model appears to be a potential and reasonably accurate tool to predict bed changes in irrigation channels due to the water flow.

Keywords: Saint Venant equations; Sediment transport; Irrigation channel; Central upwind method

Introduction

Irrigation channels that transport water in order to irrigate croplands are of significant importance in agricultural studies. The failure of irrigation channel in proper transport of water may cause substantial costs. Numerical prediction of irrigation channel erosion is of practical significance in agricultural practice. Mean while, fertilizers and nutrients essential to crops are attached to the soil surface and transported along the furrows by the irrigation flow. To optimize nutrient delivery to crops, one requires accurate predictions of flow and sediment transport in irrigation furrows [1].

The hydraulics of irrigation channel flows has been commonly calculated by kinematic wave and zero inertia models in early studies. A zero inertia model of the stream flow was presented in the context of negligible accelerations by Strelkoff and Katopodes [2]. Schwankl and Wallender [3] adopted the zero inertia furrow model with a combination of spatially-varying infiltration and steady infiltration. Clemmens [4] also used the zero inertia model for simulating irrigation channels. Elliott et al. [5] modified the zero-inertia model for continuous furrow irrigation and showed that the zero inertia model can effectively simulate the hydraulics of furrow irrigation. Oweis and Walker [6] developed a surge flow furrow irrigation model based on the zero inertia concept. Chen [7,8] used kinematic wave theory to solve the problem of irrigation flow over a wide porous bed and concluded that the kinematic wave method may only be valid for supercritical flow. Woolhiser [9] showed that the method would also be applicable to subcritical flows but may lead to poor results if water pounds at the downstream boundary and a moving backwater extended over an appreciable length of the field. Smith [10] adopted a kinematic-wave method of characteristics and the Lax-Wendroff scheme to simulate irrigation flow. Walker and Humphrys [11] developed and verified a kinematic-wave model of furrow irrigation under both continuous and surged flow management. Lu et al. [12] developed a numerical kinematic wave model to simulate the dynamic erosion process in a ridge-furrow system. Based on the kinematic wave method, Reddy and Singh [13] used moving control volume approach to model the flow in irrigation systems and used fixed control volume to model the nearly stationary phase and the runoff rate, and showed a good agreement with the experimental data.

These studies showed that the zero inertia model is applicable for surface irrigation flow with negligible errors, the kinematic wave model is more applicable for steeper slopes in irrigation furrows with a larger Reynolds number. Nonetheless, the accuracy of these models may be limited for unsteady irrigation flow due to the neglection of flow acceleration.

One-dimensional models solving St. Venant equations are considered to be accurate and rigorous for hydraulics of irrigation flow. Sakkas et al. [14] used a mathematical model based on the complete hydrodynamic equations describing open-channel flow developed for simulation of surface irrigation systems. Abbasi et al. [15] presented a system combining an overland waterflow and solute transport model and applied it to furrows/borders. The integrated model numerically solves the 1D dispersion-advection and simplified form of the Saint-Venant equations as the governing equations in a decoupled fashion. Burguete et al. [16] adopted a coupled system including solute transport and implicated the model on furrow irrigation and obtained reasonable results.

Several sediment transport models have been examined for furrow irrigation flows. Lu et al. [12] examined two concepts of the erosion process of the loose, deposited sediment and concluded that the best concept for describing erosion of this loose layer of sediment was one where the deposited sediment was eroded uniformly across the width of the channel. Strelkoff and Bjorneberg (2001) implemented Yang's and Yalin's sediment transport method on furrow erosion and concluded that furrow erosion by upstream inflow may be largely different from hillside erosion by rain-fed overland flow and predictive approaches and formulas for the latter problem may not be satisfactory for the former.

---

*Corresponding author: Xin Liu, Associate Professor, Department of Civil Engineering, University Of Ottawa, Ottawa, Canada, E-mail: majid.mohammadian@uOttawa.ca

This study focuses on the bedload sediment transport which is defined as the type of transport where sediment grains roll or slide along the bed. A lot of sediment transport models have been developed to calculate the erosion of channel bed such as the Grass equation [17], Meyer-Peter & Muller formula [18], Van Rijn equation [19-21], Nielsen formula [22], Kalinske equation [23,24], Einstein formula [25,26], Smart equation [27] etc. which are generally obtained by empirical methods. Some of these models are deterministic while others are derived based on probabilistic methods. After trials and comparison, the Einstein equation is adopted to compute the channel bed erosion in this study.

This research is aimed at developing a one-dimensional numerical model based on finite volume method for simulating flow and sediment transport. The Central-Upwind scheme is used here to calculate the numerical flux. The model is developed by combining a set of one-dimensional Saint Venant equations and an accurate sediment transport formula. The experimental data from Brush and Wolman (Brush and Wolman, 1960) are used in this research as a test case to examine the performance of the numerical model.

Hydraulic Flow and Channel Bed Erosion

For soil loss cases where the flow through the breach is usually unsteady, the hydraulics can be formulated based on continuity and momentum equations. In this study, a set of 1-D St. Venant equations based on non-cohesive sediment and mobile bed are applied focusing on the incompressible flow in channel. A coordinate system is defined where the x-axis represents the horizontal direction along the channel and the z-axis represents the vertical upward direction. In this study, the width of the irrigation channel is assumed to be constant and equal everywhere. Ignoring the flow infiltration, the governing equations are adopted based on mass and momentum conservation for each phase.

Governing equations

The 1-D St. Venant equations describing the flow through a breach can be written as:

$$\frac{\partial A}{\partial t} + \frac{\partial Q}{\partial x} = 0 \tag{2.1}$$

$$\frac{\partial Q}{\partial t} + \frac{\partial}{\partial x}\left(\frac{Q^2}{A} + 0.5gAh\right) = gA(S_b - S_f) \tag{2.2}$$

where A = the wetted cross-sectional area, $Q = uA$ is the flow discharge, u = mean flow velocity in the x-directions; t = time; h = water depth; g = gravitational acceleration. The bed slope term (S_b) is written as $S_b = -\partial z/\partial x$, where z = bed elevation.

The term S_f is friction slope term which can be expressed using the Manning formula

$$S_f = n^2 Q|Q| A^{-2} / R^{4/3} \tag{2.3}$$

where n = Manning roughness coefficient which depends on physical and natural properties, estimated from numerical calibration or experimental values and is kept constant during the modeling; R = hydraulic radius which can be approximately replaced by water depth h in this study.

Figure 1 shows the physical significance of some notations used in this paper.

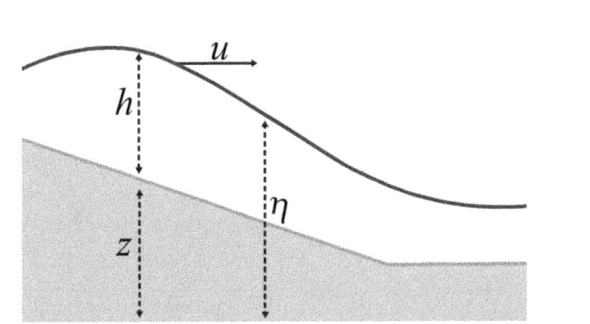

Figure 1: Schematic diagram of an unsteady flow over an irregular bottom and the corresponding notations.

Bed change model

The erosion of the soil surface can be calculated by the sediment continuity equation based on the sediment mass conservation. The continuity equation (the Exner equation) is given by

$$\frac{\partial z}{\partial t} + \xi \frac{\partial q_b}{\partial x} = 0 \tag{2.4}$$

where $\xi = 1/(1-\lambda)/B$, λ = the porosity of the soil sediment layer and B= the wetted perimeter of the movable bed. For a wide rectangular channel, the parameter B can be replaced by the channel width. The parameter q_b is the sediment transport rate per unit width which can be expressed in a general form as:

$$q_b = f(u,h,d,......) \tag{2.5}$$

where d is the particle size.

To control the stability and accuracy of the numerical modeling result, an explicit modified Lax scheme [28] is introduced as follows:

$$z_i = (1-\alpha)z_i + \alpha \frac{z_{i-1} + z_{i+1}}{2} \qquad (\mu^2 \le \alpha \le 1) \tag{2.6}$$

where α is a weighting coefficient which can be adjusted from μ^2 to 1. The subscript i indicates the nodes' positions along the x-axis. $\mu = c_b \Delta t / \Delta x$ is the Courant number of sediment transport, Cb is the bed celerity in a frictionless system which is also given by [29]:

$$c_b = \frac{u}{h(1-Fr^2)} \frac{dq_b}{du} \tag{2.7}$$

where u = flow velocity in the x-direction; h = water depth; q_b = sediment transport rate per unit width and Fr is Froude number in a rectangular channel which can be expressed as $Fr = u/\sqrt{gh}$.

Numerical Scheme

In this study, we assume the breach width to be constant (unit width). The 1-D St. Venant equations in a conservative form can be written as:

$$\frac{\partial \vec{U}}{\partial t} + \nabla \vec{F} = \vec{S} \tag{3.1}$$

where

$$\vec{U} = (h, uh)^T \tag{3.2}$$

$$\vec{F} = (uh, u^2h + 0.5gh^2)^T \tag{3.3}$$

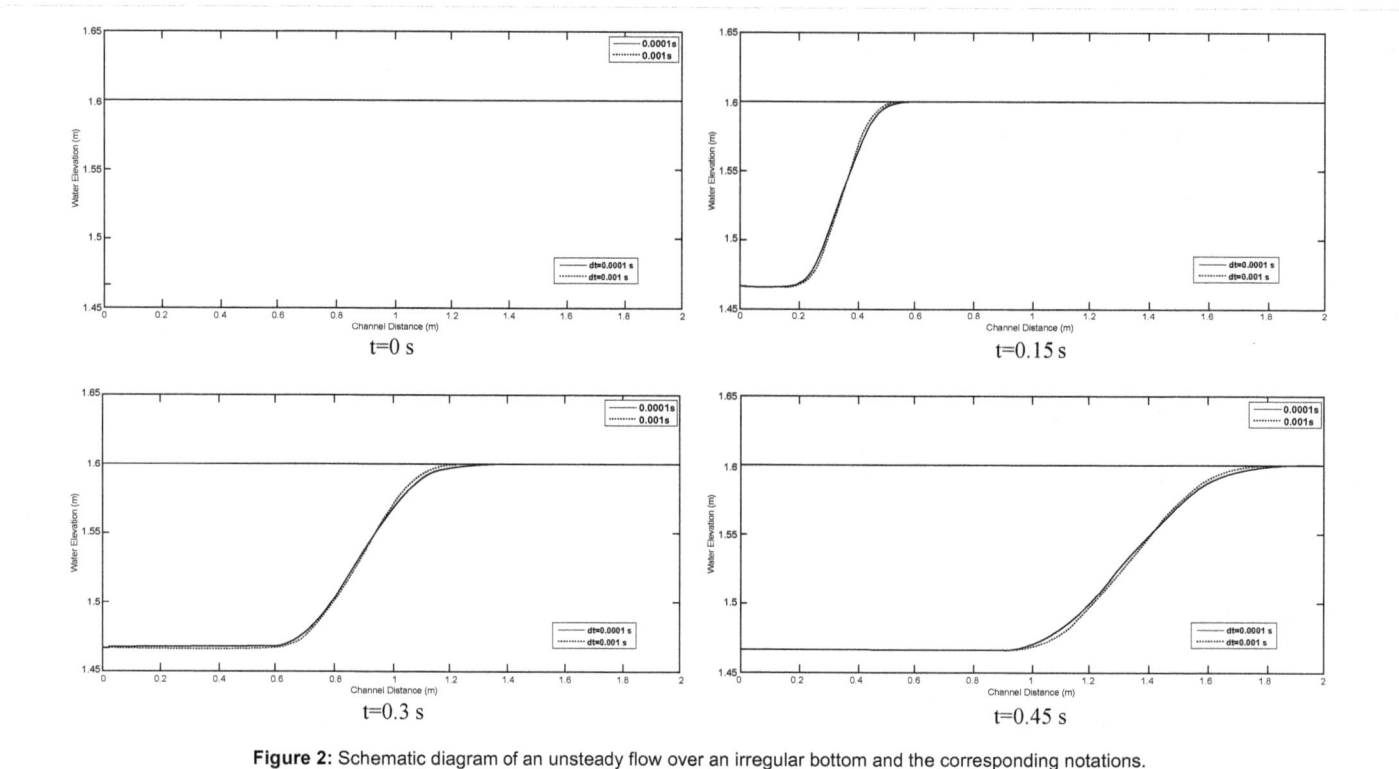

Figure 2: Schematic diagram of an unsteady flow over an irregular bottom and the corresponding notations.

$$\vec{S} = \left(0, -gh\frac{\partial z}{\partial x} - gn^2 u|u| / h^{1/3} \right)^T \tag{3.4}$$

The St.Venant formula can also be written in a non-conservative form:

$$\frac{\partial \vec{U}}{\partial t} + A\frac{\partial \vec{U}}{\partial x} = \vec{S} \tag{3.5}$$

where A is the Jacobian matrix which can be written as:

$$A = \frac{\partial \vec{E}}{\partial \vec{U}} = \begin{pmatrix} 0 & 1 \\ c^2 - u^2 & 2u \end{pmatrix} \tag{3.6}$$

and $c = \sqrt{gh}$ s the wave velocity.

The eigen values a_i corresponding to the matrix A are $a_1 = u+c$, $a_2 = u-c$ respectively.

The St. Venant equations are integrated over every control volume based on the finite volume method.

$$\frac{d}{dt}\int_{\Omega_C} \vec{U}dx + \int_{\Gamma_C} \vec{F}\cdot\vec{n}d\Gamma_C = \int_{\Gamma_C} \vec{S}\cdot\vec{n}d\Gamma_C \tag{3.7}$$

An explicit scheme for discretizing the time step integration term is used in this study. The time step is restricted by the Courant–Friendrichs–Lewy (CFL) condition. In a practical simulation, a constant time-step is usually used and the CFL condition is used to approximately determine the time-step before the simulation starts. The CFL condition is given by:

$$\frac{\Delta t}{\Delta x}(|u| + \sqrt{gh}) \le C_n \tag{3.8}$$

where $\Delta t =$ time step size; $\Delta x =$ spatial step size; C_n is the flow Courant number and its value must be less than 1 for numerical stability.

For the one dimensional case, a standard splitting algorithm is employed for discrete equation (3.8) [30]

$$U_i^{n+1} = U_i^n - \frac{\Delta t}{\Delta x}(F_{i+1/2}^n - F_{i-1/2}^n) + \Delta t \cdot \vec{S} \tag{3.9}$$

where superscript $n =$ time step index; subscript $i =$ spatial node index; $F_{i+1/2}^n$ and $F_{i-1/2}^n$ are the numerical fluxes.

The key problem in the above discretization is how to calculate the numerical fluxes $F_{i+1/2}^n$ and $F_{i+1/2}^n$ at the interface Γ_C between two neighboring cells. According to Godunov [30], the normal fluxes can be obtained by an approximate Riemann solver in the direction normal to the interface. A wide variety of reliable finite-volume methods for the St. Venant system have been developed in the past few decades. In this study, the Central upwind scheme is applied to calculate the fluxes.

$$\vec{F} = \begin{cases} \vec{F}_L \\ \vec{F}_R \\ \dfrac{S_R\vec{F}_L - S_L\vec{F}_R + S_L S_R(\vec{U}_R - \vec{U}_L)}{S_R - S_L} \end{cases} \tag{3.10}$$

$$if(S_L \ge 0)$$
$$if(S_R \le 0)$$
$$if\,(S_L < 0)\quad or\quad (S_R > 0)$$

For a 1D system with two eigen values a_1 and a_2 which have the relationship of $a_1 < a_2$,we obtain

$$a_1^L = u_L - c_L \; , \; a_1^R = u_R - c_R$$

$$a_2^L = u_L + c_L, \; a_2^R = u_R + c_R$$

S_L , S_R are the sediment concentrations on the left and right sides of the edge of control volume and are defined as:

$$S_L = \min(a_1^L, a_1^R, 0)$$
$$S_R = \max(a_2^L, a_2^R, 0)$$

The sediment continuity equation can also be discretized by using the finite difference method and we obtain:

$$z_i^{n+1} = z_i^n - \frac{\Delta t}{\Delta x} \xi (q_{bi-1}^n - q_{bi-1/2}^n) \qquad (3.11)$$

Numerical Results

In this section, two numerical experiments are performed to evaluate the performance of the model. The first case deals with a hydraulic test and the second test considers sediment transport and erosion in a channel.

Hydraulic test case

Here, we perform a numerical test to evaluate the performance of hydrodynamic solver. This test is considered to predict the water surface profile changing in an irrigation channel when the upstream flow volume rate decreases. The initial upstream flow rate is 0.6 m³/s and water depth is 0.6 m. The bed elevation is 1 m and the width is 1 m. The flow rate decreases to 0.3 m³/sover at a certain time. A time step of 0.001 s is considered here and a comparative test with a time step of 0.0001 s is also carried out with a higher resolution. Figure 2 shows the water profile changing at different time in both tests.

The computed results show that the predicted water surface profiles are smooth without any numerical oscillation or diffusion. The predicted surface changing with higher accuracy is a little flatter than the one with larger time step. The two curves are very close at different times. Thus, the changing of time step has a tiny effect on the modeling result.

Experimental Test Case

An experimental case of channel bed erosion is considered in order to test the performance of this numerical scheme.

Experiment description: Knick point is a term in geomorphology to describe a location in a channel where there is an abrupt change in the longitudinal bottom. Differential rates of erosion can result from this topography change.

A famous laboratory study of the behaviour of knick points in non cohesive sediment was carried out by Brush & Wolman (1960), which explained the migration of knick points as a result of erosion potential hSf becoming maximum at the point where the slope changes.

The experiments were conducted in a flume 15.85 m long and 1.22 m wide. The actual knick point migration was performed in a trape zoidal channel 0.21 m wide and 0.03 m deep with rounded corners molded in non-cohesive sand before starting experimental run. A short steep reach with a length of 0.3 m and a fall of 0.03 m was set at a location 10.8 m from the flume entrance. This steep fall had a slope

of 0.1 which was significantly steeper than the slope of the adjacent channel which was approximately equal to 0.00125. The channel was modeled in sand with a mean diameter of 0.67 mm. The sediment porosity was $\lambda = 0.4$. The initial water depth was $h_0 = 0.0137$ m and the upstream flow rate was $Q = 0.59$ L/s. The bed level was recorded at successive times after starting the experimental run.

Numerical modeling: In the numerical model, a 1-D control volume method was used with a fixed spatial step $\Delta x = 0.1$m and a constant time step $\Delta t = 0.1$ s. The channel was assumed to be rectangular with a fixed width of 0.21 m.

The boundary conditions for the hydraulic model were constant discharge in the upstream inflow entrance and constant outflow elevation in the downstream end.

Simulation results: Figure 3 shows the water profile under a fixed bed condition before the beginning of the morphological simulation. A small spatial step 0.01 m is adopted in order to observe more flow details. It can be shown that the Froude number Fr of the subcritical flow away from the steep reach is approximately 0.5 with a depth of around 0.015 m and a velocity about 0.2 m/s, which is in agreement with the experimental values.

The flow velocity increases when it comes to the knick point and becomes supercritical with a depth of only 0.0037 m and a maximum velocity of 0.65 m/s. A hydraulic jump was also captured at the end of the steep reach.

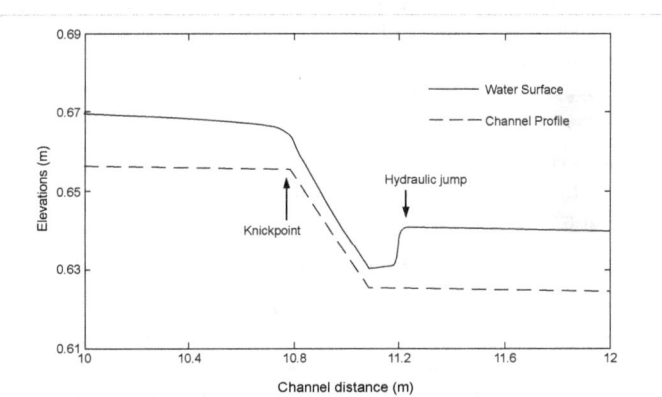

Figure 3: Initial profiles of channel bed and water surface before the beginning of morphological calculation.

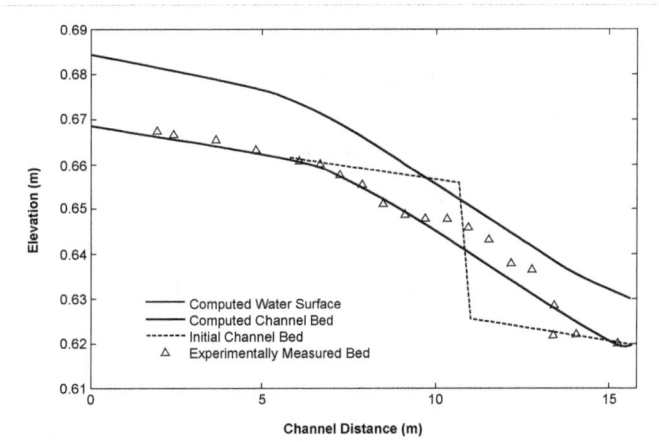

Figure 4: Experimental and computed data of channel bed development at the time of 2 hours 40 min.

Many different sediment transport equations were employed to predict the sediment discharge rate and the development of the shape of channel bed. In this numerical study, the sediment discharge rate was estimated using the bed load formula of Einstein [25] which is also used to predict the channel bed development by [31]

$$q_b = 2.15\sqrt{(s\text{-}1)gd_{50}^3}\ e^{-\frac{0.391\,(s\text{-}1)d_{50}}{hs_f}} \qquad (4.1)$$

A moderate time step of 0.1 s and a spatial step of 0.1 m were used. Both channel bed and water surface profiles are smooth without any numerical oscillations.

Figure 4 shows the comparison between the measured and computed channel bed profiles at a time of t = 160 min. It can be observed that the computed bed profile has a very good agreement in the upstream section with the measured data. It accurately predicted the knick point migration in the upstream section and agreed with the measured values very well. In the downstream section of the reach, the computed profile was smoother than the original measured data and it predicted an average value of experimental data.

Conclusions

The modeled results of irrigation channel erosion have a reasonable agreement with the measured data considering the large uncertainty of the sediment transport equation and the assumption of fixed width. From the report of Brush & Wolman (1960), the width of the channel increased downstream of the knick point by as much as ± 20% at the widest section. The increase of the downstream width may cause more sediment deposition and maybe one of the reasons accounting for the differences between the measured and the computed profiles in the downstream section.

The developed numerical model is found to be a convenient and adequate tool in predicting the bed erosion of irrigation channels, the main features of this model are:

(1) It uses a 1-dimensional grid with finite volume method.

(2) The numerical flux is calculated using the Central Upwind method which is able to simulate complex flows including sub, super and transcritical currents with a good accuracy. It shows a good numerical stability without oscillations or diffusion under a moderate spatial step of 0.1m, which is crucial for an accurate simulation.

(3) The numerical model has a good ability to capture shock waves and is able to predict the fronts of hydraulic jump accurately.

(4) The model with Einstein bed load formula can dynamically update the elevation of the channel bed during the deposition process and is in good overall agreement with experimentally measured data.

The main limitations of current model are:

(1) It ignores sediment suspension.

(2) Sediment particle size is assumed to be equal to the mean value.

(3) The 1-D model assumes a constant width of the irrigation channel.

(4) The sediment transport formula is empirical and is not suitable for all situations.

From the results of the test case, the computed values were in good overall agreement with the measured ones. Although some disagreements exist between the simulation and experimental results, these may be explained by the limitations listed above. These preliminary results show a good potential to simulate the bed erosion of irrigation channels.

References

1. Zhang S, Duan J, Strelkoff T, Bautista E (2010) Simulation of unsteady flow and soil erosion in irrigation furrows. J Irrig Drain Eng 138: 294-303.

2. Strelkoff T, Katopodes ND (1977) Border-irrigation hydraulics with zero inertia. J Irrig Drain Eng 103: 325-342.

3. Schwankl LJ, Wallender WW (1988) Zero Inertia Furrow Modeling with Variable Infiltration and Hydraulic Characteristics. Transactions of the ASABE 31: 1470-1475.

4. Clemmens AJ (1979) Verification of the Zero-Inertia Model for Border Irrigation. Transactions of the ASABE 22: 1306-1309.

5. Elliott RL, Walker WR, Skogerboe GV (1982) Zero-inertia modeling of furrow irrigation advance. J Irrig Drain Eng 108: 179-195.

6. Oweis TY, Walker WR (1900) Zero-inertia model for surge flow furrow irrigation. Irrigation Science 11: 131-136.

7. Chen CL (1966) Mathematical hydraulics of surface irrigation. Utah Water Res Lab 1-98.

8. Chen CL (1970) Surface irrigation using kinematic-wave method. J Irrig Drain Div 96: 39-46.

9. Woolhiser DA (1970) Discussion of Surface irrigation using kinematic-wave method by C. L. Chen. J Irrig Drain Div 96: 498-500.

10. Smith RE (1972) Border irrigation advance and ephemeral flood waves. J Irrig Drain Div 98: 289-307.

11. Walker WR, Humpherys AS (1983) Kinematic-wave furrow irrigation model. J Irrig Drain Eng 109: 377-392.

12. Lu JY, Foster GR, Smith RE (1987) Numerical simulation of dynamic erosion in a ridge-furrow system. Trans Am Soc Agric Eng 30: 969-976.

13. Reddy JM, Singh VP (1994) Modeling and error analysis of kinematic-wave equations of furrow irrigation. Irrigation Science 15: 113-121.

14. Sakkas JG, Bellos CV, Klonaraki MN (1994) Numerical computation of surface irrigation. Irrigation Science 15: 83-99.

15. Abbasi F, Simunek J, van Genuchten MTh, Feyen J, Adamsen FJ, et al. (2003) Overland water flow and solute transport: Model development and field-data analysis. J Irrig Drain Eng 129: 71-81.

16. Burguete J, Garcia-Navarro P, Murillo J (2009) One-dimensional conservative coupled discretization of the shallow-water with scalar transport equations. Monografias de la Real Academia de Ciencias de Zaragoza 31: 21-33.

17. Grass AJ (1981) Sediments transport by waves and currents. SERC London.

18. Meyer-Peter E, Muller R (1948) Formulas for bed-load transport. Repository Hydraulic Engineering Reports.

19. van Rijn LC (1984) Sediment Transport, Part I: Bed Load Transport. J Hydraul Eng 110: 1431-1456.

20. van Rijn LC (1984) Sediment transport, Part II: suspended load transport. J Hydraul Eng 111: 1613-1641.

21. van Rijn LC (1984) Sediment transport, Part III: bed forms and alluvial roughness. J Hydraul Div 112: 1733-1754.

22. Nielsen P (1992) Coastal bottom boundary layers and sediment transport. Advanced series on ocean engineering 4: 1-270.

23. Kalinske AA (1942) Criteria for determining sand transport by surface creep and saltation. Trans AGU 23: 639-643.

24. Kalinske AA (1947) Movement of sediment as bedload in rivers. Trans AGU 28: 615-620.

25. Einstein HA (1947) Formulas for the transportation of bed load. Trans ASCE 107: 561-597.

26. Einstein HA (1950) The bed load function for sediment transport in open 1026: 1-71.

27. Smart GM (1984) Sediment transport formula for steep channels. J Hydraul Eng 110: 267-276.

28. De Vries M (1987) Morphological computations.

29. Toro EF (2001) Shock-Capturing Methods for Free-Surface Shallow Flows. Wiley, England.

30. Godunov SK (1959) A difference method for the numerical calculation of discontinuous solutions of hydrodynamic equations. Mat Sb (N.S.) 47: 271-306.

31. Mohammadian A, Tajrishi M, Lotfi Azad F (2004) Two dimensional numerical simulation of flow and geo-morphological processes near headlands by using unstructured grid. International Journal of Sediment Research 258-277.

7

Evaluating Irrigation Scheduling Efficiency of Paddy Rice and Berseem Fodder Crops in Sandy Loam Soil

Hatiye SD[1]*, Hari Prasad KS[1], Ojha CSP[1], Kaushika GS[1] and Adeloye AJ[2]

[1]Department of Civil Engineering, Indian Institute of Technology, Roorkee, India
[2]Institute for Infrastructure and Environment, Heriot-Watt University, Edinburgh, UK

Abstract

In this study, irrigation scheduling efficiency of two field crops; paddy rice and berseem fodder, grown in unpuddled sandy loam soil for a typical existing and imposed irrigations has been evaluated using the WINISAREG water balance and irrigation scheduling model that was calibrated and validated using data collected at field experimental plot in Roorkee, India. During the 1st season of each crop, typical irrigation schedules as practiced in the farmers' field was followed while in the 2nd crop season, a reduced irrigation schedule was imposed aiming for water saving. Water balance components were monitored daily during the crop growth periods. Deep percolation was measured using drainage type lysimeters. Soil moisture content in the root zone was observed using soil moisture profile probe (PR2/6). The crops were provided with all the necessary inputs including fertilizer, pesticide and weeding operations following agronomic practices of the area. The results show that nearly 82-87% of the input water goes to deep percolation during paddy season-1 (continuous irrigation period) while 64%-70% of input water was lost through deep percolation during berseem season-1. Due to the imposed irrigation, the deep percolation has been reduced to nearly 78-80% of input water during paddy season-2 and 42-52% of input water during berseem season-2 besides large input water saving in the crop seasons. The large input water saving was due to alternative irrigation scheduling strategy whose efficiency has been significantly improved. Irrigation scheduling efficiency has been increased from 9.65% to 30.5% for paddy and 23% to 92% for berseem. In particular, comparative irrigation water saving of 64-74% in paddy season and 82-88% in berseem season was achieved with nominal yield penalty. This study shows the possibility of large volume of water saving in water intensive crops such as paddy rice and berseem fields under un-puddled sandy loam soils by considering a reduced irrigation scheduling option.

Keywords: Berseem; Deep percolation; Irrigation scheduling; Lysimeter; Paddy rice; Water balance model

Introduction

Rice (*Oryza Sativa L.*) is the most important staple food crop in Asia, providing 35-80% of the total calorie intake [1]. In Asia, irrigated agriculture accounts for about 90% of total diverted fresh water and more than 50% of this is used to irrigate rice [2]. Due to the inherent nature of water application, lowland rice is often seen as an inefficient water user [3]. This is due to the fact that large proportion of the applied water is lost through deep percolation and seepage [2,4-6]. Sizable efforts have been made to reduce deep percolation especially from rice fields including: alternate wetting and drying (AWD) [5,7-9]; aerobic rice [10]; delayed application of continuous flooding [11] and puddling [6,12]. Berseem fodder (*Trifolium alexandrinum L.*) is a forage crop widely cultivated in northern India and the other parts of the world [13]. Specifically, in India it grows in winter from October to June and offers a good rotation with other summer crops such as rice, cotton, barely and maize. The water requirement and irrigation scheduling of berseem fodder is almost similar to that of alfalfa forage crop. However, berseem is preferred to alfalfa since it provides the possibility of rotation with other crops, improves soil structure, relished by all kinds of stock and poultry, more succulent, supper in fattening stock and milk production [14].

Appropriate irrigation scheduling favours increase in crop yield, water saving, environmental protection as well as economic boost [15]. Accurate scheduling of irrigation water would limit over and/or under-irrigation situations, and thus, help to avoid deep percolation of water and stress to crops. Deep percolation phenomena, however, has become a major threat to proper irrigation scheduling particularly for surface methods of irrigation which are mostly practiced in developing countries. Most research studies considered presence of an impermeable layer (hard pan) below the bottom of rice paddy root zone to arrest deep percolating water. However, the efficiency of the

hard pan under farmer operated field conditions is not well proved to serve the purpose of impeding deep percolation. Incidences of large deep percolation under bunds and through cracks have been reported in puddled field conditions [3-5,16]. Another problem which has been documented against puddling practices in paddy fields is the interference of the puddled layer with the next crop [17]. Further, puddling operation is a costly task as it needs extra labour and cost. Therefore, nowadays farmers are escaping the puddling operation and grow both rice and berseem crops under unpuddled conditions. However, the quantity of deep percolation under such agronomic conditions is not well understood specifically applied to coarse textured soils. So far field measurement of deep percolation using drainage type lysimeters for water intensive crops was also limited probably due to large volumes of drainage below the root zone which is often difficult to monitor. Consequently, only limited studies are available with regard to deep percolation and irrigation scheduling evaluation of such water intensive crops on coarse textured unpuddled field conditions. Therefore, in this study, we carried out typical irrigation applications and timings as practiced in this particular region and imposed reduced irrigation applications for both crop periods under non-puddled coarse textured soil to study the extent of water saving

*Corresponding author: Hatiye SD, Professor, Department of Civil Engineering, Indian Institute of Technology, Roorkee, India
E-mail: samueldagalo@gmail.com

due to imposed irrigations, evaluate the irrigation scheduling efficiency of each crop seasons, test the applicability of the WINISAREG water balance model in simulating the water balance components and finally assess the efficiency of locally constructed lysimeters in metering deep percolation from the crop root zones.

Materials and Methods

Study site

The study site is located in the Uttrakhand state of India, an experimental plot situated at Department of Civil Engineering, Indian Institute of Technology, and Roorkee. Roorkee is located near the River Ganges in the geometric grid of 77°53′52″ East Longitude and 29°52′00″ North Latitude at an average altitude of 274 m above mean sea level. The climate of Roorkee is typical of north western India with hot humid summer and very cold dry winter [18]. The monthly average maximum temperature of the study area is recorded in the range of 19.33 (January) to 37.73°C (May) and monthly average minimum temperature in the range of 7.2 (January) to 25.6°C (July) according to the data from National Institute of Hydrology (NIH), at Roorkee. The average relative humidity runs from 52.2% (May) to 89.7% (January). The average annual daily sunshine duration is 7.7 hrs. The normal rainfall of Roorkee is 1060 mm per annum and out of which almost 80% is recorded during the monsoon season (June to September). The soil in the region can be classified as 'soils in old alluvial plains', which are well drained fine loamy soils on nearly level plain with sandy loam surface [18].

Field and laboratory experiments

The field experiment consisted of growing paddy rice (var. Supper

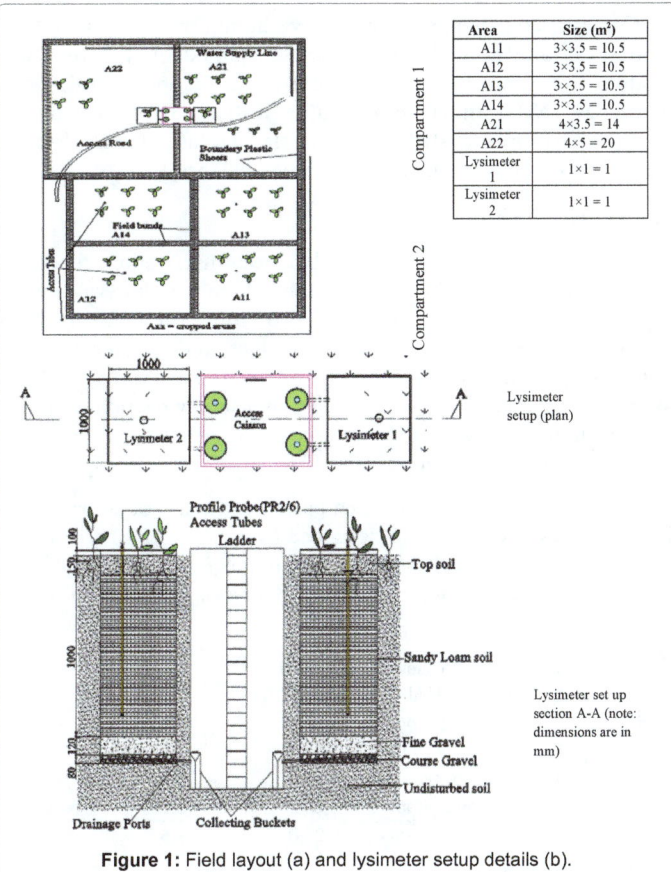

Area	Size (m²)
A11	3×3.5 = 10.5
A12	3×3.5 = 10.5
A13	3×3.5 = 10.5
A14	3×3.5 = 10.5
A21	4×3.5 = 14
A22	4×5 = 20
Lysimeter 1	1×1 = 1
Lysimeter 2	1×1 = 1

Figure 1: Field layout (a) and lysimeter setup details (b).

Basmati) in the 2013 and 2014 growing seasons with continuous saturation and intermittent application of irrigation respectively. The 1st paddy rice season hereafter called paddy season-1 was transplanted on July 23, 2013 and harvested on 02 November 2013 while the 2nd paddy rice (paddy season-2) was transplanted on 15 July, 2014 and harvested on 22 October, 2014. On the same field, berseem fodder (var. JB-1) crop was also grown in the winter seasons of 2013/14 and 2014/2015. The 1st season berseem (berseem season-1) was sown on 12 December, 2013 and finally harvested on 08 May, 2014 while the 2nd season berseem (berseem season-2) was sown on 17 November, 2014 and harvested on finally harvested on 16 April, 2015. Berseem was cut four times for green fodder in each of the seasons.

Lysimeter experiments were conducted at the experimental farm from July 2013 to April 2015 in both crop periods. The area of lysimeters is 1 m² having a depth of 1.35 m repacked soil monolith of the experimental field. A repacked soil column may take several years to duplicate the natural soil state for research with restricted irrigation (i.e., rain-fed or dry land applications) [19]. The construction of the lysimeters was took place in 2006 and hence they are considered to replicate the surrounding root zone soil environment. The lysimeters were constructed of steel metal sheets having a square shape. The soil monolith is a repacked soil material consisting of the upper 1.15 m filled with a sandy loam textured soil, moderately homogeneous throughout the profile, characterized by an organic content of 1.1 to 1.2%. The bottom 0.08 m was filled with a very coarse gravel of size more than 3 cm in diameter overlain by 0.12 m thick gravel of about 2 cm in diameter. This bottom arrangement allows drainage towards imbedded pipes which carry percolating water towards collecting buckets in the access hall (Figure 1). The lysimeters were located in the centre of the 1st compartment (plots A21 and A22) where plastic sheets were buried at field boundaries to a depth of 60 cm to impede lateral seepage out of the field. Each compartment has been further partitioned into smaller plots to manage the experimental run. The boundaries of the 2nd compartment (A11- A14) were left open to mimic actual field conditions elsewhere (Figure 1). The same experimental conditions have been maintained inside and outside the lysimeters in each of the growing periods of the crops.

During the paddy growing seasons, 21 days old seedlings were transplanted after thorough field preparation and flooding to saturate the soil. Prior saturation of the field by flooding before transplanting was made, which favoured initial conditions for the crop growth. A basal dose of Di-ammonium Phosphate (DAP) and Zinc Sulphate during transplanting and Urea after three weeks of transplanting were applied following agronomic practices of the area. Weeding was undertaken manually by hand removing all weeds from field three times during the growth periods of the crop. The crop was also protected from threatening insects by applying an insecticide commonly used in the area. The soil root zone in this particular experimental condition was left un-puddled. Similarly, in the winter season, berseem fodder crop was sown on prepared beds on the same plot. The field was soaked with water before sowing the seed to favour easy seed germination. Required does of DAP was applied for the fodder crop and weed control was undertaken in the similar way as that of the paddy rice. Irrigation was scheduled when nearly 40% of moisture depletion in the surface layer took place (mainly in berseem season-1). Additional irrigations were also provided during the winter season to ease the soil freezing effect on the crop.

The soil physical and hydraulic characteristics have been determined in the laboratory for three representative spots of the experimental plot and replicate depths from 0 to 140 cm. In this study, laboratory experimental works consisting soil physical properties such as grain size, bulk density, particle density and the soil characteristic curve of the experimental plot have been determined. The data from pressure plate apparatus (Ψ versus θ) enabled to furnish soil water characteristic curve of the experimental plot from which the wilting point and field capacity of the experimental plot have been determined. The soil bulk density was determined using the core cutter method as suggested by Trout et al. [20]. The soil particle density was obtained by conducting water Pycnometer test. The soil particle size was determined by employing mechanical sieve analysis (for coarser particles) and hydrometer (for the finer portion of the soil) methods as recommended by the American Society for Testing and Materials, ASTM. The soil properties determined are shown in the Table 1.

Irrigation water was applied for a specific area by knowing the supply line discharge and calculating time required to provide a predetermined depth of water for a given plot area. The depth of water required varies depending on pondage required (paddy rice season) or to fill up to field capacity after certain depletion of the available water has been occurred in the root zone (berseem fodder). The discharge of a permanent water supply line at a particular period was determined by measuring the time required (stop watch) to fill a known volume of container. After deciding irrigation depth, mainly based on local practices during paddy season (20-100 mm during continuous irrigation in paddy season-1), imposing reduced irrigation size (10-50 mm during paddy season-2) and based on certain deficit during the berseem season-1(30-60 mm) and berseem season-2 (4.8-18.5 cm), the time required to spend the flow in a particular plot of known area was calculated using the continuity equation. In similar fashion all the plots were irrigated. The variations of head and tail end ponding and consequently percolation are disregarded in this case since the areas of the plots and the border lengths are small (Figure 1) so that the water advances to the tail end in a very short time. Plastic hoses were used to deliver water to a particular plot from the water supply line.

During paddy season-1, the effort was to saturate the field every time to keep a saturated culture while in paddy season-2 the method similar to alternate wetting and drying (AWD) was practiced. Ponding for a long time in our case was impossible because the water quickly infiltrates.

The soil water status was monitored by using soil water probe (Profile Probe-PR2/6; Delta-T Devices, Cambridge) through access tubes installed both inside and outside the lysimeters. The probe consists of a sealed polycarbonate rod approximately 25 mm diameter, with electronic sensors arranged at fixed intervals along its length. Each of the sensors comprises a 100 MHz oscillator and transmits an electromagnetic field extending about 100 mm into the soil. The water content of the soil surrounding the rings dominates its permittivity, $\sqrt{\varepsilon}$. The permittivity of a material is a measure of its response to polarisation in an electromagnetic field. Water having a strong permittivity (≈ 81) than soil (≈ 4) and air (≈ 1) can easily be detected in an electromagnetic field. The detectors are sensitive to the different

proportions of transmission and refection, and convert them into stable voltage output that acts as a simple, sensitive measure of soil moisture content. When installed in an access tube constructed from a composite material it can measure the dielectric constant at soil depths of 10, 20, 30, 40, 60 and 100 cm. The probe can either be logged in the access tube or it can be moved from one access tube to another to make spot readings. The output of the probe is in volts which can be converted into dielectric constant, ε which is useful for describing the microscopic interaction between electromagnetic radiation and matter [21]. The probe enables to measure the soil water content in volumetric bases for different types of soils ranging from clayey to sandy soils with accuracy between ±0.04 (after soil specific calibration) and ±0.06(after generalized soil calibration in normal soils). However, it has been reported that the probe is very sensitive to air gaps which may be created due to soil cracking and installation operations [22]. Therefore, we have installed the access tubes with due care as suggested by the manufacturer of the instrument.

Deep percolation was measured twice in a day at the bottom of lysimeters early in the morning (07:00 a.m.) and evening (around 07:00 p.m.). The lysimeter rim was kept 10 cm above the ground to avoid run-on or runoff. Collecting buckets in access hall were used to collect the drainage water. The buckets were securely covered to avoid rainwater inflow. The caisson hall was also sheltered from rainfall to avoid any inflow from rain water to the buckets so that only drainage water should be collected. Drainage ports connected to the lysimeters were used to convey and discharge the percolating water to the buckets where the percolation measurement was being made. Further, data pertaining to crop specific parameters such as root depth, crop height and leaf area index were also monitored during the growth period of the crops.

Climatic data (temperature, relative humidity, pan evaporation, wind speed and rainfall) for the growth period of the crops was obtained from nearby metrological station, National Institute of Hydrology (NIH) India, located at a distance of 0.8 kilometres from the experimental station. These data were used to calculate the evapotranspiration component of the water balance.

Model description

Model inputs and the soil water balance: An irrigation scheduling and simulation model, WINISAREG, has been used to evaluate the imposed irrigation scheduling of the two crops. The model requires the soil, crop and climatic data (variables and parameters) and irrigation application options to carry out the root zone water balance computations. It also considers certain water supply restrictions which may encounter in practical field conditions. The detailed list of required inputs is documented in Fortes et al. [23].

The WINISAREG model which performs the soil water balance at field scale was developed and described by Teixeira and Pereira [24], Liu [25] and Pereira [26] The model is an integration of two different models, the EVAP56 (for computing reference evapotranspiration)

Depth below ground level (cm)	Bulk density (g/cm³)	Particle density (g/cm³)	Sand (%)	Silt (%)	Clay (%)	Soil Class (USDA)	θ_{fc} (%)	θ_{pwp} (%)	θ_{sat} (%)
0-30	1.58	2.55	73.40	22.70	2.96	Sandy loam	18.5	6.6	38
30-60	1.55	2.57	66.89	28.39	4.01	Sandy loam	24.5	6.6	40
60-80	1.54	2.56	68.57	26.54	4.33	Sandy loam	19.9	6.0	40
80-100	1.54	2.58	69.10	26.54	3.84	Sandy loam	20.2	6.3	40
100-140	1.59	2.62	68.01	27.38	4.58	Sandy loam	20.0	7.6	39

Table 1: Soil physical characteristics of the experimental plot.

and the ISAREG (to perform water balance computations) as described by Pereira [26]. WINISAREG has been tested and used in different parts of the world under varying soil, crop and irrigation management conditions [27-29]. The model specifically enables to compute deep percolation on seasonal bases which provides additional opportunity to compare model computed percolation with that of field measured percolation.

The governing equation in the WINISAREG water balance and scheduling model which can also be applied for the lysimeter water balance model is given as:

$$\theta_i = \theta_{i-1} + \frac{P_i + I_i - ET_{ci} - DP_i - R_i + GW_i}{1000Z_{ri}} \tag{1}$$

where $\theta(m^3 m^{-3})$=soil water content in the root zone; P(mm)=precipitation; I(mm)=applied irrigation; ET_c(mm)=actual evapotranspiration; DP(mm)=deep percolation of water moving out of the root zone; R(mm) is surface runoff; GW is groundwater contribution or capillary rise into the crop root zone; Z_{ri}(m) is the rooting depth in day i; i and i-1 are, respectively, the current and previous time steps (days in this study).

Measured irrigation and precipitation from field observations were supplied as inputs. Actual evapotranspiration can be estimated by the model considering available soil water content in the root zone. The reference evapotranspiration can be estimated using the EVAP56 module from climate data or supplied directly into the modelling environment. In this particular study, the reference evapotranspiration estimated using Penman-Monteith approach has been used. The reference evapotranspiration, ET_o (mm/day), according to Penman-Monteith is:-

$$ET_o = \frac{0.408.\Delta(R_n - G) + \gamma \frac{900}{273 + T} u_2(e_s - e_a)}{\Delta + \gamma(1 + 0.34u_2)} \tag{2}$$

where R_n is net radiation at the crop surface (MJ/m²/day), G is soil heat flux density (MJ/m²/day), e_s is saturation vapour pressure (kPa), e_a is actual vapour pressure (kPa), T is air temperature at 2 m height (C°), u_2 is wind speed at 2 m height (m/s), Δ is slope of vapour pressure curve (kPa/C°), and γ is psychometric constant (kPa /C°).

The potential evapotranspiration is computed using FAO recommended procedures in the model by incorporating a crop coefficient, K_c, for a specified crop growth stage. The potential evapotranspiration for non water limiting conditions may be computed using:

$$ET_c = K_c \times ET_o \tag{3}$$

where ET_c is the potential crop evapotranspiration and K_c is the crop coefficient. However, in actual field conditions, the actual evapotranspiration might be less than the potential evapotranspiration if the soil water is limiting [25]. In such conditions, the soil water stress coefficient, K_s, may be introduced in (3) to compute actual evapotranspiration.

$$ET_c = K_s \times K_c \times ET_o \tag{4}$$

Further, the coefficient K_s may also be given by:

$$K_s = \frac{TAW - D_i}{TAW - RAW} \tag{5}$$

where TAW (mm) =$1000Z_{ri} (\theta_{fc} - \theta_{wp})$ is the total available water (mm); D_i (mm) is the soil moisture depletion on day i; RAW (mm) is readily

available water that can be obtained by multiplying TAW to a depletion coefficient, p, considering the crop water stress resistance. θ_{fc} is the soil moisture content at field capacity; θ_{wp} is the soil moisture content at permanent wilting point.

The crop coefficient for the respective crop development stages of each crop has been modified for Roorkee climatic condition and further calibrated using the model for the particular field and agronomic conditions. K_s describes the effect of water stress on crop transpiration and hence introduction of K_s in the computation of actual evapotranspiration is more valid for dual crop coefficient approach than the single crop coefficient approach. WINISAREG uses the single crop coefficient for water balance calculations. In fact, reasonable estimation of ET_c is also possible without the K_s coefficient when soil evaporation from soil is not a large component of ET_c [30]. A full detail of instructions to determine the coefficients for a particular crop season, climatic and agronomic conditions are available in FAO-56 paper [30].

Runoff component of the water balance has been neglected in this study since runoff from lysimeters is only possible when rainfall depth overtops lysimeter rim level. Whenever, such intense rainfall occurs, the depth of water which goes above the lysimeter rim level is deducted to obtain the effective rainfall. Groundwater contribution through capillary rise has also been ignored since the shallow groundwater table in the area is beyond 2 m below the ground surface.

Model outputs: The model outputs for a specified irrigation schedule (an option from a list of other irrigation simulation options) mainly consist of all seasonal water balance components such as total applied irrigation, total deep percolation from irrigation, seasonal potential and actual evapotranspiration, soil moisture storage, cumulative rainfall, unused rainfall and irrigation scheduling efficiency values. Deep percolation is computed as excess amount of water from irrigation application. The excess depth of water from rainfall is accounted as unused rainfall which may contribute to either deep percolation or runoff. The soil moisture content is determined based on the input water and the supplied soil hydrologic parameters (the wilting point and the field capacity). Thus it is possible to compare field observed and model simulated water content values as a model calibration step for specific field and crop conditions. The irrigation scheduling efficiency is determined based on the applied irrigation and deep percolation.

Irrigation scheduling efficiency

The scheduling efficiency of applied irrigation, as used in the water balance model, is given by:

$$SE = \left(1 - \frac{DP}{I}\right) \times 100 \tag{6}$$

Where SE is the irrigation scheduling efficiency and other terms were defined earlier. The deep percolation is computed as excess water above the field capacity for an applied irrigation. The excess water due to rainfall is categorized as unused rainfall in the model although in actual field conditions deep percolation is contributed from both irrigation and rainfall. In principle, irrigation is scheduled when there is an occurrence of soil water deficit in the root zone, called management allowed depletion (MAD). However, 100% precision under agricultural field conditions cannot be achievable and hence water may be applied in excess of the field capacity or only satisfy some percentage of the soil moisture deficit (deficit irrigation). We are dealing here with the excess irrigation under rice paddy and berseem fodder crops where water application is intensive compared with other crops. It is not possible to avoid deep percolation in such crop fields but opportunities do exist to reduce deep percolation. The reduction in deep percolation process can be achieved by adopting a certain irrigation schedule other than traditional approaches such as puddling. In this study, different applications of water were practiced for both crop seasons and the respective scheduling efficiency have been evaluated.

Model calibration and validation

Soil water content observations made in the lysimeters during the two crop seasons were used to calibrate the WINISAREG model for the experimental site conditions. Calibration of the model was undertaken in the 1st crop seasons of each crop period to determine crop parameters (K_c and depletion fraction for no stress, p) and soil hydrologic parameters (field capacity, θ_{fc}, and wilting point, θ_{wp}, soil moisture content values). The calibration process consisted of searching K_c and p for the crop development stages and soil hydrologic parameters for different soil layers in the root zone that allowed minimizing the differences between simulated and observed values of soil water content. In the first instance, k_c and p values as suggested by FAO were plugged as trial values to make water balance in the modelling environment. The initial entries for θ_{fc}, and θ_{wp} values were extracted from the soil moisture characteristic curve (SMCC) constructed from pressure plate test data. These values were tuned step by step, by varying a given parameter at a time and fixing the others in the model until a fitting between model simulated and field observed moisture content values is achieved. The model validation was carried out using the 2nd season of each crop. Figures 2 and 3 present the variation of model simulated and field observed soil moisture contents in the crop root zone during model calibration and validation. Observed cumulative deep percolation values were also used as additional criteria to test the sufficiency of model calibration and validation.

The soil water content values computed using the WINISAREG model represent the average soil water content variation in the entire crop root zone. Therefore, the measured average soil water content values were used to compare with the simulated values during model calibration and validation in each crop season.

Statistical parameters

Selected statistical parameters were employed to assess the significance of model calibration and validation efforts. We employed two statistical parameters, coefficient of determination (R^2) and root mean square error (RMSE) to test the performance of the model as used [31]. The respective equations for the parameters are given below.

$$R^2 = \left\{ \frac{\sum_{i=1}^{n}(x_i - \overline{x})(y_i - \overline{y})}{\left[\sum_{i=1}^{n}(x_i - \overline{x})^2\right]^{0.5}\left[\sum_{i=1}^{n}(y_i - \overline{y})^2\right]^{0.5}} \right\}^2 \quad (7)$$

$$RMSE = \left[\frac{\sum_{i=1}^{n}(y_i - x_i)^2}{n}\right]^{0.5} \quad (8)$$

Where x_i and y_i are, respectively, observed and model computed values of a given variable with the respective means $\overline{x}$ and $\overline{y}$; n refers to the number of sample data points. Statistical parameters for model calibration and validation are shown in Table 2.

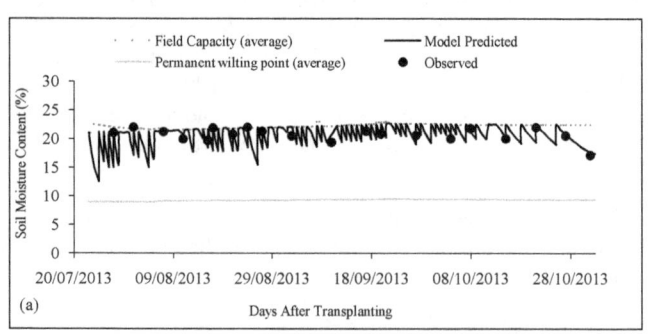

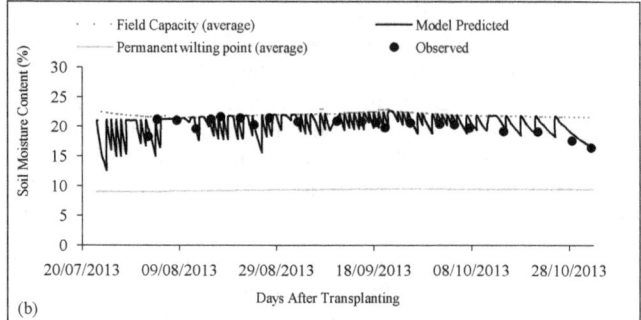

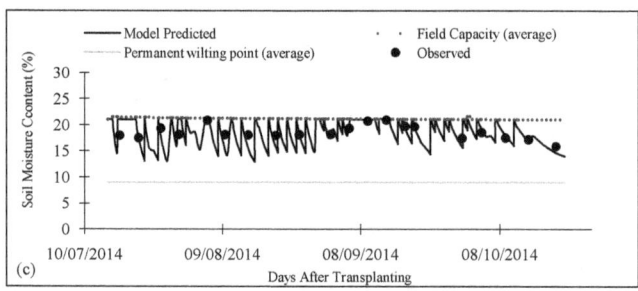

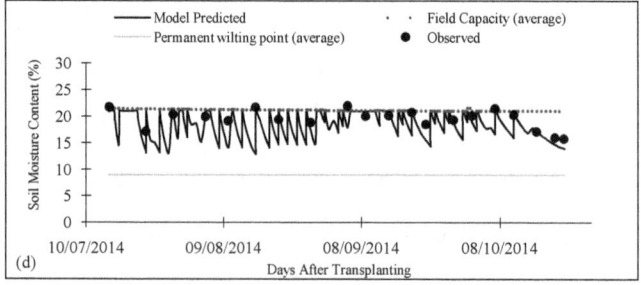

Figure 2: Model predicted and observed soil moisture content during model calibration ((a) and (b)) for paddy season-1 and model validation for paddy season-2 ((c) and (d)). Figures (a) and (c) refer to lysimeter 1 while (b) and (d) refer to lysimeter 2 conditions.

Results and Discussions

Model parameterization

Through model calibration, the parameters crop coefficient, depletion fraction for no stress, the field capacity and wilting point soil moisture characteristics have been established (Tables 3 and 4). The crop coefficient and depletion fraction for no stress are specific to crop type and growth stages of each crop.

The crop coefficients so obtained were slightly different from the FAO tabulated values [30] owing to local climatic conditions, agronomic practices and specific experimental setup. The crop coefficient value

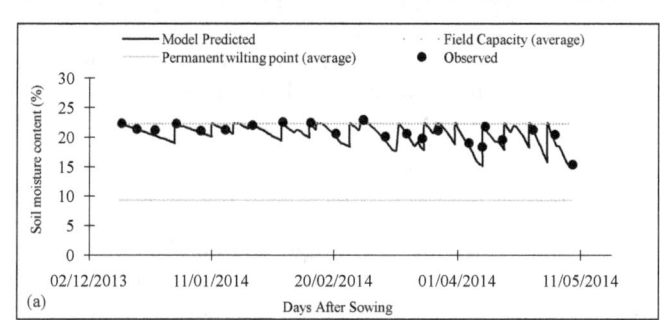

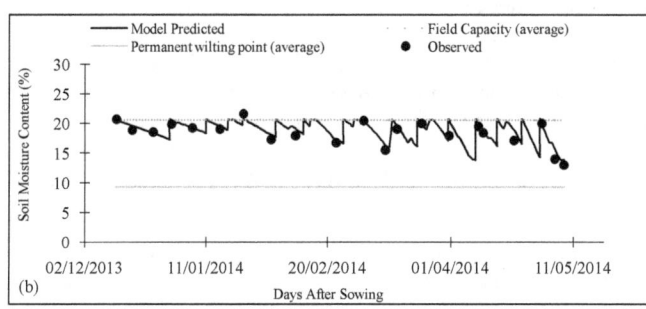

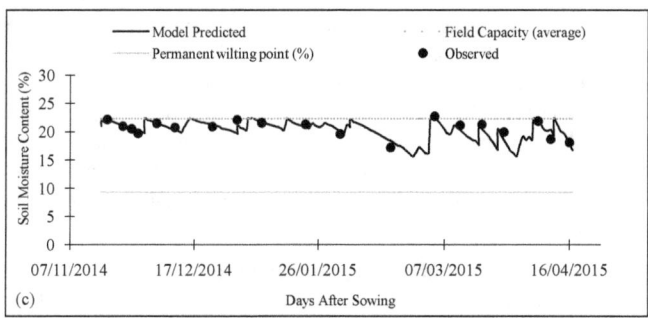

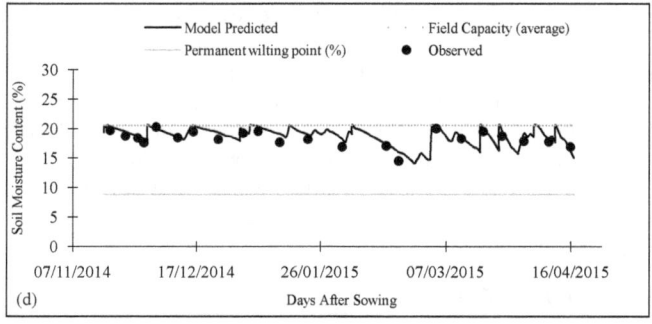

Figure 3: Model predicted and observed soil moisture content during model calibration ((a) and (b)) for berseem season-1 and model validation for berseem season-2 ((c) and (d)). Figures (a) and (c) refer to lysimeter 1 while (b) and (d) refer to lysimeter 2 conditions.

Crop season	Model process	Parameter		Lysimeter/ Locations of data
		R^2	RMSE (%)	
Paddy rice -1	Calibration	0.69	0.64	L1
		0.80	0.68	L2
Paddy rice -2	Validation	0.75	0.77	L1
		0.86	0.98	L2
Berseem-1	Calibration	0.79	0.93	L1
		0.88	0.94	L2
Berseem-2	Validation	0.77	0.66	L1
		0.83	0.88	L2

Table 2: Statistical parameters for model calibration and validation.

Crop name/ parameters	Crop growth stages			
	Initial	Development	Mid season	Late season
Paddy season -1	23/07/13-06/08/13	07/08/13-05/09/13	06/09/13-03/10/13	04/10/13 - 02/11/13
Paddy season -2	15/07/14-05/08/14	06/08/14-30/08/14	31/08/14-30/09/14	01/10/14 - 22/10/14
K_c	1.10	1.10-1.20	1.20	0.67
p	0.1	0.1	0.1	0.1
Berseem fodder	Individual Cutting Dates			
Berseem season-1	21/02/14	19/03/14	09/04/14	06/05/14
Berseem season-2	05/02/15	09/03/15	27/03/15	16/04/15
K_c	K_c initial 0.34	K_c before cut 0.85	K_c after cut 0.4	K_c end 0.95
p	0.40	0.40	0.40	0.40

Table 3: Calibrated crop characteristics used in the model.

Lysimeter (Location)	Layer (10 cm interval)	θ_{fc} (%)	θ_{wp} (%)	TAW (mm)
L1	1	21	9	120
	2	24	10	140
	3	22	9	130
	4	22	9	130
	5	22	9	130
L2	1	21	9	120
	2	24	10	140
	3	18	9	90
	4	18	9	90
	5	18	9	90

Table 4: Calibrated field soil characteristics used in the model.

grown under submerged conditions. These results at mid stage growth period are fairly at par with the calibrated values of K_c in this study although there are differences in the initial and late stages owing to differences in the agronomic practices and experimental conditions. Investigation for berseem K_c was also made in the same region earlier employing weighing type lysimeters [13]. The results reported are significantly different (0.62 -1.27) when compared with the calibrated results obtained here. This would be probably due to the differences in agronomic practices, crop variety and sowing time which are different from the current condition. In this study, berseem crop coefficients were considered for individual cutting periods.

The depletion fraction for no stress for rice is 0.1 according to FAO [30]. The average value of p equal to 0.1 has also been adopted in this study. In fact, p is more preferably applied for deficit irrigation conditions than the near saturated field conditions as in the case of rice fields. The parameter basically modifies the potential

is directly dictating the evapotranspiration component of the water balance and hence altering the deep percolation. Various studies on the crop coefficients of rice based systems were undertaken so far in the region [32,33]. Choudhury et al. [32] have reported that the Kc values of dry-seeded irrigated bed planted rice ranged from 0.62(initial) to 1.16(mid season) while for dry seeded conventional flat land it varied from 0.61(initial) to 1.42(mid-season) in the Indo-Gangetic plains of India, Karnal. Tyagi [33] have computed K_c values of 1.15, 1.23, 1.14 and 1.02, respectively for initial, crop development, reproductive and last stages using Penman-Monteith method in the same region evapotranspiration based on the soil available water. However, in paddy fields, soil moisture limitation is not a critical consideration as irrigation was applied frequently. In berseem season p has modified the evapotranspiration and hence the soil water balance since irrigation was intermittently applied during the berseem season besides reduced size of rainfall. Evapotranspiration was taking place almost at potential rate and it has a weak effect on deep percolation during the paddy season-1. However, the effect of evapotranspiration on deep percolation during paddy season-2 and berseem periods was reduced since more water was demanded by evapotranspiration from lower layers. The depletion fraction for no stress equal to 0.4 for berseem has been adopted in this study.

The soil water characteristics for the experimental field were determined for each 10 cm depth of the crop root zone (Table 4). Field observation of root growth besides observed soil moisture regime in the root zone enabled to determine the rooting depth of each crop. Accordingly, the root depths of paddy and berseem were 27 cm and 45 cm, respectively, making nearly three and five layers. The general behaviour of observed soil water content showed that the 2nd layer (10-20 cm below ground level) exhibit more water content value than the adjoining layers. Therefore, this layer was assigned with larger values of field capacity and wilting points than the other layers. Soil compaction during field preparations and cultivation of earlier crops would be responsible for the formation of a plough layer at such depths which exhibit higher water retention property than the other layers [34]. The field capacity water content is very sensitive to deep percolation. When field capacity is higher, deep percolation would be smaller and vice versa.

Climatic data and evapotranspiration

The total seasonal rainfall that fell during both crop seasons is presented in Table 5 along with other water balance components. The total numbers of growing days were almost similar for each season of both crops. Fortunately, the rainfall amount was reduced in the 2nd season of both crops during which time irrigation applications were also reduced for the purpose this study. However, rainfall was not spread over the entire growing season but concentrated in a small interval of a season in which more water goes away by runoff and/ or deep percolation losses. For example, almost half (509 mm) of the annual average rainfall in the year 2013 fell in just 15 days in the month of August in the paddy season; five of these 15 days events recorded 365.4 mm. Obviously, the rainfall which occurs during the time when the soil is near field capacity or above could not be utilized by the crops. The model computes the rainfall balance which does not take part in either soil water storage or evapotranspiration as non-used rainfall. In our field experimental plot where either run-on or run-off was controlled, the excess rainfall balance goes for augmenting deep percolation. Accordingly, more percentage of rainfall was left unused during the paddy seasons than the berseem seasons. In paddy season-1,

for example, nearly 80% of the rainfall was returned as deep percolation. During berseem season 2 and lysimeter 1 irrigation schedule, only 44% of the rainfall occurred was lost. This shows that by appropriately reducing irrigation frequency and depth, it is possible to utilize more amount rainfall for crop production.

In the paddy growing seasons, intense and continuous downpours for two to three days were not considerably contributed for crop water utilizations. Such rains have more of basin water resources importance than field scale water use as they quickly contribute to runoff or deep percolation and thus for surface water storage or groundwater aquifers. During berseem seasons, rainfall was intermittent and two to three major storms occurred in both years of growing. As these heavy storms occur after long intervals of time, most of these were returned as percolation losses due to formation of cracks and macropores in the root zone in the season [3,35]. Runoff was considered only for a heavy rainfall event occurred on August 6, 2013 in paddy season -1 when the rainfall depth overtopped the lysimeter rim level. This particular excess depth was deducted before adding the rainfall value into the modelling environment. Thus, the total rainfall recorded was taken as effective rainfall for model simulation.

In general, temperature, relative humidity and wind speed play a greater role in shaping the evapotranspiration in the area, although wind speed has comparatively less impact as it seldom appears to be more than 1 m/sec. During the growth periods, Maximum temperatures were observed in April and May (late seasons for berseem crop) and the minimum temperature values were in December (in seedling stages of berseem seasons). The average relative humidity has ranged between 100% and 29%. Therefore, larger values of evapotranspiration during periods of April through August for both crop periods were attributable to high temperature, comparatively less humidity and proportionately windy weather conditions.

The seasonal potential evapotranspiration computed in paddy season-1 and paddy season-2 were 403.80 mm and 433.3 mm respectively. During berseem season-1 the potential evapotranspiration was 260.3 mm while it was 198 mm in berseem season-2. This shows that in all the seasons, the seasonal potential evapotranspiration was higher than the actual evapotranspiration showing there was certain limitation in soil moisture in the root zone, although there was heavy irrigation and rainfall during paddy season-1, for example. Comparatively, large root zone soil moisture stress was occurred during paddy season-2 due to reduced irrigation application and high evaporative demand in the particular crop season. The seasonal actual evapotranspiration values computed were shown in Table 5.

Irrigation schedules

The total amount of applied irrigation during the crop seasons is shown in Table 5. Figures 4 and 5 also present irrigation schedules conducted in the crop seasons. In each of the crop seasons, the 1st season

Crop season	Lysimeter	Irrigation (mm)	Rainfall (mm)	Measured DP (mm)	Irrigation water saving (%)	Input water (I + P) saving (mm)	Percentage reduction in DP (%)
Paddy season-1	L1	2388.80	659.30	2668.83	Control	Control	Control
	L2	2388.80	659.30	2525.86	Control	Control	Control
Paddy season-2	L1	630.00	532.90	937.19	73.60	61.80	7.0
	L2	851.00	532.90	1069.16	64.40	54.60	5.6
Berseem season-1	L1	520.00	225.8	522.79	Control	Control	Control
	L2	520.00	225.80	478.49	Control	Control	Control
Berseem season-2	L1	63.10	220.8	148.15	88.00	61.90	17.8
	L2	91.90	220.8	132.27	82.30	58.10	21.9

Table 5: Water balance components during the crop seasons for each of the lysimeters.

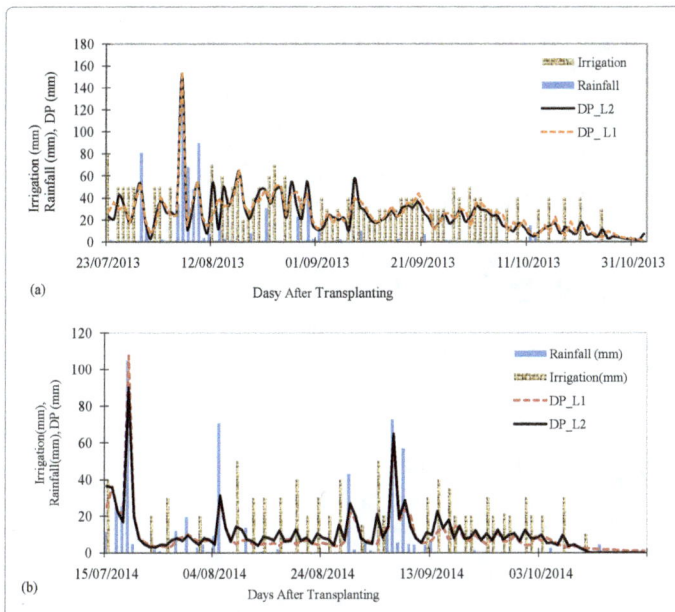

Figure 4: Irrigation schedules, rainfall and observed deep percolation in lysimeter 1 (DP_L1) and lysimeter 2 (DP_L2) in the crop paddy season-1 (a) and paddy season-2 (b).

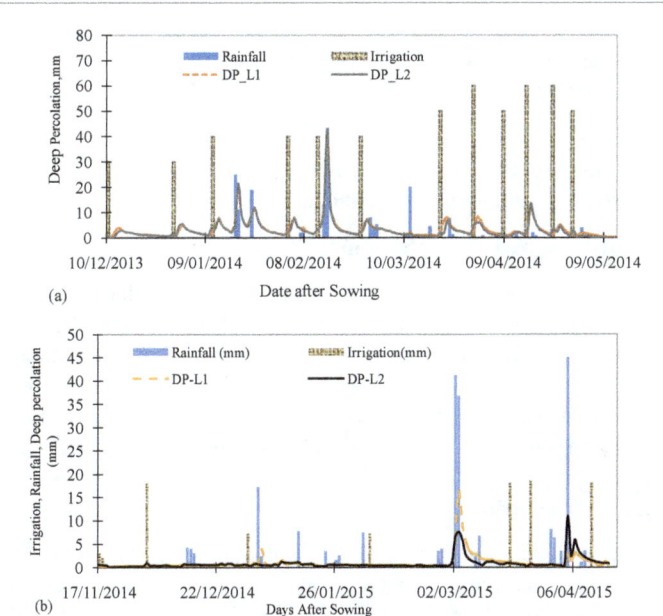

Figure 5: Irrigation schedules, rainfall and observed deep percolation in lysimeter 1 (DP_L1) and lysimeter 2 (DP_L2) in berseem season-1 (a) and berseem season-2 (b).

was used as control and resembles typical irrigation applications in the region at farmers' field while the 2nd season was presenting scenario of reduced water application. Table 6 presents the seasonal irrigation depth, percentage water saving, deep percolation and percentage reduction in deep percolation.

Average depth of applied irrigation for paddy season-1 was 41 mm (±13 mm). In this season, irrigation was applied every day, except the rainy days, in the development and mid-season growth stages while in the late season stage 2-3 days irrigation interval was imposed. During the initial growth stage, more frequent rainfalls also supplied the water demand of the crop besides irrigation. The schedule in this crop season yielded large volume of irrigation input. Such types of seasonal irrigation input for paddy rice fields were also reported in literature [4,5], although mainly concerned with puddled field conditions. During paddy season-2 average irrigation depths ranging 20 and 27 mm (±7.2 and10 mm) were imposed in lysimeters 1 and 2 respectively. The number of irrigations was halved besides reducing the depth of irrigation events in the 2nd season of both crop periods. Overall, 59 irrigations during paddy season-1 and 31 irrigations events in paddy season-2 have been applied. An average irrigation interval of nearly 2 days was practiced in paddy season-2 period. Kukal et al. [2] reported that an interval of 2 days after complete infiltration of ponded water is a recommended procedure in North-western Indian condition irrespective of soil type and irrigation depth under puddled paddy fields.

Berseem, on the other hand, needs frequent irrigation throughout its growing season because of its shallow root zone that dries quickly [13]. Average depth of irrigation equal to 47.3 mm (±10 mm) was applied during berseem season- 1 with average irrigation interval of nearly 9 days. In berseem season-2, the average depth of application was ranging between 8-11.5 mm (±4.6-7.3 mm) with an average irrigation interval of nearly 12 days for each of the lysimeters. In berseem season-1 a total of 11 irrigations were applied and only 6 irrigation events were made in berseem season-2. More frequent irrigations were applied near the end of the crop season (April-May) owing to the increased evaporative demand as responded in the crop root zone soil moisture content variation. It has been shown that large saving in irrigation as well as overall input water was achieved by imposing reduced irrigation in both crop seasons. Further, percentage reduction in percolation has also been achieved; i.e., for example the during paddy season-1 in lysimeter 1, 87.60% of input water has been turned as deep percolation. In paddy season-2, for the same lysimeter, deep percolation was 80.60%; showing a percentage reduction of 7% due to the imposed irrigation schedule.

Deep percolation

The measured and model computed deep percolation results are summarized in Table 5 and their temporal patterns during the growth

Crop season	Lysimeter	Irrigation (mm)	Rainfall (mm)	DP Irrigation (mm)	Unused Rainfall (mm)	Total DP (mm)	Measured DP (mm)	Actual ETa (mm)	Soil moisture Storage (mm)	Irrigation Scheduling Efficiency (%)
Paddy season-1	L1	2388.80	659.30	2157.90	524.10	2682.00	2668.83	378.5	-12.40	9.65
	L2	2388.80	659.30	2158.20	524.10	2682.30	2525.86	378.0	-12.20	9.67
Paddy season-2	L1	630.00	532.90	438.00	380.10	818.10	937.19	360.9	-16.10	30.48
	L2	851.00	532.90	659.00	380.10	1039.10	1069.16	360.9	-16.10	22.56
Berseem season-1	L1	520.00	225.8	364.10	145.80	509.86	522.79	260.3	-24.4	29.98
	L2	520.00	225.80	365.80	145.80	511.60	478.49	257.7	-23.50	29.65
Berseem season-2	L1	63.10	220.8	4.81	97.70	102.51	148.15	196.3	-14.91	92.38
	L2	91.90	220.8	24.62	107.30	131.92	132.27	196.0	-15.22	73.21

Table 6: Seasonal irrigation depth, percentage water saving and reduction in deep percolation in the crop seasons.

periods are shown in Figures 4 and 5. During the continuous irrigation season of paddy rice, nearly 82-87% of input water has been returned as deep percolation while in the intermittent irrigation season, the percolation loss amounted approximately 77-80% of the overall input water. For coarse textured soils, nearly same depth of percolation has been reported [4,5]. In fact, in unpuddled paddy field conditions, the percentage reduction in percolation due to reduced irrigation size is not significantly differing. This may be attributable to preferential flow through cracks and macropores during the intermittent irrigation season when such soil phenomena are prevalent (3). However, the overall water saving together with percentage reduction in percolation loss is encouraging.

In berseem season, comparatively less amount of water has been percolated as expected. Respectively, 64.2% and 70% of input water was lost through deep percolation during berseem season-1 in lysimeter 1 and 2. Due to the imposed irrigation schedule, deep percolation was limited to 52.2% and 42.3% of input water in lysimeter 1 and 2 respectively.

The deep percolation in the model was estimated as cumulative value for the whole growing season. The model predicted deep percolation well, although it separately computed percolation loss from applied irrigation and rainfall. The rainfall amount not contributing to crop growth is reported as the non-used rainfall in the model. Since runoff was controlled in our experimental field, the non used rainfall again goes on the account of deep percolation. Hence, the total seasonal percolation is the sum of the non-used rainfall and percolation amount from irrigation as computed in the model. In both crop seasons, large volume of deep percolation was attributable to rainy days and weeks in which rainstorms are intense and continuous for longer hours. With regard to growth stages of the crops, the initial stages of the crops share larger losses due to deep percolation. Therefore, irrigation schedules which consider the onset of rainy days or seasons are suggested to be implemented to limit deep percolation losses and enhance water productivity.

Percolation from an irrigated field would depend on many factors such as depth of applied water, soil and plant characteristics, groundwater depth etc. However, the depth of input water and frequency of its occurrence and/or application in sand dominated soils play a major role in transferring input water to deep percolation outflow. During both crop periods, large depths of input water due to event storms contributed to maximum daily deep percolation losses (Figures 4 and 5). Apparently, the antecedent soil moisture condition before the rainfall incident favoured more percolation during paddy crop season which was frequently irrigated compared to the berseem season. Crops during development and mid-season growth periods exhibited to withdraw more water than the other periods. It can be shown that during both crop periods that deep percolation was reduced during the development and mid-season growth stages.

The performance of the two lysimeters in metering deep percolation has also been investigated. It has been seen that the amount of deep percolation observed in both lysimeters was fairly similar showing the repacked soil monolith exhibit the same property in both lysimeters particularly during the non-storm periods. During storm periods, however, the lysimeters were observed to demonstrate variations in allowing percolation. This may be due to the fact that the lysimeters depict differences in preferential flow which is significant during rainy days. Thus we deduce from these results that locally constructed drainage type lysimeters could owe better understanding of deep percolation phenomena in an irrigated farm. Figure 6 presents

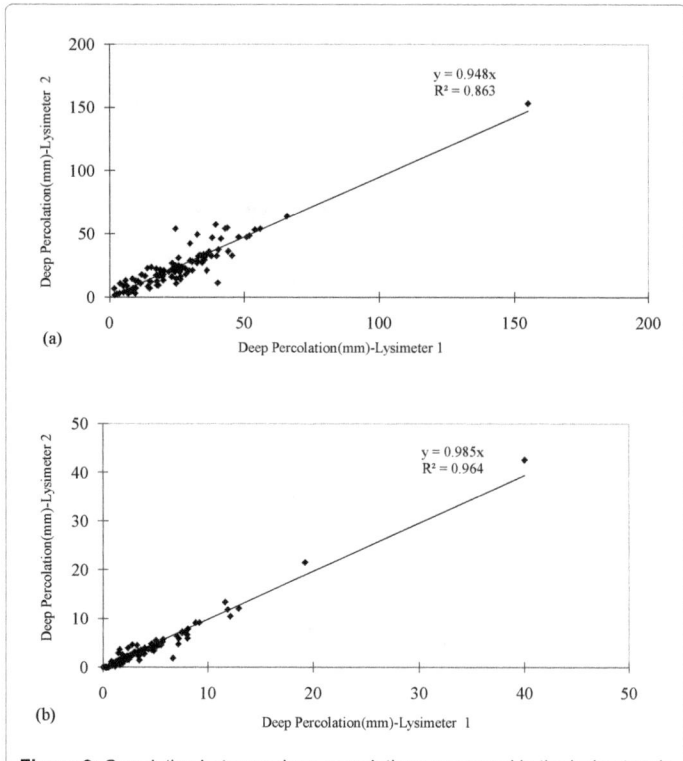

Figure 6: Correlation between deep percolations measured in the lysimeters in paddy season-1 (a) and berseem season-1 (b) growing periods.

the correlation between the measured deep percolations in the two lysimeters for paddy season-1 and berseem season-1.

Irrigation scheduling efficiency

The calculated irrigation scheduling efficiency values based on equation (4) are summarized in Table 4 above. The irrigation scheduling efficiency of the paddy season-1 was very low due to continuous flooding of the field. Again large loss of deep percolation in berseem season-1 shows that such irrigation schedules are not recommended. Due to an alternative irrigation schedule, the scheduling efficiency has been significantly improved in both paddy season-2 and berseem season-2 crop periods.

The scheduling efficiency depends mainly on applied irrigation depth and amount of deep percolation loss from irrigation. Refereeing to equation (4), it is evident that if deep percolation is zero, then the efficiency value becomes infinity. However, note that we are referring to water intensive crops growing on coarse textured soils. In fact, deep percolation can occur in even heavy soils or puddled paddy field conditions and hence negligible amount of deep percolation from surface irrigation is unlikely. For example, in berseem season irrigation was reduced eightfold and still there was deep percolation from irrigation as computed using the model. Further, the scheduling efficiency index is mainly applicable to irrigated fields than rainfed agriculture. In general, deep percolation process is the key component of water balance in sandy loam soils reducing irrigation scheduling efficiency and hence it calls for proper irrigation schedules containing less water depth and longer irrigation intervals depending on respective irrigation requirements and the nature of water consumption of a given crop under unpuddled field conditions.

Crop season	Location	Yield (tones/ha)	Water use efficiency (=Yield/input water) (Kg/m³)	Comparative yield reduction (%)	Remarks
Paddy season-1	L1	3.54	0.12	Control	Dry grain yield
	L2	3.25	0.11	Control	
Paddy Season-2	L1	2.69	0.23	24.0	
	L2	2.70	0.20	16.9	
Berseem season-1	L1	37.2	4.98	Control	Green fodder harvest (four cuts in each season)
	L2	41.2	5.52	Control	
Berseem season-2	L1	21.3	7.50	42.7	
	L2	27.9	8.90	32.3	

Table 7: Crop yield, water use efficiency and comparative yield reduction in crop seasons.

Water use efficiency and crop yield

Table 7 shows measured crop yield values for both crop periods in the lysimeters. Yield response of paddy rice can vary widely depending on rice variety, environmental factors, climatic conditions, soil characteristics and agronomic practices applied in an area. Yields ranging from 2-3.5 tonnes/ha are common as reported by Ladha et al. [36]. de Vries et al. [8] also investigated rice yield responses ranging 2-11.8 tonnes/ha in their water saving and continuous flooded rice field in the Sahelian environment. Due to reduced input water, there was yield penalty in both crop seasons. However, compared to such significant amount of water saving, the yield reduction is nominal since water use efficiency has been increased. Additionally, crop yield can be improved by employing improved agronomic practices, shifting of sowing/planting dates and adoption of appropriate technologies [37]. In fact yield reduction could have been occurred due to other reasons, but we made an effort to maintain similar crop growing and agronomic conditions except irrigation scheduling. Therefore, reductions in yield were mostly attributed to the reduced input water application in the crop seasons.

Conclusion

Two major crops, rice paddy and berseem, were grown in four seasons (two experimental runs for each crop) in an experimental field under different regimes of water application. For each crop, typical irrigation schedules as practiced in the farmers' field have been selected and conducted in the 1st growing season. Again a reduced depth and frequency of irrigation was applied in the 2nd growing period of each crop aiming for input water saving in unpuddled sandy loam field conditions. For the purpose of computing the water balance components, the WINISREG water balance and irrigation scheduling model has been employed which has been calibrated and validated for field soil and crop parameters. The field observed average root zone soil moisture content was used to calibrate and validate the model. The model was found to be adequately simulating the water balance components (deep percolation) and average root zone soil moisture content.

It has been observed that continuous application of irrigation water under non puddled paddy field of sandy loam soil resulted in quite large volume of percolation loss. Continuous irrigation in the case of paddy fields is having poor irrigation scheduling efficiency indicating very low irrigation efficiencies. On the other hand, intermittent irrigation based on soil water status would greatly reduce deep percolation and improve scheduling efficiency. Large saving in input water was achieved due to reduced irrigation applications with nominal yield penalty. In general, irrigation depths under 5 cm with 2-3 days interval for paddy irrigation and irrigation applications below 2 cm with irrigation interval of 8-12 days for berseem fodder crop resulted in percentage reduction of deep percolation in unpuddled fields besides large input water saving.

The field experiments and model results show that deep percolation is the most important component of irrigated field water balance lowering scheduling efficiency. Deep percolation was observed to mainly depend on the depth of input water and its frequency. Wetter antecedent soil moisture conditions due to irrigation favoured large deep percolation records from consecutive intense storms; in which more rainfall was observed to go unused.

Therefore, critical consideration of irrigation scheduling is suggested to reduce deep percolation to enhance irrigation scheduling efficiency and thereby increase overall irrigation efficiency for such water intensive crops. Our investigation shows that the existing irrigation schedule practiced in typical farmers' fields is by no means saving water and needs to be altered. The alternative irrigation schedule indicated in this study may be beneficial and better scheduling strategy can also be applied.

The locally constructed lysimeters were robust enough to monitor percolation loss beyond the crop root zone and can be implemented in various water management or research programs. These are affordable, can be easily constructed, maintained and provide reliable field monitoring which could be utilized in research and monitoring programs elsewhere.

References

1. Timsina J, Connor DJ (2001) Productivity and management of rice-wheat cropping systems: issues and challenges. Field Crops Res 69: 93-132.

2. Kukal SS, Hira GS, Sidhu AS (2005) Soil matric potential-based irrigation scheduling to rice (Oryza Sativa). Irrigation Sci 23: 153-159.

3. Garg KK, Das BS, Safeeq M, Bhadoria PBS (2009) Measurement and modelling of soil water regime in a lowland paddy field showing preferential transport. Agr Water Manage 96: 1705-1714.

4. Yadav S, Li T, Humphreys E, Gilla G, Kukal SS (2011) Evaluation and application of ORYZA2000 for irrigation scheduling of puddled transplanted rice in North West India. Field Crops Res 122: 104-117.

5. Bouman, Lampayan, RM, Tuong TP (2007) Water Management in Irrigated Rice: Coping with Water Scarcity. International Rice Research Institute, Los Banos.

6. Kukal SS, Aggarwal GC (2002) Percolation losses of water in relation to puddling intensity and depth in a sandy loam rice (Oryza sativa) field. Agr Water Manage 57: 49-59.

7. Belder P, Bouman BAM, Cabangon R, Lu G, Quilang EJP, et al. (2004) Effect of water-saving irrigation on rice yield and water use in typical lowland conditions in Asia. Agr Water Manage 65: 193-210.

8. de Vries ME, Rodenburg J, Bado BV, Sow A, Leffelaar PA, et al. (2010) Rice production with less irrigation water is possible in a Sahelian environment. Field Crops Res 116: 154-164.

9. Tan X, Shao D, Liu H (2014) Simulating soil water regime in lowland paddy fields under different water managements using HYDRUS-1D. Agr Water Manage 132: 69-78.

10. Nie L, Peng S, Chen M, Shah F, Huang J, et al. (2012) Aerobic rice for water-saving agriculture. A review. Agron Sustain Dev 32: 411-418.

11. Dunn W, Gaydon DS (2011) Rice growth, yield and water productivity responses to irrigation scheduling prior to the delayed application of continuous flooding in south-east Australia. Agr Water Manage 98: 1799-1807.

12. Kukal SS, Sidhu AS (2004) Percolation losses of water in relation to pre-puddling tillage and puddling intensity in a puddled sandy loam rice (Oryza sativa L.). Field Soil and Tillage Res 78: 1-8.

13. Tyagi NK, Sharma DK, Luthra SK (2003) Determination of evapotranspiration for maize and berseem clover. Irrigation Sci 21: 173-181.

14. Kennedy PB, Mackie WW (1925) Berseem or Egyptian clover (*Trifolium alexanderinum*): A preliminary report. University of California Printing Office, Berkley.

15. Mintesinot B, Verplancke H, Ranst EV, Mitiku H (2004) Examining traditional irrigation methods, irrigation scheduling and alternate furrows irrigation on vertisols in Northern Ethiopia. Agr Water Manage 64: 17-27.

16. Janssen M, Lennartz B (2009) Water losses through paddy bunds: Methods, experimental data, and simulation studies. J Hydrol 369: 142-153.

17. Mitchell J, Cheth K, Seng V, Lor B, Ouk M, et al. (2013) Wet cultivation in lowland rice causing excess water problems for the subsequent non-rice crops in the Mekong region. Field Crops Res 152: 57-64.

18. Shankar V, Hari Prasad KS, Ojha CSP, Govindaraju RS (2012) Model for Nonlinear Root Water uptake Parameter. J Irrig Drain Eng 138: 905-917.

19. Hillel D (2004) Introduction to Environmental Soil Physics. Elsevier Academic Press, Heidelberg.

20. Trout TJ, Garcia-Castillas IG, Hart WE (1982) Soil Water Engineering Field and Laboratory Manual. Academic Publishers, Jaipur, India.

21. Bohren CF, Huffman DR (1983) Absorption and scattering of light by small particles. John Wiley, New York.

22. Whalley WR, Cope RE, Nicholl, CJ, Whitmore AP (2004) In-field calibration of a dielectric soil moisture meter designed for use in an access tube. Soil Use Manage 20: 203-206.

23. Fortes PS, Platonov AE, Pereira LS (2005) GWINISAREG-A GIS based irrigation scheduling simulation model to support improved water use. Agr Water Manage 77: 159-179.

24. Teixeira JL, Pereira LS (1992) WINISAREG, an irrigation scheduling simulation model. ICID Bull 41: 29-48.

25. Liu Y, Teixeira JL, Zhang HJ, Pereira LS (1998) Model validation and crop coefficients for irrigation scheduling in the North China Plain. Agr Water Manage 36: 233-246.

26. Pereira LS, Teodoro PR, Rodrigues PN, Teixeira JL (2003) Irrigation scheduling simulation: the model WINISAREG. Kluwer Academic Publishers 161-180.

27. Popova Z, Pereira LS (2011) Modelling for maize irrigation scheduling using long term experimental data from Plovdiv region, Bulgaria. Agr Water Manage 98: 675-683.

28. Pereira LS, Paredes P, Eholpankulov ED, Inchenkova OP, Teodoro PR, et al. (2009) Irrigation scheduling strategies for cotton to cope with water scarcity in the Fergana Valley, Central Asia. Agr Water Manage 96: 723-73.

29. Liu Y, Pereira LS, Fernando RM (2006) Fluxes through the bottom boundary of the root zone in silty soils: Parametric approaches to estimate groundwater contribution and percolation. Agr Water Manage 84: 27-40.

30. Allen RG, Pereira LS, Raes D, Smith M (1998) Crop evapotranspiration: Guidelines for computing crop requirements. Irrigation and Drainage, FAO, Rome.

31. Cholpankulov ED, Inchenkova OP, Paredes P, Pereira LS (2008) Cotton irrigation scheduling in Central Asia: Model calibration and validation with consideration of groundwater contribution. Irrigation and Drainage 57: 516-532.

32. Choudhury BU, Singh AK, Pradhan S (2013) Estimation of crop coefficients of dry-seeded irrigated rice-wheat rotation on raised beds by field water balance method in the Indo-Gangetic plains, India. Agr Water Manage 123: 20-31.

33. Tyagi NK, Sharma DK, Luthra, SK (2000) Determination of evapotranspiration and crop coefficients of rice and sunflower with lysimeter. Agr Water Manage 45: 41-54.

34. Bertolino AVFA, Fernandes NF, Miranda JPL, Souza AP, Lopes MRS, et al. (2010) Effects of plough pan development on surface hydrology and on soil physical properties in south-eastern Brazilian plateau. J Hydrol 393: 94-104.

35. Tournebize J, Watanabe H, Takagi K, Nishimura T (2006) The development of a coupled model (PCPF–SWMS) to simulate water flow and pollutant transport in Japanese paddy fields. Paddy Water Environ. 4: 39-51.

36. Ladha JK, Fischer KS, Hossain M, Hobbs PR, Hardy B (2000) Improving the productivity and sustainability of rice-wheat systems of the Indo-Gangetic Plains: a synthesis of NARS-IRRI partnership research. Inst Rice Research Institute, Los Banõs, Philippines.

37. Balwinder-Singh, Humphreys E, Gaydon DS, Yadav S (2015) Options for increasing the productivity of the rice-wheat system of north-west India while reducing groundwater depletion. Part 2. Is conservation agriculture the answer? Field Crops Res 173: 81-94.

8

Mitigation of Sedimentation at the Diverstion Intake of Fota Spate Irrigation: Case Study of the Gash Spate Irrigation Scheme, Sudan

Tewedros Fikre Zenebe[1], Yasir Mohamed[1] and Haile AM[2]

[1]Faculty of Civil Engineering and Applied Geosciences, Stevinweg 1, 2600 GA Delft, The Netherlands
[2]UNESCO-IHE and Spate irrigation Network, P.O. Box 3015, 2601 DA Delft, the Netherlands

Abstract

Blocking of diversion intakes and canals by sediment deposition is a widespread problem in many spate irrigation (flood water farming) systems. In the Gash Spate Irrigation Scheme (GSIS), particularly the Fota block, sedimentation is a continuous challenge that resulted in 75% reduction of the irrigable land (2012 data). The scheme receives sediment laden flood water from the Gash River which originates from the Eritrea-Ethiopia Plateau. The GSIS is the breadbasket for the Eastern Region of Sudan with over half a million inhabitants. This research focused on sedimentation problem and its remedial measures at Fota diversion intake. The sediment deposition in front of Fota diversion intake reached up to 1.5 m depth. This deposition at the diversion intake plus sedimentation in the canal networks reduces the Fota block irrigable land by 75%. Therefore, providing remedial measures to the sedimentation problem at Fota intake would directly impact the livelihood of 100's of poor farmers. The sedimentation at the vicinity of the diversion intake was analysed using a Delft3D model. The model was calibrated and validated using observed water levels at Fota diversion intake, with a coefficient of determination (R^2) of 85% and 72% respectively. The model result under existing condition showed a 1.6 m sediment deposition at the main intake which in fact is the real situation on the ground. Alternative, sediment remedial measures based on local knowledge of farmers and technicians were modelled. Sedimentation at the intake could be reduced to almost zero if additional three Spurs (100 m, 50 m and 120 m long) are constructed on the opposite bank of the diversion at 25 m, 100 m and 200 m upstream the diversion structure respectively, and increasing the intake sill level by 1 m.

Keywords: Spate irrigation; Sedimentation; Modelling; Delft3d; River morphology; Diversion intake

Introduction

Spate irrigation is a type of water management that flood water from mountain catchments, laden with sediment concentration as much as 10%, is diverted from rivers and spread over large areas. It is unique to semi-arid environment and found in the Middle East, North Africa, West Asia, East Africa and parts of Latin America [1,2]. Spate systems are very risk-prone. The uncertainty comes from both the unpredictable nature of the floods and the frequent river bed changes due to sedimentation. Sedimentation is a real challenge for spate irrigation systems at the diversion structure from the river source. In almost all cases, flood water arrive with high sediment load, and creates serious problems at diversion intakes and canals network if it is not controlled and managed properly [3,4]. Therefore, the study of sedimentation is vital aspect for planning and design of a river basin. This study could be supported by detail analysis of flow pattern and sediment transport [5]. Advisory literature suggests several interventions to mitigate sedimentation including construction of sediment exclusion, settling basin, sediment extractor, and good design of canals and also by improving the operation and maintenance practise [6]. Field-based researches on sedimentation issues are very limited and there is still a huge knowledge gap on specific remedial measures for different magnitude of sedimentation problems under different physical, institutional and economical situations. The Gash River originates from the Eritrean-Ethiopian Plateau, and ends up in a Delta in Sudan, providing spate water for the Gash Spate Irrigation Scheme. The Gash River is characterised by flashy floods, and has uncertain flow with very high sediment concentration of up to 60,000 ppm. The river course has high sediment deposition and this seems due to braided channel aggradations and lateral migration and by both channelized and un-channelized sheet flood deposition [7]. Based on previous report, annually 5.5 -13 million tons of sediment has estimated to be brought by the River [8]. The major part of this sediment deposited in the GSIS, more noticeably at the canal and diversion intake structures. The Fota

diversion intake (a pilot study site) is one of the 7 diversion intakes in the GSIS, which is most affected by sedimentation. The sediment deposition in front of Fota diversion intake reaches up to 1.5 m. This deposition in combination with sedimentation in the canal networks reduces the Fota Block irrigable land by 75%. Therefore, providing remedies to the sedimentation problem at Fota intake would directly impact the livelihood of 100's of poor farmers who solely depends on agriculture. Primary and Secondary data such as river cross section, sediment samples, hydrological data, detailed design documents were collected. In addition, discussion with key stakeholders, namely the farmers and scheme managements were carried out. The Delft3D model was used to assess the magnitude of sedimentation problem. The model was calibrated and validated based on water levels at Fota intake and simulate alternative remedial measures. This paper encloses description of the study area in Section 2. Section 3 provides details about methods and data used and also schematization of the study area, Section 4 presents model scenario results and discussion and these are followed by conclusion under section 5.

The Study Area

The Gash Spate Irrigation Scheme (GSIS) is located in the eastern part of Sudan (Figure 1a). Spate irrigation has been practiced for more than a century for production of main crops such as Barley, Sorghum and Cotton in the area. The scheme is operated by the Gash Delta

*Corresponding author: Tewedros Fikre Zenebe, Faculty of Civil Engineering and Applied Geosciences, Stevinweg 1, 2600 GA Delft, The Netherlands
E-mail: tfikrie@yahoo.com

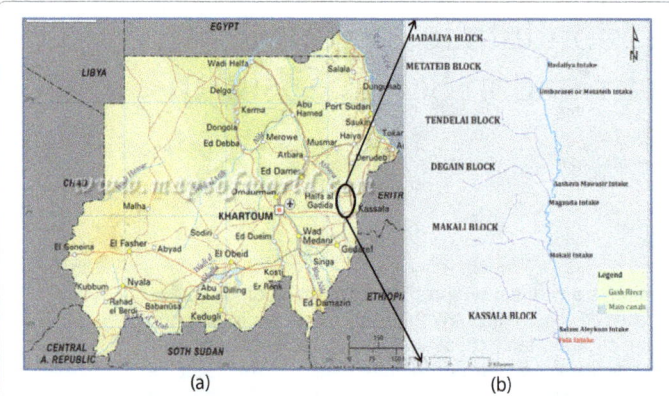

Figure 1: (a) Location map **(b)** Layout of intakes and main canal along the river.

Agricultural Corporation (GDAC), and also technically supported by the Gash River Training Unit (GRTU) of the federal ministry of Water Resources [9]. The first design of the system was developed in the 1930's and improved in the late 1950's. The GSIS area is characterized by semiarid climate with two notable seasons, winter and summer. The max temperature may exceed 45°C in the summer time, and drops to an average of 25°C in the winter [10]. The Gash River, water source of GSIS, originates from the Eritrean-Ethiopia Plateau and travels about 121 Km from the border with Eritrea down to the Gash Die (end of the delta). The total catchment area of the River is 21,000 km² [11]. It is seasonal river, flows between late June to October, with high flows occurring between July and September. The maximum annual discharge accounts 1430 million m³ recorded in 1983 and annual minimum flow of 140 million m³ was gauged in 1921 [11]. Sediment concentration varies between 3000 ppm to 30,000 ppm on average, but may exceed more than 60,000 ppm during high flood. In general, the river is characterized by large variations in annual flow and heavy silt loads [12]. The total gross area of the Gash Delta is 2,80,000 ha, of which 1,00,000 ha is used for irrigation. However, the irrigated area allocated for annual rotation is about 30,000 ha [13]. The irrigation systems includes 7 Diversion Intakes, 7 Main canals (Figure 1b), and 212 messga canals (tertiary canals). All canals receive water on the left bank of the river since the terrain slopes declines in north-west direction. The area of messgas (farm blocks), which is irrigated by one tertiary canal, varies from 300 ha to 1200 ha. The flood water is diverted from the wide river through a brick structure intake (Figure 2). The Fota intake composes of 3 openings, with dimension of 2.5 m width and 3.0 m height. The intake is closed manually by timber stop loges. The Fota main canal is an earthen canal with trapezoidal cross section and side slope of 1:0.5 /1:1.2. During the flood period, the deposited sediment at Fota diversion intake excavated by a stand-by excavator so as to keep continuous water flow to the command area. This effort has, however, often not been successful for most part of the flood season since the muddy situation around the intake prevents access for the excavator. Thus, as indicated in Section 1, the Fota command area has consistently suffered from severe shortages of irrigation water.

Materials and Methodology

Methodology

Assessment of the existing sedimentation problem at the Fota diversion intake and subsequent remedial measures were analyzed using Delft3D model, which is a three dimensional model system [14]. In the following sections, a detailed account is given on the set-

up and schematization of the model and its calibration and validation procedures.

Delft3D model setup and schematization: Two dimensional (2D) Delft3D model was developed to simulate the river morphodynamics around Fota diversion intake. Delft3D package, developed by WL/ Deltares in close cooperation with Delft University of Technology, is a software for simulating hydrodynamic flow (under the assumption of shallow water), transport of water-borne constituents, short wave generation and propagation, sediment transport and morphological changes and ecological processes and water quality parameters [15,16]. The model uses three basic equations to study the sediment transport process these are: Continuity equation, Momentum equation and Transport equation [14]. A reach of 10 km length of the Gash River was schematized in Delft3D model using 2856 grid cells (Figure 3). It starts from Kassala Bridge (upstream boundary condition, BC) and ends at Salamalikum (downstream BC). The upstream BCs are measured discharge and sediment load at Kassala Bridge station. The downstream BC is taken as the water level and sediment load measured at Salamalikum station (Section 3.2). Then the model was calibrated and validated using measured data at Fota diversion intake which is 7 km far away from Kassala Bridge (upstream BC). Based on the river cross section data (Section 3.2) and extent of the river boundary, a

Figure 2: Sedimentation affected Fota diversion intake.

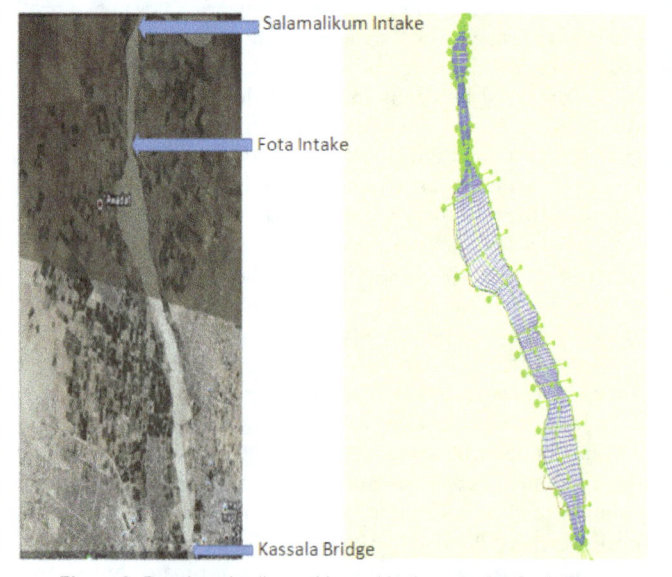

Figure 3: Developed spline, grids, and bathymetry for Gash River.

numerical grid and bathymetry of the river were established. The grid orthogonalized and smoothed to reach representative schematization of the flow pattern in the study area. The generated grids for the river have the following properties:

- Grid points in M direction is 204.

- Grid points in N direction is 14.

- Maximum and Minimum Grid size in M direction is 125 m and 1.5 m respectively.

- Maximum and Minimum Grid size in N direction is 84 m and 2 m respectively.

- Orthogonally for the whole grids ranges from 0 to 0.14, however most of, around 95%, is fall below 0.01.

- Smoothness in M direction for the whole grids ranges from1 to 3.73, however most of, around 98%, is fall below 1.14.

- Smoothness in N direction for the whole grids ranges from1 to 2.75, however most of, around 95%, is fall below 1.18.

- Aspect Ratio for the whole grids ranges from1 to 9.64, however most of, around 90%, is in the range of 1to 1.8.

The Flow module was used to enter all required data (Section 3.2) and physical parameters (Table 1) in to the Delft3D model system. The model simulation results were calibrated and validated hydrodynamicaly and then calibration of sediment transport proceeded.

Materials

Boundary condition: The River Gash has 5 gauge stations, since the model was developed within the three gauge stations, namely Kassala Bridge, Fota and Salamalikum, all available measured data from those three gauge stations were collected. Since the available data have lots of missing values, well completed daily data of year 2005 and 2006 were used to calibrate and validate the models respectively. The discharge and water level used for boundary condition for the models for 2005 and 2006 year data are shown below in Figures 4 and 5. For some missed data, a data driven graph was developed and used to fill the missing values. Sediment concentration measurement in Gash River started from 2004; however there was a lot of missing data and discontinuity in measurements. After 2008 the measurement of sediment concentration at Fota and Salamalikum station also interrupted [3]. For this study, measured silt and clay concentration is used. For the case of sand concentration it is assumed that the flow enters at the Kassala Bridge with equilibrium sand concentration. This is because of two reason: the first one is the available data on the sand concentration is very limited and the second is since the flow reaches to the Kassala Bridge travelling long distance it could attain equilibrium flow for sand concentration. Based on this, the sediment distribution of the area was studied and mitigation measures were provided. For the analysis of sediment grain size of the River, 7 deposited sediment samples from different points

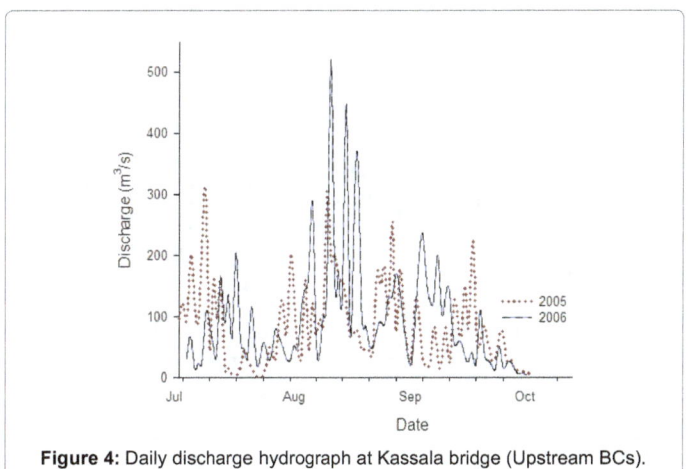

Figure 4: Daily discharge hydrograph at Kassala bridge (Upstream BCs).

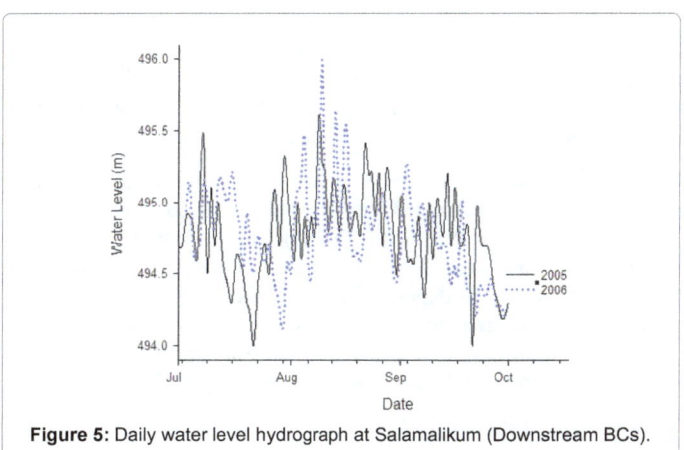

Figure 5: Daily water level hydrograph at Salamalikum (Downstream BCs).

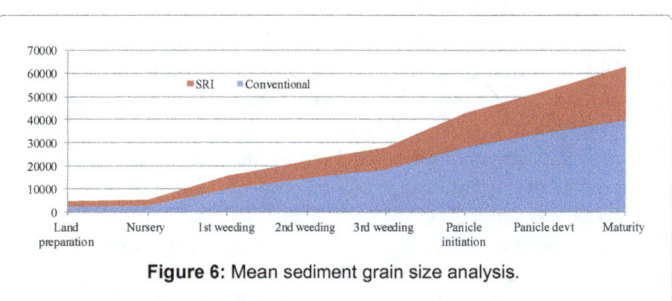

Figure 6: Mean sediment grain size analysis.

of the River bed were collected. Random sediment samples collections were done within the range of 50 m to 500 m interval between sample points. The d50 of the river bed material (after flood season) for the samples ranges from 220 μm to 280 μm (Figure 6). The sediment particle density of the sand samples in Gash varies from 2500 Kg/m³ to 2640 Kg/m³. Therefore, an average value of 2550 Kg/m³ was used. Since there was no information available on settling velocity of the Gash sediment, settling velocity from Gezira scheme which has more or less similar sediment source catchment area were used.

Topographic surveying: The 14 km length of the Gash River Cross Section was surveyed in 2004 by the GRTU. Therefore, this secondary data were used to develop the river bathymetry from Kassala Bridge to Salamalikum gauge station. In the field observation, the area around the Fota diversion intake shows a serious sedimentation problem. The elevation of the Fota intake bed level is 494.87 m.a.s.l and the crest level

Physical parameters used in the model	Values
Chezy Roughness	50 m¹ᐟ²/s
Sediment transport predictor	E-H
Morphology factor	1
Courant number	10
Time step	0.05 minutes
Horizontal eddy viscosity	1 m²/s
Horizontal eddy diffusivity	0.1 m²/s

Table 1: Model parameters.

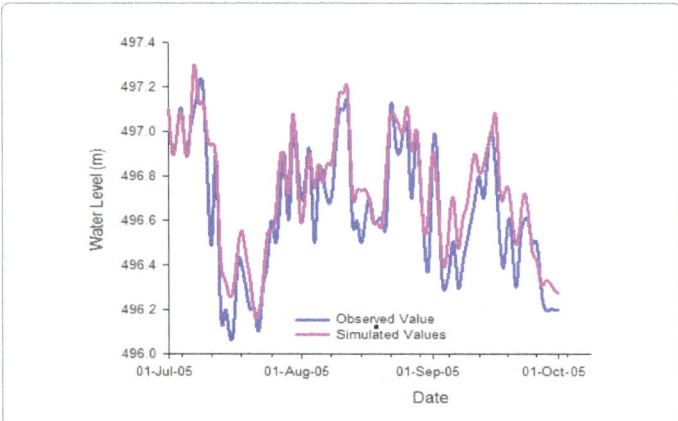

Figure 7: Daily water level for observed and simulated at Fota gauge station (2005).

of the diversion headwork/weir downstream of the intake, which is by now completely buried, has an elevation of 496 m.a.s.l. This creates about 1.13 m sediment to be deposited easily around the intake. In addition to the river cross section primary data on-length and location of existing structures/spurs was collected.

Operation and maintenance: Information's on the operation and maintenance of the Fota intake was collected from the Gash Agricultural Scheme Authority (GASA). It shows that they used a stand-by excavator near the diversion intake during irrigation period for removing sediment deposited at the intake and the most upstream part of the main canal so as to allow water flow in to the irrigation system. Due to the sediment deposition, the Fota irrigation system gets water only during high flood time.

Results and Discussions

Calibration and validation of Delft3d model

The Delft3D (hydrodynamic) model was calibrated using the available measured daily data of the Gash River water levels during 2005 at Fota Diversion Intake. The comparison of daily average water levels is shown in Figure 7. The coefficient of determination R^2 between simulated and observed water level was found to be 0.74. The observed discharge values at Fota gauge station showed a bit strange values for the period August 5, 2005 to August 15, 2005. Excluding these values led R^2 improved to 0.85. However, it couldn't be confirmed that these data were wrong. As shown in the Figure 7 there is somewhat over estimation by the model especially around August 8-20/2005 and also during pick flow. The validation results of the hydrodynamic model using 2006 year data are given in Figure 8, few outliers of the observed data (12 days) were excluded, and then R^2 found to be 0.72. As portrayed in the Figure 8, the simulated values shows poor fit for peak and low flow. This might be due to the data quality, however it follows similar trained as the observed one. Since there was no continuous sediment deposition data around Fota Diversion Intake for model calibration, qualitative calibration (sedimentation pattern) and total depth of sediment deposited on the sill of the intake (sill level is 494.87 m.a.s.l. and deposited sediment level is 496.27 m.a.s.l.) was adopted. As shown in the Figure 9 below, the sediment deposition near the intake is high and causes the water main channel to flow far away from the diversion intake (see the real condition under Figure 2 left top picture). The intake was buried by more than 1.5 m sediment deposition. This is a real situation in the area and results in limited water inflow to Fota Main Canal only during high flood time.

Comparison and sensitivity analysis

A comparison of the sediment transport equations (Van Rijn and Engelund and Hansen, E-H) were carried out to investigate the applicability of the sediment predictor equation for the Gash River. The analyses indicated that Van Rijn transport equation computed a very high sediment deposition around the diversion area than both the existing condition and E-H transport equation result (Figure 10a). Model sensitivity was also investigated for varying Chezy roughness coefficient (50 $m^{1/2}$/s and 40 $m^{1/2}$/s), as well as for different sediment particle size, d50 (225 μm and 270 μm). The result shows that the change in roughness coefficient has less effect on the sedimentation patterns. On the other hand, Chezy with 40 $m^{1/2}$/s provides relatively low sedimentation on both sides of the river banks around the diversion intake (Figure 10b). In the case of sediment particle size, the sedimentation pattern has less effect as is the case with roughness coefficient. However, with large sediment size, relatively high deposition occurs upstream of the diversion intake (Figure 10c).

Model scenario analysis

To assess the sedimentation problem in the area and suggest mitigation measures, three scenarios were evaluated:

Existing condition (baseline scenario)

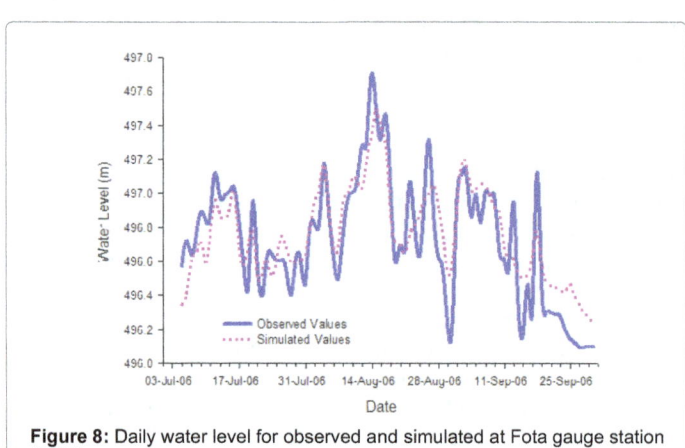

Figure 8: Daily water level for observed and simulated at Fota gauge station (2006).

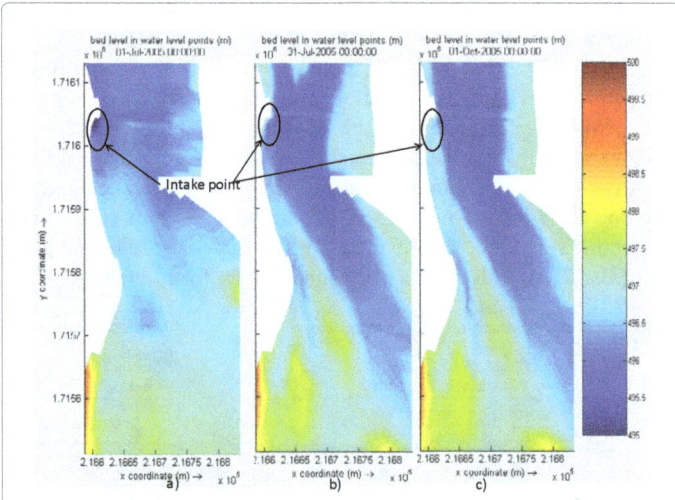

Figure 9: River morphology change at Fota diversion intake: (a) at the beginning, (b) after 1 month and (c) after one year simulation.

1. Constructing 200 m long guiding wall in the right side of the Gash River around the diversion intake area (Figure 11a)

2. Constructing 3 spurs in the right side of the Gash River around the diversion intake area and increasing the intake sill level by 1.2 m (Figure 11b)

For the existing condition (Figure12), the simulation showed high sedimentation (more than 1.5 m depth) at the diversion intake which diverts Gash flow away from the intake. The design bed level of the intake structure is 494.80 m.a.s.l, while there is a diversion weir across the Gash River with a crest level of 496 m.a.s.l. which is located at 5 m downstream of the intake structure [17]. This actually favours 1.13m sediment deposition at the intake during flood. In case of both scenarios (scenario 2 and 3) (Figures 13a and 13b), the model results showed reduction of sediment deposition around the intake. This is mainly because both interventions reduce the river flow width which leads to higher flow velocity and hence increase the sediment transport capacity. The introduction of spurs (scenarios 3) has outperformed than guiding wall (scenario 2). While the former resulted in almost zero sediment deposition, the latter could only reduce sedimentation from more than 1.5 m (under existing condition) to 0.8 m. Figure 13a, 13b and 13c provide the illustrations. The construction of spurs has

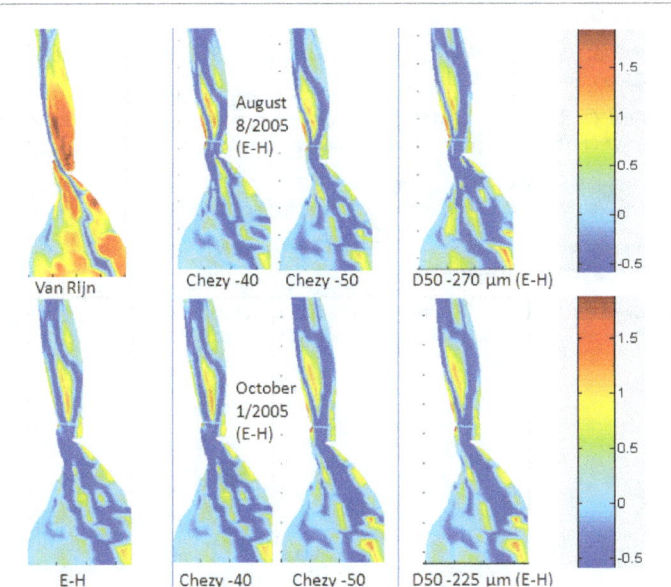

Figure 10: Comparison of sediment transport equations (a), and roughness (b) and sediment particle size (c) sensitivity.

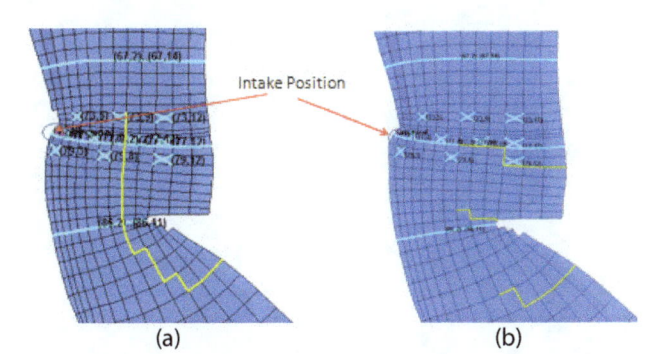

Figure 11: Proposed sediment mitigation mechanisms at the Fota intake: a) guiding wall and b) spurs.

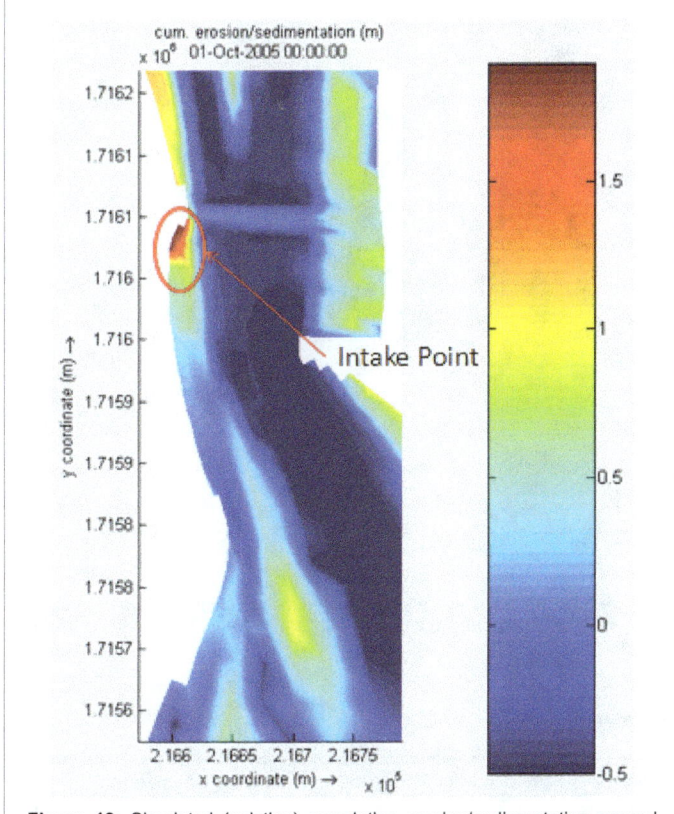

Figure 12: Simulated (existing) cumulative erosion/sedimentation around Fota intake.

additional value of creating open space on the right bank of the river for sediment deposition. As a result the downstream part of the river will not be affected by excess sediment removal from the diversion intake.

Conclusion

The research conclude that under existing condition, more than 1.5 m of sediment annually deposited and this deposition completely block the water inflow into the Fota main canal system. The deposition is attributed mainly to a weir constructed just downstream the intake, which is 1.13 m higher than the bed level of the intake, and the wide river water flowing width near to the intake. Introducing a 270 m length guiding wall reduces the river flow width near the intake, increases the sediment transport capacity thereby reducing the sediment deposition at the intake to 0.8 m depth (scenario 2). Constructing groyens/spurs (scenario 3) on the right bank of the river combination with intake modification (constructing sill with 1 m height at the intake) could almost completely avoid sediment deposition at the intake. Engelund-Hansen sediment transport predictor equation provides better result for the Gash River compared to Van Rijn equation. The sediment deposition predicted by Van Rijn equation is over estimated. It shows more than meter depositions all around the diversion weir and intake which is not the situation in the study area. Using the Engelund-Hansen sediment transport predictor equation the sensitivity of the model for change of roughness coefficient and sediment particle size was evaluated. Change of sedimentation pattern found to be less sensitive to changes of roughness coefficient and sediment particle size. However, the sediment deposition upstream of the diversion intake increases with increase of sediment particle size.

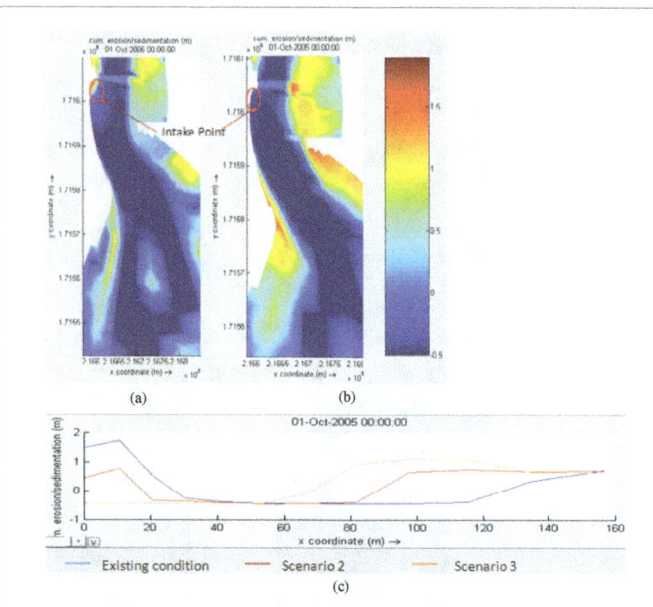

Figure 13: (a) Cumulative sedimentation/erosion for guiding wall, (b) spur/ groyens intervention and intake sill height change and (c)comparison of the three conditions.

Acknowledgement

My deep gratitude to Netherlands Fellowship program (NFP), Spate Irrigation Network, IFAD and UNDP for funding the study and article preparation.

References

1. Mehari A, Van Steenbergen F, Schultz (2010) Modernization of Spate Irrigated Agriculiure: a new approach. Irrigation and Drainage 60: 163-173.

2. Steenbergen F, Abraham Mehari Haile, Taye Alemehayu, Tena Alamirew, Geleta Y, et al. (2011) Status and Potential of Spate Irrigation in Ethiopia. Water Resour Manage 25:1899-1913.

3. Embaye TG, Beevers L, Haile AM (2011) Dealing with Sedimentation Issues in Spate Irrigation Systems. Irrigation and Drainage 61: 220-230.

4. Lawrence P (1987) Sediment Control in Wadi Irrigation Systems.

5. Osman A, Xiwu LU, LADU JLC (2012) Application of Two Dimensional Models to Simulate the Flow and Sediment Transport in the Middle Reach of Yangtze River, Renmin Island Region. Journal of American Science 8: 778-784.

6. Lawrence E, Atkinson P, Spark C, Counsell (2001) Procedure for the Selection and Outline Design of Canal Sediment Control Structures: Technical Manual, HR Wallinford and DFID.

7. Alredaisy SMA (2011) Mitigating the Catastrophic Impacts of Torrential Rivers in semi-arid Environments: a case of the Gash River in eastern Sudan. Arid Land 3: 174-183.

8. Eisa AAS (2011) Spate Irrigation in the Gash : Technical report of Gash River training unite (GRTU)

9. Kirkby J (2001) Saving the Gash Delta, Sudan. Land Degrad Dev 12: 225-236.

10. Elsheikh AEM, Khalid AEZ, Shaza AE (2011) Groundwater budget for the upper and middle parts of the River Gash Basin, eastern Sudan. Arab J Geosci 4: 567-574.

11. Anderson IM, (2011) Technical paper on Main findings and Recommendations: The Eastern Sudan Rehabilitation and Development Fund.

12. AQUASTAT (2005) FAO's Information System on Water and Agriculture -world wide web page

13. Hamid MH, Malla MM (2011) The River Gash Irrigation Scheme, Eastern Sudan: Report of GRTU

14. Delft3DUserManual (2011) Simulation of multi-dimensional hydrodynamic flows and transport phenomena, including sediments

15. Lesser GR, Roelvink JA, Van Kester JATM, Stelling GS (2004) Development and Validation of a Three-Dimensional Morphological Model.

16. Alfagemel RS (2005) Process Based Morphological Modeling of A restored Barrier Island Whiskey Island, Louisiana, USA.

17. Gismalla YA (2010) Review of the Sediment Monitoring Program in Gezira Scheme - Sudan. Science and Technology 11:1-19.

9

Effects of Trickle Irrigation System from Southern Iran District Chabahar Free Zone

Mohammad AZ*

Department of Basic Science, Faculty of Marine science, Chabahar Maritime University, Iran

Abstract

This research was carried out at steep slope area planted with trees in different elevation terraces. The experiments were conducted at the experimental site at the Chabahar Free Zone. One major disadvantage of trickle systems is the tendency for emitters to clog. A trickle irrigation system was installed in a 50 m long and 20 m wide plot. The hydraulic performance of emitters was based on water flow, uniformity coefficient, application efficiency, and water losses through deep percolation. The flow volumes along the lateral length were fairly consistent and the variation was diminutive under both types suggesting uniform distribution of water. The difference in elevation between upper and lower terraces at the area of study was about 50 m irrigated by drip irrigation system. The system of irrigation has a problem in distribution uniformity of water resulted from initial filling of the pipes and drainage of water after stopping irrigation. Therefore, the lowest terrace receives the highest while the upper terrace receives the lowest amount of water. The problem of a lateral pipe with equally emitters and uniform supply of water is investigated. The flow volumes along the lateral length were fairly consistent and the variation was diminutive under both types suggesting uniform distribution of water. The system achieved rationally high DU, CU, Ea. The CU values for randomly selected laterals with smooth emitters averaged to 81.7% and spiral emitters averaged to 87.4%. The DU values averaged to 75.4% for smooth and averaged to 81% for spiral emitters. The overall Ea achieved were 82.7% and 89.4% for smooth and spiral emitters, respectively. The higher values of CU, DU, and Ea with spiral emitters as compared to smooth emitters suggest that they performed better and could be preferred to achieve uniform water distribution. Water movement below the emission point was more pronounced in the vertical. In most cases, the wetting front followed an axially symmetric pattern. The water laterally moved to about 0.35 m while it moved to a 0.56 m depth. The root zone for many short rooted crops is located in this range hence the percolation losses would practically be negligible under such situations.

Keywords: Clogging drip; Emitter; Irrigation; Laterals; Trickle; Uniformity

Introduction

This paper describes drip or trickle irrigation system applied in Chabahar Free Zone in the Makoran area. The area of the study is near the border of Iran and Pakistan, extending south to the Gulf of Oman. The climate of the region varies from subtropical arid and semi-arid to temperate sub-humid in the plains of Sistan and Balochistan. In any system development an objective must first be looked and explained at complete different angles. As the system develops, thus the research must be known that quality or cost-effectiveness of the trickle irrigation system for that site is a direct function of how well it meets the requirements, and fulfils its basics objective. The development of a typical irrigation in such arid area in Bahokalat (Sistan and Balochistan) drip irrigation system can be a fairly straight forward process [1]. Trickle irrigation is a localized irrigation method that slowly and frequently provides water directly to the plant root zone. Emitter clogging has often been recognized as inconvenient and one of the most important concerns for trickle irrigation systems, resulting in lowered system performance and water stress to the non-irrigated plants. Partial and total clogging of emitters is closely related to the quality of the irrigation water, and occurs as a result of multiple factors, including physical, biological and chemical agents [2]. Trickle irrigation is the precise, slow application of water as discrete drops, continuous drops, small streams, or miniature sprays through mechanical devices called emitters located close to the plants. Water analysis prior to system design, a preventive maintenance program and field evaluation of clogging and uniformity are strongly recommended and proposed a classification scheme for water quality to indicate clogging potential. The evaluations have been carried out according to Mosh [1] recommendations, which have been followed in later works of authors. Irrigation is the controlled application of water artificially to the soil for the purpose of supplying the moisture needs and requirements of the crops for production and optimum performance in the field or farm. Drip irrigation is the frequent, slow application of water either directly into the land surface or into the root zone of the crops [3], according to Mosh [1], drip irrigation is a method of watering plants with a volume of water approaching the consumptive use (CU) of the plants or trees. Adequate water is vital not only for the survival of trees in Chabahar Free Zone (Sistan and Balochistan), but also the frequent intervals and application of water can boost both the quality and quantity of these trees. Drip irrigation has a higher capability for minimizing the loss of water by evaporation, runoff, and deep percolation in comparison to other irrigation systems that supply water to the soil [4]. This may include such parameters as soil conditions, topography, water quality, water quantity available in Sarbaz River where water is irrigated for that plants or trees for Bahowkalat farm in which drip irrigation is applied, although type of crop or landscaping to be planted, plant spacing and similar type parameters could be research. In this regards for Chabahar free zone arid or semi-arid area system type requirements then take this site and project conditions and transpose them into irrigation system standard [5].

*Corresponding author: Mohammad AZ, Department of Basic Science, Faculty of Marine science, Chabahar Maritime University, Iran
E-mail: mazainudini@yahoo.com

Material and Methodology

Iran with an area of 165 million hectare of arable land of which only 8 million hectare are irrigated, 6 million hectare are rain-fed, and 4.5 million hectare remain in the form of fallow land. The climate of Iran is one the greatest extremes due to its geographic location and variation in topography. The summer is extremely hot in its central deserts and fall far below zero in the West Mountains. Annual rainfall ranges from less than 50 mm in the deserts to more than 1600 mm on the Caspian Plain. The average annual rainfall is 252 mm and approximately 90% of the country is arid or semiarid. Taken as a whole, about two-thirds of the country annually receives less than 250 mm of rainfall.

This experiment was carried out at Chabahar Free Zone research center, in Makoran Chabahar, Iran. Tests were carried out in the experimental area of the Chabahar Free Zone. The irrigation system evaluated was in a subunit with trickle irrigation system, comprising 10 lateral lines, 50 m long, 3 m apart from each other, set in a hilly area. The lateral irrigation lines were of polyethylene pipes of high density, (16 mm diameter), with a screen filter at the beginning [6]. Trickle irrigation is the precise, slow application of water as discrete drops, continuous drops, small streams, or miniature sprays through mechanical devices called emitters located close to the plants. The water supply was taken from a tube well and firstly supplied into the big pound, then pumped by the tube well into the lateral line. Also the emitter's flow of a trickle irrigation system is mainly affected by hydraulic dimensions, manufacturing variations, temperature and clogging of emitter. Furthermore, in this procedure if emitter's flow is turbulent, it is less affected by temperature and, if the water taken into the system can be controlled by filters, which are essential for trickle irrigation systems, the emitter's variation will be only affected by pressure and manufacturing variations.

Individual emitters are considered in discharge and pressure estimations along the lateral and also the friction head loss between successive emitters. An important aim and objective of any trickle or drip irrigation system is a uniform distribution of water delivered through the emitters however, there would be an accurate filtering in this pipe network system. Computational of flow distribution requires knowledge of the variables such as pressure, flow rate, length of lateral, characteristics of the orifice, and frictional loss or emitters clogging. Trickle irrigation has become a well–established method for irrigation high value crops and trees in Chabahar Free Zone (Sistan & Balochistan) province where water resources are scarce and expensive.

System Consideration

During the first phase of the methodology was applied in the site the research could be analysis, the sensitive design for that area with tropical climate would be consider not only the performance and ability of the components selected, but many other parameters that influence the operation of a system. For example, the operation system will be forced by the availability of the product stated, at the time of settled in the Chabahar free zone Drip irrigation site, though experience and capability of the personnel available for its settlement and operation [1]. Increasingly, many other effects considered that come into play in determining the ultimate effectiveness of an irrigation system such as emitter clogging and physical suspended solids and chemical problem like Power hydrogen (pH) factors of water, hydrogen sulfide, dissolved solids, organic matter, temperature, bacterial growth and slime development can be influenced [6]. Emitter clogging greatly reduces the water distribution uniformity in irrigated fields [2], which negatively influences crop growth and yield. Evaluated local trickle irrigation

units and calculated average emission uniformity, average absolute emission uniformity, and system emission uniformity. According to the criteria proposed by Michael [7] there should be an accurate filter in this system to reduce sedimentation from network.

Maintenance and operating of this farm continued many years but due to miss management in this system was not play positive role in this farm thus, drip irrigation was failed in the Chabahar Free Zone farm. Because it is not only properly handle but also not suitably designed. Therefore, an increase in the initial acquisition cost is easily justified if it can be demonstrated at the Chabahar Free Zone farm that this will represent an annual savings throughout in this agricultural farm the life of a system. Of course, research create project for under initial cost. In this case it is hard to convince the client that "life-cycle" cost is a significant parameter [8]. Further more in sandy soils like Chabahar Free zone drip irrigation system, reduce the spacing between drip lines from 18" inches to 12" inches or less. Irrigation water will tend to go down through sandy soil rather than spreading out by capillary action. Therefore, in this case should be keep drip irrigation session about 45 minutes long but water more frequently. Thus, commercial growers often bury drip line to avoid damage from machinery but I could recommend keeping drip line on the surface to avoid any possibility of muddy water flowing back into the drip line during shut down the drip irrigation in Chabahar Free Zone site [3].

Effectiveness of Drip Irrigation

A drip irrigation system can be composed of filters and strainers, valves, pressure regulators, chemical injectors, emitters, hose or tubing, pipe and proper fittings to do then it could be properly management and organizing. Furthermore, the importance of this Agricultural site with this system is obviously no better than the function of its individual parts can be available in that farm. This, in turn, is a function of both the products selected and the hydraulic design of the system. Generally the measure of importance of a system is a function of how well it performs with in this farms its handle and operational scenario. From the investigation of water pumped into this system, therefore, the pumping unit was not appropriate, because the pump was applied in to this site was carried many sand and silt. That means the size and type of pump applied were not appropriate for these agricultural sites. Because it brought sand and silt into the system and clogging occur to emitters and due to this reason less water supply for trees then growth of trees were low [3]. Increasingly, in the Chabahar free zone Horticultural park PH of water is 7.4 to 7.8 which are originated from city water supply. Perhaps, the diameter of lateral pipes was not appropriate for that trees like mangoes and Chekov because these trees needs more water naturally and also due to tropical climate. However, due to less water applied to these trees the growth was failed for these trees sharply [9].

Data Analysis of Flow Meter Performance

The performance of a flow meter is defined by its characteristic (calibration) analysis curve. This curve relates flow meter response to volume of flow and volumetric flow rate. Test to find flow meter response over a range of expected flows are normally used to define the characteristic curve of a flow meter. Typical non-linear characteristic curve is shown in Figure 1a and Figures 1-5. For most nonlinear flow-meter, flow is proportional to the square root of flow meter response and the shape of the characteristic curve is similar to curve of Figure 1b. Indicated flow is that given by the flow meter or calculated from its readings while true flow is found by a high accuracy measured device used in the calibration test. Plots of such parameter as the coefficient

Figure 1a: Study area of Chabahar free zone trickle irrigation site.

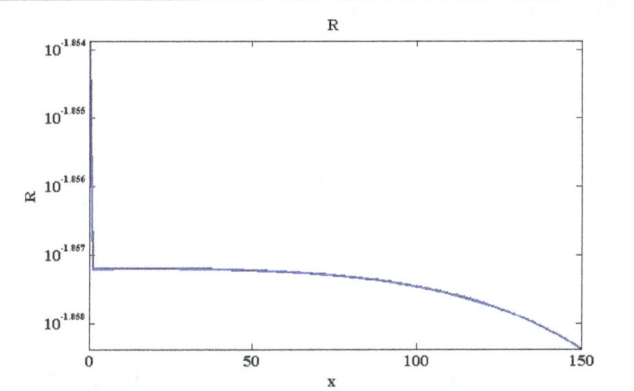

Figure 1b: Relationship between length of pipe and Reynolds number in pipe (diameter of lateral 16 mm).

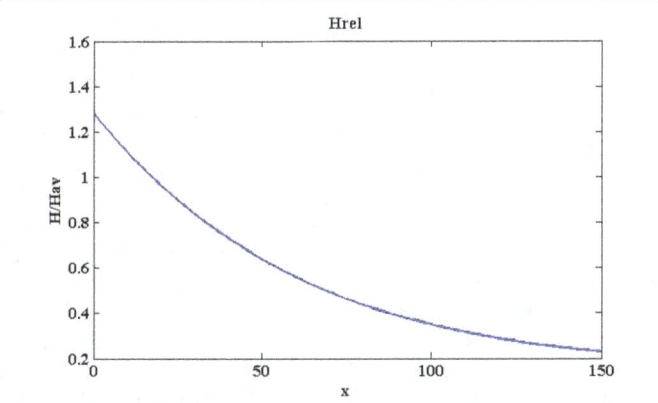

Figure 2: Relationship between emitter number and pressure head at emitter (diameter of lateral 16 mm).

of discharge, meter correction term, meter factor, and K-factor against flow or quantities such as the Reynolds number, however, describe deviations from true flow indicates such a plot in Figure 3. Therefore, Individual emitter flow for each tree of fertility soil with manifold characteristic to refer with Figures 6-8 with average irrigated water into root zone of different trees with equal space of that emitter. Considering emitters flow for each terrace amount of water flow is different for each trees according to Figures 7 and 8. It could be grateful if we have emitter with adjustable and appropriate quality of emitter, thus towards amount of flow is less than need of tree because need of each tree at least 45 liter per day per week according to Figure 8.

Individual emitters are considered in discharge and pressure estimations along the lateral and also the friction head loss between successive emitters. An important aim and objective of any trickle or drip irrigation system is a uniform distribution of water delivered through the emitters however, there would be an accurate filtering in this pipe network system. Computational of flow distribution requires knowledge of the variables such as pressure, flow rate, length of lateral, characteristics of the orifice, and frictional loss or emitters clogging. Trickle irrigation has become a well-established method for irrigation high value crops and trees in Chabahar Free Zone (Sistan and Balochistan) province where water resources are scarce and expensive.

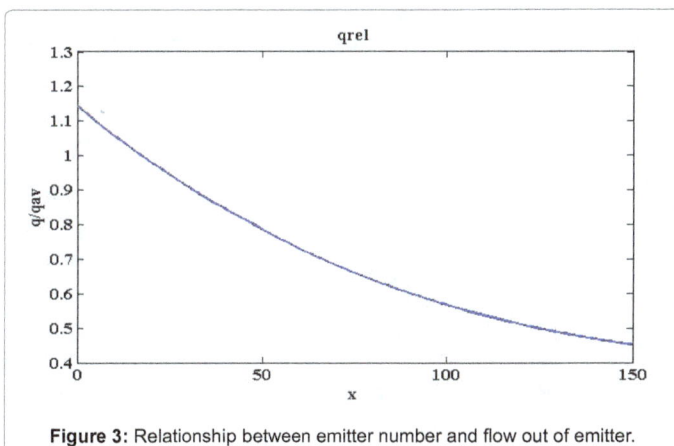

Figure 3: Relationship between emitter number and flow out of emitter.

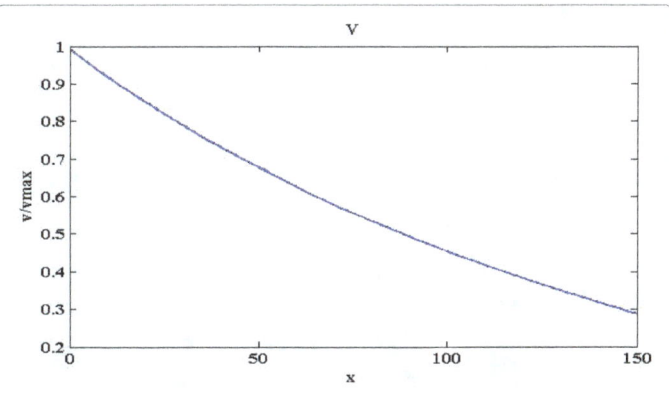

Figure 4: Relationship between length of pipe and velocity in pipe (diameter of lateral 16 mm).

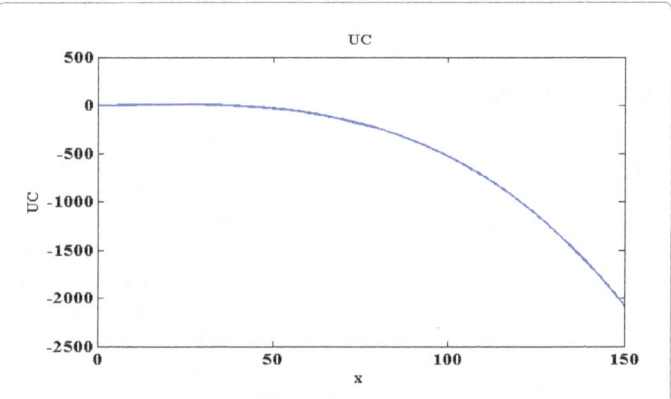

Figure 5: Relationship between length of pipe and uniformity coefficient.

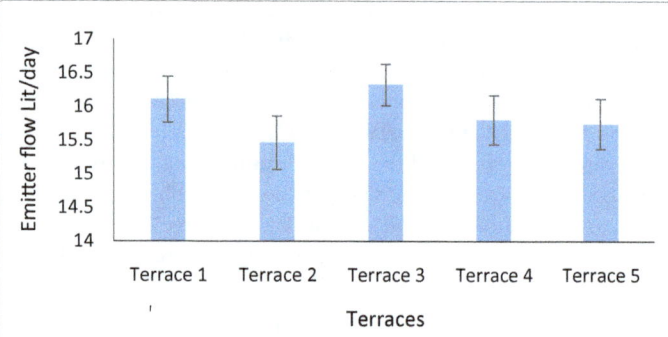

Figure 6: The amount of water for each emitter in existing design of lower slope trees.

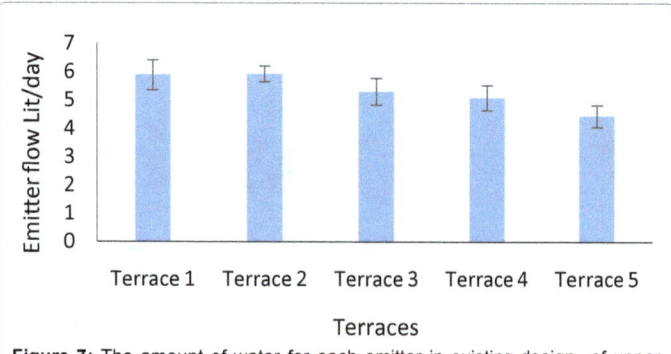

Figure 7: The amount of water for each emitter in existing design of upper slope trees.

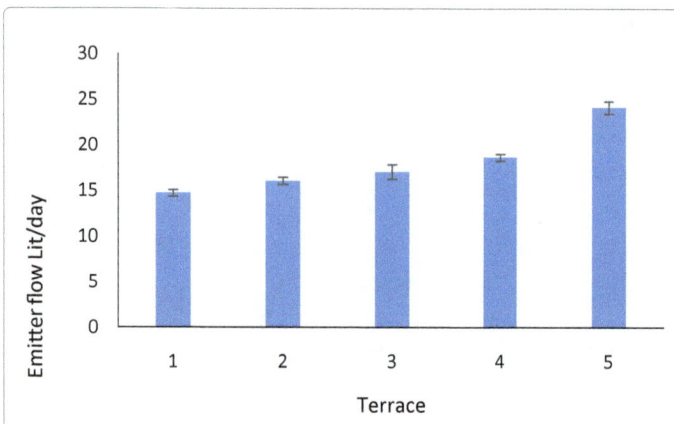

Figure 8: The amount of water for each emitter in all terraces of suggested design in the Lipar park of Chabahar free zone.

Emitter Clogging

Emitter clogging in the Chabahar Free Zone area is a major problem involved with trickle irrigation operation system. However, information is available on the causal factors; control measures are not always successful. Once, a trickle irrigation system is expensive, it must be maximized to assure a favorable cost-benefit ratio. In this way so if emitters plugged in Kahirbord agricultural farm site then a short time after their installation, reclamation procedures to correct plugging increase maintenance costs and unfortunately may not be permanent. Therefore, clogging cause to a less efficient drip or trickle irrigation method for Chabahar Free Zone regions [1].

Increasingly, emitter clogging is directly related to the quality of the irrigation water, i.e., suspended load, chemical composition, and microbial activity. Thus, these factors consequently follow the type of water treatment necessary for controlling clogging. Sometimes solution for clogging is not always available [10].

Regarding of water sources for Chabahar free zone site, trickle systems require some type of filtration to remove the bulk of suspended materials. Furthermore, it is not practical to remove all the suspended particles. Importantly, calcium or magnesium carbonate can precipitate in filters, pipe lines, or emitters when source water has PH values above 7.5 and a high degree of hardness. Though water is applied in this area of Chabahar free zone site. Sometimes it is have pH above 7.5 then and then it could be one factor for clogging of their sites for drip irrigation [3].

Physical Aspects (Suspended solids)

In the site of Chabahar Free Zone it shall be physical clogging happen by factors such as suspended inorganic particles (sand, silt, clay or plastic), organic materials (plant fragments, animal residues, fish, snails, etc.), and microbiological debris (algae, diatoms, larva) .It could be possibility organic materials and though by research confirmed there was sand and silt may carry into irrigation water supply from Sarbaz River water by open water canals or pumped from wells. Sand and silt introduced in the lateral lines during installation can cause problems unless they are flushed out before the emitters are placed on the line, that is confirmed from management and organizers this action did not to take for this Chabahar free zone site. Therefore, it is cause problematic to drip irrigation system which is applied and although that was failed due to these reasons [6].

Results and Discussion

Three factors can affect the flow variation of emitters in the drip irrigation system of steep slope land, such as: initial filling of pipe, drainage water from the pipe after stopping irrigation and pressure variation. Initial filling and drainage of water from the pipe after stopping irrigation can affect the flow variation significantly because the last lateral has the first water filling and the last one of drainage while the first line has the last filling and the first one of water drainage [3]. Drip irrigation systems, as cutting edge technology in irrigation methods has many advantages but it is associated with some problems and obstacles i.e. low water pressure at the end of lateral lines and salt accumulation. The trickle irrigation seems to have better future in the area with water scarcity. Since water is applied directly to individual plants instead of irrigating the entire area thus saves water which is otherwise lost by the use of traditional surface irrigation methods. The method is more suitable for production of orchards and high value vegetables. Results of this and previous studies suggest that over 50-75% water could be saved. Water can be provided to a plant with low pressure and at a high frequency. Closed-circuits were proposed as incorporating modification to the traditional drip irrigation system. The aims of the work were to study the effect of drip irrigation circuits (DIC) used: 1) Closed irrigation circuit with one manifold for lateral lines (CM1DIS); 2) Closed irrigation circuit with two manifolds for lateral lines (CM2DIS), 3) and Closed irrigation circuit with one manifold for lateral lines (CM1DIS), 4) and closed irrigation circuit with two manifold for lateral lines (CM2DIS), 5) as well. Traditional drip irrigation system (TDIS) as a control and lateral lines lengths (LLL): (LLL1 = 50 m, LLL2 = 50 m, LLL3 = 60 m, LLL 4 = 70 m, LLL5 = 80 m) on: flow velocity and velocity head [1].

Table 1 and Figure 6 indicated the effect of DIC and LLL on FV. The reader can deduce that the change in FV took the same trend of

Terrace	Emitter 1	Emitter 2	Emitter 3	Emitter 4	Emitter 5	Total	Mean
1	14.2	14.5	15.0	14.8	15.0	73.5	14.7
2	15.5	15.8	16.2	16.2	16.5	80.2	16.04
3	17.2	16.5	16.0	17.5	18.0	85.2	17.04
4	18.5	18.5	19.2	18.8	18.2	93.2	18.64
5	23.5	23.5	24.0	24.5	25.0	120.5	24.1

Table 1: The amount of water for each emitter in all terraces of the existing design (liters/irrigation).

PH, whereas, it was opposite to that of friction loss. The explanation for this could be due to the effect of both DIC on both PH and friction loss. Also, increasing LLL increased its discharge and de- creased the amount of water flowing along the lateral lines while, their cross section areas are constant are other reasons [2,10]. Another addition of the proposed automation system is to install the digital cameras to monitor the plant growth and overall condition of the field. In addition, the developed irrigation method partly removes the excess workload of the farmers. To identify the suitable pump with facility for maintaining certain recommended pressure in the water pipe. To identify proper sensors and monitoring device required for the farming data like soil moisture, soil temperature, soil fertilizer and chemical constituents [1,9].

Search of appropriate sensors with specifications and coordinating wireless system for acquisition of various data. To process the data based on the limits set and there by controlling the whole irrigation management. To find the economic method of drip irrigation and its technique for automation regarding short term and long term crop [1,9].

The physical, chemical and biological properties of water in the experimental areas were compared with the water quality criteria for emitter clogging proposed by Ribeiro [2]. According to Ribeir and Yavuz [2,10] the tested irrigation waters, based on properties (pH, TDS, TSS, Fe, Mn) can be classified, in general, as minor hazardous to severe hazardous in some cases. According to Ribeiro [2], the hazard rating is, in general, from minor to moderate for EC except Ghom that was severe, minor for TSS, from minor to moderate for Ca except Ghom that was severe, from minor to severe for Mg, minor for Fe and Mn. The bicarbonate values for Izeh, Damghan, Sari, Ghom, Nahavand, and Talesh was high [11,12]. Bicarbonate concentrations of more than 5 meq/l, or 305 mg/l, caused serious problems due to precipitates in the irrigation system. In Talesh, large formations of biological biofilm were observed as well as the occurrence of the same formation in the micro jet orifice [2,10]. The amounts of water (liters per irrigation) for all emitters in all terraces in the existing design are indicated in Table 1.

Conclusion

Results of the study revealed that the trickle irrigation achieved high uniformity coefficient and distribution uniformity. The *CU* of randomly selected laterals with smooth emitters ranged between 79.1 and 84.4% with an average value of 81.7%. However, it ranged between 85.9 and 89% with average value of 87.4% with spiral emitters. Similarly, the distribution uniformity ranged between 71.2 and 81.2% and averaged to 75.4% for the smooth emitters, while it ranged between 76.9 and 85% and averaged to 81% for the spiral emitters. The system achieved an overall field application efficiency of 82.7% with smooth emitters compared to 89.4% with spiral emitters. Further, the spiral emitters showed higher uniformity coefficient, and application efficiency as compared to smooth ones hence could be preferred with great degree of confidence to achieve uniform water distribution.

Flow variation of drip irrigation system of steep slope land

required a rational distribution of lateral on the manifold and careful selection of suitable length of both laterals and manifold to eliminate the problems of the initial filling, drainage of water from the pipe and reduce the pressure variation that resulted from the difference in elevation to ensure the uniformity of emitters distribution. Though the main problems associated with drip or trickle irrigation system in the Chabahar Free Zone sites was poor distribution drip system. Because that begun slowly water supply reduction to that trees. Therefore, there was no enough or lengthy loop for lateral pipe when water should be applied for that trees when they are bigger and bigger it means for many years where trees grown up and it needs more water.

Increasingly, during water supply there was sand and silt mixture with water and biological problems to this system where also cause emitters clogging occurred or less water supplied to the system. From the studies it was seen that the different chemicals are useful only to improve the relative discharge rates. Prevention is the best solution to reduce clogging of emitters. Filtration of water is absolutely essential in this pipe network. Periodical flushing of laterals line helps to minimize sediment build up in the lines. Actually water-use efficiency, which is defined as crop yield per unit volume of water applied in the trickle system excellent result in distribution and a very good system, but when it applied with best management. Therefore, trickle irrigation is a convenient and efficient means of supplying water. Of course, several problems arise in this system when it is applied due to miss management and lack of poor organizing.

Acknowledgements

The author acknowledge for his research activities, is extremely grateful to the following company for their assistance with the collection of data from Chabahar Free Zone Horticultural Organization. The author would also like to express their great appreciation to Professor Mohammad Safar Mirjat and Dr. Abdolsamad Chandio for their valuable and constructive suggestions during the planning and development of this research work.

References

1. Mosh S (2006) Guidelines for planning and design of micro irrigation in arid and semi-arid regions. Int Comm Irrig and Drain.

2. Ribeiro TAP, Paterniani JES, Coletti C (2008) Chemical treatment to unclogging dripper irrigation system due to biological problems. Sci Agric 65: 1-9.

3. Bralts VF, Gitlin HM (1981) Manufacturing variation and drip irrigation uniformity. Transaction of the ASAE 24: 113-119.

4. Alizadeh A (2006) Principles and Practices of Trickle Irrigation. Scientific Research.

5. Anyoji H, Wu IP (1998) Normal distribution water application for drip irrigation schedules. Transaction of the ASAE 37: 159-164.

6. Ella VB, Reyesand MR, Yoder R (2012) Effect of hydraulic head and slope on water distribution uniformity of a low-cost drip irrigation system. App Eng in Agric 25: 349-356.

7. Michael AM (1999) Irrigation Theory and Practice. South Asia Books.

8. Elmaloglou S, Diamantopoulos E (2011) Soil water dynamics under surface trickle irrigation as affected by soil hydraulic properties, discharge rate, dripper spacing and irrigation duration. Irrigation and Drainage.

9. Kim Y, Evans RG, Iversen WM (2009) Evaluation of closed-loop site-specific irrigation with wireless sensor network. J Irrig Drain Eng 135: 25-31.

10. Yavuz MY, Demirel K, Erken O, Bahar E, Deveciler M (2010) Emitter clogging and effects on drip irrigation system performances. African journal of Agricultural Research 5: 532-538.

11. Aali KA, Liaghat A, Dehghanisanij H (2009) The effect of acidification and magnetic field on emitter clogging under saline water application. Journal of Agricultural Science 1: 132-141.

12. Kim Y, Evans RG, Iversen WM (2008) Remote sensing and control of an irrigation system using a distributed wireless sensor network. IEEE Trans Instrum Meas 57: 1379-1387.

10

Climate Change and its Impact on Irrigation Water Requirements on Temporal Scale

Mohan S* and Ramsundram N*

Professor, Environmental and Water Resources Engineering Division, Department of Civil Engineering, Indian Institute of Technology Madras, Chennai, India

Abstract

The natural processes and man-made disturbances in the watershed have influenced the micro climate and in turn affect the hydrology of the watershed along the time scale. The increase in emission of greenhouse gases into the atmosphere might induce variation in climatic pattern in the future. In hydrological models, the climatic parameters remain to be deterministic variables in simulating the surface water or groundwater components. In the recent past, climatological cycle and its variability have been incorporated into water resources systems modeling by many researchers. In this study, an attempt has been made to study the influence of climatic variability on irrigation water requirements in an arid region on a temporal scale, which will help in the water resource planning and management of an irrigation system. A climate crop water requirement (CCWR) integrated framework has been developed to estimate the irrigation requirement in Manimuthar river basin, Tamilnadu, India, incorporating variation in climatic parameters over temporal scale. Based on the existing land use pattern and economic development prevailing in the study area, the most likely climatic scenario has been identified as A1B. From the results, it is inferred that the irrigation water requirement is likely to increase by 5% from 2010 to 2020.

Keywords: Climate change; Crop water requirement; Arid region; Scenario analysis; Manimuthar basin

Introduction

The natural processes in any watershed can be influenced by changes in climatic variables and have long term impact on economic and ecological processes [1]. In general, climate change is defined as "the difference between long-term mean values of a climate parameter or statistic, where the mean is taken over a specified interval of time, usually a number of decades" [2]. The hydrologic cycle is a part of the climate system; the interactions between the components in the system give rise to the system complexity and this system complexity has been modeled as a) empirical model, b) water balance model, c) conceptual lumped parameter model and d) process based distributed parameter models [3]. The models are then developed to replicate the hydrological cycle, the climate influence by considering the variability in the climate parameters. Intergovernmental Panel on Climate Change (IPCC) developed a global climate model (GCM) which forecast the future climate parameters. Incorporating the parameter values from GCM into hydrological cycle models will have the ability to visualize the impact of climate change on the system. The IPCC has stated that the average global surface temperature has increased by 0.45°C to 0.6°C and the average sea level has increased by 15 to 20 cm during last century [4]. The IPCC has created climate scenarios such as (A1, A2, B1, B2 and their combinations) based on economic and environmental development conditions. The potential impacts of climate change for countries with temperate climate like India are; a) current water security problems are likely to increase by 2050, b) substantial impacts on agriculture and forestry are very likely by 2050, c) the current trends of glacial melts suggest that the snow fed rivers could likely become seasonal rivers in the near future and could likely affect the economics in the region [5]. As Southern and Eastern Asia is concerned, climate change increases runoff, but this may not be very beneficial in practice because the increase tend to come during the wet season and extra water may not be available during dry season [6].

Climate parameters (precipitation, temperature and carbon dioxide levels) changes can affect the demand for water as well as supply. Increased water use efficiency attributable to higher carbon dioxide levels, this may tend to increase frequency of water application as temperature rises [7]. Changes in water demand, irrigation practice

will enhance the groundwater exploitation; this claim that climate change is likely to have dramatic impact on groundwater resources. From the analysis of rainfall data for 131 year period, it is observed that no clear role of global warming in the variability of monsoon rainfall over India [8]. Even though the above finding states that 'no impact of considering climate parameter for India', but it is worthwhile to consider the climate parameters to quantify the potential adverse impacts on water resources that may occur in future time periods. In India, precipitation has always been extremely variable, with the number of annual rainy days varying from 12 to 100 and there are rain events that have poured about 60% of the total annual precipitation within few hours. It is projected that due to climate change, the inter-annual variability of the monsoon is expected to increase. Also, the rainy days will be less with concentrated rain within a few hours and increase in dry spells [9]. The above said likely impact may create excessive runoff within a short period, thereby reducing the groundwater recharge potentials [8]. This excessive runoff in addition may influence the flooding frequency in the watershed, and also increase in dry spells may increase the frequency of drought in the watershed [7]. As the climate parameter (temperature) increases, the atmosphere air gets warmer thus the warm air accelerates the evaporation from soil moisture [8,10]. Irrigation practice or schedule depends on the frequency and quantity of irrigation water required based on the type of crop and rainfall precipitated over the time period. Evapotranspiration governs the crop

*Corresponding author: Mohan S, Professor, Environmental and Water Resources Engineering Division, Department of Civil Engineering, Indian Institute of Technology Madras, Chennai-600036, India, E-mail: smohan@iitm.ac.in

Ramsundram N, Research Scholar, Environmental and Water Resources Engineering Division, Department of Civil Engineering, Indian Institute of Technology Madras, Chennai-600036, India, E-mail: nramsundram21@gmail.com

water requirement estimation; Hargreaves et al. [11] recommended a procedure for estimating crop water requirements that only require the measurement of maximum and minimum temperatures. The most popular evapotranspiration estimation procedure is through Penman-Monteith equation, [12] evaluated the Penman-Monteith equation to develop general relationships for estimating daily average values of canopy and aerodynamic resistance parameters. Committee on irrigation water requirements of the Irrigation and Drainage Division of the American Society of Civil Engineers Jensen et al. [13] prepared a manual summarizing all available procedures for estimating irrigation water requirements. Allen et al. [14] Developed an irrigation water requirement estimation a stepwise procedure incorporating improved or modified Penman-Monteith equation considering aerodynamics and stomata opening into evapotranspiration modeling (FAO paper 56).FAO paper 56 formed the basis for estimating the crop water requirement; the Penman-Monteith equation requires information climatic parameters such as air temperature, wind speed, and humidity. The Penman-Monteith equation also requires information on meteorological parameter rainfall that precipitated during time't'. The climatic parameters are subjected to variability based on the atmospheric concentration (composition of greenhouse gases). These lead researchers to study the influence of climatic variability in crop water requirement estimation. In the recent past few research works [15-18] integrated the future climatic variations forecasted by Global Circulation Models (GCMs) within there modeling framework to study the climate change influence on crop water requirements. The developed integrated models, differs in the type of downscaling technique utilized for projecting regional climatic conditions simulated by GCMs to local scale, and also differs in the model used for estimating the crop water requirement.

One of the major findings of Mall et al. [8] is that, the influences of climate parameters in arid and semiarid climatic region are not very significant. This finding or statement might be valid at regional level (India) but may not true at local or micro scale. In this paper an attempt has been made to study the variability in climatic parameters on crop water requirement estimation in an arid region. An integrated framework has been developed and applied to case study Manimuthar basin located Tamilnadu, India.

This paper has been organized in a stepwise procedure starting with the basics of reason behind climatic change, what scenarios are, and what basis emission scenarios are developed. After the introduction of climatic scenarios, what global circulation model (GCM) has been chosen for this study and underlying reason behind the choice of GCM. To achieve the desired objective of the study an integrated climate crop water requirement (CCWR) framework has been developed and the same has been described in detail. With the description of the developed CCWR framework the same has been applied to the case study Manimuthar basin, Tamilnadu, India. The session on case study has been written to summarize the details about the study area. The developed CCWR framework has been allowed to analyze the study area database and interpretation of the outcomes have been discussed in the session on results and discussion. The final research findings and remarks are emphasized as conclusions of this study.

Climatic Scenarios

Global warming is caused primarily due to increase in carbon dioxide emission into the atmosphere from fossil fuel burning and deforestation activity. The increase or decrease in carbon dioxide concentration in the atmosphere might influence the global and local climatic conditions. Future GHG emissions are the product of very complex dynamic systems, determined by driving forces such as demographic development, socioeconomic development, and technological change [19]. Their future evolution is highly uncertain. Scenarios are alternative images of how the future might unfold and area an appropriate tool with which to analyze how driving forces may influence future emission outcomes and to assess associated uncertainties [19].The intergovernmental panel on climate change (IPCC) has developed a set of future greenhouse gas (GHG) emission scenarios known collectively as SRES (special report on emission scenarios).The IPCC has grouped future emission scenarios as four major classes or groups, namely; a) A1, b) A2, c) B1, and d) B2 based on level on economic development and environmental concern. Table 1 summaries the aspects considered in developing the future GHG emission scenarios in IPCC-TGCIA [20]. From Table 1, it can be observed that scenario group or classes A1 and A2 are concerned more about activities which will improve the economic development of the world, B1 and B2 are concerned more about environmental sustainability of the world.

An Overview on Climate Models (GCMs)

General Circulation Models (GCMs) predicts the future 100 years climatic parameters using the equation of motion. The predictions of climatic variables in GCMs are classified as eight climatic scenarios [21] based on carbon dioxide (CO_2) emission level ranging from very low to high. Very low CO_2 emission indicates that the scenario has more consideration of environment conservation and low economic development, whereas high CO_2 emission indicates maximum economic development (industrialization, urban growth) and minimum concern to environmental concern. Table 2 summarizes some of the popular GCMs developed by various research laboratories across the world. Every GCM (Table 2) has its unique grid resolution

Scenario Classes	Concerns	Remarks
A1	• Rapid economic growth • Low population growth • Rapid new technology • Concern to wealth rather than environment	• Homogenous world on economic development • Cultural convergence • No difference in per captia income
A2	• High population growth • Strengthen regional cultural identities • Less concern for economic development compared to A1	• Differential world on economic development
B1	• Dematerialization of economy • Introduction to clean technologies • Efforts for rapid technology development	• Convergent world • Global solutions for environmental and social sustainability • Improving equity
B2	• Diverse technological change • Emphasis on community initiative • Concern on environment rather than economic development	• Heterogeneous world • Local solutions for environmental and social sustainability

Table 1: Climatic Scenario Classes.

Model	Resolution(lat/lon)	Description of model
CGCM1	Atmospheric component: ~3.7° x 3.7° Ocean component: ~1.8° x 1.8°	Flato et al., 2000
HadCM2	2.5° x 3.75°	Johns et al., 1997
HadCM3	2.5° x 3.75°	Gordon et al., 2000 Pope et al., 2000
RegCM2	~50 km	Giorgi et al., 1993a, b
ECHAM4	~2.8° x 2.8°	Roeckner et al., 1996

Table 2: GCM models used for water resource problems.

and global area coverage. Among the eight reported scenarios, many of the researches prefer to observe the impact of climate change for A2 and B1 scenarios [21,22]. The selection of these two scenarios is due to 'A2' represented strong economic values and 'B1' represents strong environmental value.

The selection of a particular GCM model is based on the grid resolution that is required for modeling and the purpose for which the researcher is modeling the climatic scenario. For example, if the purpose is to estimate the irrigation requirement in the watershed over the time period, then the GCM has to be selected based on which GCM is considering the vegetation characteristics in is climatic variable prediction. The best suited GCM model for water resource will be HADCM and CSIRO which considers the vegetation characteristics. For this study climatic parameters are taken from HADCM3 model, which has a spatial resolution of 2.5×3.75 (latitude by longitude) and grid resolution of 250 km×320 km.

Climate Crop Water Requirement (CCWR) Integrated Framework

To the study the influence of variation in climatic parameters (Temperature, Wind direction, and humidity) on the irrigation water requirement on temporal scale, climate crop water requirement (CCWR) integrated framework has been developed (Figure 1). The CCWR framework integrates the crop water requirement model (CROPWAT) and spatial climate variable downscaling technique developed by Maurer et al. [23].

Climatic variable downscaling technique

Global climate model output, from the World Climate Research Programme's (WCRP's) Coupled Model Intercomparison Project phase 3 (CMIP3) multi-model dataset [24], were obtained from www.engr.scu.edu/~emaurer/global_data/. These data were downscaled as described by Maurer et al. [23] using the bias-correction/spatial downscaling method [25] to a 0.5 degree grid, based on the 1950-1999 gridded observations of Adam and Lettenmaier [26].

Irrigation demand estimation module

The irrigation requirement for various crops in the command area has been estimated using the irrigation demand estimation module (Figure 2). The data required for irrigation demand estimation module area) the precipitation that has occurred, b) prevailing climate variables (wind speed, relative humidity, maximum and minimum temperature, and sunshine hours), c) cropping pattern (time of sowing, harvest), and d) type of soil (field capacity, moisture content).

From Figure 2, it can be observed that the module begins with an estimation of excess rainfall for the rainfall that has occurred in the command area. The process is followed by estimation of the crop water requirement of the available crop types in the study area. In this research the crop water requirement for the type of crop and cropping pattern has been estimated using CROPWAT package. CROPWAT uses FAO Penman-Monteith model [14] (equation 1) for estimating reference Evapo-transpiration.

FAO Penmann-Monteith Model to estimate reference evapo-transpiration (ET_o),

$$ET_o = \frac{\left(0.408\Delta(R_n - G)\right) + (\gamma\frac{900}{T+273}u_2(e_s - e_a))}{\Delta + (\gamma(1 + 0.34u_2))} \tag{1}$$

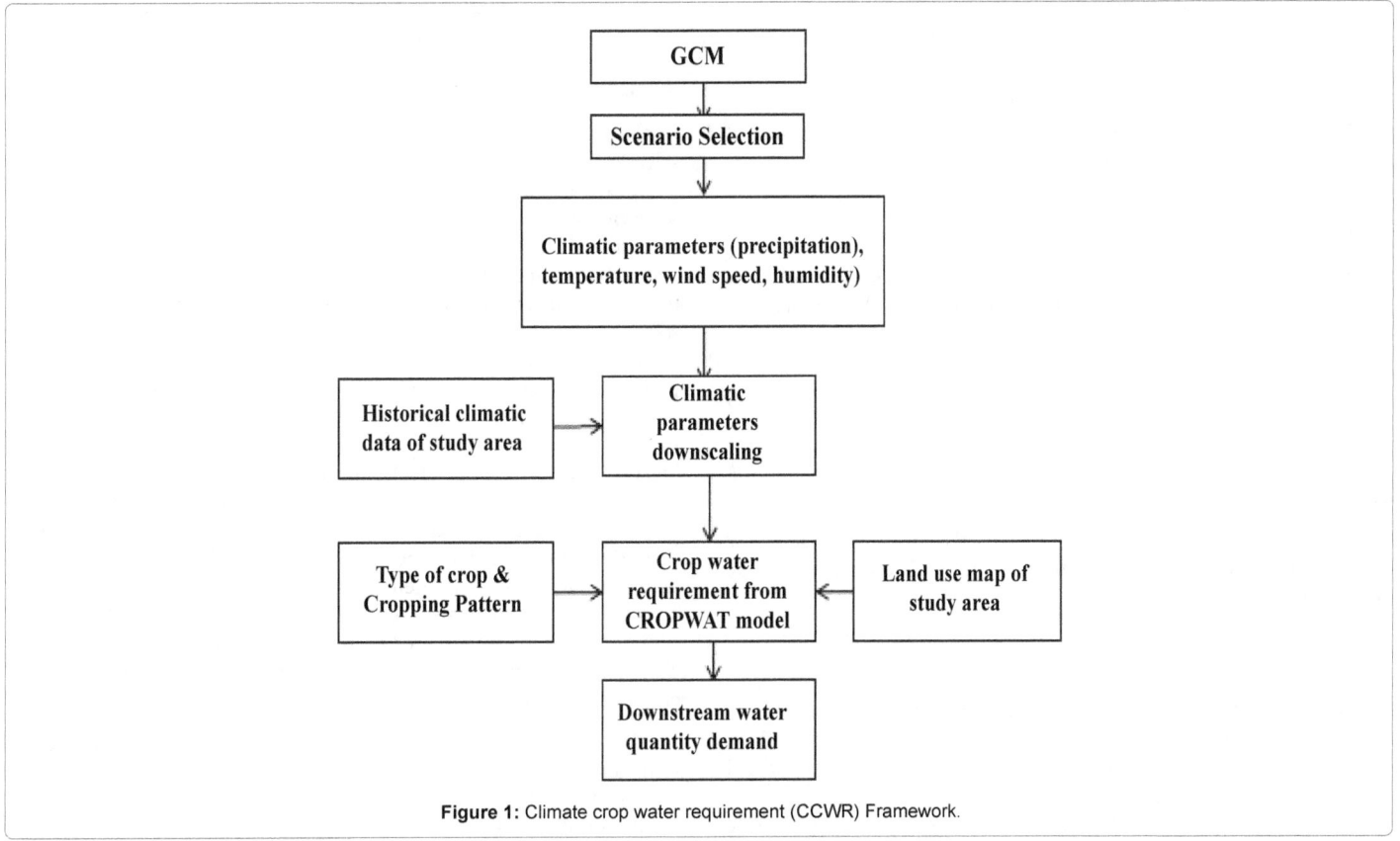

Figure 1: Climate crop water requirement (CCWR) Framework.

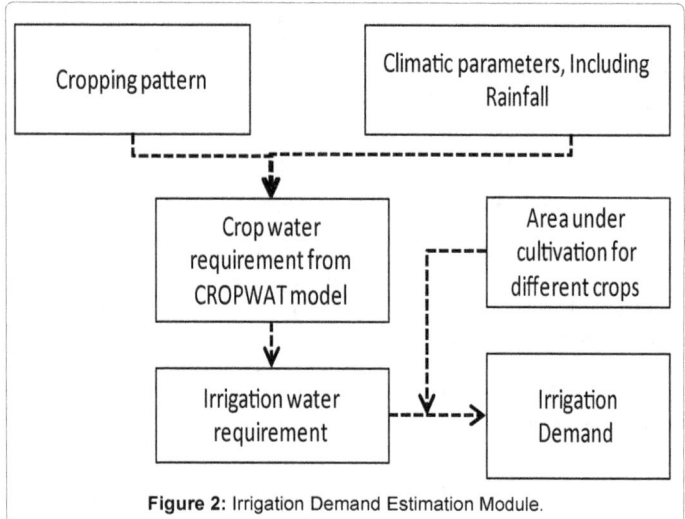

Figure 2: Irrigation Demand Estimation Module.

where,

ET_o = reference evapo-transpiration (mm / day)

R_n = net radiation at the crop surface (MJ/ sq.m /day)

G = soil heat flux density (MJ/ sq.m/ day)

T = air temperature at 2 m height (°C)

u_2 = wind speed at 2 m height (m/s)

e_s = saturation vapor pressure (KPa)

e_a = actual vapor pressure (KPa)

e_s-e_a = vapor pressure deficit (KPa)

Δ = Slope vapor pressure curve (KPa/ °C)

γ = psychometric constant (KPa/ °C)

The crop water requirement for every crop in the command area has been based on the crop coefficient (K_c) and estimated T_o as equation (2).

$$CWR_i = (kc_i ETo_i) \qquad (2)$$

Net irrigation water requirements (NIR) for a given year are thus the sum of individual crop water requirements (CWR_i) calculated for each crop 'i'.

$$NIR = \sum_{i=1}^{n} (CWR_i - P.eff) \qquad (3)$$

where, 'P.eff' is the effective precipitation after satisfying the infiltration capacity.

Gross irrigation water requirements (GIR) give the overall water requirement for the crop to grow from the time of sowing to the date of harvest.

$$GIR = \frac{1}{E} NIR \qquad (4)$$

Where, E is the efficiency of the irrigation system practiced in the command area. In the above irrigation water requirement module, among the variables temperature and precipitation are identified as the important climatic variables. The variability in temperature might affect the atmospheric gradient there by influencing the rate of evapo-transpiration. Variation in temperature is proportional to the precipitation variation in a spacial and temporal scale. From common belief, solar radiation plays a major role in influencing the global surface temperature. But, U.S. climate research program, 2001 has found that, solar radiation received by earth does not influence the global surface temperature. So, in this study net solar radiation has not been considered as a influencing variable in modeling crop water requirement.

To incorporate the temporal variation in the identified climatic parameters, the output of GCM is downscaled to the field grid level. The GCM model simulates the climatic parameters for various climatic scenarios. As described in 'Overview of GCM', the HADCM3 GCM model outputs for scenario A2 (Continuously increasing population, regionally oriented economic development), B1 (Rapid economic growth as in A1, Reductions in material intensity and the introduction of clean and resource efficient technologies) and A1B (balanced emphasis on all energy sources.) Are utilized for this study to predict the future crop water requirements.

Case Study

Tambraparani River Basin lies between geographic co-ordinates latitude 8°26'45" to 9°12'00" N and longitude 77°09'00" to 78°08'30" E. The entire basin covers an area of about 5665 sq. km and lies in the revenue districts of Tirunelveli and Tuticorin in southern Tamil Nadu. Figure 3 presents the Tambraparani river basin with multiple sub basins. For this study, Manimuthar one of the sub basin in Tambraprani river basin has been chosen. The Manimuthar originates (numeric 2 highlights Manimuthar basin in Figure 3) from the Mukkuttukal and confluence with Tambaraparani at its 36th km. The major reservoir constructed in the basin across Manimuthar is the Manimuthar reservoir. The irrigation farm fields are located at the downstream of Manimuthar reservoir. The Manimuthar reservoir has an irrigation command area of 161.61 sq.km.

Results and Discussions

From Figure 1, it can be observed that the climatic scenario selection is the first and foremost process in any climate impact analysis. In developing countries, the most commonly used climatic scenarios are; a) Scenario A2, and b) Scenario B1 representing the economic development and environmental concern / protection respectively. Modeling for these two scenarios crudely might miss lead to over or under prediction of actual. To overcome the crude way of selecting scenarios, in this study an exercise has been incorporated for understanding the change in land use pattern in the study area over a period of 30 years. From the analysis, it was observed that Manimuthar basin has a vegetative land cover and very minimum urban and industrial development. It can be observed from the geographic location that catchment of Manimuthar reservoir is majorly covered by hilly terrain. From this, it can be inferred that, there is no much change in land use pattern, so it is most suitable to choose climatic scenarios which are developed based on less CO_2 emission and more dependability on environment conservation [27,28]. The more suited scenario for this basin will be B1, in next 20 years if there is adverse economic development then it is reasonable to model for A1B scenario. To compare with the extreme condition, A2 scenario also has been chosen for modeling.

The climatic parameters of the respective scenarios are downscaled to field grid scale and coupled with crop water requirement estimation model (CROPWAT). Figure 4 shows the temperature prevailing at Manimuthar basin for various scenarios. Table 2 shows relative

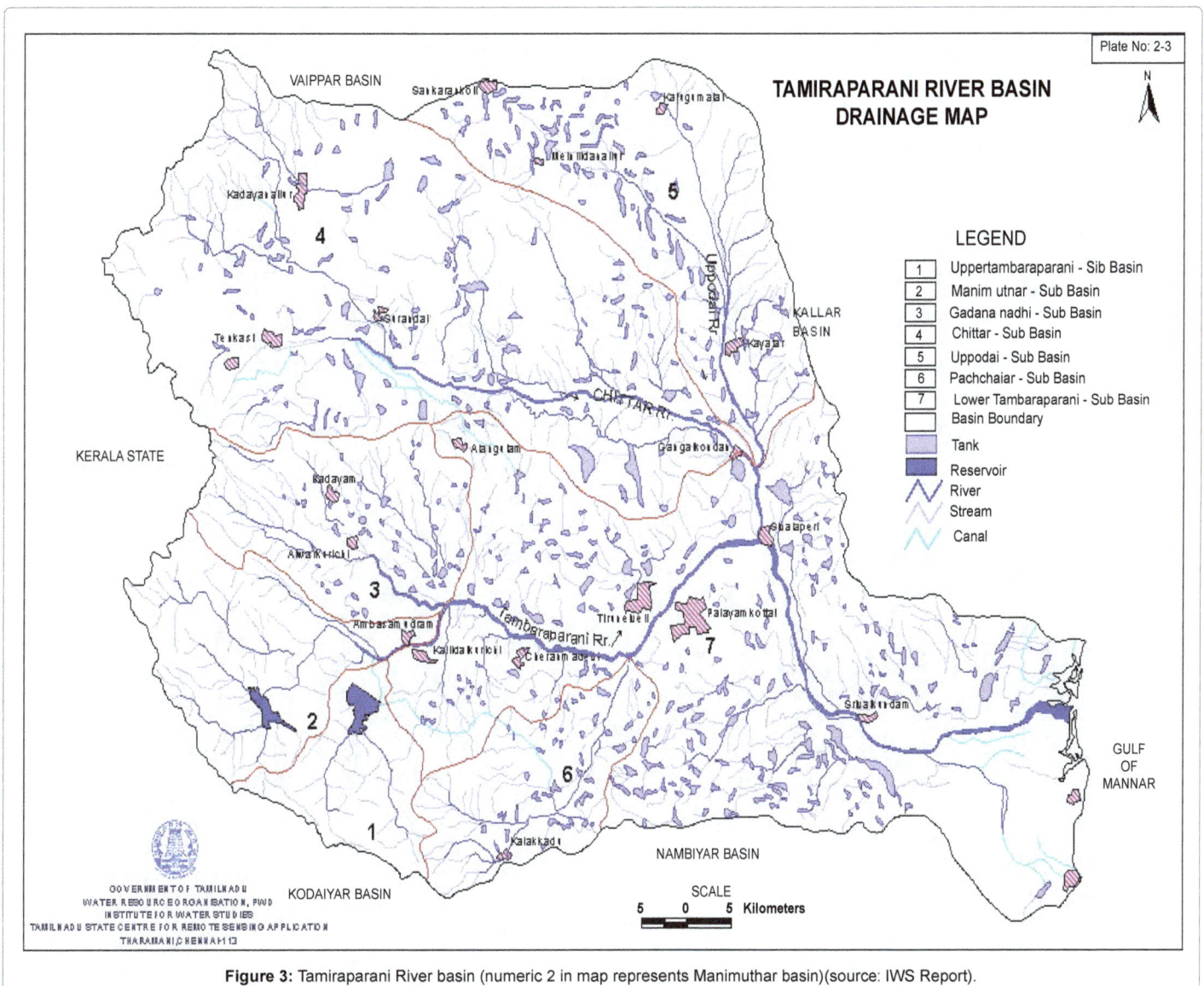

Figure 3: Tamiraparani River basin (numeric 2 in map represents Manimuthar basin)(source: IWS Report).

temperatures for year 2020 and 2050 with reference to the year 2010. From Table 2 an interesting phenomenon of scenarios behavior can be observed, i.e., scenario B1 is more about integrated world, once the population reaches 9 billion, the population will start reducing, and also B1 introduces clean and energy efficient technologies this phenomenon can be observed from the annual average temperature rises by 0.86°C relatively in year 2020 and then the temperature reduces to 0.66°C relatively in 2050 as emission reduces (scenario assumption). Similarly, assumptions of scenario's A2 and A1B are resembled in Table 3.

When we look at the precipitation corresponding to temperature variations for various scenarios, Table 3 might mislead the interpretation at annual scale. In Table 3, the comparative analysis of temperature and corresponding precipitation of scenario B1 and A1B mislead to conclude that as the temperature decreases the precipitation increases. Thus the inferred concept is not logical, as the atmospheric gradient increases as temperature increases and condensation occurs at a faster rate thereby resulting in an increase in precipitation quantity. To have insight into the above concept, the seasonal variations for various scenarios are summarized in Table 4. In Table 4, during summer

season of Scenario A2, as temperature increases precipitation also increases. But, in southwest monsoon season, the inverse phenomenon can be seen, i.e., as temperature decreases, precipitation increases. Similar kind of mechanism can be seen in the other two scenarios too [29]. From basic hydrological concepts and atmospheric science, it is understood that wind direction is also a major component in creating atmospheric gradient (Table 4).

The climatic and meteorological parameters are used in irrigation requirement estimation module to predict the future crop water requirements. Figure 5 shows the irrigation water requirement for the Manimuthar command area for various climatic scenarios. From Figure 5a, it can be observed the variation in irrigation water requirement for A2 (extreme economic development) scenario for different months in a year. Similarly, Figure 5b and 5c show the irrigation water requirement for various months in a year for scenario B1 and A1B respectively. From the comparative analysis of water requirements of A2 and B1 (extreme scenarios), it can be inferred that during the year 2050 i.e., when industrial development is assumed to be attained its saturation that is maximum CO_2 emission the irrigation water requirement increases

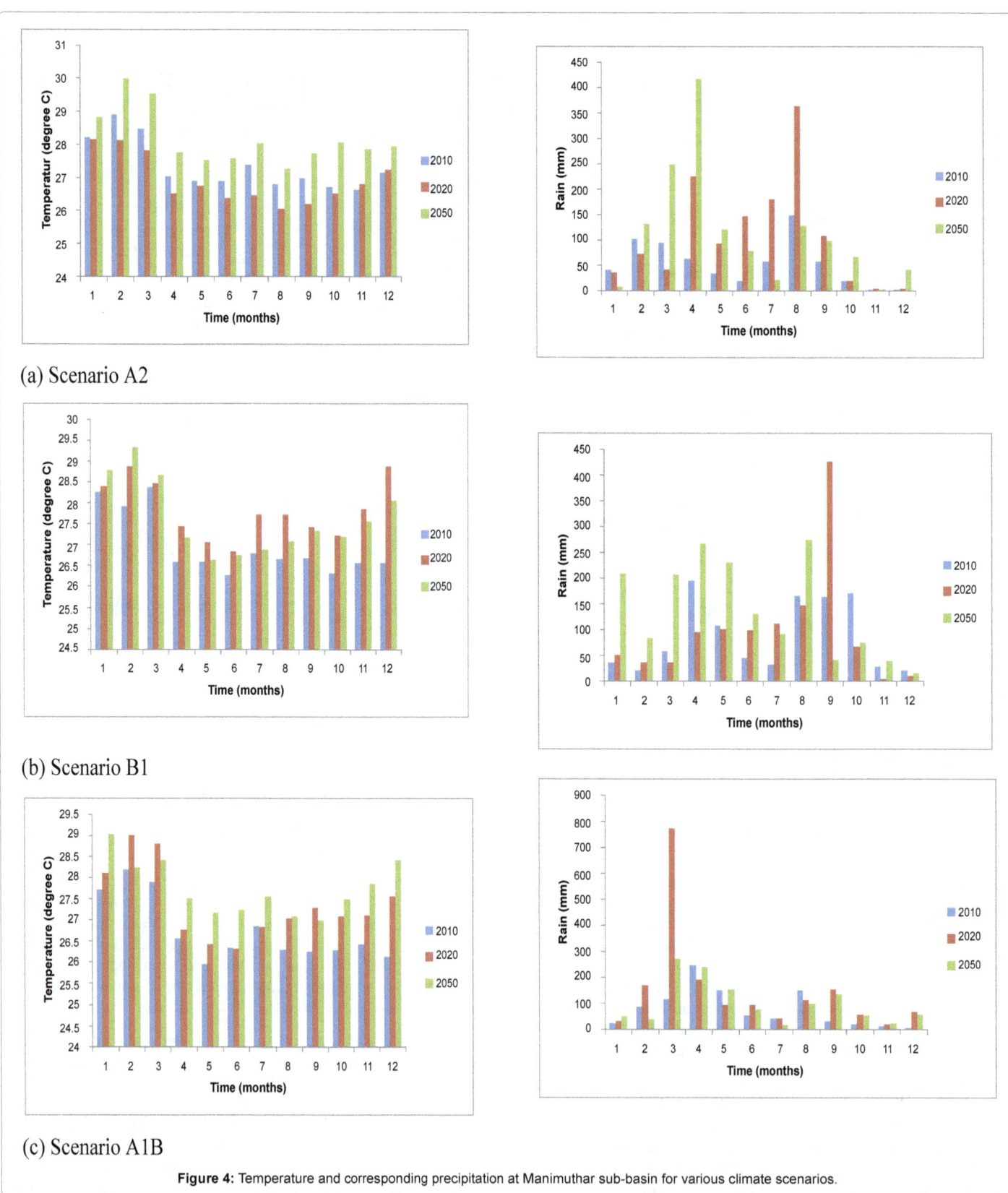

(a) Scenario A2

(b) Scenario B1

(c) Scenario A1B

Figure 4: Temperature and corresponding precipitation at Manimuthar sub-basin for various climate scenarios.

adversely compared to irrigation water requirement of scenario B2. With the increase in irrigation water requirement during year 2050 of scenario A2 is contributed by a decrease in rainfall during the

respective time periods (Figure 4a). The neutral scenario A1B, where there is balance between economic development and environmental concern, except during the summer season in all other months the

Scenario\ Year	Temperature relative to 2010 °C)		Precipitation Relative to 2010 (%)	
	2020	2050	2020	2050
A2	-0.42	0.84	100.35	110.93
B1	0.86	0.66	13.51	59.94
A1B	0.61	1.00	94.36	30.01

Table 3: Annual relative variation of climatic parameters with reference to year 2010.

Scenario	Monsoon/ Year	Avg. Rel. 2020 temp(°C)	Avg. Rel. 2050 temp(°C)	Cuml. Prcp.2020 (mm)	Cuml.Prcp.2050 (mm)
A2	Winter	-0.42	0.84	107.10	139.43
	Summer	-0.45	0.79	359.76	785.24
	South West	-0.74	0.66	798.02	324.84
	North East	0.02	1.14	27.89	111.56
B1	Winter	0.54	0.97	85.01	292.21
	Summer	0.48	0.32	231.09	704.18
	South West	0.82	0.42	782.48	537.49
	North East	1.50	1.13	81.98	129.53
A1B	Winter	0.59	0.67	198.82	85.96
	Summer	0.52	0.88	1052.15	663.64
	South West	0.43	0.79	403.07	321.45
	North East	0.96	1.64	143.55	131.35

Table 4: Seasonal relative variation climatic parameters with reference to year 2010.

change in irrigation water requirement is almost same during the year 2020 and 2050.

Table 5 summarizes the relative change in irrigation water requirement across temporal scales for various scenarios compared to irrigation water requirement during year 2010. It can be observed from Table 5 that if there is adverse development of industries in the Manimuthar basin (A2 scenario) then that might increase irrigation water demand by 25% during year 2020 and 15 % during year 2050 compared to the present condition (year 2010). If there is Manimuthar basin has not been subjected to any industrial revolution on urban migration, and continues to be a vegetative land use pattern by the year 2050 there might be an increase in irrigation demand by 20 % compared to present irrigation water requirement. This additional water requirement might be due to increase in vegetative growth. During balance or neutral scenario A1B, the irrigation water requirement achieves a saturation demanding an increase of 5% in irrigation water requirement of the year 2010.

In general, it can be inferred that crop water requirement is inversely proportional to the quantum of rainfall precipitated across the basin. From the results, it can be inferred that A2 scenario reflects the extreme or peak water requirements in the Manimuthar basin. As highlighted in the scenario selection, identification of proper scenario with respect to field condition is more essential for increasing the dependability on the predictions. From the geographic location and existing development activities in Manimuthar basin, it can be that certain that probability of extreme industrial development is very less. Similarly, chance of no urbanization or economic development is also very less. It is worthwhile to depend on the predictions made for climatic scenario A1B.

Conclusions

Increasing population and fast economic development increase the emission of CO_2 in the atmosphere, thereby accelerating global warming and its adverse effects on climatic conditions. In the recent past, one of the most research areas of interest among researchers is

to study the influence of climate change on natural resources and to develop strategic mitigation measures. As a part of the above research direction, the current study has been conducted to analyze the impact of

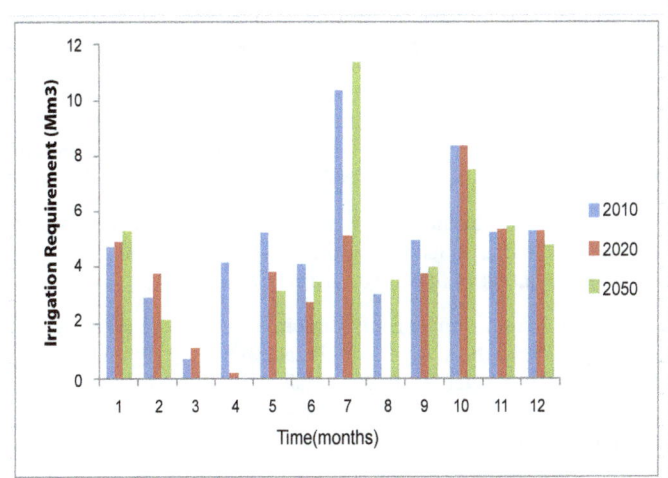

(a) Scenario A2

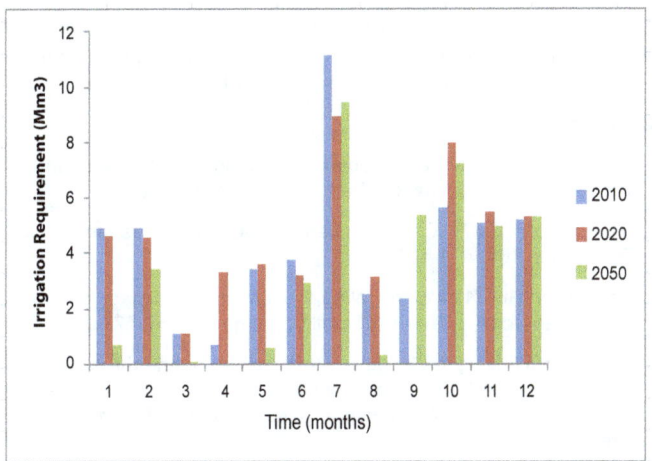

(b) Scenario B1

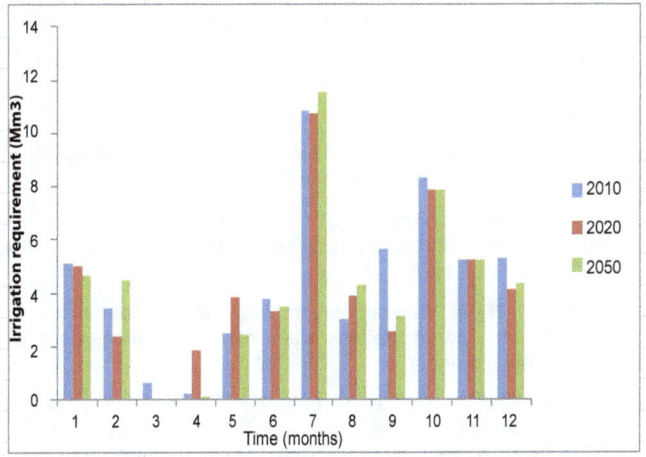

(c) Scenario A1B

Figure 5: Irrigational requirement for Manimuthar command area.

Scenario/Year	Annual Irr. Req relative to 2010 (%)	
	2020	2050
A2	24.72	14.16
A1B	5.98	4.63
B1	-0.90	20.46

Table 5: Annual relative irrigational requirement with reference to year 2010.

climate change in arid regions on the crop water requirement over time period. In general, it is observed that wind direction also plays a major role as like air temperature on the spatial variability of precipitation across a basin. From case study application, it can be concluded that;

a)   Climatic Scenario A1B will be more suitable scenario for Manimuthar basin considering the existing land use and recent development activities that are occurring in the basin.

b)   Generalization of results on climate change based on regional modeling may not be suitable for climate modeling. From the results it is inferred that generalization by Mall et al. [8] is not valid for local scale, as the change in climatic variables has contributed to increase in crop water requirement in future years (2020 and 2050).

It is concluded that climate change studies have to be conducted to assess the risk associated with water resources based on future variability in climate and meteorological variables. This assessment has to be done based on proper selection of climatic scenarios based on the land use pattern and most probable economic development that may occur in the study area for which assessment has to be made.

Reference

1. USEPA (2004) The ozone report: measuring progress through 2003. USEPA, Research Triangle Park, North Carolina, USA.

2. Askew AJ (1987) Climate change and water resources, IAHS Publication 168: 421-430.

3. Leavesley GH (1994) Modeling the effects of climate change on water resources–a review. Climatic Change 28: 159-177.

4. Houghton JT, Ding Y, Griggs DJ, Nogues M, Linden PJV, et al., (2001) Climate change 2001: The scientific basis. IPCC Third Assessment Report of the Intergovernmental panel on Climate Change, GRID-Arendal.

5. Parry M L, Canziani OF, Paultikof JP, Linden PJV, Hanson CE (2007) Contribution of working group II to the Fourth Assessment Report of the Intergovernmental panel on Climate Change. Cambridge University Press, Cambridge, United Kingdom.

6. Arnell NW (2004) Climate change and global water resources: SRES emissions and socio-economic scenarios. Global Environmental Change 14: 31-52.

7. Frederick KD (1997) Adapting to climate impacts in the supply and demand for water. Climate Change 37: 141-156.

8. Mall RK, Gupta A, Singh R, Singh RS, Rathore LS (2006) Water resources and climate change: an Indian perspective. Current Science 90: 1610-1626.

9. Mukerrji R (2009) Vulnerability and adaptation experiences from Rajasthan and Andhra Pradesh: water resource management. SDC V&A Programme, India.

10. Chattopadhyay N, Hulme M (1997) Evaporation and potential evapotranspiration in India under conditions of recent and future climate change. Agricultural and Forest Meteorology 87: 55-73.

11. Hargreaves G, Hargreaves G, Riley J (1985) Irrigation water requirements for senegal river basin. Journal of irrigation and Driange Engineering 111: 265-275.

12. Allen RG, Jensen ME, Wright JL, Burman RD (1989) Operational estimates of reference evapotranspiration. Agronomy Journal 81: 650-662.

13. Jensen M E, Burman RD, Allen RG (1990) Evapotranspiration and Irrigation water requirements. Manual of practice No.70, ASCE.

14. Allen RG, Pereira LS, Raes D, Smith M (1998) crop evapotranspiration – Guidelines for computing crop water requirements – FAO Irrigation and Drainage paper 56. Food and Agricultural Organization of the United Nations.

15. Doll P (2002) Impact of climate change and variability on irrigation requirements: A global perspective. Climate Change 54: 269-293.

16. Jones RN (2000) Analysing the risk of climate change using an irrigation demand model. Climate Research 14: 89-100.

17. Puma MJ, Cook BI (2010) Effects of irrigation on global climate during the 20th century. Journal of Geophysical research 115: D16120.

18. Silva CSD, Weatherhead EK, knox JW, Rodriguez-Diaz JA (2007) predicting the impacts of climate change-A casestudy of paddy irrigation water requirements in Srilanka. Agricultural Water Management 93: 19-29.

19. IPCC (2000) IPCC special report: Emission Scenarios, Intergovernmental panel on climate change.

20. IPCC-TGCIA (1999) Guidelines on the use of scenario data for climate impact and adaptation assessment. Intergovernmental panel on Climate Change & Task group on scenarios for climate impact assessment, Version-1: 69.

21. Mujumdar PP, Ghosh S (2008) Modeling GCM and scenario uncertainty using a possibilistic approach: Application to the Mahanadi River, India. Water Resources Research 44: W06407.

22. Rodriguez Diaz JA, Weatherhead EK, Knox JW, Camacho E (2007) Climate Change impacts on irrigation water requirements in the Guadalquivir river basin in spain. Regional Environmental Change 7: 149-159.

23. Maurer EP, Adam JC, Wood AW (2008) Climate model based consensus on the hydrologic impacts of climate change to Rio Lempa basin of Central America. Hydrological and Earth System Sciences 5: 3099-3128.

24. Meehl GA Covey C, Delworth T, Latif M, Avaney B, et al. (2007) The WCRP CMIP3 multi-model dataset: A new era in climate change research. Bulletin of the Ameican Meteorological Society 88: 1383-1394.

25. Wood AW, Leung L, Sridhar V, Lettenmaier DP (2004) Hydrologic implications of dynamical and statistical approach to downscaling climate model outputs. Climate Change 62: 189-216.

26. Adam JC, Lettenmaier DP (2003) Adjustment of global gridded precipitation for systematic bias. Journal of Geophysical Research 108: 1-14.

27. Sacks WJ, Cook BI, Buenning N, Levis S, Helkowski JH (2009) Effects of global irrigation on the near surface climate. Climate Dynamics 33: 159-175.

28. Shibao Y, Jarsjo J, Destauni G (2007) Hydrological responses to climate change and irrigation in the Aral sea drainage basin. Geophysical Research Letters 34.

29. Fischer G, Tubiello F N, Van Velthuizen V, Wiberg D A (2007) Climate change impacts on irrigation water requirements: Effects of mitigation, 1990-2080. Technological Forecasting and Social Chang 74: 1083-1107.

11

Effect of Different Deficit-Irrigation Capabilities on Cotton Yield in the Tennessee Valley

A.H. AbdelGadir[1]*, M. Dougherty[2], J.P. Fulton[3], L.M. Curtis[4], T. W. Tyson[5], H.D. Harkins[6] and B.E. Norris[7]

[1]Biosystems Engineering Department, Auburn University, Research Fellow, 200 Corley Bldg, Auburn, AL 36849-5417
[2]Biosystems Engineering Department, Auburn University, Associate Professor, 200 Corley Bldg, Auburn, AL 36849-5417
[3]Biosystems Engineering Department, Auburn University, Associate Professor, 200 Corley Bldg, Auburn University, AL 36849-5417
[4]Biosystems Engineering Department, Auburn University, Emeritus Professor, 200 Corley Bldg, Auburn, AL 36849-5417
[5]Biosystems Engineering Department, Auburn University, Professor, 200 Corley Bldg., Auburn, AL 36849-5417
[6]Tennessee Valley Research & Extension Center, Associate Director, P.O. Box 159, Belle Mina, AL 35615
[7]Tennessee Valley Research & Extension Center, Director, P.O. Box 159, Belle Mina, AL 35615

Abstract

Fluctuations in cotton (Gossypium hirsutum, L.) yield in the Tennessee Valley of Alabama are common and usually related to drought or irregular rainfall. A sprinkler irrigation study was established from 1999 to 2004 to evaluate the minimum design flow rate to produce optimum cotton yields and economic gain. A replicated randomized block design consisting of four irrigation treatments ranging from one inch every 12.5 days (equivalent to 1.5 gpm acre^{-1} design flow rate or system capability) to one inch every 3.1 days (6.0 gpm acre^{-1}) and a control, rainfed treatment. Daily plant water requirement was determined using soil moisture sensors and a spreadsheet-based scheduling program (MOISCOT) developed by Alabama Cooperative Extension engineers. Significant yield differences between irrigated and rainfed cotton were noted during the study period, with rainfall variability and treatment effects accounting for most of the yield response. The minimum design flow rate (1.5 gpm acre^{-1}) increased mean seed cotton yield by more than 500 lb acre^{-1} over rainfed yields. The most economically efficient design flow rate (4.5 gpm acre^{-1}) increased mean seed cotton yield by more than 996 lb acre^{-1}. A positive relationship was observed between cotton yield and total seasonal irrigation depth during dry years. Across all six years of the study, irrigated treatments produced significantly higher yields than rainfed cotton. The highest six-year cotton lint yield and net economic returns were obtained with the 4.5 gpm acre^{-1} irrigation treatment. This result provides a rule of thumb for estimating the extent of irrigated area based on available water supply rate.

Keywords: Cotton; Deficit irrigation; Irrigation system capacity; Design flow rate; Economic return

Introduction

Increased demand for limited water resources worldwide mandates that agricultural sectors explore increased water use efficiency for irrigation while striving for optimum economic crop productivity. Excessive irrigation aggravates water scarcity and can result in leaching and/or runoff of nutrients and pesticides. As a result, excessive irrigation can lead to increased costs for production and environmental protection. This research identifies the minimum level of irrigation for economic crop yield over a multiyear time span in a humid subtropical climate. Several studies in this humid region showed that cotton response to irrigation during seasons with insufficient rainfall [1-3]. Thus, an attempt was made here to study the response of cotton to deficit irrigation as a mean to conserve irrigation water while maintaining an economic yield.

Deficit irrigation has been reported by numerous authors as a method to improve water use efficiency in plants [4-7]. Bordovsky et al. [8] observed that deficit irrigation of short-season cotton using a LEPA system not only improved lint yield, but conserved groundwater on the Texas Southern High Plains. Likewise, Kirda et al. [6] reported that deficit irrigation was effective in saving irrigation water and increasing water use efficiency but did not decrease cotton seed yield. On the contrary, Steger et al. [9] reported that water stress caused by delayed post-planting irrigation reduced cotton lint yield. Similarly, in field studies conducted under rainfed and irrigated conditions, Pettigrew [10] found that moisture deficit reduced cotton lint yield by 25% in rainfed cotton. Moreover, DeTar [11] showed that deficit irrigation of cotton on a sandy soil reduced yield. The decline in yield as a result of moisture deficit in cotton plants is due to physiological impacts such as

reduced root growth, decreased leaf area index, lower photosynthesis, and decreased flowering and fruiting [12-20].

Northern Alabama has abundant water for crop production based on average annual rainfall (52 inches), however the region has large inter-annual variability in rainfall with low historic rainfall during the growing season (Figure 1). Sporadic convective rainfall during the growing season makes rainfed agriculture a poor competitor to the efficiency of irrigated agriculture [21].

This research originated from the broad body of knowledge related to rain-fed and irrigated crop production and to irrigation management, especially deficit irrigation practices. Earlier work by Tyson et al. [22] led to the development of a cotton scheduling procedure entitled MOISCOT (Moisture Management and Irrigation Scheduling for Cotton). This approach utilizes long-term average crop water use data, soil moisture monitoring and precipitation data to schedule cotton irrigation timing and quantity of water applied. In this experiment, the MOISCOT scheduling procedure was interrupted to incorporate a deficit irrigation component in order to simulate various design capacities, in terms of gallon per minute per acre (gpm acre^{-1}) available

*Corresponding author: A.H. AbdelGadir, Biosystems Engineering Department, Auburn University, Research Fellow, 200 Corley Bldg, Auburn, AL 36849-5417
E-mail: aha0001@auburn.edu

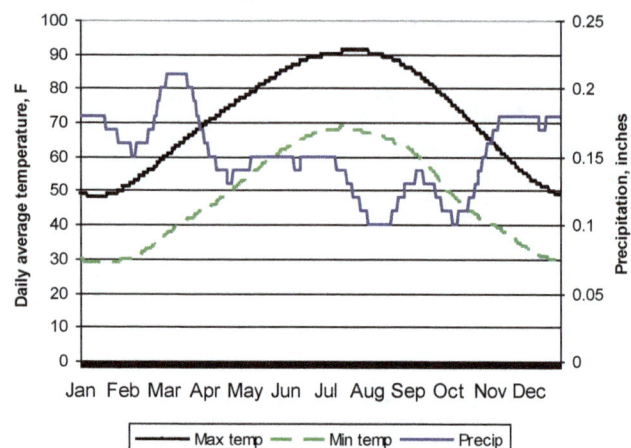

Figure 1: Long-term daily maximum/minimum temperature and precipitation, Belle Mina, Alabama, 1971-2000 (39).

for pivot irrigation of cotton. Recommended scheduling by MOISCOT was necessarily delayed when deficit irrigation capability treatments were unable to provide irrigation applications when the MOISCOT scheduling procedure called for irrigation. When extended dry periods occurred, MOISCOT would recommend irrigation but some of the capability treatments did not allow irrigation to occur because sufficient time had not passed since the last irrigation application. In terms of a center pivot system designed with a flow rate that did not meet the peak evapotranspiration rate of the cotton crop, a delay would occur from the time irrigation was called for by MOISCOT and when the system could make an application. In this case, only sufficient rainfall could return the available soil moisture in a field irrigated by the pivot to field capacity.

Because irrigation water supplies are limited on many farms, this research was designed to determine if satisfactory yields could be achieved over a number of years using irrigation systems that could not provide adequate water to replace crop evapotranspiration during peak water demand periods. The design flow rate delivered to center pivot systems is sometimes described in terms of gallons per minute per acre (gpm acre^{-1}). The desired or optimum flow rate in gallons per minute (gpm) delivered to a pivot is determined based on the anticipated crop(s) to be grown, soil type and water holding capacity, the peak water use of the crops grown, and the acres irrigated by the pivot. Dividing this flow rate by the acres irrigated determines the gpm acre^{-1}. In areas with similar climatic conditions, soil types and crops produced, this term can be used to quickly make an estimate of the flow rate needed for any size pivot. The number of acres irrigated by the center pivot multiplied by the gpm acre^{-1} is the fixed or design flow rate that will be delivered from the water source to the pivot. A higher gpm acre^{-1} flow rate may allow the pivot to match or exceed the evapotranspiration rate during peak water use periods and is preferred. A lower gpm acre^{-1} may fail to supply the peak evapotranspiration rate and thus is not preferred but may be necessary where the water supply is insufficient to provide the higher flow rate. Determining crop yield and economic benefits to a range of flow rates from low to optimum over several years is the major objective of this study.

Thus, the gpm acre^{-1} treatment levels used in this study reflect the irrigation capability of a system with the lowest gpm acre^{-1} treatment providing substantially less than the peak water demand and the

higher gpm acre^{-1} treatment providing application amounts near peak. Lower gpm acre^{-1} treatments reflect the most extreme case of deficit irrigation design. A fixed flow rate is required for a specific center pivot system design. Different flow rates can be specified for a system but not changed after initial design without major design changes. Lower or deficit flow rates might be selected because of limited water supply. Thus a 100-acre system would have continuous pumping capacities throughout the growing season of 150 gpm, 300 gpm, 450 gpm and 600 gpm for each of the treatment levels.

Therefore, a system design capacity experiment was established in 1999 with the goal of determining the minimum design capacity, gpm acre^{-1}, for center pivot irrigation in Northern Alabama to produce optimum economic cotton yield. Specific objectives of the study were to 1) compare sprinkler irrigated cotton yields to rainfed with different in-season rainfall levels and distributions, 2) determine the minimum design capacity for sprinkler irrigation without impacting cotton yield, and 3) identify the economic return for varying irrigation capacities along with non-irrigated cotton.

Materials and Methods

The research presented in this paper is located in northern Alabama in the Tennessee Valley, an area of widespread cotton production. This study was conducted on a Decatur silt loam soil (fine, kaolinitic, thermic, Rhodic Paleudults) at the Tennessee Valley Research and Extension Center located in Belle Mina, Alabama, during 1999-2004.

During the six years of this study, growing season precipitation and evaporation fluctuated across a wide range, providing representative wet and dry years for comparison (Figure 1).

Treatments included four sprinkler irrigation system (Hunter pop-up rotors, Hunter Industries Inc., San Marcos, California) capacities and a control, rainfed treatment. Irrigation was managed using soil moisture sensors and a spreadsheet-based scheduling method. The irrigation system capacities tested were (1) one inch every 12.5 days, (2) one inch every 6.3 days, (3) one inch every 4.2 days, and (4) one inch every 3.1 days. The one-inch amount represented the maximum irrigation depth applied during the number of days indicated. One inch represented a typical application that could be applied by center pivot systems to the soils in this region with minimum runoff. These four irrigation capabilities were equivalent to 1.5, 3.0, 4.5 and 6.0 gpm acre^{-1} respectively. The application amount was scheduled for one inch with an electronic controller that controlled the sprinkler run time for each plot. The actual application amounts were determined based on field measurement with rain gauges placed in each treatment plot. The actual amount of water applied throughout the six-year study ranged from 0.98 to 1.15 inches. This variability reflected the mechanical and hydraulic characteristics of the sprinklers as well as wind or drift effects. Hence, an average of one inch was applied each time the MOISCOT scheduling program called for irrigation, providing that sufficient time had elapsed between irrigations for each treatment. Thus, MOISCOT might call for irrigation, but irrigation may have been delayed until the design capacity time limitation for that treatment was met. In some cases, rainfall might occur within the waiting period that would satisfy the crop requirement for ET. Thereby, the experiment provided a realistic simulation of different center pivots with different pumping capacities and flow rates irrigating a cotton field under identical rainfed conditions, but with different availability of water for irrigation. For example, a center pivot system with a lower pumping capacity per acre would require a longer period of time to apply one-inch than a system with a higher pumping capacity.

In order to develop a sprinkler plot layout to simulate different center pivot capacities, 39 feet x 39 feet square sprinkler research plots were designed and installed. The plots were designed to deliver water with head to head coverage in each plot area. Each sprinkler was adjusted to apply water in a quarter circle so all water applied by the four sprinklers was placed in the designated plot. The irrigation system controller for all plots had a cycle and soak feature that allowed the application of one inch in an ON-OFF cycle to each plot to ensure that applied water infiltrated in the designated plot without runoff.

The planted plot size within each square irrigated plot was 26.7 feet x 39.0 feet, equivalent to eight 40-inch cotton rows, each 39 feet long. The middle four rows within each eight row plot served as data rows and the two outside rows within each plot served as guard rows. The excess width within each plot was planted in fescue and this perennial turf grass utilized irrigation overthrow outside the area planted to cotton.

Individual plots were arranged in a randomized complete block design of five treatments. From 1999 to 2000, three replications of each treatment were used. In 2001 and thereafter, a fourth replication was added when an adjacent space became available (Figure 2).

Moisture management and irrigation scheduling was accomplished using Watermark™ soil moisture sensors (Irrometer Company Inc. Riverside, California) and the spreadsheet-based MOISCOT irrigation scheduling program developed by Alabama Cooperative Extension System [22]. The MOISCOT program was designed to use data from individual farm fields to calculate anticipated soil moisture deficits in the future and to calculate the future date when irrigation should be applied to replenish an acceptable soil moisture deficit. This program required a one-time information entry into a spreadsheet program on the irrigation system type, crop, planting date, and soil characteristics of the irrigated fields, two times per week data entry of soil moisture readings at 9- and 18-inch depth, and daily entry of irrigation and rainfall inputs. The program then calculated a date in the future to replace a projected one-inch soil moisture deficit. The Watermark™ soil moisture sensors were installed according to manufacturer's recommendations in each plot at 9- and 18-inch depths. Wedge-shaped rain gauges were installed under the sprinkler irrigation system

within each plot to measure irrigation applied and another rain gauge installed adjacent to the study site to measure rainfall.

All plots were conventionally tilled from 1999-2002. In the fall of 2002 and 2003 wheat was planted as a cover crop. All treatment plots were converted to no-till in 2003-2004. In all experimental plots, KCl (0-0-60) and lime were applied as preplant at the rate of 60 lb K_2O and 2000 lb limestone per acre as per soil test recommendations. From 1999 to 2002, preplant nitrogen (Urea-NH_4NO_3, 32%N) was applied at 75 lb N acre^{-1} and sidedress nitrogen was applied at 30 lb N acre^{-1}. In 2003-2004, preplant nitrogen was applied at 100 lb N acre^{-1} and sidedress nitrogen at 30 lb N acre^{-1}. Nitrogen sidedressing was carried out 4 to 6 weeks after planting. In 2003, one additional ton of limestone per acre was applied, per soil test recommendations. Other cultural practices were carried out according to Tennessee Valley Research and Extension Center's practices.

Cotton (*Gossypium hirsutm*, L.) varieties selected for each year were DPL 33B (1999), DPL 428B (2000 and 2001), and DPL 451BR (2002 and 2004). Change in cotton variety during the study was required due to changes in seed technology and availability.

Cotton was planted in the second or third week of April each year using a 4-row planter on 40-inch row spacing with a seeding rate of 4-5 seeds per foot. Cotton was chemically defoliated 10 to 14 days prior to harvest by spraying the chemicals Finish (1.33 pt/acre) plus Ginstar (3.0 oz/acre). The four yield rows were harvested between the third week of September and the first week of October using a 2-row cotton picker. Each plot was harvested separately and weighed using a boll buggy (John Deere, Moline, Illinois) equipped with scales to provide accumulated mass which was divided by the harvested area to compute seed cotton yield. Turnout of lint was determined as average seasonal batch from a bulk seed cotton samples in local gin. The average turnout of lint from seed cotton for 1999-2001 seasons was 38% and for 2002-2004 was 35%. An economic analysis was conducted to evaluate irrigated cotton income gains over rainfed cotton using yield and total irrigation data per season for each irrigation capability. The sale price of $0.55/pound lint including a resale value of $200/ton seed; total annual irrigation system ownership costs of $87.95 per acre; and irrigation operating costs of $9.39 per acre-inch for a 140-acre pivot were used for the economical evaluation [23].

Yield data were analyzed statistically with a general linear model (GLM) using the LSD method for means separation at P ≤ 0.05 [24].

Results and Discussion

Table 1 presents total amount of irrigation water applied per treatment per acre in each season. Table 2 shows average seed cotton yields per treatment per season.

In 2004, rainfall was plentiful throughout the growing season, and rainfed and irrigated yields were not statistically ($P = 0.05$) different (Table 2). In 2003, rainfall was near optimum through much of the growing season, but a 26-day dry period occurred between August 7 and September 4. A total of only 0.61 inches of rain occurred during this period, and this rainfall was measured in seven minor rainfall events (25). Three timely one-inch irrigation applications during this period boosted irrigated yields significantly ($P = 0.05$), with more than 451 additional pounds of seed cotton per acre on the highest irrigation treatments (3.0, 4.5 and 6.0 gpm acre^{-1}). The lowest irrigation treatment was not significantly different from the rainfed cotton yield (Table 2).

In 2002, irrigated yields were significantly ($P = 0.05$) higher than

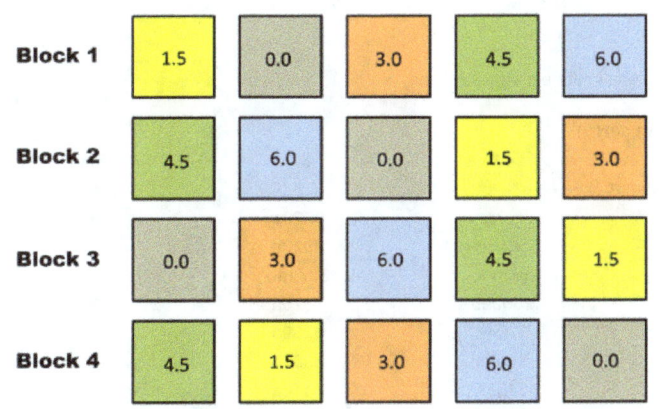

Figure 2: Sprinkler plot layout, beginning of 2001 growing season (0, denotes rainfed treatment, and 1.5, 3.0, 4.5, and 6.0 gpm acre^{-1} represent irrigated treatments). Plot size = 39' x 39'.

non-irrigated yield, but the highest yields were less than in other years for most irrigated treatments and were less than the 6-year means (Table 2). The reason for this reduced seed cotton yield was attributed to the very dry conditions late in 2002 growing season when the maximum application rate was not applied to meet peak water demand due to pumping problems resulting in reduced yields in all treatments. Non significant yield differences were noted in 2001 between rainfed and all irrigated treatments, except for the 4.5 gpm acre[-1] treatment, the highest yielding treatment. Significant yield differences were measured between rainfed and all irrigated treatments in 1999 and 2000. Although 2000 and 2002 seasons had similar rainfall (Table 1), the higher yield obtained during 2000 may be related to the greater water depth applied during this dry growing season (Table 1). Rainfall variability and treatment effects accounted for the wide range of yield responses for each of these years [25]. In drier seasons, treatments with higher irrigation gave more yields than lower irrigation treatments whereas in wet seasons little or no response was observed.

Yields in the lowest irrigation design flow, 1.5 gpm acre[-1] (1 inch every 12.5 days) were not significantly different from rainfed yields during three relatively wet seasons (Table 2). However, it is the lowest deficit irrigation design that boosted yield significantly ($P = 0.05$) during the dry years 1999, 2000, and 2002. The next highest irrigation design flow rates, at 3.0 gpm acre[-1] (1 inch every 6.3 days) did not have yields significantly different from 1.5 gpm acre[-1] in four seasons, but had an average 6-year yield significantly higher than 1.5 gpm acre[-1] and rainfed cotton. The highest irrigation design flow rates, 3.0, 4.5, and 6.0 gpm acre[-1] (1 inch every 6.3, 4.2 and 3.1 days, respectively) produced statistically similar yields in most of the years and resulted in 6-year average yields significantly higher than both rainfed and 1.5 gpm acre[-1] treatments (Table 2).

When correlating seasonal rainfall with annual treatment yields, the correlation coefficient increased with decreasing irrigation capability

design (Table 2), as would be expected. Similarly there was a positive relationship between cotton yield and irrigation capabilities except in seasons having sufficient rainfall (Table 2). Cotton yield responses to irrigation observed in most seasons of this study confirm similar results reported by others [1, 3, 26-31]. The results in this study stress the importance of irrigation to beneficially offset insufficient growing season rainfall. Nevertheless, other studies [27, 32, 33] reported no response to irrigation in cotton and attributed that to either insufficient irrigation applied or restricted root growth caused by soil compaction. In the present study, the absence of response to irrigation treatments observed during the 2001 and 2004 seasons is likely related to adequate rainfall during these seasons (Table 1, 2). In a similar study under similar conditions, Balkcom et al. [34] and Balkcom et al. [35] testing different irrigation regimes, found that irrigation increased seed cotton yield. Similarly, Howell et al. [36] in a thermally and rainfall limited environment such as the North Texas High Plains, found that deficit irrigation doubled cotton yield over rainfed yields. In contrast, Enciso et al. [37] reported that irrigation intervals ≤ 16 days did not influence cotton lint yield and quality using subsurface drip irrigation in medium to fine textured soils under limited water conditions. DeTar [11] also showed that deficit irrigation of cotton on sandy soils could greatly reduce yield. In a simulation study Jalota et al. [5] also showed that by reducing the amount of irrigation below the economic level, both yield and evapotranspiration of cotton were reduced to varying degrees depending on soil texture, precipitation and irrigation regimes.

Table 3 shows increasing seasonal operating costs for irrigation as depth of irrigation increased with corresponding increasing irrigation capability. Higher operating costs were associated with drier seasons (1999, 2000 and 2002) where total seasonal irrigation depths were higher (Table 1). Gross receipts and estimated net income gain above rainfed control for different irrigation capability treatments are given in Table 4. Gross receipts for lint yields above rainfed control for different

Year	Rainfall (in)	Irrigation applied per treatment (in)				
		0.0	1.5	3.0	4.5	6.0
1999	10.3	0.0	4.0	7.0	10.0	12.0
2000	7.1	0.0	6.1	11.1	12.6	13.5
2001	16.2	0.0	4.3	5.4	7.0	6.4
2002	7.1	0.0	4.9	9.9	9.9	9.4
2003	12.5	0.0	3.0	5.3	5.2	5.4
2004	15.9	0.0	3.3	4.4	5.3	6.3

Table 1: Total rainfall and irrigation per season in the experimental site.

Year	Irrigation capability (gpm acre[-1])					R[a]
	0	1.5	3.0	4.5	6.0	
	----------------- Seed Cotton Yield (lb acre[-1]) -------------					
1999	1700c	2637b	2984b	3708a	3920a	0.99
2000	1236c	2444b	3688a	3603a	3626a	0.98
2001	3061b	3387ab	3466ab	3595a	3371ab	0.93
2002	1759c	2531b	2871a	2853ab	2925a	0.98
2003	3288b	3579ab	3802a	3764a	3739a	0.99
2004	3530a	3300a	3208a	3505a	3367a	-0.37
Mean	2490c	3002b	3331a	3486a	3470a	--
R[b]	0.89	0.86	0.19	0.43	0.08	--

Means with the same letter in each row are not significantly different using LSD at $P = 0.05$.
R[a], R[b] = Correlation coefficient for irrigation and rainfall with yield, respectively.

Table 2: Yearly and average seed cotton yields for different irrigation capability treatments.

Irrigation capability	Total owner and operating costs ($ acre⁻¹)*					
(gpm acre⁻¹)	1999	2000	2001	2002	2003	2004
0.0	0	0	0	0	0	0
1.5	130	136	131	133	128	128
3.0	138	150	134	146	134	131
4.5	147	154	138	146	134	134
6.0	152	153	137	145	134	137

*Ownership costs = $119.35 acre⁻¹; Operating costs = $2.73 acre⁻¹-in.
Estimated costs include a 60-ac pivot system, with pump and motor.

Table 3: Total ownership and operating costs for irrigation capability treatments (23).

Irrigation capability	Gross receipts ($ acre⁻¹)*						Net income gain over rainfed ($ acre⁻¹)						Net profit
(gpm acre⁻¹)	1999	2000	2001	2002	2003	2004	1999	2000	2001	2002	2003	2004	($)
0.0	--	--	--	--	--	--	--	--	--	--	--	--	--
1.5	196	253	68	148	56	-45	66	117	-63	15	-72	-173	-110
3.0	269	513	84	214	99	-62	131	363	-50	68	-35	-193	284
4.5	420	495	111	210	92	-5	273	341	-27	64	-42	-139	470
6.0	464	500	65	224	87	-32	312	347	-72	79	-47	-169	450

* Gross receipts $0.55 /pound lint (includes resale of $200/ton seed). Gross receipt = $0.55 x Turnout (Treatment yield- Control yield)

Table 4: Gross receipts and net income gain above rainfed for different irrigation capability treatments (23).

irrigation capability treatments were calculated based on a sale value of $0.55/pound lint including a resale value of $200/ton seed [23]

Net income gains over rainfed control for overhead sprinkler irrigation capabilities were estimated when the estimated ownership and operating costs were charged against corresponding gross receipts. During seasons with sufficient rainfall (2001, 2003 and 2004), sprinkler irrigation capabilities result in a negative net income gain over rainfed indicating that irrigation added unnecessary (unrecovered) costs. However, during drier seasons, cotton producers with adequate irrigation capabilities realized significant yield increases (500-1000 lb acre⁻¹) and positive net income gain (60-360$ acre⁻¹). Durham [38] reported that cotton irrigation returned high net profit even during the wetter season of 2004. Over the six-year study period, a cumulative net profit of $470 per acre was realized with an irrigation capability of 4.5 gpm acre⁻¹. Results from this study indicate that when growing season rainfall is below 12 inches, cotton producers in the Tennessee Valley of Alabama with adequate irrigation capability can realize significant yield increases along with positive net returns over rainfed cotton production. Results provide a rule of thumb of approximately 4.5 gpm acre⁻¹ for estimating the extent of irrigated area based on available water supply rate.

Conclusions

In all treatments, irrigation was found to significantly increase seed cotton yield in seasons with inadequate rainfall. Data from this study indicate that the minimum design flow rate needed to produce optimum economic yields in irrigated cotton is 4.5 gpm acre⁻¹ which is equivalent to approximately one inch every 4.2 days. This information can be used to optimize the design of pivot irrigation pumping plants by matching pump and storage facility size to the total area irrigated in soil type typical of the Tennessee Valley and is not necessarily applicable to other areas or soil types. Thus, cotton producers in the

northern Alabama in the Tennessee Valley region with adequate irrigation capabilities can realize significant seed cotton yield increases and positive economic net returns. Results provide a rule of thumb of approximately 4.5 gpm acre⁻¹ for estimating the extent of irrigated area based on available water supply rate.

Acknowledgement

The authors would like to thank the technical personnel at TVREC for support on this project. Partial funding for this project was provided through grants from the Alabama Cotton Commission and The Tennessee Valley Authority.

References

1. AbdelGadir AH, Fulton JP, Dougherty M, Curtis LM, van Santen E, et al. (2011) Subsurface drip irrigation placement and cotton irrigation water requirement in the Tennessee Valley. Online Crop Management.

2. AbdelGadir AH, Dougherty M, Fulton JP, Burmester CH, Norris BE, et al. (2009) Sub-surface drip irrigation-fertigation for site-specific, precision management of cotton. Irrigation Conference Proceedings, , San Antonio, Texas.

3. Dougherty M, AbdelGadir AH, Fulton JP, van Santen E, Burmester CH, et al. (2009) Subsurface drip irrigation and fertigation for North Alabama cotton production. J Cotton Sci 13: 227-237.

4. Sepaskhah AR, Akbari D (2005) Deficit irrigation planning under variable seasonal rainfall. Biosyst Eng 92: 97-106.

5. Jalota SK, Sood A, Chahal GBS, Choudhury BU (2006) Crop water productivity of cotton (Gossypium hirsutum L.)-wheat (Triticum aestivum L.) system as influenced by deficit irrigation, soil texture and precipitation. Agric Water Manage 84: 137-146.

6. Kirda C, Topcu S, Cetin, M, Dasgan HY, Kaman, H, et al. (2007) Prospects of partial root zone irrigation for increasing irrigation water use efficiency of major crops in the Mediterranean region. Ann Appl Biol 150: 281-291.

7. Chaves MM, Santos TP, Souza CR, Ortuño MF, Rodrigues ML, et al. (2007) Deficit irrigation in grapevine improves water-use efficiency while controlling vigor and production quality. Ann Appl Biol 150: 237-252.

8. Bordovsky JP, Lyle WM, Lascano RJ, Upchurch DR (1992) Cotton irrigation management with LEPA systems. Trans ASAE 35: 879-884.

9. Steger AJ, Silvertooth JC, Brown PW (1998) Upland cotton growth and yield response to timing the initial postplant irrigation. Agron J 90: 455-461.

10. Pettigrew WT (2004) Moisture deficit effects on cotton lint yield, yield components, and boll distribution. Agron J 96: 377-383.

11. DeTar WR (2008) Yield and growth characteristics for cotton under various irrigation regimes on sandy soil. Agric Water Manage 95: 69-76.

12. Guinn G, Mauney JR (1984) Fruiting of cotton. I. Effects of moisture status on flowering. Agron J 76: 90-94.

13. Guinn G, Mauney JR (1984) Fruiting of cotton. II. Effects of plant moisture status and active boll load on boll retention. Agron J 76: 94-98.

14. Radin JW, Mauney JR, Kerridge PC (1989) Water uptake by cotton roots during fruit filling in relation to irrigation frequency. Crop Sci 29: 1000-1005.

15. Carmi A, Plaut Z, Heuer B, Grava A (1992) Establishment of shallow and restricted root systems in cotton and its impact on plant response to irrigation. Irrig Sci 13: 87-91.

16. Carmi A, Plaut Z, Sinai M (1993) Cotton root growth as affected by changes in soil water distribution and their impact on plant tolerance to drought. Irrig Sci 13: 177-182.

17. Gerik, TJ, Faver KL, Thaxton PM, El-Zik KM (1996) Late season water stress in cotton. I. Plant growth, water use, and yield. Crop Sci 36: 914-921.

18. Plaut Z, Carmi A, Grava A (1996) Cotton root and shoot responses to subsurface drip irrigation and partial wetting of the upper soil profile. Irrig Sci 16: 107-113.

19. Pace PF, Cralle HT, El-Halawany SHM, Cothren JT, et al. (1999) Drought-induced changes in shoot and root growth of young cotton plants. J Cotton Sci 3: 183-187.

20. Pettigrew WT (2004) Physiological consequences of moisture deficit stress in cotton. Crop Sci 44: 1265-1272.

21. Dougherty M, Bayne D, Curtis LM, Reutebuch E, Seesock W (2007) Water quality in non-traditional off-stream polyethylene-lined reservoir. J Environ Manage. 85: 1015-1023.

22. Tyson TW, Curtis LM, Burmester CH (1996) Using MOISCOT: moisture management and irrigation scheduling for cotton. Circular ANR-929. The Alabama Cooperative Extension System, Auburn and Alabama A&M Universities, AL.

23. Tyson TW, Curtis LM (2007) Alabama Irrigation and Water Resources. Timely Information: Agriculture, Natural Resources & Forestry. Biosystem Engineering Series, Publication No. BSEN-IRR-07-01, May 2007. Alabama Cooperative Extension System, Auburn University, AL.

24. Analytical Software (2003) Statistix 8. Analytical Software, Tallahassee, FL.

25. Curtis L, Burmester CH, Harkins DH, Norris BE (2004) Sprinkler irrigation water requirements and irrigation scheduling, Tennessee Valley Research Extension Center. Alabama Agric Exp Station. Cotton Research Report, Research Report No. 26: 25-26.

26. Camp CR, Thomas WM, Doty CW (1994) Drainage and irrigation effects on cotton. Trans ASAE 37: 823-830.

27. Camp CR, Bauer PJ, Hunt PG (1997) Subsurface drip irrigation lateral spacing and management for cotton in the southeastern coastal plain. Trans ASAE 40: 993-999.

28. Bronson KF, Onken AB, Keeling JW, Booker JD, Torbert HA (2001) Nitrogen response in cotton as affected by tillage system and irrigation level. Soil Sci Soc Am J 65: 1153-1163.

29. Pringle III HC, Martin SW (2003) Cotton yield response and economic implications to in-row subsoil tillage and sprinkler irrigation. J Cotton Sci 7: 185-193.

30. Sorensen RB, Bader MJ, Wilson EH (2004) Cotton yield and grade response to nitrogen applied daily through a subsurface drip irrigation system. Appl Eng Agric 20: 13-16

31. Kalfountzos D, Alexiou I, Kotsopoulos S, Zavakos G, Vyrlas P (2007) Effect of subsurface drip irrigation on cotton plantations. Water Resour Manage 21: 1341-1351.

32. Bauer PJ, Hunt PG, Camp CR (1997) In-season evaluation of subsurface drip and nitrogen-application method for supplying nitrogen and water to cotton. J Cotton Sci 1: 29-37.

33. Camp CR, Bauer PJ, Busscher WJ (1999) Evaluation of no-tillage crop production with subsurface drip irrigation on soils with compacted layers. Trans ASAE 42: 911-917.

34. Balkcom KS, Reeves DW, Shaw JN, Burmester CH, Curtis LM (2006) Cotton Yield and Fiber Quality from Irrigated Tillage Systems in the Tennessee Valley. Agron J 98: 596-602.

35. Balkcom KS, Shaw NJ, Reeves DW, Burmester CH, Curtis LM (2007) Irrigated Cotton Response to Tillage Systems in the Tennessee Valley. J Cotton Sci 11: 2-11.

36. Howell TA, Evett SR, Tolk JA, Schneider AD (2004) Evapotranspiration of full-, deficit-irrigated, and dryland cotton on the Northern Texas High Plains. J Irrig Drain Eng 130: 277-285.

37. Enciso JM, Unruh BL, Colaizzi PD, Multer WL (2003) Cotton response to subsurface drip irrigation frequency under deficit irrigation. Appl Eng Agric 19: 555-558.

38. Durham S (2005) Multicrop rotation and irrigation study for optimal water use in the southeast. USDA-ARS, GA. Agricultural Research Magazine 53:18-20.

39. AWIS (2012) Weather Services, Inc., P.O. Box 3267, Auburn, AL 36831-3267.

12

Study on Water Requirement of Selected Crops under Tarikere Command Area using CROPWAT

Nithya KB* and Shivapur AV

Water and Land Management, Visvesvaraya Technological University, Belagavi, India

Abstract

A study was carried out to determine the crop water requirement of few selected crops for the command area in Tarikere taluk in Karnataka state, India. The crops include areca nut, coconut, and cotton, banana for two seasons, sweet pepper, onion, potato, rice, pulses, mango, and cotton, sugarcane and millet (ragi). Crop water requirement for each crop was determined by using 30-year climatic data in CROPWAT. Reference crop evapotranspiration (ET_0) was determined using the FAO Penman Monteith method. For all the crops considered, three decades: decades I, II, and III and seven crop growth stages: nursery, nursery/land preparation, land preparation, initial stage, development stage, mid-season and late season stage were considered. The study shows that reference evapotranspiration (ET_0) varies from 2.5 to 3.36 mm/day for the area under study. The gross water requirement was 342.42 mm/year with an application efficiency of 70% and hence the entire crop area of 4466 ha requires 16 MCM. Thus the dam can conveniently supply the water required for irrigation in the area.

Keywords: Crop water requirement; Peak water requirement; Reference evapotranspiration; Crop evapotranspiration; Climatic data

Introduction

Water is important for plant for its growth as well as for food production. There is a competition between municipal, industrial and agriculture users for the water available in reservoirs. Estimating irrigation water requirements is prerequisite for water project planning and management [1]. The primary objective of irrigation is to apply water to soil to meet crop evapotranspiration (ETA) requirement when rainfall is insufficient to raise crops till harvesting. Hess defined crop water requirements as the total water needed for evapotranspiration, from planting to harvest for a given crop in a specific climate regime, when adequate soil water is maintained by rainfall and/or irrigation so that it does not limit plant growth and crop yield [2].

Net irrigation water requirements (NIWR) in a specific scheme for a given year are thus the sum of individual crop water requirements (CWR) calculated for each irrigated crop. Multiple cropping (several cropping periods per year) is thus automatically taken into account by separately computing crop water requirements for each cropping period. By dividing the area of the scheme (S. in ha), a value for irrigation water requirements is obtained and can be expressed in mm or in m^3/ha (1 mm = 10 m^3/ha). FAO, Smith et al. and Smith reported that CROPWAT is meant as a practical tool to help agro meteorologists, agronomists and irrigation engineers to carry out estimation for evapotranspiration and crop water use studies, and more specifically the design and management of irrigation schemes [3]. Recommendations for improved irrigation practices, the planning of irrigation schedules under varying water supply conditions, and the assessment of production under rain-fed conditions or deficit irrigation can be derived from this. Broner and Schneekloth opined that water requirements of crops depend mainly on environmental conditions. Different crops have different water use requirements, under the same weather conditions [4].

Crops will transpire water at the maximum rate when the soil water is at field capacity. Broner found that knowing seasonal crop water requirements is crucial for planning your mixed crop planting especially during drought years [4]. Adequate data on irrigation water requirements of most crops is not available in developing nations of the world. This is one of the reasons why for the failure of large scale irrigation projects in most developing countries of the world. The objective of this study was to determine crop water requirements of arecanut, coconut, cotton, and banana for two seasons, sweet pepper, onion, potato, rice, pulses, cotton, mango, sugarcane and millet (ragi).

Study area

Tarikere is situated about 40 km south of Chikkamagaluru on the Bangalore-Honnavar road, and 45 km from Chikkamagaluru city. Tarikere is located at latitude of 13.72°N and longitude of 75.82°E at an average elevation of 698 m with annual average rainfall of 963 mm. The south west monsoon starts normally from 1st week of June and peak precipitation will occur during September.

Bhadra Dam located around 50 km from Tarikere; (Bhadra Reservoir) has been the life-saving water source for irrigating over 4,466 hectares of land spread over villages around Tarikere (Figure 1).

Stages of growth

For the present study three decades and seven stages of plant growth were used, the decades include I, II and III while the crop growth stages include nursery, nursery/land preparation, land preparation, initial stage, development stage, mid-season stage and late season stage.

Estimation of Water Requirement

The term crop water requirement is defined as the "amount of water required to compensate the evapotranspiration loss from the cropped field". "Although the values for crop evapotranspiration and crop water requirement are identical, crop water requirement refers to the amount of water that needs to be supplied, while crop evapotranspiration refers

*Corresponding author: Nithya KB, Water and Land Management, Visvesvaraya Technological University, India, E-mail: nithyakrishna2391@gmail.com

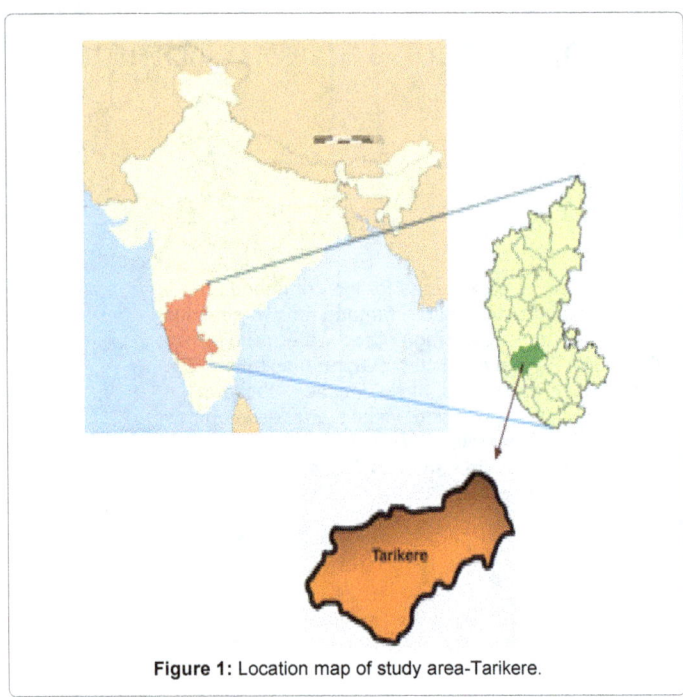

Figure 1: Location map of study area-Tarikere.

to the amount of water that is lost through evapotranspiration". The crop ET (ET_c) was estimated by FAO Penman-Monteith equation:

$$ET_c = K_c (ET_o) \qquad (1)$$

where ET_o = Reference crop (mm/day) and it is determined by:

$$ET_o = 0.408\Delta(R_n - G) + \gamma \left[\frac{900}{T + 273} \right] u_2 \frac{(e_a - e_d)}{\Delta} + \gamma(1 + 0.34u_2) \quad (2)$$

Where,

ET_0 = Reference evapotranspiration [mm day^{-1}]

R_n = Net radiation at the crop surface [MJ m^{-2} day^{-1}]

G = Soil heat flux density [MJ m^{-2} day^{-1}]

T = Mean daily air temperature at 2 mm height [°C]

U_2 = Wind speed at 2 m height [m s^{-1}]

E = Actual vapour pressure [KP$_a$]

e_s-e_a = Saturation vapour pressure deficit [KP$_a$]

D = Slope vapour pressure curve [KP$_a$C^{-1}]

g = Psychometric constant [KP$_a$C^{-1}]

CROPWAT

The CROPWAT programme (version 5.7) developed for the FAO Penman-Monteith method was utilized for estimating the crop water requirement of the crops studied. To ensure the integrity of computations, the weather measurements were made at 2 m (or converted to that height) above the surface of green grass, shading the ground [5]. The climatic data used for the calculations were obtained from a meteorological station located at Tarikere (Tables 1-4).

Conclusion

The estimation of actual irrigation requirement of Tarikere command area was carried out as shown in Tables 1-4. The net irrigation water requirement is 292.7 mm/year (Tables 1-4). This is summation of the NIR2 values from January to December. Using an irrigation application frequency of 70%, the gross water requirement of 342.42 mm/year was obtained. Therefore the entire land area of 4466 ha requires 16 MCM. The reservoir capacity is 71.50 MCM. Thus the capacity of 71.5 MCM is sufficient to irrigate the irrigation water requirement for the entire area under the command area, which is 16 MCM. The results show that the dam can conveniently supply the water required for irrigation in the area. The results obtained from the study can be used as a guide by farmers for selecting the amount and frequency of irrigation water for the crops studied under consideration.

Country: India		Station: Tarikere		Altitude: 693 m		Lat: 13.43°N		Long: 75.49° E
Month	Min Temp (°C)	Max Temp (°C)	Humidity (%)	Wind (Km/day)	Sun (hrs)	Rad (MJ/m²/day)	ET₀ (mm/day)	
January	15	32	67	1.4	0.6	8.4	2.74	
Febraury	16	34	60	1.4	0.6	9.2	3.15	
March	18	37	57	1.1	0.6	10	3.33	
April	21	36	66	1.1	0.6	10.4	3.24	
May	21	36	68	1.1	0.5	10.3	3.26	
June	20	32	72	1.7	0.3	9.9	3.19	
July	20	30	74	2.8	0.2	9.8	3.36	
August	20	29	76	2.8	0.2	9.8	3.15	
September	19	31	73	2.2	0.4	9.8	3.18	
October	20	34	73	1.4	0.5	9.2	2.91	
November	18	30	70	1.1	0.5	8.4	2.5	
December	16	31	66	1.4	0.5	8	2.67	
Average	18.7	32.7	69	1.6	0.5	9.4	3.06	

Where, ETo = Reference Crop Evapotranspiration computed using the FAO Penman-Monteith Method.

Table 1: Reference crop evapotranspiration.

Crop	Planting Date	ETc(mm/dec)	Eff rain(mm/dec)	Irr. Req.(mm/dec)
Sweet pepper	01-Nov	309.2	67.7	247.6
Mango	20-Mar	1034.5	743.7	400.6
Banana(Summer)	01-Jun	786	670	407
Banana(Winter)	01-Jan	849.6	478.9	416.6
Cotton	01-Aug	428.6	389.3	178.3
Potato	01-Oct	318.7	176.5	210.9
Ragi	15-Jul	256.3	429.4	12.1
Rice	01-Jan	336.1	541.9	182.5
Pulses	10-Jun	244.1	404.6	12
Sugacane	01-Jan	1162.4	743.7	460.7
Onion	01-Nov	234.6	63.2	174.7
Coconut	01-Dec	1097.3	741.6	455
Arecanut	01-Jun	1113.5	743.7	467

Table 2: Evapotranspiration and irrigation requirement for major crops.

S.No	Crop	Jan	Feb	Mar	Apr	May	Jun	Jul	Aug	Sep	Oct	Nov	Dec
1	Sweet pepper	86.2	83.9	14.1	0	0	0	0	0	0	0	5.4	53.3
2	Mango	73.5	57.1	57.4	37	24.7	9.4	0	0	21.6	0.2	37.7	75.2
3	Banana1	85.3	93.2	104.5	47.3	0	0	0	0	0	0	16.2	62.1
4	Banana2	85.7	86	99.1	59.9	51.7	32.2	0	0	0	0	0	0
5	Cotton	52.9	0	0	0	0	0	0	0	0	2.9	44.6	79
6	Potato	84.8	15.5	0	0	0	0	0	0	0	0	27.8	81.1
7	Grains	0	0	0	0	0	0	0	0	10.5	0	1.4	0
8	Rice	0	0	0	0	0	93.1	90	1.3	30.7	0	5.7	0
9	Pulses	0	0	0	0	0	0	1.2	0	0.3	0	0	0
10	Sugarcane	31.4	50.1	95.5	70.1	58.8	39	0	3.3	35.4	0	28.7	55
11	Onion	85.3	8	0	0	0	0	0	0	0	0	12.1	67.5
12	Coconut	80.5	80.6	89.4	37.5	31.8	13.8	0	0	10.1	0	27.7	67.1
13	Areca nut	85.3	93.2	104.5	47.3	0	0	0	0	0	0	16.2	62.1
14	NIR 1	1.5	1	1	0.6	0.4	0.8	0.5	0	0.4	0	0.5	1.2
15	NIR 2	45.2	29.4	31.5	16.6	11.6	22.8	16.3	0.4	12.8	0.1	14.8	38.2
16	NIR3	0.17	0.12	0.12	0.06	0.04	0.09	0.06	0	0.05	0	0.06	0.14
17	IA (%)	59	58	37	36	31	49	25	24	73	26	94	56
18	IR$_a$	0.29	0.21	0.32	0.18	0.14	0.18	0.24	0.01	0.07	0	0.06	0.25

where, NIR 1 = Net Water Requirement (mm/day), NIR 2 = Net Water Requirement (mm/month), NIR 3 = Net Water Requirement (l/s/h), IA = % of the total area that is actually irrigated, IR$_a$ = Net Water Requirement for Actual Irrigated Area (l/s/h)

Table 3: Irrigation scheming.

Crops	ETc (mm)		Irrigation requirement (mm)
	Minimum	Maximum	
Sweet pepper	1.5	3.27	0-32.7
Mango	1.79	3.39	0-27.2
Banana1(summer)	1.6	3.72	0-35.6
Banana2(Winter)	2.8	3.39	0-33.9
Cotton	1.1	3.26	0-27.5
Potato	1.5	3.1	0-31.5
Ragi	0.99	3.06	0-8.2
Rice	0.292	2.48	0-90.3
Pulses	1.1	3.64	0-6.1
Sugarcane	1.9	4.13	0-35.5
Onion	1.84	2.9	0-31.1
Coconut	2.6	3.27	0-31.4
Arecanut	2.56	3.36	0-31.7
Total	21.572	42.97	422.7

Table 4: Minimum and maximum values of evapotranspiration and irrigation requirements for various crops.

References

1. Michael AM (1999) Irrigation Theory and Practice. Vikas Publishing House, India.

2. Hess T (2005) Crop Water Requirements, Water and Agriculture, Water for Agriculture.

3. Smith M (1992) CROPWAT: A Computer Program for Irrigation Planning and Management. Food and Agriculture Organization, Italy.

4. Broner I, Schneekloth J (2003) Seasonal Water Needs and Opportunities for Limited Irrigation for Colorado Crops. Colorado State University Extension.

5. Food and Agriculture Organization (FAO) (2005) Irrigation water requirements, In: Irrigation Potential in Africa - A Basin Approach. FAO Corporate Document Repository, Rome.

13

Comparison of Yield and Water Productivity of Rice (*Oryza sativa* L.) Hybrids in Response to Transplanting Dates and Crop Maturity Durations in Irrigated Environment

Akhter M[1], Ali M[1], Haider Z[1]*, Mahmood A[2] and Saleem U[1]

[1]Rice Research Institute, Kala Shah Kaku, Lahore, Pakistan
[2]Ayub Agricultural Research Institute, Faislabad, Pakistan

Abstract

Water scarcity, due to abruptly accruing phenomenon of climate change, is perilously disturbing agricultural crops such as rice as well as its quality in many countries of the world. It is an acute threat to livelihood of residents of those countries where water resources are already a limiting factor to agriculture. Therefore, an experiment was conducted to ascertain and compare yield and water productivity of rice (*Oryza sativa* L.) hybrids in response to transplanting dates and cultivar duration in irrigated sub-tropical regions of Punjab, Pakistan. The experiment was conducted in experimental fields of Rice Research Institute, Kala Shah Kaku. It was determined that the water productivity was increased with the shifting of transplanting date towards shorter water demand period and variety to shorter life duration. Water stress is more damaging to those varieties or hybrids that have longer life cycle as compared to early maturing hybrids and varieties. Conclusively, same method may be used to test other rice varieties and hybrids to ascertain their minimum water requirements for maximum yield returns.

Keywords: Rice; Water management; Irrigation; Water productivity; Water saving technology

Introduction

Rice is the second largest staple food crop of Pakistan and is also an exportable item. It accounts for 3.2% in the value added in agriculture of Pakistan and 0.7% of GDP. During July-March 2014-15, rice export earned foreign exchange of US$1.53 billion. During 2014-15, rice was sown on an area of 2.89 Million hectares of Pakistan showing an increase of 3.6% over last year's area of 2.79 Million hectares. Rice recorded highest ever production at 7.01 Million tonnes, showing a growth of 3.0% over corresponding period of last year's production which was 6.79 Million tonnes [1].

Water scarcity is becoming a major problem in the agriculture sector, especially in the case of rice which is the staple food of half of the world's population. Per capita water availability has declined tremendously in many countries of Asia [2,3]. The production system of the rice crop is requiring higher water availability than other crops like cereals, fruits or vegetables. Rice is transplanted in early June on large scale during which peak evaporative demand contributes majorly to water table decline in these regions [4]. Furthermore, due to abrupt increase in urban and industrial sectors, agriculture's share of fresh water has declined by 8-10% [5,6]. Rice production in Asia is increasingly constrained by water limitation and therefore there is an increasing pressure to reduce water use in production of irrigated rice. Already declining quantity as well as quality of ground water and poor infrastructure systems is threatening the sustainability of the irrigated rice-based production system [7-10]. Exploring ways to produce more rice with less water is essential for food security and for a sustainable environment; however its confirmation still stands uncertain [11].

In Pakistan, most of the rice hybrids and cultivated varieties are late maturing. Late maturing hybrids and varieties have longer life duration and thus also show increased water requirements due to higher evapotranspiration (ET) demands crop water use efficiency (WUE)/water productivity (WP) are the most important criterion to consider where available water resources are limited or diminishing. It has been reported earlier that WP can be increased by adopting water-saving management practices e.g., improved irrigation management

technologies Bouman and Tuong [12], growing early maturing/short duration hybrids and varieties and synchronizing the crop growing cycle with the days of lower evaporative demands [13,14]. Therefore, it has now become a dire need to develop short duration rice hybrids and varieties capable of producing more grains using less water. The present study was conducted to contour methods of screening out the genotypes (hybrids/varieties) that show higher yield potential with minimum water requirements and to find out their optimum time of transplanting at maximum water productivity (WP).

Materials and Methods

A field experiment was conducted at the research fields of Rice Research Institute Kala Shah Kaku, Lahore, Pakistan during Kharif 2015. Two rice hybrids i.e., Arize Swift and INH10008 and one check variety (KSK 133) were used in the experimental study. The experiment was laid out in split plot design with three replications. Main plot treatments were consisted of three transplanting dates. 30 days old rice seedlings were transplanted on 1st, 16th July and 31st July.

Irrigations were applied when 50% plants reached to score 7 (IRRI: SES, 2002). Quantity of irrigation water was measured using water meter and water productivity computed by dividing the economic yield (kg ha-1) with amount of irrigation water applied. Important dates of crop cycle such as date of sowing, date of transplanting, Date of 10% flowering, Date of finished flowering, Date of maturity, Date of harvest were noted on regular basis, and Days of flowering period, Days to maturity and Days to harvest were calculated. Furthermore, other important morphological and agronomical traits such as Number of

---

*Corresponding author: Haider Z, Rice Research Institute, Kala Shah Kaku, Lahore, Pakistan, E-mail: z.haider.breeder@gmail.com

Productive tillers/ square meter, Number of filled spike lets per panicle, Number of sterile spike lets per panicle, seed set (%), harvest per plot, moisture Content of grain at harvest were collected. Stress related traits such as quantity of irrigation water, adjusted yield per hectare at 14% moisture content and water productivity was computed as yield per ha/ water using the following formula given.

Water Productivity (WP) (kg/Litre)=grain yield (kg)/irrigation water (Litres)

The data for each character was statistically analysed using Analysis of Variance (ANOVA) technique and significant means were separated by using Least Significant Difference (LSD) test or comparing the means of treatments as given by Gomez and Gomez [15], Singh et al. [16].

Results and Discussion

Rice yields in different treatments as influenced by transplanting date, variety and irrigation regime are presented in Tables 1a and 1b. Results depicted in Tables 1a and 1b clearly show that both the hybrids and check variety showed highly significant different performances under stress treatment for a number of parameters that include spikelet fertility, water productivity, yield per hectare and yield per plot; while for 1000-grain weight they were significantly different. On the other hands, there was no varietal difference for other traits such as tillers per meter square, spikelet fertility and seed setting. The water stress had remarkable influence on growth and yield components of rice (Tables 1a and b). It was observed that yield, filled spikelet, 1000 grain weight, plant height and number of tillers reduced with duration of water stress cycle. These results are in consistent with the results of Chahal et al.

[14] who found a significant relation between rice yield and number of days under stress during the period of post-transplanting.

Similarly, transplanting dates had significant effects on performance of hybrids and check variety. More the duration or period of stress, less yield was recovered. Therefore, average yield was fewer in first date of transplanting as compared to second date for all the studied hybrids and variety due to the reason that genotypes had to suffer stress for a longer duration that resulted in remarkable differences in yields of hybrids transplanted at different dates.

At the probability level of 0.01, both the hybrids and check variety showed highly significant different performances in terms of spikelet fertility, water productivity and yield as well as irrigation water required. Whereas, 1000 grain weight, spikelet fertility was seed setting were significantly different at probability level of 0.05. However, tillers per meter square were not significantly different for different genotypes and transplanting dates.

Averaged over genotypes, the paddy yields per hectare of hybrids INH10008 and Swift were 3192 and 3377 kg/ha, which were statistically at par as depicted in Table 2a. However, average yield of check variety KSK 133 was significantly different (p<0.01) from both the hybrids under stress treatment in all transplanting dates. When we compare the effect of transplanting dates on average yield performance of all the three genotypes, it becomes clear that there exist a significant effect of transplanting these genotypes at different dates (p<0.01) as given in Table 2b. Average yields of both the hybrids and one check variety were at par. Hybrid Swift produced highest paddy yield (3501 kg/ha) followed by INH10008 (2873 kg/ha) and KSK 133 (2539 kg/ha) under water stress treatment (Table 2c) (Figure 1).

Source of variation	Spikelet fertility	1000 grain weight (kg)	Tillers per m²	Spikelet sterility	Seed set (%)
Replications	25.04	4.97	746.26	338.48	99.97
Dates	6814.37**	10.54*	6739.15	2318.81*	118.43*
Varieties	9786.26**	24.78*	3080.70	393.59	75.52
Dates* Varieties	2876.65*	4.81	2002.20	338.15	250.69*
*, **Significant level P=0.01					

Table 1a: Mean squares of different parameters/traits.

Source of variation	Water productivity (kg-ha/Lit.)	Yield (kg/ha) (14% M.C.)	Yield (kg/plot) (14% M.C.)	Irrigation per hectare (Lit.)	Irrigation per plot (Lit.)
Replications	5.642E-08	2643	0.132	3.549E+10	60018
Dates	2.556 E-07**	2147370**	2.073**	1.229 E+12**	2077263**
Varieties	5.687 E-07**	2740890**	5.291**	1.274 E+11	215701
Dates* Varieties	1.352 E-07**	1605876**	2.200**	4.930 E+10	83370
*, **Significant level P=0.01					

Table 1b: Mean squares of different parameters/traits.

	Source of variation	Water productivity		Yield per hectare		Yield per plot	
		Between hybrids	Between dates	Between hybrids	Between dates	Between hybrids	Between dates
1	KSK 133	9.07 E-04 B	9.90 E-04 B	2343.0 B	2539.3 C	3.1533 B	3.7680 AB
2	INH 10008	1.25 E-03 A	1.27 E-03 A	3192.9 A	2873.0 B	4.1853 A	4.5483 A
3	Swift	1.39 E-03 A	1.28 E-03 A	3377.8 A	3501.3 A	4.6517 A	3.6740 B

Table 2a: Pair-wise comparisons (at 0.01 probability) of means of varieties transplanted at different dates.

	Source of variation	Spikelet fertility		1000 grain weight (kg)		Irrigation per hectare	
		Between hybrids	Between dates	Between hybrids	Between dates	Between hybrids	Between dates
1	KSK 133	84.11 B	151.11 A	23.589 AB	22.200 A	2.80 E+06 A	2.97 E+06 A
2	INH 10008	127.22 A	110.44 B	24.933 A	24.322 A	2.59 E+06 A	2.77 E+06 A
3	Swift	148.89 A	98.67 B	21.633 B	23.633 A	2.60 E+06 A	2.25 E+06 B

Table 2b: Pair-wise comparisons (at 0.01 probability) of means of varieties transplanted at different dates.

	Source of Variation	Irrigation per plot		Spikelet sterility		Tillers per m²	
		Between hybrids	Between dates	Between hybrids	Between dates	Between hybrids	Between dates
1	KSK 133	3643.8 A	3862.8 A	43.000 A	67.000 A	254.00 A	264.11 AB
2	INH 10008	3369.9 A	3601.3 A	49.333 A	35.444 B	235.89 A	223.33 B
3	Swift	3381.8 A	2931.3 B	56.222 A	46.111 AB	272.89 A	275.33 A

Table 2c: Pair-wise comparisons (at 0.01 probability) of means of varieties transplanted at different dates.

	Name of variety	Filled spiklets	1000-Grain weight (g)	Tiller/m²	Yield at 14% M.C. (kg/ha)	Quantity of irrigation water/ha (Lit.)	Water productivity (kg-ha/ Lit.)
D1	KSK 133	80	22.00	280	982	3062882	0.00046
	INH10008	193	23.00	240	3458	2795508	0.00127
	Swift	180	21.67	272	3178	3055539	0.00124
D2	KSK 133	88	25.10	225	3428	2947974	0.00117
	INH10008	95	26.83	224	3274	2796000	0.00118
	Swift	148	21.03	221	3803	2566256	0.00148
D3	KSK 133	84	24.11	257	2619	2397384	0.00109
	INH10008	94	24.56	244	2847	2185231	0.00130
	Swift	118	22.24	325	3153	2182154	0.00147

Table 3: Average yield performances of three genotypes at all the three transplanting dates along with their some yield important yield components.

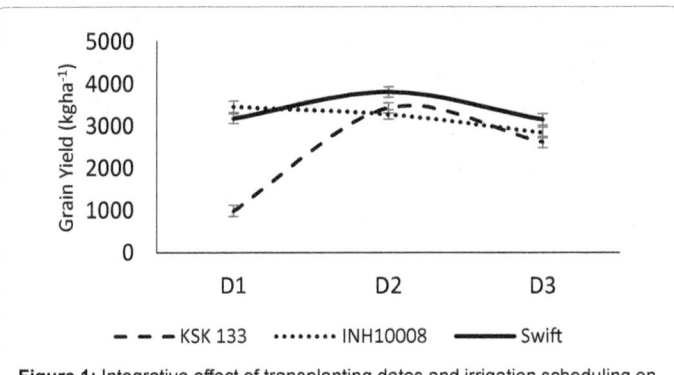

Figure 1: Integrative effect of transplanting dates and irrigation scheduling on grain yield of rice hybrids.

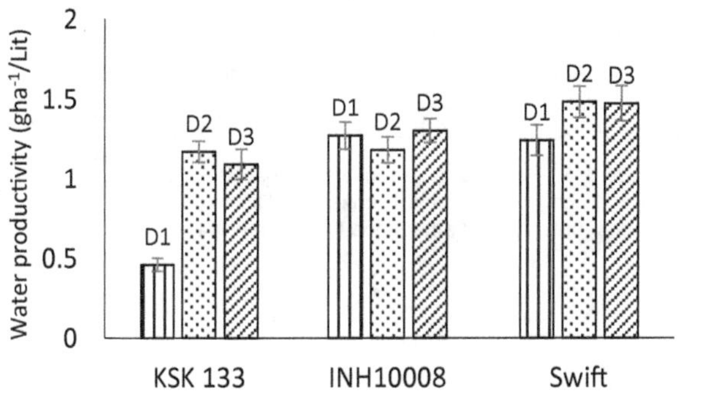

Figure 2: Integrative outcome of transplanting dates and irrigation scheduling on water productivity.

Apparent crop water productivity under different date of transplanting and genotypes are given in Table 3. The values ranged from 0.00046 to 0.00148 kg-ha/ Lit, which are close to the measured data in an independent experiment conducted by Singh et al. [16]. Keeping in view the results elaborated in Table 3, it becomes obvious that first transplanting date (D1) gave least yield due to the fact that both the hybrids and the check variety had to suffer water stress for a longer duration as compared to late of transplanting. For early transplanting, hybrid INH10008 gave highest yield (3458 kg/ha), followed by Swift (3178 kg/ha) and KSK 133 (982 kg/ha) as given in Table 3. Similarly, in second and third transplanting dates, Swift hybrid was the best yielder with average yield of 3803 and 3153 kg/ha respectively. Likewise, Swift showed maximum/significantly highest water productivity values in second and third dates of transplanting, while KSK 133 showed the lowest. However, at D1, INH10008 showed highest that was at par with Swift. Considering these results, it can be emphasized that Swift hybrid may be considered as best among studied hybrids under water stress conditions.

Comparing the average yields of all the three genotypes at three transplanting dates, D1 was the lowest yielding as compared to other two dates. Likewise, D2 gave highest yield while the yield was again reduced when further delayed to D3. However, the date of transplanting showed a significant interaction with the genotypes. Average paddy yield in the D3 treatment declined significantly compared to D2. It might be due to the shorter period of growth till flowering than required for optimum vegetative growth that in turn contributes to final yield (Figure 2).

The amount of total irrigation water applied (Table 3) in D3 was 2.25 million litres, that is 0.516 and 0.201 million litres less than D1 and D2 treatments, respectively. Comparatively in short duration variety irrigation water applied was much less than that in long duration varieties. Therefore, shorter duration varieties and hybrids should be evolved for saving more water required.

Considering the obtained results, it can be concluded that the water productivity were increased with the shifting of transplanting date towards lower evaporative demand period and variety to short duration. Swift hybrids showed maximum value for water productivity (0.00148 and 0.00147 kg-ha/L at D2 and D3 respectively) as compared to other hybrid and KSK 133 (check). At D1, both hybrids were at par for their water productivity.

Conclusion

Considering the obtained results, it may be concluded that the water productivity were increased with the shifting of transplanting date towards shorter water demand period and variety to short duration. Keeping in view the results, hybrid Swift may be considered best in terms of its yield performance under water stress condition as well as its high value of water productivity. Water stress is more damaging to those varieties or hybrids that have longer life cycle as compared to shorter duration hybrids and varieties. KSK 133 was least performing under water stress due to its longer life cycle as compared to shorter life durations of other hybrid varieties. Swift showed highest water productivity and paddy yield due to its shorter life cycle as compared to INH10008 hybrid.

Acknowledgement

This research project was funded by Bayer crop science (pvt.) Ltd.

References

1. Rice: Agriculture, Pakistan Economic Survey (2014) Pakistan Bureau of Statistics. Ministry of Finance, Pakistan, p: 28.

2. Gleik PH (1993) Water Crisis: A guide to world's fresh water resources, Pacific Institute for Studies in Development, Environment and Security. Oxford University Press, New York, pp: 1-34.

3. Rijsberman, Frank R (2006) Water scarcity: Fact or Fiction? Agricultural Water Management 80: 5-22.

4. Chauhan BS, Mahajan G, Sardana V, Timsina J, Jat ML (2012) Productivity and sustainability of rice–wheat cropping system in the Indo-Gangetic Plains of Indian sub-continent: Problems, opportunities, and strategies. Advances in Agronomy 117: 315-369.

5. Seckler D, Molden D, Barker R (1998) Water scarcity in the twenty-first century. Sri Lanka: International Water Management Institute, pp: 105-107.

6. Tuong TP, Bouman BAM (2003) Rice Production in Water-scarce Environments. In: Kijne JW, Barker R, Molden D (eds.), Water Productivity in Agriculture: Limits and Opportunities for Improvement. CAB International, pp: 53-67.

7. Harrington LW, Fujisaka S, Morris ML, Hobbs PR, Sharma HC et al. (1993) Wheat and rice in Karnal and Kurukshetra Districts, Haryana, India: Farmers practices, problems, and an agenda for action P: 44.

8. Sharma HC, Dhiman SD, Singh VP (1994) Rice-wheat cropping system in Haryana: Potential, possibilities and limitations. In: Proceedings of symposium on sustainability of rice-wheat system in India. CCS Haryana Agriculture University, India, pp: 27-39.

9. Sondhi SK, Kaushal MP, Singh P (1994) Irrigation management strategies for rice–wheat cropping system. In: Dhiman SD (ed.), Proc of the Symposium on Sustainability of Rice–Wheat Systems in India. CCS Haryana Agricultural Univ, India, pp: 95-104.

10. Hira GS, Jalota SK, Arora VK (2004) Efficient Management of Water Resources for Sustainable Cropping in Punjab. Punjab Agricultural University, India, p: 20.

11. Humphreys E, Meisner C, Gupta R, Timsina J, Beecher HG et al. (2005) Water saving in rice wheat systems. Plant Prod Sci 8: 242-258.

12. Bouman BAM, Tuong TP (2001) Field water management to save water and increase its productivity in irrigated lowland rice. Agric Water Manage 49: 11-30.

13. Bennett J (2003) Opportunity for increasing water productivity of CGIAR crops through plant breeding and molecular approaches. In: Kijne JW, Barker R, Molden D (eds.), Water Productivity in Agriculture: Limits and Opportunities for Improvement, pp: 103-126.

14. Chahal GBS, Sood A, Jalota SK, Choudhury BU, Sharma PK (2007) Yield, evapotranspiration and water productivity of rice (*Oryza sativa* L.) wheat (*Triticum aestivum* L.) system in Punjab-India as influenced by transplanting date of rice and weather parameters. Agric Water Manage 88: 14-27.

15. Gomez AK, Gomez AA (1984) Statistical Procedures of Agricultural Research, New York P: 704.

16. Singh CB, Aujla TS, Sandhu BS, Khera KL (1996) Effect of transplanting date and irrigation regime on growth, yield and water use in rice (Oryza sativa) in northern India. Indian J Agric Sci 66: 137-141.

14

Evaluation of Drip Irrigation Emitters Distributing Primary and Secondary Wastewater Effluents

Mike Rowan[1]*, Karen M. Mancl[1] and Olli H. Tuovinen[2]

[1]*Department of Food, Agriculture and Biological Engineering, Ohio State University, 590 Woody Hayes Dr., Columbus, OH 43210, USA*
[2]*Department of Microbiology, Ohio State University, 484 West 12th Avenue, Columbus, OH 43210, USA*

Abstract

Drip irrigation is a reliable and efficient way to deliver water to the soil; however, drip emitter clogging is a major concern when irrigating treated wastewater. Four types of drip irrigation emitters from three manufacturers were analyzed over a one-year period to monitor the incidence of clogging and its effect on irrigation uniformity. A controlled laboratory experiment was conducted using two different types of pressure compensating emitters designed for reclaimed wastewater, one type of non-pressure compensating emitter designed for reclaimed wastewater, and one type of non-pressure compensating agricultural emitter designed for potable water applications. Emitters of each type distributed tap water, primary treated septic tank effluent, and secondary treated sand filter effluent. Emitter flow rates were measured each month to identify clogged or flow restricted emitters. Some clogging was seen in each type of emitter over the course of the experiment and emitter flow rates fluctuated over time, suggesting that clogging was gradual and often incomplete. Many of the emitters exhibited a cyclical flow rate indicating that clogging was reversible. The emitters distributing septic tank effluent exhibited the most significant reduction in flow. The most severely clogged emitter experienced a reduction of 63% after one year of irrigation with septic tank effluent. Secondary treatment using the sand filter showed the least clogging in all four types of emitters. One of the reclaimed wastewater emitter types experienced an average reduction in flow of 1% while the other two actually increased in flow by 1% and 4% after one year of irrigation with effluent from a sand filter. Water quality appeared to have a more pronounced effect than did emitter type. The effect of wastewater type on emitter discharge was +3.3% for tap water, -9.4% for septic tank effluent, and -0.3% for effluent from secondary sand filtration. While the agricultural drip emitters experienced a significant negative impact after one year of operation the three drip emitters designed for distributing septic tank effluent and reclaimed wastewater showed little clogging and a high degree of uniformity.

Keywords: Drip irrigation; Emitter clogging; Micro irrigation; On-site wastewater treatment; Trickle irrigation

Introduction

Drip irrigation has been utilized for wastewater distribution and reuse where soil conditions prohibit traditional types of wastewater dispersal such as leach fields and mound systems. Drip irrigation offers a solution where other soil treatment systems are inappropriate due to a seasonally high water table, shallow dense soil layer, vegetative cover, space constraints, or other site limitations. The goals of drip dispersal are to attain unsaturated flow, encourage lateral movement through capillary action rather than gravitational flow, and distribute the effluent over the entire application area to promote physical, chemical and biological processes of the soil. Many advantages of distributing treated wastewater with drip irrigation have been established including water conservation, nutrient uptake, ground water protection, pathogen reduction and public safety [1-3]. Clogging of emitters is a major concern in drip irrigation systems because of high levels of suspended solids, organic matter, and nutrients (N and P especially) in treated wastewater effluents. Previous studies have sought to determine the causes and prevention of emitter clogging [4-10]. Causes of clogging can be divided into three main categories: (1) physical, caused by suspended solids; (2) chemical, caused by precipitation reactions; and (3) biological, caused by growth and metabolism of microorganisms; i.e., biofilm formation [11]. Emitter clogging is usually the result of two or more of these processes working in concert [12]. In response to previous research findings, the manufacturers of drip systems have made numerous modifications to emitter design and other system components to prevent emitter clogging. A filtration system to remove suspended solids from the effluent prior to dispersal is now a required component of all drip systems. These are in line systems with 120, 150 or 200-mesh screen. The frequent automatic flushing of these filters is essential for proper operation. Filtration alone will not, however,

adequately prevent clogging of the emitters. Scanning electron micrographs indicate that particles small enough to pass through a 120-micron filter were trapped by biofilm growing in the emitter flow pathway [13]. They found accumulation of the small particles lead to the formation of agglomerates of cells and solid particles and the eventual clogging of the emitters. Drip emitters are now designed to produce turbulent water flow inside the chamber. This reduces particle settling and discourages biofilm attachment to the interior walls. A flushing velocity of 0.30-0.61 m sec-1 (1-2 ft sec-1) must be maintained or periodically introduced in the piping network to flush out settled particles and remove slimes and biofilms [14].

In spite of these improved strategies, emitter clogging continues to be a concern with wastewater irrigation. The hydraulic loading regimen of the drip irrigation system is designed based on the assumption that each emitter is distributing the same amount of water per dosing cycle. Emitter clogging is the main cause of discharge variation within the irrigation system [15]. The coefficient of uniformity (CU) shown in equation 1, can be used to describe the spatial uniformity of the drip irrigation system [16,17].

*Corresponding author: Mike Rowan, Department of Food, Agriculture and Biological Engineering, Ohio State University, 590 Woody Hayes Dr., Columbus, OH 43210, USA, E-mail: rowan.7@osu.edu

$$CU = \left(1 - \frac{\sum_{i=1}^{n} |q_i - q_{ave}|}{\sum_{i=1}^{n} q_i}\right) 100 \qquad (1)$$

where q_i = individual emitter flow rate, q_{ave} = mean emitter flow rate, and $|q_i-q_{ave}|$ = absolute deviation from the mean. The CU classification is ranked as >89% excellent, 80-89% good, 70-79% fair, and <70% poor. Even a few clogged emitters can greatly reduce the uniformity of water application [18,19] and a drip irrigation system with non-uniform water distribution may fail to operate as designed. Water quality has been identified as the main factor associated with emitter clogging. Capra and Scicolone [20] found a strong correlation between emitter discharge and pH, total suspended solids (TSS), and Biochemical Oxygen Demand (BOD5). Increasing treatment levels to reduce suspended solids and organic matter can improve emitter function. Sand filters can be effective at reducing BOD5 and total suspended solids [21], transforming ammonia to nitrate, removing nitrogen via denitrification [22], and neutralizing pH [23]. Sand filters have been used to treat domestic sewage, food processing waste and industrial waste [24]. Chlorination has also been shown to reduce emitter clogging and is recommended to help prevent development of slimes and biofilms [25].

The purpose of this research was to determine the relationship between wastewater quality and the performance of drip irrigation emitters. Four types of emitters were evaluated. Three types of emitters were designed for use with treated wastewater and the fourth was designed for agricultural irrigation with potable water. The emitters were tested by tracking the flow rates as they discharged tap water, septic tank effluent, and effluent from a sand filter over a one-year period.

Experimental

This study used a laboratory scale drip irrigation setup with 12 lines of drip tubing, each 3.7 m in length. Four different types of wastewater drip tubing were selected for analysis and two sources of wastewater, (i) primary septic tank effluent, and (ii) secondary wastewater effluent from sand filtration, were discharged through the drip emitters. A third source, tap water, was used as control. Three laboratory scale septic tanks were established using a protocol described by Peeples and Mancl [26]. The effluent characteristics, analyzed at the start of the experiment, are given in Table 1.

Two septic tanks were loaded at 24 hr cycles with approximately 57 L tap water, 500 mL primary sludge and 250 mL of 0.363 M ammonium chloride. The primary sludge was acquired from the Southerly Wastewater Treatment Plant (Columbus, Ohio). The third tank matched all aspects of the protocol except that the primary sludge and NH4Cl were not added. This tank served as the control. BOD5 and TSS were analyzed using Standard Methods 5210 B and 2540 D [27]. Ammonia was measured using Quickchem Method number 12-107-06-2-A.

At 24-hour intervals, the effluent from one of the septic tanks was discharged into a sand filter for secondary treatment. The sand filter consisted of a cylindrical polyethylene container, 0.79 m high and 0.55 m in diameter filled with 76 cm of sand. The sand had an effective size of 0.5 mm with a uniformity coefficient of 4. After passing through the sand filter, the effluent was transferred to a dosing tank that was identical to the septic tank in size and design. The effluents from the

second septic tank and the tap water control tank were transferred to their respective dosing tanks directly, with no further treatment. Each dosing tank contained a ¼ h.p. (0.18 kW) submersible pump that delivered 0.06 L s-1 of effluent through 0.03 m clear flexible tubing to a Netafim Low Volume Control Zone (LVCZ) unit. The LVCZ combines a 24 V solenoid controlled valve, a 100-micron disc filter and a 138 kPa (20 psi) pressure regulator into a single unit. Each LVCZ unit was connected to a 0.10 m PVC manifold with four 0.01 m male adapters to direct the wastewater into four separate drip lines, one drip line for each of the four different types of emitters (Figure 1). The pressure at the PVC manifold was 90 kPa (13 psi). Drip irrigation tubing, 3.7 m in length, was connected over the 0.01 m adapters to another identical manifold where the remaining effluent was collected and returned to the dosing tanks. The velocity through the drip line tubing was 0.46 m s-1 (1.5 ft s-1). The irrigation water temperature was maintained at ambient room temperature, fluctuating between 23 and 25°C.

Four different types of drip irrigation tubing and emitters were examined.

Type 1: Netafim Bioline; a pressure compensating drip line for wastewater: A pressure compensating diaphragm emitter spaced 0.61 m (24 in) on center and impregnated with the biocide Vinyzene. The nominal flow rate is 3.4 l/h at 7-60 psi.

Type 2: Geoflow PC Wasteflow; a pressure compensating drip line for wastewater: A pressure compensating turbulent flow emitter spaced 0.61 m (24 in) on center with the root intrusion preventing ROOTGUARD (Treflan), and the bactericide Ultra Fresh DM-50 in the dripper line. The nominal flow rate is 2.0 l/h at 7-60 psi.

Type 3: Geoflow NPC Wasteflow; a non-pressure compensating drip line for wastewater: A non-pressure compensating turbulent flow emitter spaced 0.61 m (24 in) on center with the root intrusion preventing ROOTGUARD (Treflan), and the bactericide Ultra Fresh DM-50 in the dripper line. The nominal flow rate is 3.9 l/h at 20 psi with a CV of 0.05.

Type 4: Eurodrip GR: a non-pressure compensating drip line with emitter spaced 0.3 m (12 in) on center for traditional agricultural and horticultural irrigation applications. This drip line was not designed to distribute wastewater. The nominal flow rate is 4.0 l/h at 15 psi with a CV of 0.03.

A Netafim Miracle AC 12 Irrigation Controller maintained the irrigation schedule. This unit controlled the solenoid valves in each Netafim LVCZ unit and the dosing pumps. The controller was programmed to sequentially irrigate all emitter types with each effluent for 10 min, four times each day, seven days a week.

To evaluate emitter performance, an initial flow rate was measured for each emitter in the system prior to beginning the experiment. The flow rate was determined by measuring the volume of effluent over a fixed unit of time. After the irrigation dose was initiated, plastic cups were placed directly under each emitter and effluent was collected for 120 s. This procedure was repeated each month during the one-year experiment. The emitter flow rate data were analyzed using the Microsoft Excel analysis of variance (ANOVA) statistical program. The

Wastewater Source	Characteristics			
	BOD$_5$ (mg/L)	TSS (mg/L)	Ammonia (mg/L)	pH
Septic Tank Effluent	147	55	18	6.5
Sand Filter Effluent	0.5	3	1	7.2

Table 1: Wastewater Effluent Characteristics.

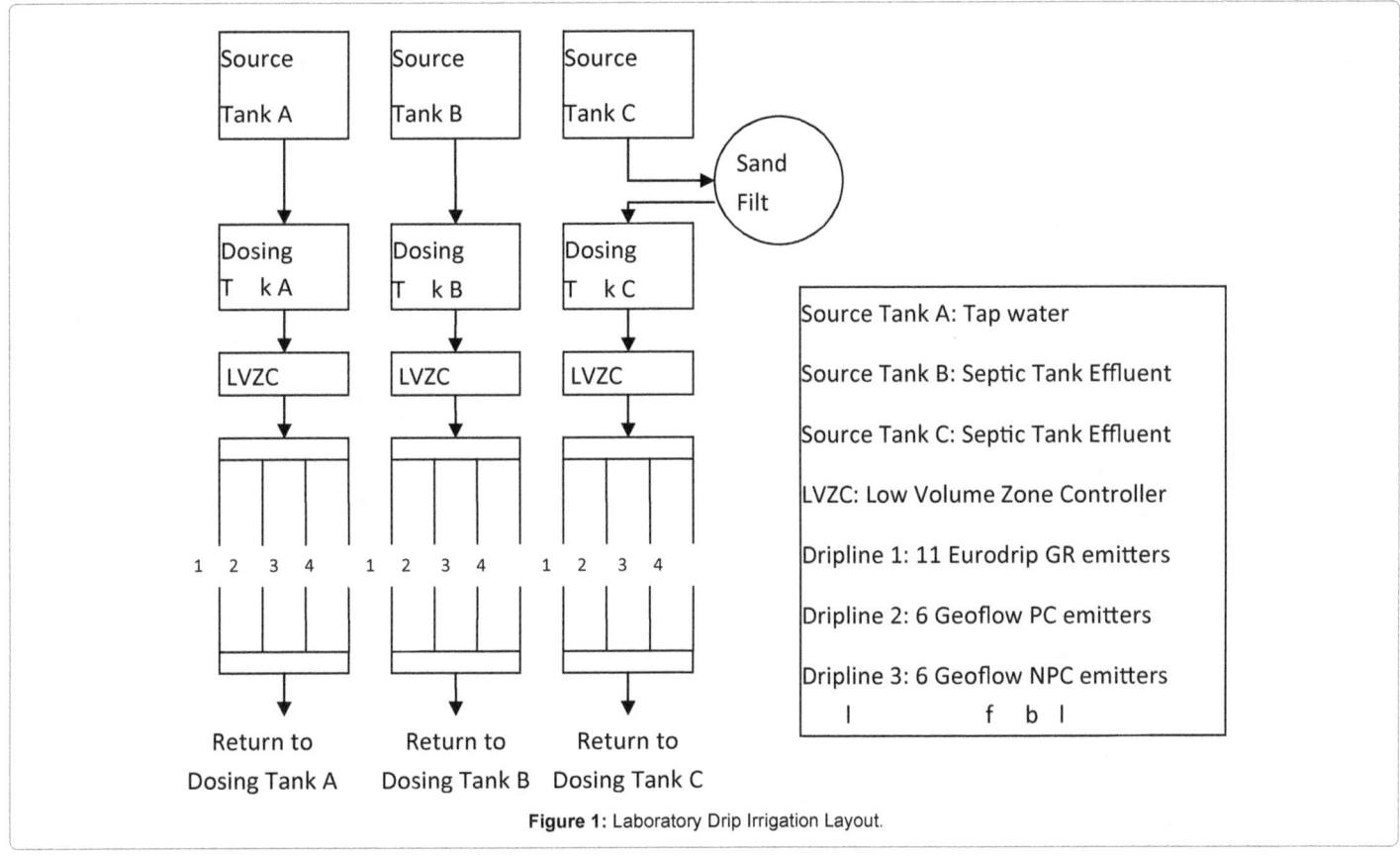

Figure 1: Laboratory Drip Irrigation Layout.

variables were flow rate change over time versus influent wastewater quality.

Results and Discussion

Eurodrip gr emitters

The initial and final flow rates and relative changes over a one-year period for each emitter type are listed in Tables 2-6. The emitter position indicates the sequence of emitters in the drip tubing. Emitter position was not considered a variable in this study. Water pressure and velocity were considered constants because of the short length of each lateral. After one year of continuous irrigation the control emitters distributing tap water showed no signs of clogging. The flow rates of emitters distributing septic tank effluent were reduced by 16% overall, with the most severe reduction of 63% for emitter 7. The flow rates of the emitters distributing sand filter effluent were reduced by an average of 3%. The standard deviation of the overall reduction in flow was greatest for the Eurodrip GR emitters distributing septic tank effluent.

Geoflow pc emitters

The Geoflow PC emitters distributing tap water showed no overall signs of clogging. However, emitter #2 experienced a flow reduction of 7%. The emitters distributing sand filter effluent showed no significant clogging with the exception of emitter 2. All of the emitters distributing septic tank effluent showed reduced flow rates with an average reduction of 12%. The standard deviation of the emitters distributing septic tank effluent was also the highest of this group.

Geoflow npc emitters

The Geoflow NPC emitters distributing tap water and sand filter effluent showed no signs of clogging overall. The flow rates of emitters distributing septic tank effluent were reduced by an average of 6%. The standard deviation of average reduction in flow for emitters distributing septic tank effluent was nearly double the other two emitter types in this group.

Netafim bioline emitters

The Netafim emitters showed little sign of clogging after one year of continuous operation with the exception of emitter 5, distributing septic tank effluent. The average flow rate change for Netafim emitters was positive regardless of wastewater quality. As with the other three groups, the standard deviation of the average flow reduction was highest for those emitters distributing septic tank effluent.

Recovery of clogged emitters

A gradual reduction in flow rate followed by a partial recovery was observed in emitters of each type (Table 7). The maximum flow reduction occurred after eight to ten months for the Eurodrip GR emitters 4, 5 and 6 distributing septic tank effluent. For the emitters Geoflow PC 1 and 3, Geoflow NPC 1 and 3, and Netafim 4, all distributing septic tank effluent, the maximum flow reduction occurred between six and ten months. A partial or complete recovery of emitter flow rate was transiently observed in most cases. The recovery of partially clogged emitters has been observed in other laboratory studies [28] as well as in field applications using lagoon wastewater [3]. It is plausible that biofilm growth inside the emitter constricted the passageway, trapped debris that would otherwise pass through the emitter, and resulted in the observed flow reduction. The start of each irrigation cycle experienced a surge velocity and pressure as the irrigation pumps were turned on, which may have caused biomass and trapped debris to be flushed out, thus resulting in the partial recovery of emitter flow rates.

Emitter Position	Tap Water Initial (L/h)	Final (L/h)	% Change	Septic Tank Effluent Initial (L/h)	Final (L/h)	% Change	Sand Filter Effluent Initial (L/h)	Final (L/h)	% Change
1	2.52	2.64	5	2.54	2.7	2	2.7	2.61	-3
2	2.64	2.85	8	2.52	2.7	7	2.7	2.79	3
3	2.52	2.64	5	2.7	2.79	3	2.7	2.34	-13
4	2.64	2.76	5	2.55	1.95	-24	2.64	2.61	-1
5	2.46	2.61	6	2.37	2.25	-5	2.79	2.85	2
6	2.61	2.79	7	2.46	1.68	-32	2.79	2.73	-2
7	2.55	2.7	6	2.58	0.96	-63	2.91	2.67	-8
8	2.64	2.79	6	2.55	1.56	-39	2.82	3	6
9	2.58	2.7	5	2.46	1.86	-24	2.82	2.43	-14
10	2.67	2.7	1	2.46	2.49	1	2.88	2.85	-1
11	2.67	2.79	4	2.61	2.4	-8	2.85	2.76	-3
Average 2.59 2.72			5	2.53	2.12	-16	2.78	2.69	-3
Std. Deviation			1.7			20.9			6

Table 2: Initial and Final Flow Rates of Eurodrip GR Emitters for One Year. Three sets of 11 emitters with one set receiving tap water, one set receiving septic tank effluent, and one set receiving effluent from a sand filter.

Emitter Position	Tap Water Initial (L/h)	Final (L/h)	% Change	Septic Tank Effluent Initial (L/h)	Final (L/h)	% Change	Sand Filter Effluent Initial	Final	% Change (L/h) (L/h)
1	1.32	1.32	0	1.2	0.99	-18	1.35	1.35	0
2	1.35	1.26	-7	1.32	1.26	-5	1.32	1.14	-14
3	1.29	1.32	2	1.23	1.11	-10	1.41	1.41	0
4	1.29	1.32	2	1.26	1.23	-2	1.35	1.41	4
5	1.26	1.35	7	1.23	0.99	-20	1.29	1.23	-5
6	1.23	1.32	7	1.38	1.14	-17	1.35	1.41	4
Average 1.29 1.32			2	1.27	1.12	-12	1.35	1.33	-1
Std. Deviation			4.7			6.8			6.2

Table 3: Initial and Final Flow Rates of Geoflow PC Emitters for One Year. Three sets of 6 emitters with one set receiving tap water, one set receiving septic tank effluent, and one set receiving effluent from a sand filter.

Emitter Position	Tap Water Initial (L/h)	Final (L/hr)	% Change	Septic Tank Effluent Initial (L/h)	Final (L/h)	% Change	Sand Filter Effluent Initial	Final	% Change (L/h) (L/h)
1	2.04	2.01	-1	1.95	1.77	-9	1.89	1.95	3
2	1.89	1.92	2	1.92	1.68	-13	1.92	2.04	6
3	1.86	1.89	2	1.83	1.71	-7	2.01	2.13	6
4	1.86	1.92	3	1.92	1.89	-2	2.01	2.13	6
5	1.98	2.1	6	1.92	1.77	-8	2.01	2.01	0
6	1.98	2.07	5	1.92	1.92	0	2.04	2.13	4
Average 1.94 1.99			3	1.91	1.79	-6	1.98	2.07	4
Std. Deviation			2.3			4.3			2.2

Table 4: Initial and Final Flow Rates (ml/min) of Geoflow NPC Emitters for One Year. Three sets of 6 emitters with one set receiving tap water, one set receiving septic tank effluent, and one set receiving effluent from a sand filter.

Emitter Position	Tap Water Initial (L/h)	Final (L/h)	% Change	Septic Tank Effluent Initial (L/h)	Final (L/h)	% Change	Sand Filter Effluent Initial	Final	% Change (L/h) (L/h)
1	3.18	3.3	4	3.09	3.3	7	3.39	3.36	-1
2	3.21	3.18	-1	3.12	3.21	3	3.21	3.24	1
3	3.18	3.24	2	3.06	3.21	5	3.33	3.45	4
4	3.15	3.15	0	3.09	3.12	1	3.3	3.33	1
5	3.24	3.24	0	3.21	3.06	-5	3.33	3.45	4
6	3.21	3.36	5	3.15	3.3	5	3.3	3.3	0
Average 3.20 3.25			2	3.12	3.2	3	3.31	3.36	1
Std. Deviation			2.2			3.9			1.9

Table 5: Initial and Final Flow Rates (ml/min) of Netafim Emitters for One Year. Three sets of 6 emitters with one set receiving tap water, one set receiving septic tank effluent, and one set receiving effluent from a sand filter.

Emitter Type and Position	Flow Rates (ml/min)						
	Month 0	Month 2	Month 4	Month 6	Month 8	Month 10	Month 12
Eurodrip GR #4	43	47	42	26	21	8	16
Eurodrip GR #5	40	44	37	26	24	7	26
Eurodrip GR #6	41	43	36	19	15	28	31
Geoflow PC #1	20	21	20	19	18	7	16
Geoflow PC #3	21	23	19	10	8	19	18
Geoflow NPC #1	33	34	30	29	28	6	30
Geoflow NPC #3	31	32	29	19	26	13	29
Netafim #4	52	55	51	46	50	52	52

Table 6: Changes in Flow Rates of Septic Tank Effluents over 12 Months.

Emitter Type	Mean Flow Rate	Final CU (%)	Rating
	% Change		
Eurodrip GR	-9.5	86.9	Good
Geoflow PC	-7.1	90.5	Excellent
Geoflow NPC	-1	90	Excellent
Netafim	2	96.13	Excellent

Table 7: Flow Rate Changes and Coefficient of Uniformity Values for Each Emitter Type.

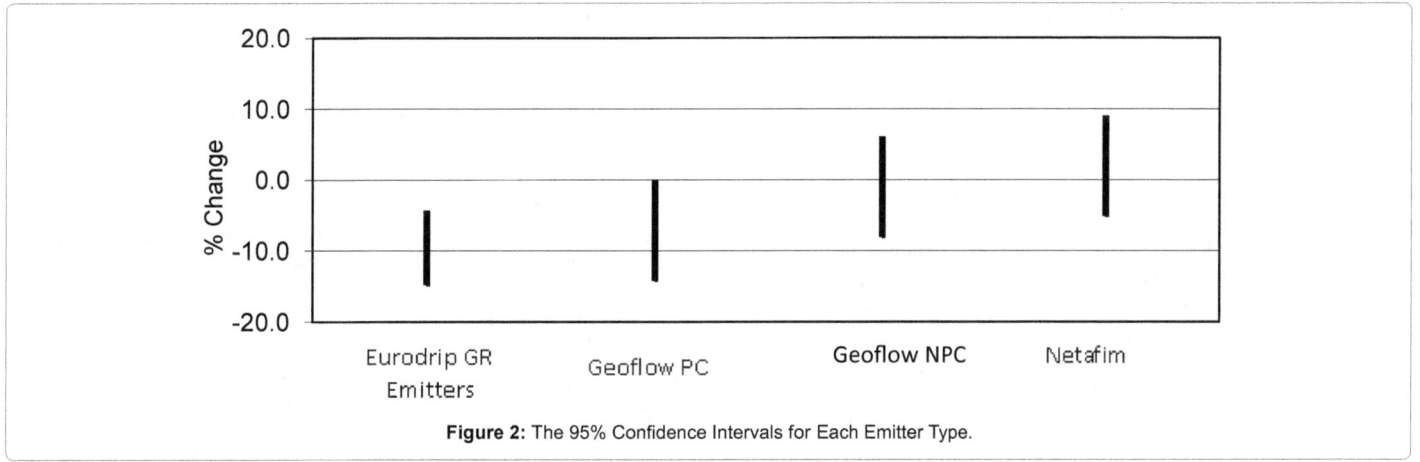

Figure 2: The 95% Confidence Intervals for Each Emitter Type.

Overall emitter performance evaluation

The initial and final flow rate changes of each emitter type for all three effluent types are listed in Table 7. The Eurodrip GR emitters show the greatest reduction in flow while the three wastewater emitters flow rates change ranged from -7% to + 2% of their initial values.

The CU values calculated for the four emitter types, after one year of service, are shown in Table 7. The CU for the agricultural emitters was good (80-89%), while the CU for the wastewater emitters was excellent (>89). The extent of clogging in certain emitters was greater than in others; however, each type of wastewater emitter maintained a high level of uniformity overall. This indicates that clogging in the Geoflow PC, Geoflow NPC and Netafim emitters did not negatively impact the overall performance of the irrigation system.

Analysis of variance for emitter type

The results of the ANOVA analysis for emitter type based on initial and final flow rates indicated that emitter type is a significant variable in determining emitter clogging at the 95% confidence level. Using the initial and final flow rate data, the single factor ANOVA analysis returned a P-value of 0.037, an F critical value of 2.77 and an F calculated value of 3.01. The 95% confidence intervals for each emitter type show that the mean flow was more reduced in the Eurodrip GR as compared to Geoflow PC, Geoflow NPC, and Netafim emitters (Figure 2).

Analysis of variance for wastewater type

The influence of wastewater type on emitter discharge was +3.3% for tap water, -9.4% for primary septic tank effluent, and -0.3% for secondary sand filter effluent. The results of the ANOVA analysis for effluent type based on initial and final flow rates indicated that effluent type was also a significant variable in determining emitter clogging at the 95% confidence level. Using the initial and final flow rate data the single factor ANOVA analysis returned a P-value of 0.000129, an F critical value of 3.1 and an F calculated value of 12.9. The septic tank effluent was detectably different from the tap water and sand filter effluent. The tap water and sand filter effluent were not detectably different from each other. The 95% confidence intervals for each effluent type show the clear distinction between the mean of the septic tank effluent and the mean of the other two effluent types (Figure 3).

Conclusions

The purpose of this study was to analyze four different drip emitters with respect to their resistance to clogging and to determine the relationship between effluent treatment levels and emitter performance. This experiment utilized a bench scale irrigation system to simulate a full-scale field application. The drip lines, emitters and filters used in this study were identical to those of a full sized system; however, due to space limitations some design modifications and adjustments were necessary. After one year of dispersal, individual emitters experienced

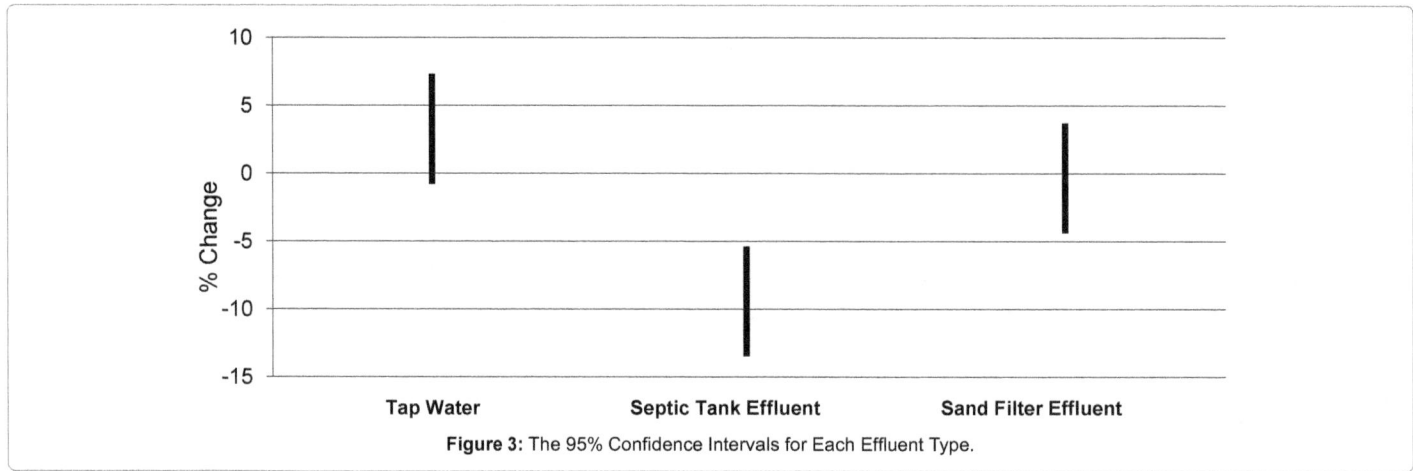

Figure 3: The 95% Confidence Intervals for Each Effluent Type.

discharge variations from slight increases to severe reductions of up to 63%. The agricultural emitters showed the greatest extent of clogging. The emitters specifically designed for discharging wastewater showed less clogging and maintained the highest CU rating after one year of continuous operation. Secondary treated wastewater from a sand filter provided the best performance for all emitter types. Primary septic tank effluent was also successfully distributed with drip irrigation emitters that were designed for use with treated wastewater. The emitters designed for traditional agricultural/horticultural irrigation applications, performed satisfactorily with sand filter effluent but experienced a high degree of clogging with septic tank effluent. The analysis of variance between emitter type and change in emitter flow rate (Figure 2) showed that emitter type is a significant variable in emitter clogging incidence, but there was no detectable difference between the three wastewater emitters at a 95% confidence level. The analysis of variance between wastewater type and emitter flow rate (Figure 3) showed that wastewater type (i.e., wastewater quality) is a more significant variable than emitter type. The extent of emitter flow reduction was highest for the primary septic tank effluent while the secondary treated effluent was not detectably different from tap water at the 95% confidence level. This study suggest that using drip irrigation emitters that are specifically designed for distribution of wastewater and treating the wastewater using a sand filter will reduce the extent of emitter clogging and improve overall distribution uniformity.

The clogging process of the drip emitter was not linear over time or unidirectional. Emitter flow rates fluctuate with time and emitters have the ability to self clean during normal startup and dose cycles. This self-correcting ability was most evident with the three wastewater drip emitters distributing the higher quality wastewater effluents. The traditional agricultural drip emitters distributing septic tank effluent showed the least ability to self clean after clogging had begun.

Acknowledgements

Support for this project was provided by The Ohio State University Extension, Ohio Agricultural Research and Development Center Site. The project was financed in part through a grant from the Ohio Environmental Protection Agency and the United States Environmental Protection Agency, under the provisions of Section 319(h) of the Clean Water Act.

References

1. Gushiken EC (1995) Irrigation with reclaimed water through permanent subsurface drip irrigation systems. Proceedings of 5th International Micro irrigation Congress, Hawaii

2. Trooien TP, Lamm FR, Stone LR, Alam M, Clark GA, et al. (2000) Subsurface drip irrigation using livestock wastewater: Dripline flow rates. Appl Eng Agr 16: 505-508.

3. Lamm FR, CR Camp (2006) Subsurface drip irrigation. Dev Agric Eng 13: 473-539.

4. Oron G, Shelef G, Turzynshki B (1979) Trickle irrigation using treated wastewaters. J Irrig Drain Div 105:175-186.

5. Adin A (1987) Clogging in irrigation systems reusing pond effluents and its prevention. Wat Sci Technol 19: 323-328.

6. Adin A, Sacks M (1991) Dripper clogging factors in wastewater irrigation. J Irrig Drain Eng 117: 813-826.

7. Ravina EP, Sofer Z, Marcu A, Shisha A, Sagi G (1992) Control of emitter clogging in drip irrigation with reclaimed wastewater. Irrig Sci 13: 129-139.

8. Ravina EP, Sofer Z, Marcu A, Shisha A, Sagi G, et al. (1997) Control of clogging in drip irrigation with stored treated municipal sewage effluent. Agr Wat Manag 33: 127-137.

9. Tajrishy M, Hills D, Tchobanoglous G (1994) Pretreatment of secondary effluent for drip irrigation. J Irrig Drain 120: 716-731.

10. Taylor HD, Bastos RKX, Pearson HW, Mara DD (1995) Drip irrigation with waste stabilization pond effluents: Solving the problem of emitter fouling. Water Sci Technol 31: 417-424.

11. Bucks DA, Nakayama FS, Gilbert RG (1979) Trickle irrigation water quality and preventive maintenance. Agr Wat Manag 2: 149-162.

12. Nakayama F, Gilbert R, Bucks D (1978) Water treatments in trickle irrigation systems. J Irrig Drain Div 104: 23-34.

13. Yan D, Yang P, Rowan M, Ren S, Pitts D (2010) Biofilm accumulation and structure in the flow path of drip emitters using reclaimed wastewater. Trans ASABE 53:751-758.

14. Smajstrla AG, Boman BJ (2000) Flushing Procedures for Microirrigation Systems. Florida Cooperative Extension Service, Bulletin 333, University of Florida, USA.

15. Ghaemi A, Chieng ST (1997) Impacts of emitter clogging on hydraulic characteristics in microirrigation: field evaluation. CSCE and CSAE Joint Meeting A: 162-171.

16. Nakayama FS (1986) 3.2-Water treatment. In Trickle irrigation for Crop Production: Design, Operation and Management 9:164-187.

17. Camp CR, Sadler EJ, Busscher WJ (1997) A comparison of uniformity measures for drip irrigation systems. Trans ASAE 40:1013-1020.

18. Bralts VF, Wu IP, Gitlin HM (1981) Manufacturing variations and drip uniformity. Trans ASABE 24:113-119.

19. Nakayama FS, Bucks DA (1981) Emitter clogging effects on trickle irrigation uniformity. Trans ASABE 24: 77-80.

20. Capra A, Scicolone B (2004) Emitter and filter tests for wastewater reuse by drip irrigation. Agr Wat Manag 68: 135-149.

21. Wigrig D, Peeples J, Mancl K (1996) Intermittent sand filtration for domestic wastewater treatment: Effects of filter depth and hydraulic parameters. Appl Eng Agr 12: 451-459.

22. Gold AJ, Lamb BE, Loomis GW, Boyd BJ, Cabelli VJ (1992) Wastewater renovation in buried and recirculating sand filters. J Environ Qual 21: 72-725.

23. Xi J, Mancl KM, Tuovinen OH (2005) Carbon transformation during sand filtration of cheese processing wastewater. Appl Eng Agr 2: 271-276.

24. Kang YW, Mancl KM, Tuovinen OH (2007) Treatment of turkey processing wastewater with sand filtration. Bioresour Technol 98: 1460-1466.

25. Cararo DC, Botrel TA, Hills DJ, Leverenz HL (2006) Analysis of clogging in drip emitters during wastewater irrigation. Appl Eng Agr 22: 251-257.

26. 26. Peeples J, Mancl K (1998) Laboratory scale septic tanks. Ohio J Sci 48: 75-79.

27. American Public Health Association (1998) Standard Methods for the Examination of Water and Wastewater, 20th ed, Washington DC, USA.

28. Yan DZ, Bai ZH, Rowan M, Gu LK, Ren SM, Yang PL (2009) Biofilm structure and its influence on clogging in drip irrigation emitters distributing reclaimed wastewater. J Environ Sci 21: 834-841.

15

Optimizing Cropping Pattern Using Chance Constraint Linear Programming for Koga Irrigation Dam, Ethiopia

Kassahun Tadesse Birhanu*, Tena Alamirew, Megerssa Dinka Olumana, Semu Ayalew and Dagnachew Aklog

Natural Resources Management, Debre Markos University, Ethiopia

Abstract

Optimal cropping pattern decisions without consideration to water supply uncertainty would result in yield/benefit that is less than expected and probability of system failure in meeting a given irrigation demand. In this study, a chance constraint linear programming (CCLP) model was used for optimizing cropping pattern for major crops grown at Koga Irrigation scheme, Ethiopia. The model incorporated uncertainty of inflow at exceedance probability of 90%, 80%, 70%, 60% and 50%. The model objectives were yield and benefit maximizations subject to land and water availability constraints. Each objective function has four scenarios. The models were solved using LINGO14. The cropping patterns under yield and benefit maximization models were found to be identical under all scenarios. However, the cropping patterns of each model varied among scenarios. The study showed that the possibility of irrigating 5904.3 to 8051.0 hectares of land at 80% by optimizing cropping patterns at irrigation efficiency of 48%. This could increase the yield by 108 to 153%, benefit by 153 to 208% and physical water productivity by 132% to 186% and economic water productivity by 205% to 241% of the actual values. In conclusion, the irrigated land in 2012/13 was below the optimal value and the irrigation water was mismanaged. Therefore, with optimal crop planning and water management, the design command area of 7000 ha could be irrigated. Finally, a study should be made to determine optimal levels of crop water deficit that maximize water productivity.

Keywords: Cropping pattern; Exceedance probability; LINGO; Optimization

Introduction

The world's readily available fresh water resources are becoming increasingly scarce due to higher demands by municipal, industrial, recreational, and agricultural sectors. This is mostly because of population increase and higher standards of living in many areas, but also due to changes in land use and global climate change as a result of rapid development [1]. Irrigation accounts for 70% of total freshwater withdrawals globally, with the industrial and domestic sectors accounting for the remaining 20% and 10%, respectively [2]. With expected increases in population by 2030, food demand is predicted to increase by 50% (70% by 2050) [3]. Without improved efficiencies, agricultural water consumption is expected to increase by about 20% globally by 2050 [4]. Irrigation accounts for more than 40% of the world's production on less than 20% of the cultivated land [4]. Globally, irrigated crop yields are 2.7 times those of rain fed farming; hence irrigation will continue to play an important role in food production [4]. However, the increasing competition for water usage in different sectors is making this resource more scarce and valuable. Hence, today, the agricultural sector around the world is under more pressure for limiting its water use, not only because of the increasing water demand, but also because of climatic changes and more frequent droughts [5]. Water scarcity and decreasing availability of water for agriculture constrain irrigated production overall, and particularly in the most hydrologically stressed areas and countries [2]. These call for increasing production per unit area and per unit water. However this needs more scientific utilization of the water resources and their optimal allocation to achieve maximum returns. Improving the irrigation efficiency, revising crop patterns, and optimizing water allocation [5]. Optimal irrigation system planning and operation are amongst many measures. The success of irrigation system operation and planning depends on the quantification of supply and demand and equitable distribution of supply to meet the demand if possible, or, to minimize the gap between the supply and demand [6]. For this purpose, optimization models are required to select optimal solutions, systematically, under agreed-upon objectives and constraints [7].

Ethiopia has vast cultivable land of 30 to 70 M ha, but only 15 M ha of land is under cultivation, with current irrigation schemes covering about 640,000 ha out 5.3 M ha of total potential irrigable land [8]. In Ethiopia, due to lack of water storage infrastructure and large spatial and temporal variations in rainfall, there is not enough water for most farmers to produce more than one crop per year [8]. Hence, Ethiopia is increasingly investing in irrigation sector in order to exploit the agricultural production potential of the country: (i) to achieve food self sufficiency at the national level, (ii) to generate foreign currency from export earnings, and (iii) satisfy the raw material demand of local industries [9]. One of the investments is Koga Dam and Irrigation Scheme located south of Lake Tana, in the Upper Blue Nile Basin. The scheme was designed to irrigate 7000 ha command areas [10]. According to MacDonald [10], the potential irrigable area is 7572 ha. However, the maximum actual irrigated area was 5123 ha in 2011/12 and 5144.36 ha in 2012/13. This is 73.5% of the design command areas. This implies that either the reservoir water was mismanaged or too small to irrigate design command areas. Multiple cropping patterns are also the common practice at Koga and other irrigation schemes in Ethiopia. An economically efficient cropping pattern defines the optimal crop area and water allocation for seasonal, annual and perennial crops, subject to constraints on land and water availability [11]. The cropping pattern of Koga Irrigation Scheme varies year to year depending upon farmers' preferences, socioeconomic factors and government directives. This affects the amount and timing of irrigation water demand, the size of cropping area, the yield /benefit,

*Corresponding author: Kassahun Tadesse Birhanu, Professor, Natural Resources Management, Debre Markos University, Ethiopia
E-mail: kbirhan@gmail.com

and reservoir operation policy. Irrigation managers and/or decision makers always face a problem of optimally allocating available land water resources to multiple crops at the beginning of every irrigation season to maximize the yield or benefit from irrigation projects. Under these circumstances, optimization techniques are required to balance water supply and demands of multiple crops. Among the available optimization techniques such as Linear Programming (LP), Dynamic Programming (DP) and Genetic Algorithm (GA), it is LP model that is more popular because of the proportionate characteristic of the allocation problems [12]. Moreover, LP model can handle a large number of constraints and thus, are an effective tool to aid in the optimization process [13]. Linear programming based optimization methods are popularly used to derive the policies and are found as an effective tool in dealing with the allocation of resources during irrigation planning [14]. The LP is also easy to apply with the problem of irrigation planning using several available programs [15]. However, neither of optimization approaches was used to define optimal cropping patterns during designing nor operation phase of Koga irrigation project. Moreover, none of optimization methods to date has been widely adopted in irrigation sector, Ethiopia. Nevertheless, in the past years, various optimization methods have been used in irrigated agriculture world-wide to define optimal cropping patterns. However, no single optimization method or algorithm can be applied efficiently to all problems [16]. The Linear programming (LP) model developed to select the optimal cropping patterns in, Zimbabwe were found to be more superior to traditional methods in maximizing profit while achieving other goals such as food security [17]. LP models were also used to improve annual benefits in multi objective crop planning [18]. A linear programming model was used to maximize the total gross margin for the delivery of water to agricultural areas that cover an irrigation network over a planning horizon in Iran [19]. A weekly LP model was formulated for determining the optimal cropping pattern and reservoir water allocation for an existing storage based irrigation system in India [20]. LP model based on simplex method was used for optimal utilization of water and land resources [21]. A linear programming model was developed for the optimal land and water resources allocation to maximize net annual returns from a command area [12]. Multiregional LP model was applied for managing cropping patterns of agricultural crops in Iran [22]. LP model was used for optimization of irrigation water management in Pakistan [23]. A Linear programming technique was applied to determine the optimum land allocation to 10 major crops in India [24]. Uncertainty is always an important factor that affects the management of an agricultural water resources system owing to the existence of spatial and temporal variations [25]. Two possible results of decisions without consideration to uncertainty are the creation of a net benefit that is less than expected and probability of system failure in meeting a given demand or other system constraint [26]. For this reason, the application of chance constraint programming that considers uncertainty of reservoir inflow for optimizing cropping pattern is imperative. This study was conducted with the objectives of finding optimal cropping patterns for yield and benefit maximizations. Therefore, this research would help the irrigation managers to make concrete decisions on how much land and water should be allocated to each crop to get more out of the stored and incoming water during irrigation season.

Material and Methods

Description of the study area

The study area is Koga Irrigation scheme located south of Lake Tana in the Upper Blue Nile River Basin, Ethiopia (Figure 1). The Koga

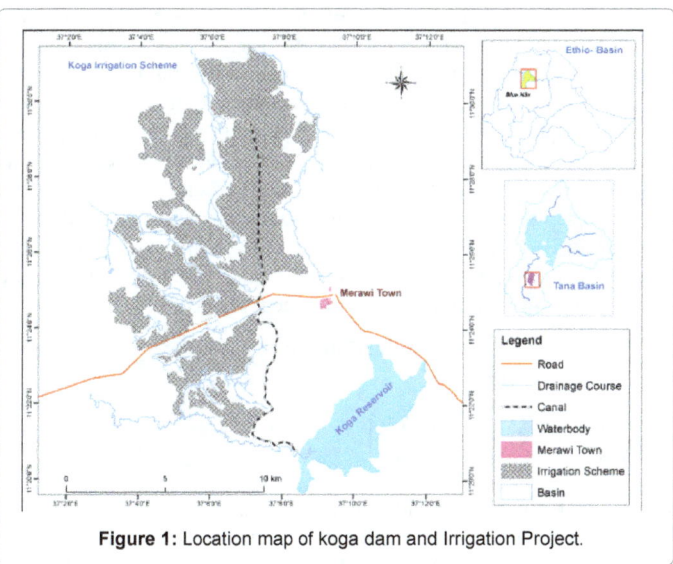

Figure 1: Location map of koga dam and Irrigation Project.

catchment lies in the Tana Basin between 11°10' and 11°22' North Latitude and 37°02' and 37°17' East Longitude. It covers an area of 22,000 hectares at dam site (37°08' E and 11°20' N) that drains into Koga River. The Koga River is a tributary of the Gilgel Abay River in the headwaters of the Blue Nile. The Gilgel Abay flows into Lake Tana. The monthly flow characteristics of the Koga River follow the rainfall pattern. Minimum flow occurs in April. Flow begins to increase in May, in response to the early rains, reaching a peak in August. About 70% of the runoff occurs in the three months from July to September. Its average annual rainfall is 1578 mm. The rainfall has a uni-modal characteristic that extends from May to October.

Data collection

Hydrological data were collected from Koga irrigation project office and Ministry of Water and Energy in 2012/13. Crop area, yield, cost of production and farm get prices of crops and reservoir water release, were collected in 2012/13 from office of Koga irrigation project. The soil texture within the command area is uniformly clay and the total available water (TAW) is 160 mm/m [27,28]. The soil moisture content at field capacity, permanent wilting point and the initial available soil moisture in December were 46.71%, 30.93%, and 15.26 mm/m [28].

Model development

Chance constraint linear programming (CCLP) models were developed by incorporating uncertainty of inflows at exceedance probabilities (ρ) of 90%, 80%, 70%, 60% and 50%.

These objective functions were maximization of total yield (Equation 4) in tons and total benefit (Equation 7) in Ethiopian Birr (ETB). Then the decision variables, irrigated land and water for each of five major crops in the project area: maize (*Zea mays L.*), wheat (*Triticum aestivaum*), potato (*Solanum tuberosum L.*), onion (*Allium cepa L.*) and pepper (*Capsicum spps.*) were determined by maximizing the developed model subject to land and water resources availability constraints. Also included in the model was a constraint of limits on the irrigated lands for each crop. This was imposed to allow for diversity and prevent the domination of one crop over the others. Constraint containing variable of stochastic inflow was transformed in to its deterministic programming before optimization [29]. Then the model was solved using Language for Interactive General Optimization (LINGO) version 14. Four scenarios were suggested through this study.

For scenario I and II, the total sum of crop areas were made less or equal to 7000 ha. Irrigated land constraints for each crop was made less or equal to its maximum allowed limit in scenario I (3290, 1260, 1120, 840 and 490 ha for maize, wheat, onion and pepper respectively) but made greater or equal to the maximum limit of each crop except for pepper in scenario II. In scenario III and IV, the sum of total crop areas was less or equal to the potential irrigable area of 7572 ha. Land constraints for each crop was made less or equal to its maximum allowed limits in scenario III (3558.8, 1363,1211.5, 908.64 and 530.04 ha for maize, wheat, onion and pepper respectively) but made greater or equal to the maximum allowed limit of each crop except for pepper in scenario IV.

The following water production function [30] was used to derive objective function.

$$\frac{Ya_c}{Ym_c} = 1 - Ky_{cs}\left(1 - \frac{ETa_{cs}}{ETm_{cs}}\right) \tag{1}$$

where, Ya_c is the actual yield of the crop with the available water (kg/ha), Ym_c is the maximum yield (kg/ha) that could be obtained when there is no limitation of water (Allen et al., 1998). ETa_{cs} and ETm_{cs} are the actual and potential crop seasonal evapotranspiration (mm/season), and Ky_{cs} is the seasonal yield response factor of crop 'c' for the full growing season [30]. Actual and potential crop evapotranspirations were estimated using crop-water simulation model (CROPWAT) version 8.0 for Windows. Seasonal actual evapotranspriration (ETa_{cs}) in mm was estimated as the sum of net irrigation applied and effective rainfall.

$$ETa_{cs} = \sum_s\left(\frac{R_{cs}\eta}{A_c} \times 10^5 + Pef_s\right) \tag{2}$$

where, R_{cs} is the irrigation release to crop 'c' in season 's' (Mm³), η is the irrigation efficiency. A_c is cultivated area under crop 'c'. The adopted canal lining in the irrigation scheme has a conveyance efficiency of 78%, field application efficiency of 62.5% and the overall irrigation efficiency of 48.75% [10]. Pef_s is seasonal effective rainfall (mm) calculated from 80% dependable rainfall of Merawi Meteorological Station using USDA Soil Conservation Service Method in CROPWAT. Seasonal potential evapotranspiration of crop was estimated as

$$ETm_{cs} = \sum_{g=1}^{NG}\left(Kc_g * ETo_g\right) \tag{3}$$

where, Kc_g is crop coefficient of crop 'c' in growth stage 'g' [31] and ETo_g is the reference evapotranspiration in growth stage 'g'. ETo was estimated from long-term monthly averaged values of meteorological data using CROPWAT. Combining equations 1 to 3, seasonal yield maximization objective function (Equation 4) was developed.

$$Ya_{cs} = \sum_{c=1}^{N}\left[A_c\left(Ym_{cs}\left(1 - Ky_{cs} + \frac{Ky*Pef}{ETM_{cs}}\right)\right) + Ym_{cs}\eta*10^5\left(\frac{Ky_{cs}}{ETM_{cs}}\right)R_{cs}\right] \tag{4}$$

The net benefit for each crop (NB_c) was estimated as:

$$NB_c = P_cYa_{cs}A_c - C_cA_c \tag{5}$$

where P_c is the market price in Ethiopian Birr (ETB) per ton, Ya_c is the yield per unit area (kg/ha), C_c is the cost of cultivation per unit area (ETB/ha) and A_c is area of crop 'c' in hectare. Total annual net benefit from all command areas of the project was given as:

$$NB_{Tot} = \sum_{c=1}^{N}NB_c \tag{6}$$

Combining Equation 4 to 6, the following objective function for total net benefit was obtained.

$$NB_{Tot} = \sum_{c=1}^{N}\left[A_c\left(P_cYm_{cs}\left(1 - Ky_{cs} + \frac{Ky*Pef}{ETM_{cs}}\right) - C_c\right) + P_cYm_{cs}\eta*10^5\left(\frac{Ky_{cs}}{ETM_{cs}}\right)R_{cs}\right] \tag{7}$$

where, Ya_c and Ym_c are the actual and maximum yields in kg /ha,

The objective functions (Equations 4 and 7) were maximized subject to the following constraints.

$$A_{c1} + A_{c2} + A_{c3} + A_{c4} + A_{c5} \leq A_{Tot} \tag{8}$$

$$A_{min} \leq A_c \leq A_{max} \tag{9}$$

$$R_{c1} + R_{c2} + R_{c3} + R_{c4} + R_{c5} \leq V_{cs} \tag{10}$$

$$R_{cs\,max} = \left(\frac{ETm_{cs} - Pef_s}{\eta*10^5}\right)A_c \tag{11}$$

$$R_{cs\,min} = P_c\left(\frac{ETm_{cs} - Pef_s}{\eta*10^5}\right)A_c \tag{12}$$

$$S_{t+1} - S_t - P_t + R_t + ER_t + SP_t + EVP_t = I_t^\rho \tag{13}$$

$$S_{min} \leq S_t \leq S_{max} \tag{14}$$

$$R_{cs} \geq 0, A_c \geq 0, EVP_s \geq 0, ER_s \geq 0 \tag{15}$$

where, Equation 8 is land resources availability constraint. Total area allocated to each crop cannot exceed total available land (A_{Tot}). A_{c1} to A_{c5} are seasonal area allocated to maize, wheat, potato, onion and pepper, respectively. Equation 9 is maximum and minimum crop area limit imposed during project designing [10]. The cropping pattern used to derive the crop water requirements during irrigation dam designing was 47% maize (3290 ha), 18% wheat (1260 ha), 16% potato (1120 ha), 12% onion (840 ha) and 7% pepper (490 ha) out of 7000 ha area [10]. Equation 10 is water availability constraint. The sums of seasonal water allocated to each crop should not exceed reservoir water supply for irrigation (V_s). R_{c1} to R_{c5} are seasonal water released to maize, wheat, potato, onion and pepper, respectively. Equation 11 is maximum irrigation release to each crop (R_{c5max}). R_{csmax} was restricted by its seasonal potential irrigation requirement ETm_{cs} (Equation 3). Equation 12 is minimum irrigation release to each crop (R_{csmin}) restricted to management allowed depletion (P_c) of the seasonal irrigation requirement. Equation 13 is water balance of reservoir during the monthly time interval 't' in irrigation season estimated by chance constraint form of reservoir storage continuity equation, where, S_{t+1} is storage at the end of time period t, S_t is storage at the beginning of time period t, P_t is 80% dependable rainfall of Merawi station during time period t , R_t is release volume at time period t+1, ER_t is environmental (compensation) release at time period t according to MacDonald [32]. SP_t is the spilled or over flow water. I_t^ρ is the expected monthly inflows (I_t) into the reservoir at different ρ values of 90%, 80%, 70%, 60% and 50% which was estimated from the distributions fitted using Cumulative Frequency Program. Transpiration and seepage are seen as a minor contribution to the losses factors [33] and therefore not included in the water balance equation.

EVP_t is the evaporation rate at time period t estimated by Linacre [34] method. Monthly reservoir evaporation (Mm³) was estimated by multiplying evaporation rate (mm) and reservoir surface area for each month. Reservoir surface area on which evaporation would take place was determined from the best fitted capacity area curve derived using Koga reservoir area, volumes and stage relationships data from MacDonald [35]. The units of all parameters in Equation 13 are in million cubic meters (Mm³). Equation 14 is minimum and maximum storage volumes. S_{min} and S_{max} were minimum and maximum storage limits, respectively. The dead storage volumes of 4.8 Mm³ and full capacity of storage (83.1 Mm³) were used as minimum and maximum storage volume limits, respectively. Equation 15 is non-negativity constraint in which all decision variables were made greater or equal to zero.

Water productivity

Physical water productivity is calculated in terms of yield per volume of irrigation water released (Equation 16), and economic water productivity is calculated in terms of net income per volume of irrigation water released (Equation 17). Finally, the calculated values of water productivities were compared with the actual water productivity values for 2012/13 irrigation season.

$$W_p = \frac{Y}{R_T} \tag{16}$$

$$W_e = \frac{I}{R_T} \tag{17}$$

where, W_p is physical water productivity (ETB/m³), Y is total yield (kg) and R_T is total irrigation water released (Mm³), W_e is economic water productivity (ETB/m³) and I is net income (ETB).

Results and Discussion

Irrigation demand reservoir water supply

Gross irrigation (mm) and 100% of irrigation water demands (Mm³) for different scenarios of cropping pattern are presented in Table 1. As it was shown in the table, the maximum water demand occurs in March. According to MacDonald [10], the 7000 ha irrigated area is governed by the 80% reliability yield per annum from the dam. This study has also confirmed that the design reservoir yield could be achieved at all exceedance probability levels of 80% and less when the reservoir is empty at the end of irrigation season (May). Cumulative reservoir inflow was 77.67 Mm³ at ρ 80%.

Equation 18 is the best fit reservoir surface area–capacity curve (R²=0.99) for Koga irrigation reservoir. Estimated monthly evaporation

rate (mm) is shown in Table 2.

$$A_t = -0.002(V_t - 30.4429)^2 + 0.26 * V_t + 2.68 \tag{18}$$

where, A_t is reservoir area (km²) at month t and V_t is reservoir volume (Mm³) at month t.

Actual irrigation practice

The actual yields, potential yields, production costs and farm gate crop prices in the Koga irrigation scheme for the year 2012/13 are shown in Table 3. It is the farmer who decides the types of crops grown on his farm. As a result, thirteen types of crops were sown in the irrigation project in 2012/13. Out of 7000 ha irrigable lands, a total of 5144.36 hectares were cultivated, 37,920.6 tons of yield was harvested from all crops and net benefit of 113,343,614 ETB was obtained. Sensitivity test showed that the actual values of yield and benefit would not be changed if environmental flow had been permitted for downstream environment. Cultivated areas of major crops: maize, wheat, potato, onion and wheat were 1967, 511, 1318, 83 and 32 ha, respectively. The cultivated areas of these crops except for potato were less than their maximum allowed land limits.

Actual reservoir operation during 2012/13 irrigation season is shown in Table 4. Total seasonal irrigation water released was 67.7 Mm³.

Optimal land and water allocations

The optimal results for irrigated land and water allocated to each crop, the total yield and benefits gained at 80% exceedance probability (ρ) of reservoir inflows under each scenarios are described in the following sections. Under each crop column (Table 5), the first values are land in ha and the values under bracket are allocated water in Mm³.

Scenario I: The global optimal solution under scenario I for total yield and benefit maximizations of 7000 ha of land is shown in Table 5. Both yield and benefit optimization models allocated 97%, 115%, 131%, 132% and 134% of the actual irrigated lands in 2013 irrigation season at ρ 90%, 80%, 70%, 60% and 50%, respectively. The actual yield harvested and the benefit gained during 2012/13 irrigation season was 37,920.6 tons and 113,343,614 ETB, respectively from 5144.36 ha of

Types of Irrig. Demand	Dec	Jan	Feb	Mar	Apr	May	June	Total
Gross Irrigation	235.50	834.80	1453.80	1736.50	1501.9	634.50	64.85	
Scenario I	7.75	15.27	19.00	19.44	13.31	1.31	–	76.08
Scenario II	7.75	16.43	22.14	23.55	16.79	1.81	–	88.47
Scenario III	8.38	17.20	21.54	22.42	15.48	1.42	–	86.44
Scenario IV	8.38	17.77	23.95	25.47	18.16	1.96	–	95.69
Scenario V	7.75	16.43	21.61	23.43	16.97	2.69	0.32	88.88

Table 1: Gross irrigation (mm) and irrigation demand (Mm³) for different scenarios.

Jun	Jul	Aug	Sep	Oct	Nov	Dec	Jan	Feb	Mar	Apr	May
123.9	94.2	85.8	87.2	88.2	72.1	63.7	70.5	86.8	125.7	145.6	148.2

Table 2: Monthly reservoir evaporation in mm.

Crop type	Max yield (ton/ha)	Actual yield (ton/ha)	Average production cost (Birr/ha)	Average farm gate price(Birr/ton)
Wheat	6	5.5	8700	4800
Maize	6	2.3	8476	7500
Potato	22	20	12900	3500
Onion	35	16.5	14380	4000
Pepper	2.3	2.2	8730	4000

Table 3: Maximum yield [36] production cost and farm gate price of crops.

Months	Stage (m.a.s.l*)	Volume (Mm³)	Release (Mm³)
Oct	2015.10	81.00	1.53
Nov	2014.88	76.00	4.54
Dec	2014.25	66.00	12.82
Jan	2013.25	51.00	15.17
Feb	2012.00	35.00	15.64
Mar	2010.63	21.00	11.94
Apr	2010.00	15.50	5.33
May	2009.50	12.00	0.72
Total			67.7

*m.a.s.l is meters above sea level

Table 4: Actual reservoir operation for the year 2012/13.

Scenario	Maize	Wheat	Potato	Onion	Pepper	Land (ha)	Yield (ton)	% actual yield	Benefit (ETB)	% actual benefit
I	3290 (32.0)	654.3 (7.3)	1120 (13.6)	840 (8.4)	0.0 (0.0)	5904.3	78734.47	108	314072100	177
II	3290 (32.0)	1260 (7.7)	1120 (13.6)	1330 (13.3)	0.0 (0.0)	7000	92161.64	143.	343355800	203
III	3558.8 (34.6)	257.1 (2.9)	1211.5 (14.7)	908.6 (9.1)	0.0 (0.0)	5936.1	82405.86	117	349179700	208
IV	3558.8 (23.8)	1363.0 (8.3)	1211.5 (14.7)	1438.7 (14.4)	0.0 (0.0)	7572	95828.70	153	348946300	208
Actual values						5144.36	37920.6		113,343,614	

Values in the parenthesis are seasonal water (Mm³) allocated to each crop.

Table 5: Optimal land and water allocation under all scenarios at excedance probability (ρ) of 80%.

ρ value	Released water	Scenario I		Scenario II		Scenario III		Scenario IV	
	(Mm³)	W_p	W_e	W_p	W_e	W_p	W_e	W_p	W_e
90%	51.3	1.4	5.7	1.6	5.6	1.5	6.1	1.6	5.5
80%	61.3	1.3	5.1	1.5	5.6	1.3	5.7	1.6	5.7
70%	70.3	1.2	4.6	1.4	5.4	1.3	5.2	1.5	5.8
60%	71.3	1.2	4.6	1.4	5.4	1.2	5.2	1.5	5.8
50%	72.2	1.2	4.5	1.4	5.3	1.2	5.1	1.4	5.8
Actual	67.74	0.56	1.67	0.56	1.67	0.56	1.67	0.56	1.67

Table 6: Water productivity under four scenarios.

land. The use of optimization model increased the yield by 93%, 108%, 119%, 119% and 120%, and the benefit by 156%, 177%, 188%, 188% and 189% as compared to the actual yield at ρ 90%, 80%, 70%, 60% and 50%, respectively. No land was allocated to wheat at ρ 90% and pepper at ρ 90% and 80%. The land allocated to wheat at ρ 80% was 51.93 % and it was 100% of their maximum allowed land of 1260 ha at ρ 70%, 60% and 50%. The land allocated to maize was 92.32 % of its maximum allowed land of 3290 ha at ρ 90%, and 100% of its maximum allowed land at ρ 80%, 70%, 60% and 50%. But the land allocated to pepper was 41.69 %, 60.08% and 78.47 % of its maximum allowed land of 490 ha at ρ 70%, 60% and 50%, respectively. The land allocated to potato and onions were 100% of their maximum allowed lands of 1120 and 840 ha, respectively at all ρ values. The slacks or unirrigated lands in scenario I, during both yield and benefit maximizations, were 20130.5, 1095.7, 285.6, 195.6 and 105.5 hectares at ρ 90%, 80%, 70%, 60% and 50%, respectively. These unirrigated lands would make some of the land holders disadvantaged from the irrigation scheme. Under this scenario, 100% of irrigation demand could be met for wheat, potato and onion and 89% for maize. The 11% water deficit for maize is very much less than its management allowed depletion of 55%. Therefore this deficit would not result into significant yield reduction.

Scenario II: The global optimal solutions under scenario II for total

yield and benefit maximizations is shown in Table 5. Both yield and benefit optimization models allocated 7000 ha of land for irrigation. The land allocated to all crops except for pepper was 100% of their maximum permitted areas at all ρ values. But no land was allocated to pepper. For example, the optimal amount of areas allocated to maize, wheat, potato and onion at ρ 80% were 3290, 1260, 1120 and 1330 ha, respectively and their corresponding allocated water was 26.7, 7.7, 13.6 and 13.3 Mm³. In general, the CCLP model improved the yield by 121%, 143%, 159%, 161% and 162%, and the benefit by 153%, 103%, 137%, 139% and 141% of the actual at ρ 90%, 80%, 70%, 60% and 50%, respectively. Under this scenario 100% of irrigation demand could be met for potato and onion, 89% for maize and 63% for wheat ρ 90%. The 37% water deficit for wheat is very much less than its management allowed depletion of 55%. Therefore, the deficit water for maize and wheat would not result into significant yield reductions.

Scenario III: The global optimal solutions for scenario III by relaxing the constraint of total command area up to 7572 ha is shown in Table 5. The total areas that could be irrigated under this scenario using yield and benefit maximization models were 96%, 115%, 131%, 133% and 135% of the 5144.36 ha of actual irrigated area in 2012/13 at ρ 90%, 80%, 70%, 60% and 50% respectively. The CCLP model improved the yield by 102%, 117%, 130%, 132% and 133%, and the benefit by 176, 208

Scenario	Irrigation water released					Irrigated area				
	RHS	Incr-ease	Decr-ease	Min	Max	RHS	Incr-ease	Decr-ease	Min	Max
I	61.3	6.7	7.3	54.0	68.0	7000	Infinity*	1095.7	5904.3	-
II	61.3	5.3	9.0	52.2	66.6	7000	905.4	490	6510.0	7905.4
III	61.3	12.3	2.9	58.4	73.6	7572	Infinity*	1635.9	5936.1	-
IV	61.3	10.8	4.8	56.5	72.1	7572	478.9	530.0	7042.0	8051.0

*Irrigated area coefficient that can be increased indefinitely

Table 7: The right-hand side (RHS) ranges of constraints at ρ 80%.

%, 223%, 224% and 226% of the actual in 2012/13 at ρ 90%, 80%, 70%, 60% and 50%, respectively. The allocated lands of potato and onion at all ρ values were 100% of their maximum allowed lands of 1211.52 and 908.64 ha, respectively. The lands allocated to maize were 2824.3 ha at ρ 90% and 100% of its maximum allowed lands of 3558.84 ha at ρ 80% to 50%. No land was allocated to wheat at ρ 90%. However, 19%, 78%, 85% and 91% of maximum allowed lands of 1362.96 ha of wheat were allocated at ρ 80%, 70%, 60% and 50%, respectively. On the other hand, no land was allocated to pepper at all exceedance probabilities. The slack or unirrigated lands out of 7572 ha of potential irrigable lands were 2627.5, 1635.8, 826.2, 736.2 and 646.2 ha at ρ 90%, 80%, 70%, 60% and 50%, respectively during yield and benefit maximizations. Under this scenario 100% of irrigation demand could be met for wheat, potato and onion, 89% for maize at ρ 80%. Thus, the deficit water for maize would not result into significant yield reduction.

Scenario IV: The global optimal solutions of scenario IV for total yield and benefit maximizations were shown in Table 5. Both yield and benefit optimization CCLP models allocated 100% of the maximum permitted lands to maize, wheat, potato and onion crops but, no land was allocated to pepper at all ρ values. The CCLP model improved the yield by 176%, 208%, 223%, 224% and 226%, and the benefit by 149%, 208%, 258%, 264% and 269% of the actual at ρ 90%, 80%, 70%, 60% and 50%, respectively. Under this scenario 100% of irrigation demand could be met for potato and onion, 61% for maize and 63% for wheat at ρ 80%. Water deficits for maize and wheat would not result into significant yield reduction. Yield and benefit improvements remain almost constant at ρ 70%, 60% and 50% under all scenarios. This is because water supply is greater than irrigation demand at these probabilities of reservoir inflows.

Productivity

Table 6 shows physical water productivity (W_p) and economic water productivity (W_e) under scenario I, II, III, IV and actual conditions at different exceedance probabilities (ρ) of water availability. For ρ 90% to 50%, physical and economic water productivities range from 1.2 to 1.4 kg/m³ and 4.5 to 5.7 ETB/m³, respectively under scenario I, 1.4 to 1.6 kg/m³ and 5.3 to 5.6 ETB/m³, respectively under scenario II, 1.2 to 1.5 kg/m³ and 5.1 to 6.1 ETB/m³, respectively under scenario III, and 1.4 to 1.6 kg/m³ and 5.5 to 5.8 ETB/m³, respectively under scenario IV.

The physical and economic water productivities of the actual irrigation practice were 0.56 kg/m³ and 1.67 ETB, respectively. The W_p and W_e improvements at ρ values of 90% to 50% were 114% to 150% and 169 % to 241%, respectively under scenario I, 150 % to 186% and 217% to 235%, respectively under scenario II, 114% to 168% and 205% to 265%, respectively under scenario III, and 150% to 185% and 229% to 247%, respectively under scenario IV. Optimal cropping pattern improved W_p and W_e by 132 to 205, 168 to 235, 132 to 241 and 186 to 241 under scenario I, II, III and IV, respectively at ρ 80%. Hence minimum and maximum W_p were 132 and 186, respectively and whereas minimum and maximum W_e were 205 and 241 of the actual

values, respectively at ρ 80%. Generally, W_p and W_e remained constant from ρ 70% to 50% under all scenarios. Whereas W_p and W_e decreased from ρ 90% to 80% under scenario I and III. Under scenario II, W_p decreased and W_e remained constant. W_p remained constant and W_e decreased at scenario IV from ρ 90% to 80%.

Optimal ranges of water and land allocations

Table 7 shows the right-hand side (RHS) ranges of water and land constraints in which the optimal solutions of yields and benefits are unchanged. For example, at scenario II, the optimal yield of 92161.64 tons and 343355800 ETB could be gained from 7000 ha of land irrigated at ρ 80% using 61.3 Mm³ of water (Table 5). In this scenario, the reservoir water released for irrigation and the cultivated areas can change between 52.2 to 66.6 Mm³ and 6510 to 7905.4 ha, respectively (Table 7), but the current optimal values of yield and benefit remain optimal in these intervals or ranges. Among the four scenarios, the minimum amount of land that could be irrigated at ρ 80% was 5904.3 ha at scenario I and the maximum was 8051.0 ha at scenario IV. The optimal irrigated areas under scenario I and III were the minim values of optimal ranges of area allocations for these scenarios. Whereas that of scenario II and IV were within the ranges of optimal area allocations. Therefore, optimal cropping pattern would increase the irrigated area by 15% to 56%, the yield by 108 to 153% and the benefit by 153 to 208% of the actual values at ρ 80%. This implies that the actual irrigated land in 2012/13 was below the optimal values and water released for irrigation was mismanaged.

Conclusion

This study optimizes cropping patterns of five major crops: maize, wheat, potato, onion, and pepper using chance constraint linear programming (CCLP) model under four scenarios of cropping pattern. The yield, benefit, water productivity and irrigated areas under all scenarios were greater than that of the actual values in 2012/13 irrigation season. The study showed that the possibility of irrigating 5904.3 to 8051.0 hectares of land at ρ 80% by optimizing cropping patterns at irrigation efficiency of 48%. This could increase the yield by 108 to 153%, benefit by 153 to 208% and physical water productivity by 132% to 186% and economic water productivity by 205% to 241% of the actual values. Optimal water allocations to each crop, under all scenarios of cropping pattern, would not result into significant yield reduction. Thus, it was concluded that the actual irrigated land (5144.36 ha) in 2012/13 was below the optimal values, and water released for irrigation was mismanaged. Therefore, with optimal crop planning and water management, it is possible to irrigate the design command area of 7000 ha. Therefore, the developed CCLP models could be valuable tools for decision making in selecting optimum combination of crops that maximize the benefit/yield of an irrigation project. Knowledge of optimum crop combination is important in determining irrigation demand on which reservoir operation is based. Finally, a study should be made to determine optimal levels of crop water deficit that maximize water productivity and optimization of cropping pattern using computational intelligence techniques.

References

1. Goodarzi E, Ziaei M, Hosseinipour EZ (2014) Introduction to optimization analysis in hydrosystem Engineering. Springer International Publishing, Switzerland.

2. WWAP (United Nations World Water Assessment Programme) (2014) The United Nations World Water Development Report 2014: Water and Energy. Paris, UNESCO.

3. Bruinsma J (2009) The Resource Outlook to 2050: By How Much do Land, Water and Crop Yields Needto Increase by 2050? Prepared for the FAO Expert Meeting on 'How to Feed the World in 2050', 24-26 June 2009, Rome.

4. WWAP (World Water Assessment Programme) (2012) United Nations World Water Development Report 4: Managing water under uncertainty and risk. Paris, UNESCO.

5. Homayounfar M, Lai SH, Zomorodian M, Sepaskhah AR, Ganji A, et al. (2014) Optimal Crop Water Allocation in Case of Drought Occurrence, Imposing Deficit Irrigation with Proportional Cutback Constraint. Water Resources Management 28: 3207-3225.

6. Singh A (2012) Optimalal location of resources for the maximization of net agricultura lreturn. Journal of Irrigation and Drainage Engineering 138: 830-836.

7. Wang LK, Yang CT (2014) Modern Water Resources Engineering. Humana Press, New York.

8. Awulachew SB, Teklu E, Namara RE (2010) Irrigation potential in Ethiopia: Constraints and opportunities for enhancing the system. International Water Management Institute, Ethiopia.

9. MoWR (Ministry of Water Resources) (2002) Water sector development program 2002-2012. Main report volume I. Addis Ababa, Ethiopia.

10. MacDonald (2006b) Koga dam and irrigation project design report. Part 2: Irrigation and drainage. Ministry of water resources, Addis Ababa, Ethiopia.

11. Mujumdar PP (2002) Mathematical tools for irrigation water management- An Overview. Journal of Water International 27: 47-57.

12. Matanga GB, Marino MA (1979) Irrigation planning 1. Cropping pattern. Water Resource Research 15: 672-678.

13. Singh DK, Jaiswal CS, Reddy KS, Singh RM, Bhandarkar DM, et al. (2001) Optimal cropping pattern in a canal command area. Agricultural Water Management 50: 1-8.

14. Regulwar DG, Pradhan VS (2013) Irrigation planning with conjunctive use of surface and ground water using fuzzy resources. Journal of Water Resource and Protection 5: 816-822.

15. Sivanpheng O, Kangrang AM, Lamom A (2009) A Varied-Utilized Soil Type in Linear Programming Model for Irrigation Planning. American Journal of Engineering and Applied Sciences 2: 133-138.

16. Reddy MJ (2006) Swarm intelligence and evolutionary computation for single and multi objective optimization in water resource systems (un published PhD thesis), Department of Civil Engineering, Faculty of Engineering, Indian Institute of Science, Bangalore -560012, India.

17. Felix M, Judith M, Jonathan M, Munashe S (2013) Modeling a small farm livelihood system using linear programming in Bindura, Zimbabwe. Research Journal of Management Sciences 2: 20-23.

18. Rani Y R, Rao T (2012) Multi Objective Crop Planning for Optimal Benefits. International Journal of Engineering Research and Applications 2: 279-287.

19. Sabouni MS, Mardani M (2013) Application of robust Optimization Approach for Agricultural Water Resource Management under Uncertainty. Journal of Irrigation and Drainage Engineering 139: 571-581.

20. Prasad AS, Umamahesh NV, Viswanath GK (2011) Optimal irrigation planning model for an existing storage based irrigation system in India. Irrigation and Drainage Systems 25: 19-38.

21. Mutnuru RA, Ahsan N, Hassan Q (2013) Application of optimization modeling in sustainable agricultural and water resources planning: a case study. International Journal of Civil, Structural, Environmental and Infrastructure Engineering Research and Development (IJCSEIERD) 3: 31-40.

22. Aghajani A, Bidabadi FS, Joolaei R, Keramatzadeh A (2013) Managing cropping patterns agricultural crops of three Counties of Mazandarnprovince of Iran. International Journal of Agriculture and Crop Sciences 5: 596-602.

23. Qureshi AL, Khero ZI, Lashari BK (2012) Optimization of irrigation water management: a case study of secondary canal, Sindh, Pakistan. Sixteenth International Water Technology Conference, IWTC, Istanbul, Turkey.

24. Wankhade MO, Lunge HS (2012) Allocation of agriculturall and to the major crops of Saline Track By linear programming approach: A case study. International Journal of Scientific and Technology Research 1: 21-25.

25. Guo P, Chen X, Tong L, Lib J, Lia M, et al. (2014) An optimization model for a crop deficit irrigation system under uncertainty. Engineering Optimization 46: 1-14.

26. Watkins DW, McKinney DC (1997) Finding robust solutions to water resources problems. Journal of Water Resources Planning and Management 123: 49-58.

27. MacDonald (2008) Koga irrigation scheme manual: operation and maintenance part C. Irrigation and drainage system, Volume 1. Ministry of Water Resources, Addis Ababa, Ethiopia.

28. Abebaw A (2012) Evaluation of stage-wise deficitfurrow irrigation application on maize production at Koga irrigation scheme, West Gojjam (unpublished M.Sc. thesis), Haramaya University, Ethiopia.

29. Mays LW, Tung YK (1992) Hydrosystems Engineering and Management. McGraw-Hill Inc, Singapore.

30. Doorenbos J, Kassam AH (1979) Yield response towater. FAO Irrigation and DrainagePaper 33, Food and Agricultural Organization of United Nations, Rome, Italy.

31. Allen RG, Pereira LS, Raes D, Smith M (1998) Cropevapo transipiration: guidelines for computing water requirements. FAO Irrigation and Drainage Paper 56, Food and Agriculture Organization of the United Nations, Rome, Italy.

32. MacDonald (2006) Koga dam and irrigation project, design report, Part 1: Koga dam. Ministry of Water Resources, Addis Ababa, Ethiopia.

33. Tukimat NNA, Harun S (2014) Optimization of Water Supply Reservoir in the Framework of Climate Variation. International Journal of Software Engineering and Its Applications 8: 361-378.

34. Linacre ET (1993) Data-sparse estimation of lakeevaporation using a simplified Penmanequation. Agricultural and Forest Meteorology 64: 237-256.

35. MacDonald (2004) Kogadam and irrigation project contract KDIP 3: Kogairrigation and drainage system hydrology factual report. Ministry of water resources, Addis Ababa, Ethiopia.

36. FAO (2013) Crop Water Information. Natural Resources and Environment Department: Water Development and Management Unit.

16

Influence of Drip Irrigation and Plastic Mulch on Yield of Sapota (Achraszapota) and Soil Nutrients

K N Tiwari*, Mukesh Kumar, Santosh D T, Vikas Kumar Singh, M K Maji and A K Karan

Precision Farming Development Centre, Agricultural & Food Engineering Department, Indian Institute of Technology, Kharagpur- 721 302(W.B), India

Abstract

The experiment was carried out to study the response of Sapota (Achraszapota) crop under drip irrigation and plastic mulch. Three levels of irrigation water applied through drip, ring basin irrigation method in combination with plastic mulch were experimented with five replications on Sapota plants. Reference evapotranspiration was estimated using FAO-56 Penman Monteith approach. The Sapota crop water requirement was estimated using reference evapotranspiration data and crop co-efficient for different crop growth stages. The irrigation water was applied at 60%, 80% and 100% of the crop water requirement. Irrigation intervals were at 2 and 5 days respectively in drip and ring basin irrigation treatments. The water requirement of Sapota crop varies between 2.14 mm (10.71 L) per day per plant in winter season and 6.89 mm (34.44 L) per day per plant in summer season for 100% water requirement treatment at peak growth stage. To investigate the effect of plastic mulch on soil, the physico-chemical analysis of soil was performed for the soil samples collected from three different depths (0-30, 30-60, 60-90 cm). The soil chemical analysis indicated increase in organic carbon, organic matter, humic acid, microbial count, available potassium, available phosphorus, total nitrogen content and C:N ratio for the soil covered with the plastic mulch treatment. The pH and available nitrogen was found to decrease in the soil covered with plastic mulch. The biometric observations (canopy, height, girth, no. of branches) of Sapota plants showed positive influence of the irrigation and plastic mulch treatments on growth of Sapota crop. Due to mulch alone the increase in Sapota yield varied from 7.62% to 41% in different treatments. Yield of Sapota crop was found to increase by 21.05% due to drip in comparison to ring basin irrigation.

Keywords: Sapota; Drip irrigation; Water requirement; Plastic mulching; Soil nutrient

Introduction

Optimum moisture level in the soil near the root zone of the crop is critical to agriculture and plantation crops. Drip irrigation is frequent application of water directly on or below the soil surface near the root zone of plants. Drip irrigation is one of the irrigation methods which can help to increase irrigation water potential and crop yield. Crops yield get adversely affected due to excess or deficit water supply. Crop yield can be considerably increased by optimal water supply. Drip irrigation can be helpful if water is scarce or expensive because evaporation, runoff, and deep percolation are reduced and irrigation application efficiency is improved. Strategically deficit water supply through drip irrigation can save water and energy input. In general, water management assumes paramount importance to reduce the wastage of water. It is also necessary to increase the Water Use Efficiency (WUE) and ensure equitable water distribution.

Drip irrigation in combination with plastic mulch research studies carried out at Precision Farming Development Centre, IIT Kharagpur, India on vegetable and fruit crops showed increase in yield, saving in water, higher water use efficiency and net increase in profit [1-3]. Growth and production of peach trees were monitored under furrow, surface and subsurface drip and micro jet irrigation systems for different irrigation scheduling. Higher water use efficiency, yield and larger trees growth were reported under surface and subsurface drip as compare to micro jet irrigation systems and furrow irrigation [4]. Saving in irrigation water and greater net profit due to drip irrigation in banana production has been reported by Kanannavar and Pawar [5,6]. Gunduz et al. [7] investigated the effect of amount of irrigation, irrigation interval using drip irrigation on yield and quality of peach. Amount of irrigation water application to peach crop was estimated using pan evaporation data with different pan coefficient. The amount of irrigation water application had significant influence on peach yield.

Mulch is a protective cover placed over the soil surface. Mulch can play an important role for sustainable fruit production. Beneficial aspects of plastic mulch include conservation of moisture, controls weeds and moderate soil temperature for better root growth and higher crop yield [8]. The use of plastic mulch alters soil temperature. Dark opaque plastic mulches and clear mulches applied over the soil intercept sunlight that warm the soil allowing earlier planting as well as encourage faster growth and early crop production. White mulch reflects solar radiation from the sun effectively reduces soil temperature. This reduction in temperature may help to establish plants in mid-summer when cooler soil might be required. Plastic mulches reduce the amount of water lost from the soil due to evaporation. This means less water will be needed for irrigation. Plastic mulches also aid in evenly distributing moisture to the soil which reduces plant stress. Plastic mulches prevent sunlight from reaching the soil which can inhibit most annual and perennial weeds. Clear plastic mulch does not prevent weed growth as light passes through the film and reaches soil and weed plant. In black plastic mulch holes in the mulch for plants tend to be the only pathway for weeds to grow [9].

The use of drip irrigation in conjunction with plastic mulch allows conservation of water and fertilizers. Using drip irrigation for orchard crop eliminates the use of ring basin irrigation that applies large quantities of water to the soil which in turn tends to leach nitrogen and

*Corresponding author: K N Tiwari, Precision Farming Development Centre, Agricultural & Food Engineering Department, Indian Institute of Technology, Kharagpur-721 302(W.B), India; Email: kamlesh@agfe.iitkgp.ernet.in

other nutrients to depths below the root zone. Drip irrigation applies lower amount of water with fertilizers frequently to the root zone as and when needed. This reduces the amount of fertilizer application as compared to broadcasted method of fertilization. Drip irrigation in combination with plastic mulch achieves additional benefits of water and fertilizer saving besides greater yield.

Sapota juice is good sources of sugar, proteins, ascorbic acid, phenolics, carotenoids and minerals like iron, copper, zinc, calcium and potassium [10].There are various nutrients needed by the crop to normal functioning of its metabolic activities and to be disease free. Trees must be healthy to produce good quality fruit. Weak or diseased trees produce either poor quality fruit or no fruit at all. Fruit weight, volume, and peel-pulp ratio increases with the optimum irrigation water supply as water availability influences cell division more than cell expansion but no influence on fruit shape [11].

Drip method of irrigation requires fixed capital investment for installation of drip system. The amount of investment depends upon the kinds of crop, its spacing, type and discharge capacity of the dripper and the distance from water source. Wide-spaced crops require relatively low capital cost. Besides capital investment there are investments such as operating cost, cost of fertilizer used, water used, skilled labors requirement, etc. It is very important for the Sapota crop growers to know the amount of money to be invested before cultivation of the crop. The investment made by the grower on the crop must get adequate profit to adopt drip and plastic mulch. The yield and quality of the crop produce should be high so as to overcome the investment made on the crop cultivation. Any new technology would be acceptable only if crop production give greater Benefit-Cost (B.C.) ratio.

In this research paper an attempt is made to study the response of combined effect of drip irrigation and plastic mulch with different levels of irrigation on Sapota crop grown in sub- humid and sub-tropical climate of Kharagpur. This study is also aimed to investigate influence of plastic mulch on soil properties and Sapota yield, kept above soil surface for long time.

Materials and Methods

Description of study area

The study area is located at Precision Farming Development Centre, experimental farm of Agricultural and Food Engineering Department, IIT Kharagpur, India. It is situated at 22°20' N latitude and 87°20' E longitude and at an altitude of 48 m above the mean sea level. The climate of the region is sub-humid, with an average annual rainfall of about 1400 mm. The minimum temperature varies from 9.6°C to 27°C and maximum temperature ranges from 27.2°C to 41.8°C during winter and summer seasons respectively. The maximum and minimum relative humidity varies respectively from 79 to 99 % and 19 to 78 % throughout the year.

Crop details

Field experiments were conducted on Sapota crop belongs to Sapotaceae family Kalipatti variety. The crop was planted on 4th October 2005 in the PFDC farm which is now six years old. Being a tropical fruit crop it can be grown from sea level up to 1200 m above MSL. It needs warm (10-38°C) and humid climate (70% relative humidity) for growth and can be cultivated throughout the year. Alluvial, sandy loam, red laterite and medium black soils with good drainage are ideal for cultivation of Sapota. The fruit is a large ellipsoid berry, 4–8 cm in diameter, very much resembling a smooth-skinned

potato and containing 2-5 seeds. Inside, its flesh ranges from a pale yellow to an earthy brown color with a grainy texture. The fruit has a high latex content and ripens after maturity and picked from plants.

Field layout and experimental details

The experimental field of sapota is rectangular in shape with 40 m long and 25 m wide and plant to plant and row to row distance is 5m x 5m. There are total of forty numbers of plants in the field. The topography of the land area is plain and it has sandy-loam texture of soil. The total length of 12 mm dia lateral used in the field was 150 m with 70 drippers of different discharge combinations. The distribution of water in different lateral was controlled by gate valves provided at the entry end of each lateral. The operating pressure of about 1 kg/cm² was maintained to obtain design dripper discharge. The layout of the field and the division of plots along with the laterals fitted with drippers is shown in Figure 1.

The combination of different levels of irrigation with drip alone and drip with black plastic mulch ring basin with plastic and ring basin alone (control) were considered as treatments of experiment. Total eight treatments had eight rows with five plants in each row having different levels of irrigation water application.

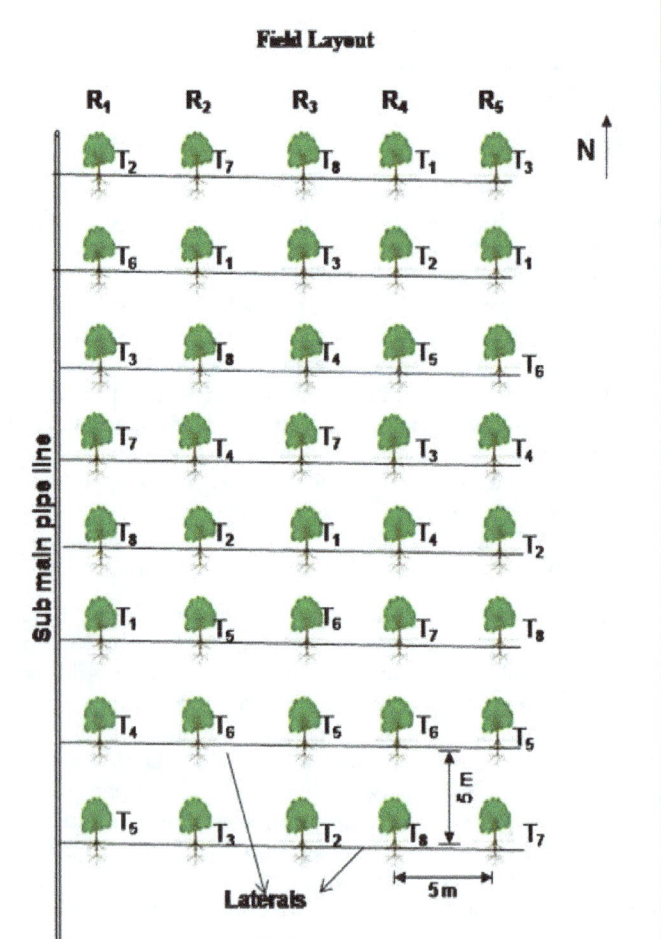

T_1=VD, T_2=VDM, T_3=0.8VD, T_4=0.8VDM, T_5=0.6VD, T_6=0.6VDM, T_7=RBM, T_8=RB

Figure 1: Experimental field layout of different irrigation levels and plastic mulch treatments.

The treatments followed for the study were as stated below:

T 1: 100% of irrigation requirement met through drip irrigation (VD)

T 2: 100% of irrigation requirement met through drip irrigation with plastic mulch (VDM)

T 3: 80% of irrigation requirement met through drip irrigation (0.8VD)

T 4: 80% of irrigation requirement met through drip irrigation with plastic mulch (0.8VDM)

T 5: 60% of irrigation requirement met through drip irrigation (0.6VD)

T 6: 60% of irrigation requirement met through drip with plastic mulch (0.6VDM)

T 7: 100% of irrigation requirement met through ring basin irrigation with plastic mulch (RBM)

T 8: 100% of irrigation requirement met through ring basin (RB)

Randomized block design was used to supply water to plants at different irrigation levels. The five of plants in each treatment were set randomly for irrigation either by drip irrigation or ring basin. Treatments T1 and T2 had combination of two drippers of 4 lph and one dripper of 2 lph discharge. Treatments T3 and T4 had two drippers of 4 lph discharge and Treatments T5 and T6 had combinations of one dripper of 4 lph and one dripper of 2 lph discharge.

Estimation of irrigation water requirement

The daily irrigation water requirement for the sapota crop was estimated by using the following relationship

$$WR = ET_o x K_c x W_p x A \qquad (1)$$

Where,

WR = Crop water requirement (L d-1)

ETo = Reference evapotranspiration (mm d-1)

Kc = Crop coefficient

Wp = Wetting fraction

A = Plant area (m2)

Net irrigation water requirement was estimated by using Equation 2.

$$IR = ET_0 x Kc x W_p x A - Re x W_p x A \qquad (2)$$

Where,

IR=Net irrigation requirement (L d- 1)

Re=Effective rainfall (mm d- 1)

Daily reference evapotranspiration (ET0) was estimated using FAO-56 Penman Monteith Equation [2] and using estimated ET_0 values for Kharagpur by Singh [12] and Gontia [7]. The crop co-efficient for different growth stages were considered based on the available local studies from unpublished literature and similar crop information given in Allen [13]. Wetting per cent (Wp) was decided based on average canopy growth of the plant (which was 40 % of the area of each plant). As the canopy area increases the wetting fraction also increased [14].

In case of ring basin irrigation, the wetted area was more than the drip irrigation due to larger volume of water supply at 5 days interval. Rainfall occurred during non monsoon months (October-May) was considered as effective rainfall. Effective rainfall during monsoon months was estimated using the guidelines proposed by Doorenbos and Pruitt [15]. Irrigation during monsoon months (June-September) was given when dry spell exceeded more than 7 days.

The irrigation system was operated to meet the irrigation water requirement for plant under drip, ring basin, as well drip and ring basin with plastic mulch. The time of operation of the drip system was determined based on emitter discharge and volume of water to be delivered in different treatments. Two days irrigation interval was made for the plants under the treatments T1 to T6 in case of drip irrigation and for the irrigation treatments T7 and T8 (ring basin) water was applied in five days interval directly to the basin.

Measurement of biometric response of crop and analysis of physico-chemical properties of soil

Biometric observations on plant canopy diameter, height, girth and number of branches were measured at three months interval for the plants under different treatments in order to monitor the influence of irrigation treatments and plastic mulch on crop growth.

The soil samples were collected from 0-30, 30-60 and 60-90 cm depths using screw auger from the experimental plots of Sapota crop for analyzing the physico-chemical properties of soil. The soil samples collected were kept for air drying and then it was grinded manually and passed through 2 mm sieve. Some selected chemical properties such as soil pH, carbon content, available nitrogen, available phosphorus and potassium content which have direct influence on the fertility status of soil were evaluated for all the treatments. The soil moisture content, bulk density and porosity of soils of different treatments were determined using standard methods.

Results and Discussion

Water requirement of sapota crop

Figure 2 shows average weekly variation in Crop EvapoTranspiration (ETc) for 52 weeks. The maximum weekly evapotranspiration was found as 21.26 mm for 18th week and minimum value as 6.62 mm in 51st week.

The crop water requirement was estimated for 52 weeks for different treatments. Water requirement of Sapota crop varies as evapotranspiration varies throughout the year and raises maximum during summer months and minimum during winter months. The maximum value of crop water requirement was found as 6.89 mm (34.44 L) per day in 18th week (May month) and minimum value as 2.14 mm (10.71 L) per day in 51st week (December month) for 100% crop water requirement. The irrigation water requirement supply varied as per the treatments i.e. 100%, 80 % and 60%.

Soil temperature

The soil temperature was measured at 15, 30 and 90 cm of soil depth both for the control (without mulch) and black plastic mulch (100μ). It can be seen from the observations presented in Table 1 that average temperature at 15 cm soil depth during winter season varied between 13.5°C to 27.5°C under control plot where as temperature varied between 15 to 31.5°C under black plastic mulch condition. During summer months (March to May) temperature at different depth ranged

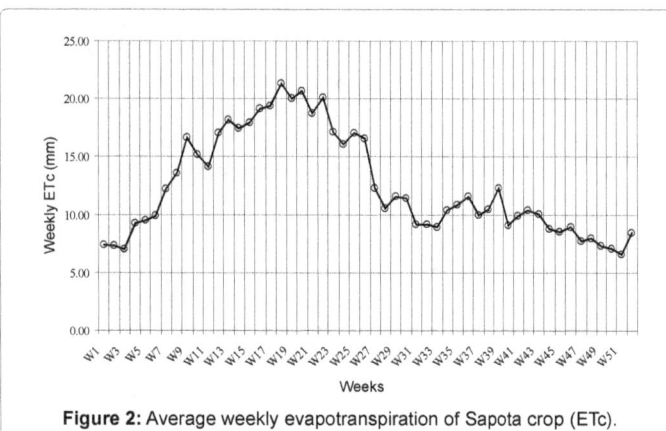

Figure 2: Average weekly evapotranspiration of Sapota crop (ETc).

from 14.5 to 32.5°C in control plot and 15 to 35°C under black plastic mulch. In general rise in the soil temperature was observed under plastic mulch condition. Increase in soil temperature has been reported under plastic mulch condition by Wu and coworkers [16]. They also reported faster growth of rice. Rise in the soil temperature during winter season under plastic mulch might have influenced for increase in the yield of Sapota crop due to increased activity of microorganisms which transforms nutrients. Soil treated with plastic mulch caused accumulation of more organic matter and humic acid which is the store house of plant nutrients that supplies nutrient to plant.

Biometric response of Sapota crop with drip irrigation and plastic mulch

Treatment-wise biometric observations of the crop were recorded from April 2006 to April 2011. Table 2 shows the pooled values of biometric attributes (canopy, height, girth and no. of branches) under different treatments.

From the Table 2, it is revealed that the drip irrigation with plastic mulch has the significant influence on plant growth and yield in comparison to basin irrigation (T7) and without mulch (T8). All the vegetative attributes and yield at 80% and 100% irrigation water requirement met with drip irrigation plastic mulch (T2 & T4) were statistically at par and non significant for canopy development. Results showed highest value of plant height, girth, number of primary branches under the treatment T2 and lowest under the treatment T8 except for number of primary branches under treatment T5.

The yield data presented in Table 2 and Figure 3 shows that the yield of Sapota crop was also statistically significant under different treatment combinations. The Sapota yield was found to decrease as the amount of irrigation water supply was reduced from 100 % to 60 % of irrigation requirement. Maximum yield of 16.1 t ha-1 was found in treatment T2. With the same level of irrigation water application between two treatments, the yield was always greater in case of plastic mulch treated plants. This could be due to greater nutrients and water availability to plants as compared to non mulched condition. These results are also supported by studies conducted on rice crop in China by Wu [16] and Liu Wu [17].

With 100 % irrigation water supply through drip system the yield of Sapota was estimated to be 21.05 % more than the conventional ring basin irrigation. This may be due to better soil water environment in root zone because of reduction in bulk density and greater porosity due to drip irrigation. With 20 per cent reduced irrigation water supply

through drip, 17.89 per cent greater yield was found over conventional ring basin irrigation, with 40% reduced water supply through drip the yield was reduced marginally by 0.32 per cent over ring basin irrigation (RB), which is statistically at par. Hence in case of water scarcity the drip irrigation is a viable option to adopt.

Physico-chemical properties of soil

To study the effect of drip irrigation and plastic mulch on soil, the soil samples were collected from the experimental plots for evaluating the physical and chemical properties of the soil of the Sapota crop root zone. The treatment-wise average value of results of the analysis is shown in Table 3.

Table 3 shows the effect of drip irrigation and plastic mulch on bulk density and porosity of the surface soil (0-30 cm) of Sapota crop. From the analysis of the results it was revealed that the bulk density has reduced and porosity has increased in six years for all the plastic mulch treated soils having same amount of water application. The maximum value of bulk density and minimum value of porosity was found as 1.72 g/cm³ and 35.1% respectively for plots with ring basin irrigation without mulch (T8). The minimum value of bulk density and maximum value of porosity was found as 1.59 g/cm³ and 40.0% respectively for the plots within the drip irrigation and plastic mulch (T2). This analysis revealed that soil gets compacted due to surface sealing in basin irrigation where as plots with drip irrigation with plastic mulch, the bulk density was found to reduce and corresponding increase in the porosity Similar results are reported by Khan [18].

Table 4 shows the changes in the chemical properties of the soil samples collected from Sapota crop field due to different treatments. It is observed that soil has become acidic, pH of the surface soil (0-30 cm) ranged from 5.52 to 5.98. In the sub-surface soil (30-60 cm) pH ranged from 5.62 to 5.98 and in the subsoil (60-90 cm) it ranged from 5.94 to 6.27.

In natural condition, due to enzymatic oxidation carbon dioxide is evolved on soil surface but in mulch condition carbon dioxide evolution is restricted in soil. The accumulated carbon dioxide in soil reacts with soil moisture and formed carbonic acid. Therefore pH of soil covered with plastic mulch is marginally decreased compared to soil without plastic mulch. Soil pH value generally increased in subsoil as base material like Ca, Mg, Na, K, etc. leaches down to subsoil.

The organic carbon in surface soil (0-30 cm) was found to vary from 2.50 g.kg⁻¹ to 4.79 g.kg⁻¹. In sub surface soil it ranged from 0.66 g.kg⁻¹ to 1.73 g.kg⁻¹ and in subsoil (60-90 cm) it ranged from 0.48 g.kg⁻¹ to 1.24 g.kg⁻¹. The organic matter was found to vary from 4.31 to 8.25 g.kg⁻¹ in surface soil and further it was found to decrease in subsurface soil. In the mulched soil the organic carbon and organic matter content is preserved and found greater as compare to the non-mulched soils.

The statistical analysis revealed that drip irrigation treatments with and without mulch has significant influence on soil bulk density, organic carbon and soil pH. However at 60% water requirement met with drip and ring basin irrigation had no significant influence of plastic mulch on soil pH.

Table 5 shows the effect of plastic mulch on changes of available nutrients (N, P and K) and soil moisture. From the Table 5 it was also revealed that with the same volume of irrigation water applied to plants in treatments T1 and T2, the soil moisture content measured after two days of irrigation was found to be greater in the soil covered with plastic mulch. It is also found that the plastic mulch had greater

Season	Soil temperature in Control (°C)			Soil temperature in plastic mulch (°C)		
	15 cm	30 cm	90 cm	15 cm	30 cm	90 cm
Winter (Dec-Feb)	13.5-27.5	14.5-29.0	16.5-26.0	15.0-31.5	17.5-30.5	17.0-28.0
Summer (March-May)	17.0-30.5	14.5-30.5	17.5-32.5	16.5-35.0	15.0-32.0	17.5-33.5

Table 1: Temperature variation in control and plastic mulched plots at different soil depth.

Treatment	Plant height (m)	Girth (cm)	No. of primary branches	Canopy (East-West) (m)	Yield (t/ha)
T_1 (VD)	4.55	37.30	13.40	4.12	11.50
T_2 (VDM)	4.82	39.10	14.00	4.34	16.10
T_3 (0.8VD)	3.90	38.30	11.00	3.85	11.20
T_4 (0.8VDM)	4.00	39.00	12.60	3.85	15.60
T_5 (0.6VD)	4.81	36.20	10.80	3.83	9.20
T_6 (0.6VDM)	3.95	37.40	11.20	3.89	11.04
T_7 (RBM)	3.92	30.90	11.60	3.98	11.04
T_8 (RB)	3.75	30.00	11.20	3.65	9.50
S.Em (±)	0.242	2.057	0.704	0.403	0.903
CD (P=0.05)	0.603	5.122	1.753	NS	2.248

Table 2: Effect of plastic mulch and drip irrigation on biometric attributes and Sapota yield.

Treatments	Bulk density (g/cm³)	Porosity (%)
T_1 (VD)	1.69	36.2
T_2 (VDM)	1.59	40.0
T_3 (0.8VD)	1.71	35.4
T_4 (0.8VDM)	1.60	39.6
T_5 (0.6VD)	1.70	35.8
T_6 (0.6VDM)	1.61	39.2
T_7 (RBM)	1.65	37.7
T_8 (RB)	1.72	35.1

Table 3: Treatment-wise bulk density and porosity of surface soil (0-30 cm) of Sapota crop.

Treatments	Depth (cm)	pH	Organic carbon (g kg⁻¹)	Organic matter (g kg⁻¹)
T_1 (VD)	0-30	5.98	3.32	5.72
	30-60	5.99	0.66	1.14
	60-90	6.27	0.65	1.12
T_2 (VDM)	0-30	5.57	4.79	8.25
	30-60	5.62	1.40	2.41
	60-90	6.09	0.74	1.28
T_3 (0.8VD)	0-30	5.90	3.55	6.12
	30-60	5.92	1.73	2.98
	60-90	6.18	0.74	1.28
T_4 (0.8VDM)	0-30	5.6	4.38	7.55
	30-60	5.98	1.73	2.98
	60-90	6.27	1.24	2.14
T_5 (0.6VD)	0-30	5.69	3.81	6.57
	30-60	5.90	1.21	2.09
	60-90	5.98	0.55	0.95
T_6 (0.6VDM)	0-30	5.58	4.09	7.05
	30-60	5.79	1.15	1.98
	60-90	5.95	0.58	1.00
T_7 (RBM)	0-30	5.52	2.95	5.06
	30-60	5.69	1.01	1.74
	60-90	5.94	0.48	0.83
T_8 (RB)	0-30	5.64	2.50	4.31
	30-60	5.78	0.95	1.64
	60-90	6.02	0.52	0.90

Table 4: Effect of drip irrigation and plastic mulch treatments on chemical properties of soils of experimental plots with different treatments.

Treatment	Depth (cm)	Moisture content (%)	Available N (mg kg⁻¹)	Available P (mg kg⁻¹)	Available K (mg kg⁻¹)
T_1 (VD)	0-30	11.62	80	52.7	75
	30-60	13.12	76	24.8	91.2
	60-90	14.95	68.5	17.6	112.5
T_2 (VDM)	0-30	13.10	78.5	77.8	85.7
	30-60	13.98	73	25.5	95.2
	60-90	14.98	49	22.5	118.4
T_3 (0.8VD)	0-30	11.23	91	22.8	68.7
	30-60	12.65	90	22.3	75.9
	60-90	14.76	80.5	2.7	102.5
T_4 (0.8VDM)	0-30	12.10	68.5	26.2	87.2
	30-60	13.45	66	14.9	90.5
	60-90	14.65	39.2	3.2	116.5
T_5 (0.6VD)	0-30	9.35	88	43.7	83.7
	30-60	11.11	64.3	25	97.5
	60-90	13.62	52.1	8.7	111.2
T_6 (0.6VDM)	0-30	10.55	81.6	52.2	85.7
	30-60	12.12	63	32.7	98.5
	60-90	13.44	47.5	11.6	110.7
T_7 (RBM)	0-30	10.45	44.0	18.4	80.7
	30-60	12.89	32.9	16.8	86.8
	60-90	13.40	32.1	15.2	100.6
T_8 (RB)	0-30	8.90	55.9	17.3	62.5
	30-60	11.23	43.8	15.8	75.2
	60-90	13.34	25.9	10.2	88.7

Table 5: Effect of drip and plastic mulch treatments on soil moisture content and available soil nutrients (N, P, K) at different soil depths.

Treatment	Organic carbon (g kg⁻¹)	Total nitrogen (g kg⁻¹)	C:N ratio	Humic acid (g kg⁻¹)	Total microbial count (cfu/g)
T_1 (VD)	3.32	0.37	8.9:1	0.46	35×10^4
T_2 (VDM)	4.79	0.45	10.6:1	1.20	25×10^5
T_3 (0.8VD)	3.55	0.41	8.6:1	0.80	5×10^5
T_4 (0.8VDM)	4.38	0.46	9.5:1	1.32	6×10^5
T_5 (0.6VD)	3.81	0.42	9.1:1	0.62	15×10^4
T_6 (0.6VDM)	4.09	0.44	9.3:1	0.72	16×10^4
T_7 (RBM)	2.95	0.36	8.3:1	0.42	12×10^4
T_8 (RB)	2.50	0.33	7.6:1	0.38	20×10^3

Table 6: Effect of drip irrigation and plastic mulch treatments on total nitrogen and C:N ratio in surface soil (0-30 cm).

influence on conserving moisture content of surface soil (0-30 cm) than the subsurface soil. Similar trend was found for other treatments. This may be due to the fact that plastic mulch prevents water to evaporate and conserves moisture within the soil.

The available nitrogen content in surface soil (0-30 cm) was found to vary between 44 mg.kg⁻¹ and 91 mg.kg⁻¹. In subsurface soil (30-60 cm) it varied from 32.9 mg.kg⁻¹ to 91 mg.kg⁻¹ and in subsoil (60-90 cm) from 25.9 mg.kg⁻¹ to 80.5 mg.kg⁻¹. In plastic mulched soil, nitrogen mineralization is restricted due to lack of exchange of gases like oxygen for micro-organisms with the evolution of carbon dioxide. It seems soil covered with the plastic mulch contains less available nitrogen for plant compared to soil without plastic mulch.

The treatment with plastic mulch was found to contain more available phosphorous for plant compare to non mulched soils. Surface soil contained more available phosphorous than subsurface soil. The maximum available phosphorous in surface soil was found to contain 77.8 mg.kg⁻¹ in treatment T2 where as in treatment T8 it was found to be minimum as 17.3 mg.kg⁻¹. Available phosphorous in water soluble and exchangeable form could be more in 100% of irrigation requirement met through drip in comparison to ring basin irrigation

it is expected that Al, Fe and Mn will be more available in ring basin irrigation treatment (T8) which makes complexes with these metals.

The plastic mulched soil was found to contain more available potassium for the Sapota plants compared to plants grown in bare soil (without plastic mulch). Subsoil had more available potassium than surface soil as available potassium for plant generally leached down to subsoil horizon. Maximum value in surface soil was found as 87.2 mg.kg⁻¹ in treatment T4 and minimum value as 62.5 mg.kg⁻¹ in treatment T8. Due to leaching the potassium in subsoil increased to 118.4 mg.kg⁻¹ in treatment T2 and minimum as 88.7 mg.kg⁻¹ in treatment T8. This results in consistent with other research studies [19,20]. The increase in soil available P and K content may be due to greater temperature in mulched soil. It is hypothesized that plastic mulch might have changed soil moisture content and aeration conditions, which in turn resulted in the changes of soil microbial communities and redox potential, thus affected the soil phosphorus status.

Table 6 shows the effect of drip irrigation and plastic mulch on changes in organic carbon and total nitrogen in surface soil (0-30 cm). From the chemical analysis it was revealed that the carbon and total nitrogen percentage contents were greater in black plastic mulch

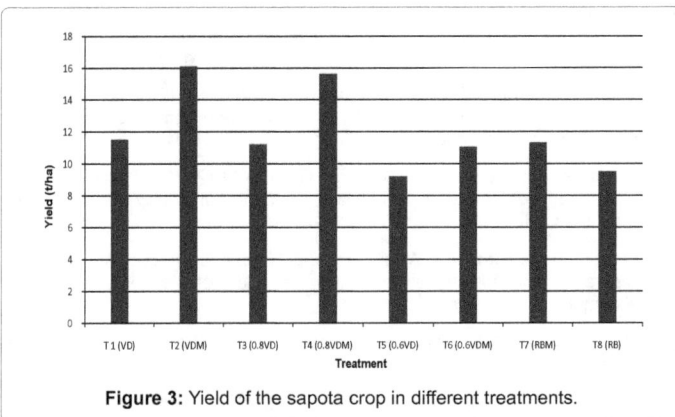

Figure 3: Yield of the sapota crop in different treatments.

treated plots compared to non-mulch. The maximum value of total nitrogen was found as 0.46 g.kg⁻¹ in treatment T4 (0.8VDM) and minimum as 0.33 g.kg⁻¹ in treatment T8 (RB). Range of C:N ratio in the surface soil was found to vary between 7.6:1 and 10.6:1. The availability of nitrogen to plant was observed highest in the treatment T2 followed by treatment T4 as 10.6 and 9.5 respectively and relatively lower in rest of the treatments. Higher C:N ratio for plant growth was observed in soil covered with black plastic mulch compared to non-mulch. The C:N ratio availability for growth of the plant was estimated to be normal in all the treatments. The C:N ratio of the organic material added to the soil influenced the rate of decomposition of organic matter and this results in the release (mineralization) or immobilization of soil nitrogen. If the added organic material contains more nitrogen in proportion to the carbon, then nitrogen is released into the soil from the decomposing organic material. On the other hand, if the organic material has a less amount of nitrogen in relation to the carbon then the microorganisms will utilize the soil nitrogen for further decomposition and the soil nitrogen will be immobilized and will not be available to plants.

The humic acid concentration was always found greater in the mulched soil than without mulch. Maximum humic acid concentration was found as 1.32 g.kg⁻¹ in treatment T4 and lowest as 0.38 g.kg⁻¹ in treatment T8. Humic acid in soil is amorphous dark brown color lignin and protein body of enzyme complexes. These enzymes remaining in microbial cell transform nutrient in ionic form and exchange by the plant root. Humus is known as store house of crop plant nutrient. The plants with drip irrigation and black plastic mulch check loss of carbon and carbon dioxide from soil. Carbon in lignin with protein body of enzyme in microbial cell which make complexes of humic acid this is expected to contain higher in plastic mulched soil with and drip irrigation due to optimum moisture content as compared to without mulch and surface irrigation.

The total microbial count was found to be greater in the plastic mulched soil than the soil without mulch. Maximum microbial count was found as 25 x 10⁵ cuf/gm in the plants for treatment T2 and lowest as 20 x 10³ cfu/gm in treatment T8, which shows that plastic mulched soil with drip irrigation increases the microbial activities in the soil.

Table 7 shows the N, P, K concentration in Sapota plant leaf. Nitrogen content in Sapota plant was always greater in mulch treated plants compared to without mulch. Highest concentration of nitrogen in leaf sample was found as 1.407 per cent in treatment T4, whereas lowest as 1.260 per cent in treatment T8.

Phosphorous content in Sapota plant was greater in plastic

mulched treatments compare to without mulch. Highest concentration of phosphorous in Sapota leaf was found as 0.0648 per cent in the treatment T6 and lowest in treatment T8 as 0.0383 per cent.

Potassium content in Sapota plant leaf was also greater in mulch treated plants compared to without mulch. Highest concentration of potassium was found as 0.78 per cent in treatment T2 and lowest as 0.53 per cent in treatment T8.

From above analysis of results, it revealed that phosphorous and potassium content was higher in mulched plants which may be due to accumulation of more organic matter and humic acid. Black plastic mulch enhances soil temperature due to restricted CO_2 evolution from soil surface. Increased temperature of soil solution the nutrient concentration like P & K also increased in the root zone. The basal ATPase and P & K activities in the presence of K+ were related with the root levels of thiscations. Similar results are also reported by Ruiz [12]. Soil temperature and moisture remain at optimum level under black plastic mulch condition during winter season. During winter the plant metabolism reduces that causes low uptake of P. The available soils P remain in the soil for residual use. This is more under mulch condition. Similar results are reported by Almeida [21].

Water use efficiency was found maximum for treatment T4 as 6.71 kg.ha⁻¹mm⁻¹ containing drip irrigation and plastic mulch in which 80% irrigation water was met through drip. The lowest value of water use efficiency was found in treatment T8 as 3.27 kg.ha⁻¹mm⁻¹ in which 100% water requirement was met through ring basin irrigation method. In ring basin irrigation there is huge loss of water is found in terms of evaporation, deep percolation, seepage loss, etc which lead to less water use efficiency than irrigation methods like drip and sprinkler.

Economic analysis of the project was carried out to determine the economic feasibility of using drip irrigation and plastic mulch in cultivation of Sapota crop. It allow us to compare different treatments containing drip only, drip with plastic mulch, ring basin with mulch, and ring basin alone. For this purpose costs for installation cost of drip, cost of mulching, labour cost, cost of fertilizer applied, cost on water and electricity, etc are considered. Based on present study all these costs and benefits obtained were taken annually and for one hectare of land. The economic analysis of different treatment in terms of Benefit-Cost ratio revealed that highest Benefit-Cost ratio was 3.59 for the treatment T2 (100% VDM) followed by 3.55 for the treatment T4 (80% VDM). The lowest benefit cost ratio of 1.84 was found under the treatment T8 (Ring Basin irrigation without mulch).

Conclusions

The maximum daily water requirement of the Sapota plant was estimated as 6.89 mm during 30th April to 6th May (18th week) and minimum as 2.14 mm in winter season during 17 to 23 December (51st week).

Treatment	N (%)	P (%)	K (%)
T₁ (VD)	1.092	0.0590	0.77
T₂ (VDM)	1.365	0.0640	0.78
T₃ (0.8VD)	1.386	0.0455	0.66
T₄ (0.8VDM)	1.407	0.0465	0.70
T₅ (0.6VD)	1.302	0.0460	0.73
T₆ (0.6VDM)	1.318	0.0648	0.76
T₇ (RBM)	1.302	0.0425	0.59
T₈ (RB)	1.260	0.0383	0.53

Table 7: Concentration of N, P, K in Sapota leaf.

Due to surface covering water is not able to escape therefore plastic mulch treatment conserved about 17.41% greater soil moisture than without mulch. Soils covered with plastic mulch and drip irrigation was found to remain soft and well aerated which is favorable for plant growth. The porosity of soil was found to be enhanced by 4.2% due to plastic mulch. The pH value of the soil covered with plastic mulch marginally decreased due to formation of carbonic acid. Organic carbon in surface soil due to plastic mulch was found to increase from 7.35 to 44.27 % in different irrigation levels and the similar trend was observed for organic matter content in the soil. Available potassium increased to vary between 2.39% and 26.9% and available phosphorus content increased to vary between 14.91% and 47.62% in soil covered with plastic mulch as compared to non mulced soil. Available nitrogen in mineralized form was generally found to decrease from 7.27% to 24.72% in soil covered with black plastic mulch as compared to bare soil due to depletion of oxygen for normal functioning of micro-organism. Total nitrogen content was found to be increased from 4.76% to 21.62% in plastic mulched soils than without mulch. It revealed that due to slow mineralization the release of nitrogen is restricted in soil covered with plastic mulch which reduces the wastage of nitrogen thereby it can be used in sustainable manner. Hence, C:N ratio was found good in plastic mulched soil than non-mulched soil. Humic acid and microbial counts were found greater in plastic mulch treated plots. Evolution of carbon dioxide from soil surface is restricted and leaching loss of nutrients is checked due to rainfall in mulched treated plots. In long duration crop like Sapota the black plastic mulch cover on soil causes cumulative accumulations of the slow organic matter from structural carbon with high lignin and plant nutrients conserve organic matter. So nutrients supplies to plant in slowly available form in sustainable manner for longer period. The analysis of N, P, K concentration in Sapota plant leaf was estimated to be more in soil plastic mulched plots than non-mulched plots. The yield of Sapota crop was found to be increased which varies from 7.62% to 41% in different irrigation treatments with plastic mulch. Yield of Sapota crop was found to increase by 21.05% due to drip in comparison of ring basin irrigation.

This study shows that the drip irrigation in combination of plastic mulch enhances Sapota yield and improves soil properties.

Acknowledgement

Authors are thankful to the National Committee on Plasticulture Application in Horticulture (NCPAH), Department of Agriculture and Cooperation, Ministry of Agriculture, Government of India for providing necessary funds to conduct this research studies.

References

1. Tiwari KN, Mal PK, Singh RM, Chattopadhyay A (1998) Response of Okra (Abelmoschusesculentus (L.) Moench.) to drip irrigation under mulch and non-mulch condition. Agr Water Manage 38: 91-102.

2. Tiwari KN, Mal PK, Singh RM, Chattopadhyay A (1998) Feasibility of drip irrigation under different soil covers in tomato. J AgricEngg 35: 41-49.

3. Tiwari KN, Singh A, Mal PK (2003) Effect of drip irrigation on yield of cabbage (Brassica oleracea L. var. capitata) under mulch and non-mulch conditions. Agr Water Manage 58: 19-28.

4. Bryla DR, Thomas JT, James EA, Johnson RS (2003) Growth and production of young peach trees irrigation by furrow, microjet, surface drip or subsurface drip systems. HortScience 38: 1112-1116.

5. Kanannavar PS, Kumathe SS, Premanand BD, Kawale N (2009) Water saving and economics of banana production under drip irrigation in north eastern dry zone of Karnataka. Birasa Agricultural University, Ranchi: 39-42.

6. Pawar BR, Landge VV, Deshmukh DS, and Yeware PP (2010) Economics of banana production in drip irrigated and flood irrigated gardens. International Journal of Commerce and Business Management 3: 88-91.

7. Gunduz M, Korkmaz N, Asik S, Unal HB, Avci M (2011) Effects of various irrigation regimes on soil water balance, yield and fruit quality of drip-irrigated peach trees. J Irrig Drain Eng 137: 426-434.

8. Ramakrishna A, Tam HM, Wani SP, Long TD (2006) Effect of mulch on soil temperature, moisture, weed infestation and yield of groundnut in northern Vietnam. Field Crop Res 95: 115–125.

9. Tarara JM (2000) Micro climate modification with plastic mulch. Hort Science 35: 169-180.

10. Kulkarni AP, Policegoudra RS, Aradhya SM (2006) Chemical and antioxidant activity of Sapota (AchrasSapota) fruit. J Food Biochem 31: 399-414.

11. Proietti P, Antognozzi E (1996) Effect of irrigation on fruit quality of table lives (Oleaeuropaea), cultivar 'Ascolanatenera'. New Zealand Journal of Crop and Horticultural Science 24: 175-181.

12. Ruiz MJ, Hernandez J, Castilla Nicolas, Romero L (2002) Effect of soil temperature on K and Ca concentrations and on ATPase and pyruvate kinase activity in potato roots. Hort. Science 37: 325-328.

13. Allen RG, Pereira LS, Raes D, Smith M (1998) Crop evapotranspiration-guidelines for computing crop water requirements. Irrig and Drain Paper No 56, FAO, Rome, Italy.

14. Cetin O, Uygan D (2008) The effect of drip line spacing, irrigation regimes and Planting geometries of tomato on yield, irrigation water use efficiency and net return. Agr Water Manage 95: 949-958.

15. Doorenbos J, Pruitt WO (1977) Guidelines for predicting crop water requirements.

16. Wu LH, Zhu ZR, Liang YC, Shi WY, Zhang LM (1999) A high-yielding, water-saving and fertilizer-saving cultivation technique for rice mulched by plastic film under dry land condition. J Zhejiang AgricUniv 25: 41-42.

17. Liu M, Wu LH (2003) Study on changes of soil fertility in rainfed paddy soils with mulching plastic film. Acta Agriculture Zhejiangensis 15: 8-12.

18. Khan AR(2002) Mulching effects on soil physical properties and pea nut production. Ital J Agron 6: 113-118.

19. Yong SL, Wu LH, Li MZ, Xing HL, Qiao LF, Fu SZ (2007) Influence of continuous plastic film mulching on yield, water use efficiency and soil properties of rice fields under non-flooding condition. Soil and Tillage Res93: 370-378.

20. Liang YC, Hu F, Yang MC, Zhu XL, Wang GP et al. (1999) A study on high-yielding and water-saving mechanisms of upland rice mulched by plastic film. SciAgricSinica 31: 26-32.

21. Almeida DO, Fillho OK, Almeida HC, Gebler L, Felipe AF (2011) Soil microbial biomass under mulch types in as integrated apple orchards from Southern Brazil. SciAgric 68: 217-222.

17

Performance Evaluation and Development of Daily Reference Evapotranspiration Model

Hashem A[1,2]*, Engel B[1], Bralts V[1], Radwan S[2] and Rashad M[2]

[1]Department of Agricultural and Biological Engineering, Purdue University, USA
[2]Department of Agricultural Engineering, Suez Canal University, Egypt

Abstract

Using agricultural water wisely in irrigated fields is very important, especially with water scarcity in arid and semi-arid countries globally. An accurate irrigation water requirement calculation is required to determine real time irrigation scheduling, in order to apply the specific amount of irrigation water at the right time, and avoid crop growth stress which leads to reduced crop production. The main objective of this paper is to develop a mathematical model to accurately calculate daily Reference Evapotranspiration (ETo) as a first step for the accurate calculation of irrigation water requirements. Also, the model output was compared to ETo estimated using CROPWAT, an irrigation software program used for ETo calculation and irrigation scheduling. The reference evapotranspiration model was built using the Food and Agricultural Organization FAO-56 Penman-Monteith equation with the SIMULINK tool in MATLAB software. The model was validated by comparing daily estimates of evapotranspiration with Class A pan and evapotranspiration gauges in the United States. The results indicated a good fit between daily ETo calculated by the model and that observed from Class A pan and evapotranspiration gauge. There were some discrepancies between measured, modeled and CROPWAT ETo. This model is the first step to calculate accurate irrigation water requirements.

Keywords: Reference evapotranspiration; FAO Penman Monteith; Modeling; Irrigation scheduling; Irrigation water requirement

Introduction

Efficient management of irrigation water involves precise irrigation scheduling. To achieve this, an accurate crop water requirement calculation is required. Irrigation is a practice to apply water to the root zone of a crop to reach field capacity. Water use efficiency is driven by three factors; the specific amount of water applied, the timing of the application, and the efficiency of the irrigation method. Irrigation scheduling aims for yield maximization, high irrigation efficiency, and crop quality improvement by adding the appropriate amount of water to the crop in order to bring the soil moisture to the desired level. Crop water requirement is the aggregate volume of water needed to satisfy the evapotranspiration from a specific crop. Crop water requirement varies in two dimensions, spatial and temporal. Reference evapotranspiration is the proportion of evapotranspiration from a uniform reference crop with a crop height 0.12 m from an extensive surface of a green grass of uniform height, well irrigated, actively growing, and completely covering the soil. Reference ET is a major factor required for irrigation water requirement calculations and crop irrigation scheduling. Mathematical modeling is an essential tool to estimate ET and crop water requirements for best water management practices, and further, it is important for irrigation scheduling and irrigation water management.

The objective of this research was to develop a tool to: (1) simulate daily reference ET (ET_o) using real time climatological data, rather than using historical climate data such as that in CLIMWAT, and (2) calibrate it to accurately calculate daily ET_o as a first step for accurate calculation of irrigation water requirements. This study contains two parts, the first, to build the reference evapotranspiration model using the United Nations Food and Agricultural Organization Penman-Monteith (FAO56-PM) equation. This was done using the SIMULINK tool in MATLAB software. The model was validated by comparing daily ET_o calculated by the model versus evapotranspiration using a Class A evaporation pan and evapotranspiration gauges in the United States. The second step is a *comparison* of monthly ET_o estimated from the model using daily data obtained from weather stations with both ET_o measured from the evaporation pan and ET_o calculated using CROPWAT.

Background

Evapotranspiration is the primary consumer of irrigation water and rainfall from an agricultural field. A correlation between evapotranspiration and crop yield has been published for different ET levels and their effects on crop yield [1]. Evapotranspiration is a driving factor for both hydrological and climatological research, in addition to irrigation management [2]. ET determination is commonly preceded by estimation of ET_o [3].

ET model validation requires measurements of evapotranspiration. ET models are often used due to the difficulty and cost of ET measurement. There are different ways for directly measuring evapotranspiration, for example weighing lysimeters and eddy covariance. Indirect measurement includes soil water balance and surface energy balance, using conservation of mass and energy balance [4]. With advances and technology improvement in data acquisition and measurement, improvement of ET estimation is possible, especially with measurement of near vegetation surface climate elements and surface energy exchange [5].

ET_o estimation from weather data has been used in different applications of crop water requirement and irrigation water management calculations. In developing nations, where there is a shortage of direct measurements of ET using lysimeters or soil moisture balances, most irrigation consultants estimate ET_o based on meteorological data. The Penman–Monteith FAO 56 (PMF-56) equation is recommended for

*Corresponding author: Hashem A, Department of Agricultural and Biological Engineering, Purdue University, USA, Department of Agricultural Engineering, Suez Canal University, E-mail: ahashem@purdue.edu

the estimation of reference evapotranspiration and provides reliable ETo values under different climate conditions [6,7].

The Penman–Monteith FAO 56 is recognized worldwide as a reasonable ET_o estimator in comparison with other methods [8]. Most irrigation planners, climatologists, hydrologists and agronomists use it in research field applications [9]. The PMF-56 has a major disadvantage as it needs multiple meteorological elements, and this is not applicable in developing countries [10,11]. There are several models to estimate ET_o, such as Ref-ET and CROPWAT. CROPWAT primarily imports weather data from CLIMWAT, which is a database containing historical climate data. Ref-ET is software, but it must be purchased to obtain its full capabilities. Ref-ET also contains a variety of equations that can be used to estimate ET_o thereby facilitating comparison of different ET estimation methods at a location.

ET_o can be estimated using weather station data, measured by Bellani plate evapotranspiration gauges, or obtained from the evaporation pan multiplied by K_{pan} factor [12]. Evapotranspiration estimation models require input data that are field observations and derived or assumed parameters. Field measurement of meteorological variables is a critical part of the evaporation estimation process. Measurements and recording errors in field variables result in ET estimation errors [4]. There has been significant progress in the capability of near surface meteorological variable measurement such as temperature, precipitation, wind speed, solar radiation, and humidity using automated climate stations [13]. This has the effect of simplifying ET model usage.

The Bellani plate evapotranspiration gauge (atmometer) is another way to measure ET_o by using a plate to simulate water evaporation from a green surface to match short canopy reference evapotranspiration. The ET measured using a Bellani gauge is inaccurate, especially in humid climates, where poor performance occurs on rainy days. The ET_o estimated using ET gauges is 27% lower than the FAO56-PM ET_o. The correction factor between the evaporation rate (E_A) and ET_o was 0.84 as expressed in the following $E_A\big/0.84 = ET_o$ [14].

Models for Irrigation Planning

CROPWAT

CROPWAT uses the Penman-Monteith equation [15] for computing reference evapotranspiration. The reference evapotranspiration is used to calculate crop water requirement and irrigation scheduling [15,16].

CROPWAT has a user friendly interface with input and output menus. The input data consists of the following: monthly weather data to estimate ET_o, monthly rainfall data, cropping pattern and crop coefficient data, and soil type. The irrigation schedule is calculated based on the input data. Different methods are used in CROPWAT to calculate irrigation scheduling; once an appropriate method is selected, the irrigation dates and amounts will be calculated [17]. CROPWAT provides results at a monthly time step, which is not accurate enough for real time irrigation management. CROPWAT can provide outputs with a daily time step, but the data must be entered manually, which is time consuming and prone to errors.

CLIMWAT is a meteorological dataset used to export the input files to CROPWAT to calculate the crop water requirement and irrigation scheduling for different crops for more than 5000 stations worldwide. CLIMWAT exports the following climate elements: Monthly maximum and minimum temperature (°C), wind speed (km/day), relative humidity (%), solar radiation (MJ/m²/day), sunshine

hours per day, monthly rainfall (mm month⁻¹), effective rainfall (mm month⁻¹) and calculated reference evapotranspiration (mm day⁻¹).

The CLIMWAT historical monthly data typically is not accurate enough to calculate reference evapotranspiration, which leads to inaccurate estimates of ET_o, causing stress on plants due to insufficient irrigation or over irrigation, resulting in yield losses or crop failure. Irrigation water requirement calculated based on daily weather data is more accurate than average monthly data because the actual need for plants is determined. All required weather elements are not available in each CLIMWAT station, and many weather stations merely measure air temperature and precipitation. As a result, the information in such datasets should never replace the actual data [18].

ET_o observed from pan evaporation

In many regions, evaporation pans are widely used because of the simplicity of the method, as well as being inexpensive in comparison with ET measurement and its application. Evaporation pans are useful in some locations, where no weather data is available. In Egypt for example, agronomists used the evaporation pan for Egyptian clover and maize irrigation scheduling in Kafr El-Sheikh and Giza 1 in Giza [19].

The depth of water evaporated from the pans is easy to measure by subtracting the new depth of water from the initial water depth. The pan measurement is a combination of different climatological factor effects on a free open water surface, including wind, radiation, humidity, and temperature. In recent years, the evaporation rate from pans has been the subject of much debate. However, there are other considerations which contribute significantly to water loss from open water surfaces rather than from crop surfaces. The pan side heat transfer affects the energy balance, and the pan heat storage, which evaporates water throughout the night. Also, turbulence variances, air temperature, and relative humidity differ beyond the water and crop surface [20]. Validated a model of evapotranspiration based on the Penman-Monteith method at two locations southern Italy and southern France in Europe, using soybean datasets, permanently stressed, planted in the Mediterranean weather, with a semi-arid and a semi-humid weather, respectively. The model provided good results for the two sites with hourly, daily, and seasonal time scales [21]. Validated an evapotranspiration model using meteorological and lysimeter evapotranspiration hourly data sets at Davis, California, and daily time steps at Policoro, Southern Italy. The model output was validated with the ET estimated using the FAO Penman-Monteith method, and the model reference ET estimate is reasonable on two time steps hourly and daily.

Materials and Methods

In this research, an ET_o model was developed to investigate estimation of daily reference evapotranspiration using meteorological data. To validate the model, ET_o data from the class A evaporation pan at Dubois, Indiana and evapotranspiration gauges at Purdue Center for Research and Education (ACRE), West Lafayette, Indiana, USA, were compared with ET_o estimated by the model in both locations. This model uses the FAO PM-56 as this method fits different locations globally with the same inputs, in addition to having a user friendly interface.

The Simulink tool in MATLAB was used to build the ETo model using the FAO Penman-Monteith equation expressed by [18]. The main inputs of the model are the daily averages of climate elements: maximum and minimum air temperature, air humidity, wind speed, and solar radiation as shown in Figure 1. Also, the latitude, longitude, and altitude are required.

The Penman Monteith-FAO 56 equation:

$$ET_o = \frac{0.408\Delta\left(R_n - G\right) + \gamma\dfrac{900}{T+273}U_2\left(e_s - e_a\right)}{\Delta + \gamma\left(1 + 0.34U_2\right)} \tag{1}$$

Where ET_o is reference evapotranspiration (mm day^{-1}), R_n is net radiation at crop surface (MJ m^{-2} day^{-1}), G is soil heat flux density (MJ m^{-2} day^{-1}), T is mean daily air temperature at 2 m height (°C), U_2 is wind speed at 2 m height (m s^{-1}), e_s is saturation vapor pressure (kPa), e_a is actual vapor pressure (kPa), e_s-e_a is saturation vapor pressure deficit (kPa), Δ is slope vapour pressure curve (kPa °C^{-1}) and γ is psychrometric constant (kPa °C^{-1}).

For as and bs, average values (as =0.25, bs =0.50) as recommended by FAO were used [18]. The ET_o model produces daily reference evapotranspiration (mm day^{-1}).

For this research, data was obtained from the NOAA database website and Wunderground database website for Dubois S IN forage farm, IN, USA (Station ID: GHCND: USC00122309) located at 38.46° N and 86.69° W, with 210.3 m elevation above sea level from May 17th to July 31st 2006, May 5th to October 14th 2010, April 12th to September 30th 2011, and May 4th to October 31st 2012. For this data set the missing data was not replaced. For the ACRE site, the data were collected from the Indiana State Climate Office website. Data at the ACRE site was obtained using a Bellani plate evapotranspiration gauge (atmometer). ACRE is located at 40.47° N and 86.99° W, with 214 m elevation above sea level for the growing season (May 1st to October 31st, June 1st to October 31st) for 2010 and 2011.

The principle weather parameters considered were maximum and minimum air temperature, air humidity, wind speed, and solar radiation. According to FAO 56, the equations for a Class A evaporation pan with green fetch are indicated as [18]:

$$ETo = K_p \times E_{pan} \tag{2}$$

Where ET_o is the reference evapotranspiration (mm day^{-1}), K_p is pan coefficient, and E_{pan} is the pan evaporation (mm day^{-1}).

Under some conditions, the K_p coefficients may need some adjustment where tall crops surrounded the evaporation pan. The daily average relative humidity, wind speed (U_2) and the upwind fetch distance of the evaporation pan location are factors affecting the Pan coefficient [18,22].

CROPWAT was also used to calculate ET_o for the two sites in Indiana, USA using monthly historical data. The input data was climate, crop, soil and planting dates. The CLIMWAT data set was based on weather station data. With the humid weather in Indiana, solar radiation and other climatic factors are affected by cloud cover. This leads to uncertainty in evapotranspiration estimation occurs.

The Nash-Sutcliffe coefficient (NS) for model performance accuracy was used in the study to validate the ET_o model by comparing predicted and observed ET_o. The Nash-Sutcliffe coefficient is a sign of the model's capability to predict about the 1:1 ratio between experimental and estimated data. Nash–Sutcliffe can be a value from negative infinity to one, efficiency of 1 means an exact ET_o values estimated by the modeled to the measured data, efficiency of 0 means the model forecasts are no

more accurate than the mean of the measured data, and efficiency less than zero means the measured mean is better than the model [23].

Results and Discussion

Model validation on a daily basis

The daily ET_o data calculated by the ET_o model for West Lafayette and Dubois, IN USA was compared with the pan evaporation and ET_o gauge observed ET_o values. The results are presented graphically in Figure 2, and the correlation coefficient (R^2) and Nash-Sutcliffe coefficients (NS) are shown. The R^2 and NS for the model and evaporation pan differ by location and year. The figure shows that

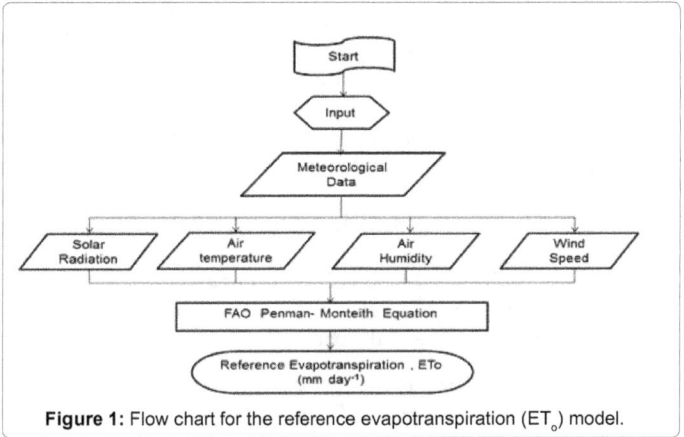

Figure 1: Flow chart for the reference evapotranspiration (ET_o) model.

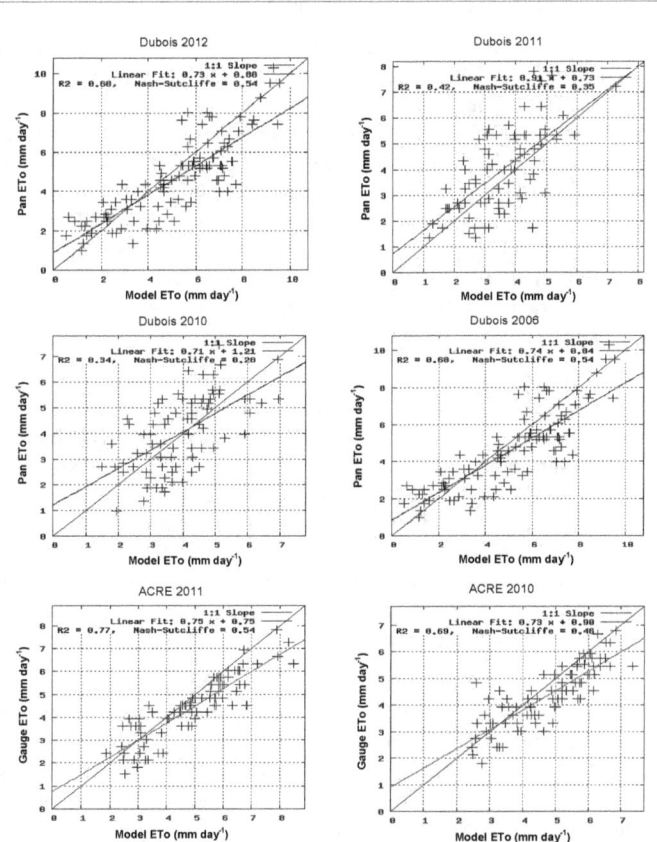

Figure 2: Reference evapotranspiration model validation between measured and simulated daily evapotranspiration from evaporation pan& gauge with model ET_o.

the ET$_o$ values calculated by the ET$_o$ model are in the range of those obtained by pan evaporation and ET$_o$ gauges for most days.

In Dubois, the relationship between ET$_o$ estimated from the model and ETo observed from the pan is linear, with differing R^2 and NS coefficients between different years. The R^2 and Nash-Sutcliffe coefficients were equal to 0.68 and 0.54 in 2012, 0.42 and 0.35 in 2011, 0.34 and 0.28 in 2010 and 0.68 and 0.54 in 2006 between the ET$_o$ model and ET$_o$ pan. The R^2 and Nash-Sutcliffe coefficients for ACRE were 0.77 and 0.54 in 2011 and 0.69 and 0.47 in 2010 between the ET$_o$ model and ET$_o$ gauge. The R^2 coefficient is better for ACRE rather than Dubois and the NS is similar between the two sites in different years.

For the Dubois site, the Nash-Sutcliffe coefficients were 0.54 in 2012, 0.35 in 2011, 0.28 in 2010 and 0.54 in 2006 as shown in Figure 2. For the ACRE site, the NS was 0.54 in 2011 and 0.47 in 2010. In 2012, there was a drought in Indiana, which meant higher temperatures and lower relative humidity than in a typical year. For the drought year, the model performance was good, as the ET$_o$ values from the model were close to the ET$_o$ values estimated from the evaporation pan and gauges. However, in 2010 the average temperature was much lower and humidity was much higher than 2012 and 2006, which appears to impact model performance in those years. Lower temperature and high humidity results in reduction of the evaporation rate from the pan and gauges, which leads to increases in the differences between the ET$_o$ estimated from the model and the pan.

For hydrology related model performance, the NS values larger than 0.4 and R^2 values greater than 0.5 are considered acceptable model performance. Satisfactory models achieving a NS coefficient higher than 0.5 and a R^2 higher than 0.6 specify acceptable model [24].

The minimum and maximum differences between the calculated from the model and measured from the pan and gauge based on daily values are in the range of -3.96 to 5.11 mm with an absolute average of 0.56 mm in 2012, for 2011 in the range -4.94 to 5.8 mm with an absolute average of 0.10 mm, for 2010 in the range -1.49 to 4.46 mm with an absolute average of 0.66 mm, and for 2006 in the range -0.61 to 5.09 mm with an absolute average of 1.34 mm for Dubois. However, in ACRE, the differences between the ET$_o$ estimated from the model and determined by the gauge ranges from -1.24 to 2.31 mm with an absolute average of 1.24 mm in 2011 and for 2010 the range was from -2.24 to 1.90 mm with an absolute average of 0.51 mm.

The results of this work indicated the ET$_o$ model provided reasonable estimates of ET$_o$ as shown in Figure 2. There is a slight variance between ET$_o$ estimated from the model and ETo obtained from the evaporation pan. The model provides higher ET$_o$ than the pan, likely due to the humid weather and the cloud cover in the study area; these results agree with the findings of [25]. In Indiana, the weather is humid and this may be the key reason that the model performance in humid years was not as accurate as performance in dry years. The high humidity reduces the evaporation from the pan, which means there is a lower evaporation from the pan.

At ACRE, the required dataset was obtained using one source, which is the weather station located in the center of the site. However, in Dubious the required dataset was obtained using two different sources - NOAA and Wunderground. The use of two sources means the use of different locations and instrumentation for each source, potentially leading to different accuracies and measurement approaches.

Model performance on a monthly basis

In order to compare the monthly performance of the model versus the evaporation pan, gauge and CROPWAT software. The daily ET$_o$ data was being averaged on a monthly basis for the evaporation pan, the model, and compared with CROPWAT ET$_o$ in DUBOIS site, and for ACRE site, the evaporation gauge, the model, and compared with CROPWAT ET$_o$ software as shown in Figure 3 .There are differences between the monthly ET$_o$ from pan, gauge, model, and CROPWAT software for both of the locations. The daily ET$_o$ estimated from the model is mostly higher than ET$_o$ from evaporation pan and gauge, and the monthly average is nearer the average of ET$_o$ pan than the monthly ET$_o$ from CROPWAT. These results prove a better performance of the ET$_o$ model with pan evaporation. The model provided a more accurate estimate with an evaporation pan data than CROPWAT.

As shown in Figure 3, in Dubois for 2012, the model provided a good estimate of ET$_o$, and there was a peak for the ET$_o$ model in July as there were high temperatures, which increased the predicted ET by the model. The ET$_o$ estimated from the model is higher than the ET$_o$ obtained from the pan and ET$_o$ calculated from CROPWAT from May to July, especially during July as the air temperature is higher than May and June, and the model sensitivity is much higher to the climate elements than the pan. However, ET$_o$ estimated from the model in September and October is closer to the ET$_o$ obtained from the pan than CROPWAT. For 2011, the ET$_o$ simulated from the model was higher than ET$_o$ from the pan and less than CROPWAT from May until July, and then the model gives higher estimates in August and then returns in September to be closer to CROPWAT than the pan. In 2010, the ET$_o$ estimated from the model is higher than both ET$_o$ estimated from the pan and calculated from CROPWAT from May through September. These results are due to higher wind speed than previous years. Then, a decline occurred in the ET$_o$ estimated by the model in October to levels approximately the same as ET$_o$ from the pan and CROPWAT. Finally, in 2006, the ET$_o$ estimated from the model is nearer ET$_o$ obtained from CROPWAT than the pan. However, the ET$_o$ estimated from the model is less than ET$_o$ CROPWAT values in June, although it is higher than ET$_o$ calculated from CROPWAT in May and July.

In ACRE for 2011, the model estimated a higher ET$_o$ than the gauge and CROPWAT from June to August, then the model estimates declined, with values relatively similar to ET$_o$ measured by the gauge. In 2010, the model estimated values were larger than those for the gauge, except in June and October when the values of the ET$_o$ model and the gauge were similar.

With respect to use of CLIMWAT and CROPWAT software with average monthly meteorological data, there are differences between monthly ET$_o$ calculated from CROPWAT, pan observations, and the gauge. Significant underestimation of ET$_o$ with similar models was detected in analyses for arid and semiarid sites under Mediterranean climate conditions [20,21,26,27].

This could result in an incorrect irrigation water requirement calculation when using CLIMWAT and CROPWAT software for an estimated ET$_o$. Over-irrigation results in an excess of water, which is priceless for many arid nations, with additional potential for increasing of groundwater level and unwanted wetness of the root zone. Under-irrigation during the growing season causes plants to wilt. Extended periods of under-irrigation may result in yield loss or crop failure.

Figure 3 illustrates the relationship between monthly reference evapotranspiration (ET$_o$) measured from evaporation pan and gauge, simulated by the reference ET$_o$ model and calculated by CROPWAT. There are differences between monthly ET$_o$ due to the use of old meteorological data in CLIMWAT. This result agrees with [28].

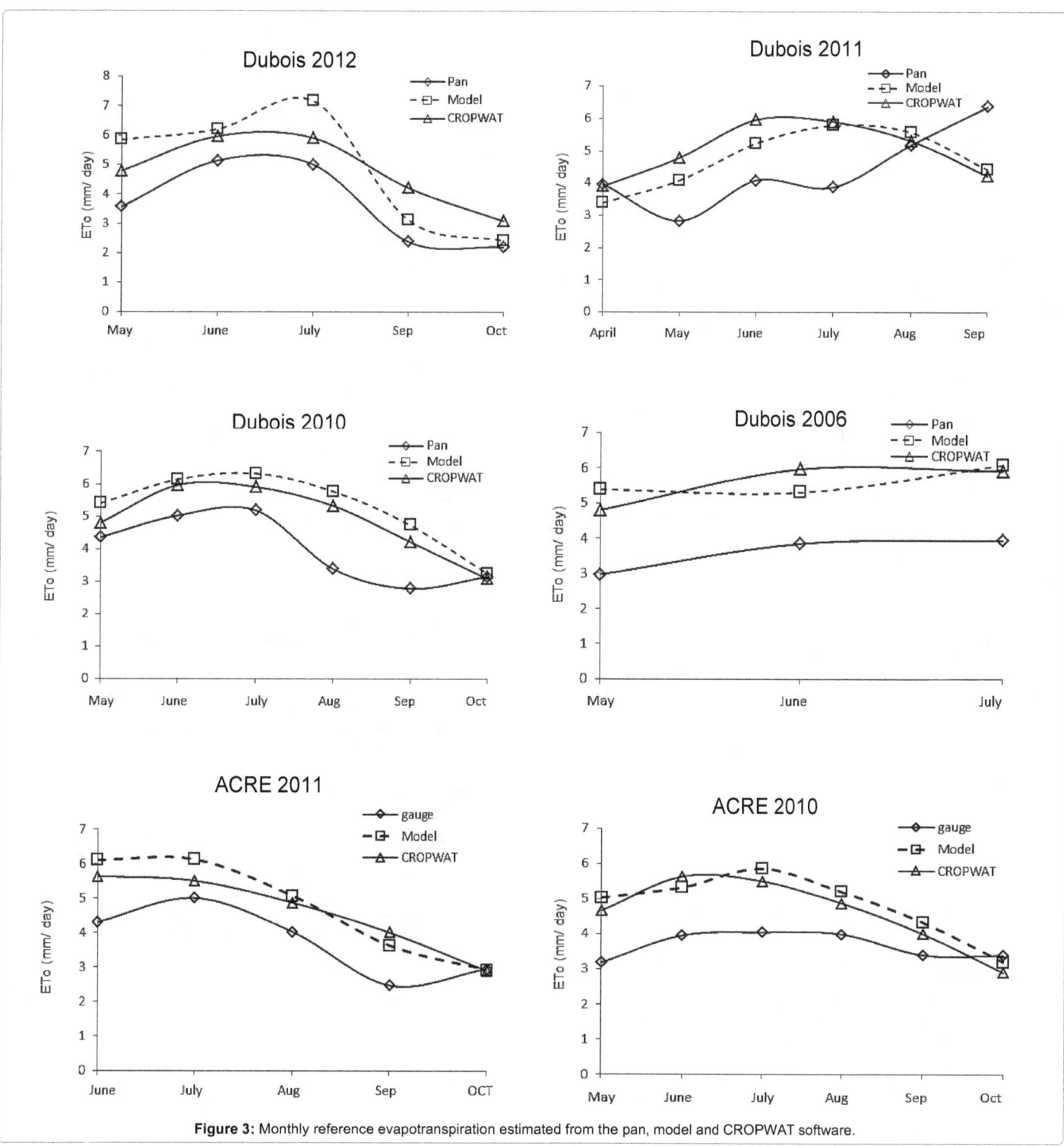

Figure 3: Monthly reference evapotranspiration estimated from the pan, model and CROPWAT software.

CLIMWAT is a reasonable meteorological dataset that contains data from 3262 climatological locations globally. In this study, CLIMWAT data was exported to CROPWAT to calculate ET_o, then was compared with ET_o estimated from the model and ET_o obtained from the gauge as shown in Figure 3. In this case, CROPWAT underestimated ET_o values as cloud affect was neglected on CLIMWAT, which reduced solar radiation. However, it may be precise when using an existing weather dataset. These results should use for preparatory applications because mean monthly data only used in this approach. These results agree with those of [18].

Summary and Conclusions

For accurate irrigation water requirement calculations, a mathematical model was built to estimate daily reference evapotranspiration from meteorological data. The model was built using the Food and Agricultural Organization Penman-Monteith equation with the SIMULINK tool in MATLAB software. The model was validated for two locations in the USA.

The process of developing the proposed model is based on the equations presented by the FAO Penman-Monteith method. The ET model uses public climatic variables measured beyond the crops. The model uses daily temperature (max, min, dew), sunshine hours and wind speed to estimate ET_o. The ET_o model simulates the daily reference ET amount from a short, green grassland. Then, the model was validated by comparing daily data between the ET_o model with ET_o pan evaporation and ET gauge in the USA.

The results of the analyses comparing model ET_o estimate with pan evaporation demonstrate that the model performed well in estimating daily ET_o from meteorological data. The model gives accurate estimates based on a daily and monthly basis, which lead to improved accuracy in ET_o estimation compared with using old weather data such as the CLIMWAT dataset. The model performance was more accurate in ACRE than Dubois, based on daily calibration between ET_o estimated from the model versus ET_o obtained from the evaporation pan and gauge, respectively. The CROPWAT estimate is typically lower than the estimate from the model created in this study and measured ET_o.

Finally, the model is a useful tool for calculating reference ET, which is needed for the accurate calculation of irrigation water requirements. Nonetheless, more calibration of this model is necessary to evaluate its appropriateness for diverse regions beyond the study areas of the United States when applied to irrigation scheduling.

Acknowledgment

We thank Prof. Mahmoud Hany Ramadan (passed away) for his feedback and suggestion to improve the data set analysis and the text. This research was supported by a governmental general mission scholarship administrated by the Egyptian Cultural and Education Bureau, Washington, DC, and by the Department of Agricultural and Biological Engineering at Purdue University.

References

1. Burt C, Mutziger A, Allen RG, Howell TA (2005) Evaporation Research: Review and Interpretation. J Irrig Drain Eng 131: 37-58.

2. Sentelhas PC, Gillespie TJ, Santos EA (2010) Evaluation of FAO Penman–Monteith and alternative methods for estimating reference evapotranspiration with missing data in Southern Ontario, Canada. Agric Water Manage 97: 635-644.

3. Lopez-Urrea R, Martin de Santa Olalla F, Fabeiro C, Moratalla A (2006) Testing evapotranspiration equations using lysimeter observations in a semiarid climate. Agric Water Management 85: 15-26.

4. Abtew W, Melesse A (2013) Evaporation and Evapotranspiration Measurements and Estimations, New York: Springer.

5. Farahani H, Howell TA, Shuttleworth W, Bausch W (2007) Evapotranspiration: Progress In Measurement and Modeling In Agriculture. Transactions of the ASABE 50: 1627-1638.

6. Allen RG, Clemmens A, Burt C, Solomon K, O'Halloran T (2005) Prediction accuracy for projectwide evapotranspiration using crop coefficients and reference evapotranspiration. J Irrig Drain Eng ASCE 131: 24-36.

7. Allen RG, Pruitt WO, Wright JL, Howell TA, Ventura F, et al. (2006) A recommendation on standardized surface resistance for hourly calculation of reference ETo by the FAO56 Penman–Monteith method. Agric Water Manage 81: 1-22.

8. Cai J, Liu Y, Lei T, Pereira LS (2007) Estimating reference evapotranspiration with the FAO Penman–Monteith equation using daily weather forecast messages. Agric For Meteorol 145: 22-35.

9. Alexandris S, Kerkides P, Liakatas A (2006) Daily reference evapotranspiration estimates by the 'Copais' approach. Agric Water Manage 82: 371-386.

10. Gocic M, Trajkovic S (2010) Software for estimating reference evapotranspiration using limited weather data. Comput Electron Agric 71: 158-162.

11. Tabari H, Hosseinzadeh TP (2011) Local calibration of the Hargreaves and Priestley–Taylor equations for estimating reference evapotranspiration in arid and cold climates of Iran based on the Penman-Monteith model. J Hydrol Eng ASCE 16: 837-845.

12. Xing Z, Chow L, Meng F, Rees HW, Monteith J, et al. (2008) Testing reference evapotranspiration estimation methods using evaporation pan and modeling in Maritime region of Canada. J Irrig Drain Eng ASCE 134: 417-424.

13. Snyder RL, Brown PW, Hubbard KG, Meyer SJ (1996) A guide to automated weather station networks in North America. Advances in Climatology 4: 1-61.

14. Irmak S, Dukes M, Jacobs J (2005) Using Modified Bellani Plate Evapotranspiration Gauges to Estimate Short Canopy Reference Evapotranspiration. ASCE 131: 164-175.

15. Smith M (1992) CROPWAT, a computer program for irrigation planning and manegment. Food & Agriculture Org, Italy.

16. Clarke D (1998) CropWat for windows: user guide. FAO.

17. Ali MH (2011) Practices of irrigation & on-farm water management. Springer.

18. Allen RG, Pereira L, Raes D, Smith, M (1998) Crop evapotranspiration (guidelines for computing crop water requirements). FAO Irrigation and Drainage, Italy.

19. Abdel-Wahed MH, Snyder RL (2008) Simple Equation to Estimate Reference Evapotranspiration from Evaporation Pans Surrounded by Fallow Soil. J Irrig Drain Eng 134: 425-429.

20. Rana G, Katerji N, Mastrodli M, El Moujabber M, Brisson N (1996) Validation of a model of actual evapotranspiration for water stressed soybeans. Agricultural and Forest Meteorology 1996: 86: 215-224.

21. Todorovic M (1999) Single- layer evapotranspiration model with variable canopy resistance. Journal of Irrigation and Drainage Engineering 125: 235-245.

22. Doorenbos J, Pruitt WO (1977) Crop water requirements. FAO irrigation and drainage paper 24. Land and Water Development Division, FAO, Rome pp: 144.

23. Burkey J (2007) Nash-Sutcliffe Model Accuracy Metric. Math works.

24. Santhi C, Arnold JG, Williams JR, Dugas WA, Srinivasan R (2001a) Validation of the SWAT Model on a Large River Basin With Point and Nonpoint Sources. Journal of the American Water Resources Association 37: 1169-1188.

25. Benli B, Bruggeman A, Oweis T, Üstün H (2010) Performance of Penman-Monteith FAO56 in a Semiarid Highland Environment. J Irrig Drain Eng 136: 757-765.

26. Caliandro A, Catalano M, Rubino P, Boari F (1990) Research on the suitability of some empirical methods for estimating the reference evapotranspiration in Southern Italy. 1st Congr of the Eur Soc of Agronomy, Rome.

27. Steduto P, Caliandro A, Rubino P, Ben Mechlia N, Masmoudi M, et al. (1996) Penman-Monteith reference evapotranspiration estimates in the Mediterranean region. Proceedings of the International Conference on Evapotranspiration and Irrigation Scheduling, USA.

28. Cornejo C, Haman D, Espinel R, Jordan J (2006) Irrigation Potential of the TRASVASE System Santa Elena Peninsula, Guayas, Ecuador. J Irrig Drain Eng 132: 453-462.

18

Influence of Concentration and Type of Clay Particles on Dripper Clogging

Oliveira FC[1], Lavanholi R[1], Camargo AP[1]*, Frizzone JA[1], Ait-Mouheb N[2], Tomas S[2] and Molle B[2]

[1]Department of Biosystems Engineering, Luiz de Queiroz College of Agriculture, University of São Paulo, Piracicaba, São Paulo, Brazil
[2]National Research Institute of Science and Technology for Environment and Agriculture, UMR G-EAU Montpellier, France

Abstract

The leading causes of emitters' clogging are known, although the processes involved are seldom studied. The present research is based on the hypothesis that the susceptibility of drippers to clogging is influenced by the emitter discharge, the type of clay, and the concentration of clay in the irrigation water. The objective of this study was to analyse the susceptibility of drippers to clogging caused by water containing suspended clay particles. The susceptibility of the drippers to the clogging was analysed with respect to the following factors: the concentration of suspended clay in water, the discharge of emitters of same labyrinth geometry, and the type of clay particles in suspension. We used four concentrations of kaolinite and montmorillonite (500, 750, 1,000, and 2,000 mg L^{-1}) and two drip line models with similar labyrinth geometries, one model having a lower flow rate (0.6 L h^{-1}) than the other (1.7 L h^{-1}). The concentration of suspended clay particles affected the flow rate of the drippers, particularly at concentrations above 1000 mg L^{-1}. The drip line model with the lower flow rate was more susceptible to variations in the flow rate than the higher-flow rate model. The type of clay had no significant effect on the dripper clogging.

Keywords: Plugging; Emitter; Clay particles; Flow rate; Micro irrigation

Introduction

Clogging of emitters has been identified as the main limitation of microirrigation systems. In addition to reducing the durability of equipment [1], it can compromise the water application uniformity [2,3]. The uniformity of distribution is an important parameter for the performance of irrigation systems, as it expresses the variations in volume of water applied at different points on a given surface. In particular, in areas where fertigation occurs, a high uniformity of distribution is essential to ensure that plants receive equivalent amounts of nutrients [4].

Clogging can be caused by different factors (physical, chemical, or biological) [5]. Clogging caused by physical processes has been identified as the most common type and is due to particles in suspension. Generally, these particles are part of the soil and can be classified according to their mean diameter, as follows: sand (2 to 0.05 mm), silt (0.05 to 0.002 mm), and clay (less than 0.002 mm) [5,6].

Clay particles are usually too small to clog drippers. However, small-diameter particles can pass through the filtering system and reach the drippers [7]. Inside the drippers, depending on the characteristics of the flow in the labyrinth, particles can be deposited in the vortices and stagnation zones, increasing the potential of clogging [8,9].

Although the main causes of clogging are known, thorough studies on the processes involved are lacking. The present study is based on the hypothesis that the susceptibility of drippers to clogging is influenced by the emitter flow rate, the type of clay, and the concentration of clay in the irrigation water. To further our understanding of the clay clogging process, the objective of this study was to analyse the susceptibility of drippers to clogging caused by water containing suspended clay particles. We analysed the performance of drippers exposed to water containing clay particles in suspension. To study the susceptibility of the drippers to clogging, the following factors were analysed: the concentrations of clay suspended in water, the flow rate of drippers having the same labyrinth geometry, and the types of clay particles in suspension (kaolinite and montmorillonite).

Materials and Methods

The experiment was performed on a workbench designed for clogging tests, at the ESALQ/USP Irrigation Material Testing Laboratory in Piracicaba, SP.

The test bench was equipped with a 250 L water reservoir, a pump, a mixer, a manifold with symmetrical bifurcations, a set of collectors and an automated system for the continuous flow rate monitoring of 32 drippers. The water flow was equally distributed among eight lateral lines of equal length (2.8 m), set up in parallel.

The test pressure (100 kPa) was monitored using a digital pressure gauge installed at the top of one of the lateral lines. The collectors were equipped with pressure transducers (Motorola/Freescale-MPX5010DP) connected to a data-acquisition system, which were used for monitoring the variations of the water level and the subsequent calculation of the flow rates of the drippers. Pinch solenoid valves were installed in the bottom part of the collectors, for drainage. The water drained by the collectors was channelled to the reservoir through gutters, perpetuating water recirculation through the system.

Two types of clay compounds were used in this study: (1) kaolinite (Kt), clay type 1:1, density of 2419.9 kg m^{-3}, 2.651-μm average diameter of particles, commercial brand named kaolin extracted from a mineral source located at Pântano Grande, Rio Grande do Sul, Brazil; (2) montmorillonite (Mt), clay type 2:1, density of 2364.2 kg m^{-3}, 2.296 μm average diameter of particles, commercial brand named Brasgel provided by the company Bentonit União Nordeste S.A. (Brazil). Kt was selected because of its abundance in surface waters, and Mt was

*Corresponding author: Camargo AP, Department of Biosystems Engineering, Luiz de Queiroz College of Agriculture, University of São Paulo, Piracicaba, São Paulo, Brazil, E-mail: apcpires@usp.br

chosen for its sensitivity to the physicochemical and hydrodynamic conditions that may lead to aggregation phenomena in irrigation systems.

The flow rate of the drippers was initially determined using distilled water, without adding particles, to avoid the risk of clogging. This flow rate is referred to as the initial flow (qi). Further testing was subsequently performed with clay particles in suspension, resulting the final flow (qf).

The effect of the concentration of clay particles on the susceptibility of the drippers to clogging was studied intermittently over 40 h, by alternating between 8-h operation periods and 16 h of rest. From one concentration to the other, the drip lines were replaced with new ones, and the bench was cleaned in order to avoid residual effects on future tests.

We attempted to maintain the pH of the water at approximately 6 to reduce the risk of ions precipitation and the biological activity. pH corrections were performed when necessary by using hydrochloric acid (HCl).

The sensitivity of drippers to clogging was assessed based on the average values of the relative flow rate (average of 15 drippers). The average value was obtained over an 8-h testing period, corresponding to 384 automated flow rate measurements. The relative flow rate (qr) was expressed as a percentage and it depends on the initial flow rate (qi) and final flow rate (qf) of the drippers (equation 1).

$$q_r = 100 \frac{q_f}{q_i} \tag{1}$$

The results were presented in graphs, indicating the average relative flow rate as a function of the elapsed testing time. Values of the average relative flow rate are written above whiskers. Whiskers indicate standard deviation ranges. Additionally, the mean values are followed by a letter used to indicate if the average values are significantly different from each other. Mean values followed by the same letter does not present significant difference in statistical terms. The experimental design was completely randomized, and a Tukey's test (p<0.05) was performed to compare the relative flow rate averages obtained for a given tested parameter over the same period of time. The Turkey's test is a statistical test that can be used to find whether mean values are significantly different from each other or not.

Four concentrations of each clay compound (Kt and Mt) were analysed. The concentrations used were C1 (500 mg L^{-1}), C2 (750 mg L^{-1}), C3 (1000 mg L^{-1}), and C4 (2000 mg L^{-1}).

Two non-pressure compensating flat drippers were tested. Fifteen drippers of each model were evaluated simultaneously. Drip lines of the Naan Dan Jain® brand and the Tal drip model were selected because the emitters have similar geometries, with flow rates of 0.6 L h^{-1} (model A) and 1.7 L h^{-1} (model B) (Table 1).

Results and Discussion

Figure 1 shows the effect of the kaolinite (Kt) concentration on the flow rate of drip line model A with respect to the testing time. At the concentration C1, there was a progressive decrease in the relative flow rate, with a reduction of 2% after 40 h of testing. The C2 concentration results a flow rate reduction of 7% after 24 h of testing, and subsequently there was a small increase in the flow rate (1%), which was then maintained until the end of the 40-h testing period. The same trend was observed for the concentration C3, with an initial reduction of 5% after 16 h and a subsequent increase (1%), followed by the flow rate remaining relatively constant until the end of the test. The higher concentration (C4) yielded a greater fluctuation of the relative flow rate values. On the first day of testing (8 h), the reduction reached 2%; then, a progressive increase occurred, with the flow rate returning to the initial value after 24 h of testing. Finally, after 40 h of testing, there was a reduction of 9% in the relative flow rate.

Analysis of the concentrations at each testing time step reveals that no significant difference occurred on the first day (8 h). Significant differences began to appear after 16 h. At 16 h, the concentration C3 led the smallest reduction (5%). At 24 h, the concentration C2 yielded the smallest reduction (7%). After 32 and 40 h of testing, the concentrations C2 and C4 were similar and differed from the concentration C1.

The increase in Kt concentration did not significantly influence the flow rates of the model B drippers. The flow rate reductions over time were not significant, even after 40 h of testing. The largest reduction (4%) was obtained for the highest concentration of particles (C4). However, there was no significant difference between the average discharge at each concentration and at any testing time, except at 16 h, when there was a difference between the concentrations C1 and C2 (Figure 2).

Manufacturer/Model	NaanDanJain/ Taldrip	NaanDanJain/ Taldrip
Nominal discharge (L h^{-1})	1.7	0.6
Average discharge under 98.1kPa (L h^{-1})	1.55	0.55
Coefficient of variation - CVq (%)	1.7	2.0
Design		
Labyrinth dimensions (top view / units=mm)		
Labyrinth depth (mm)	0.71	0.42

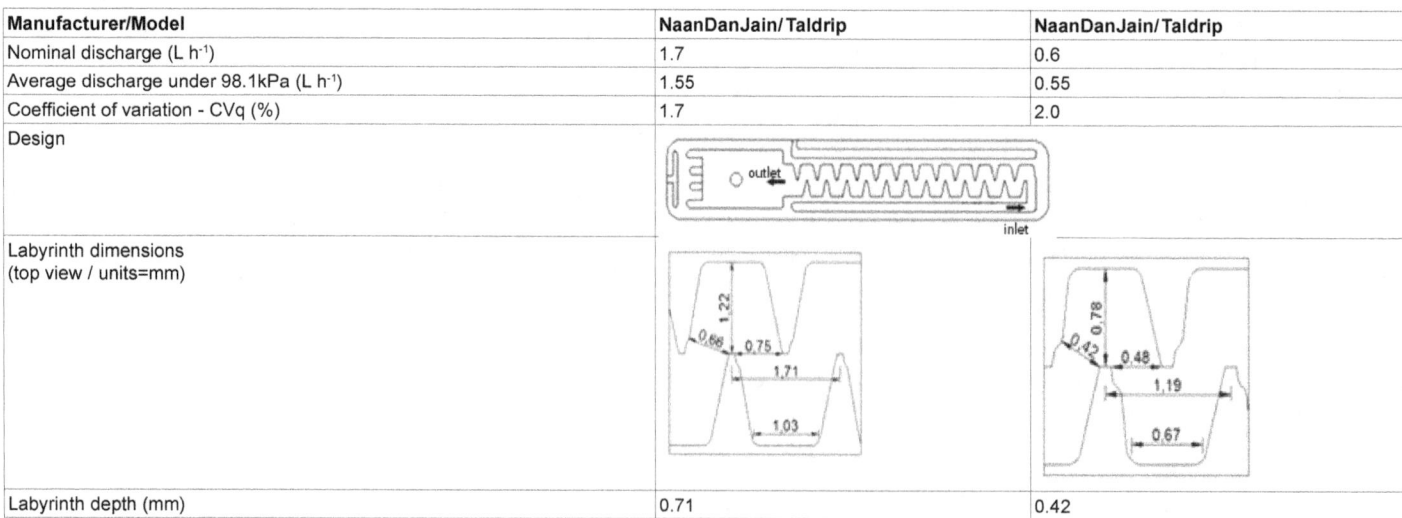

Table 1: Characteristics of the evaluated dripline models.

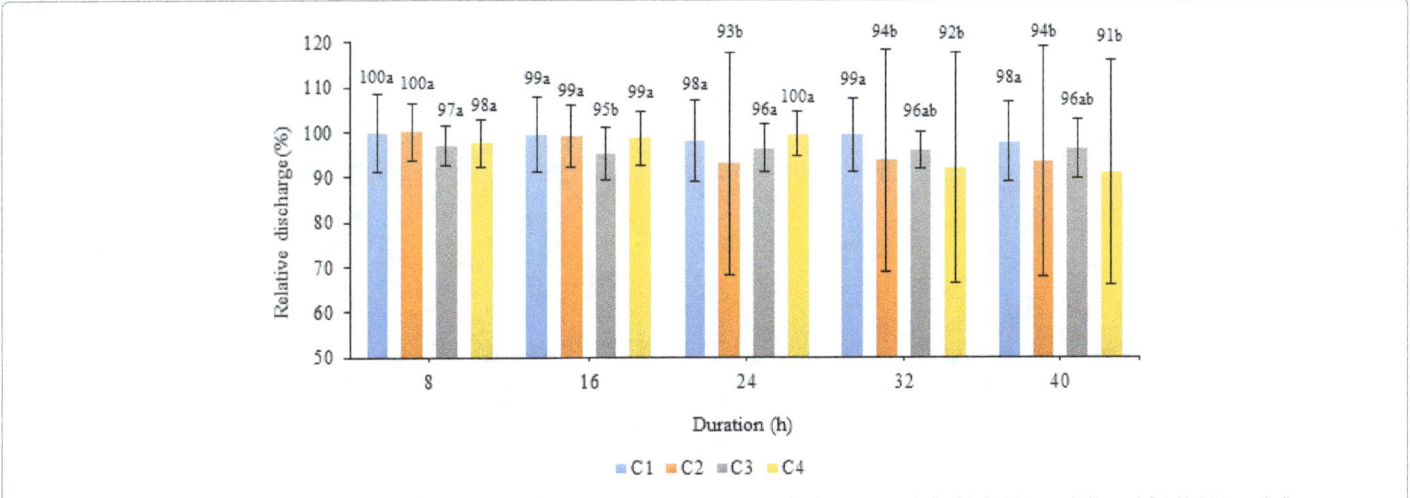

Figure 1: Effect of the concentration of Kt particles in the suspension — C1 (500 mg L⁻¹), C2 (750 mg L⁻¹), C3 (1,000 mg L⁻¹), and C4 (2,000 mg L⁻¹) — on the clogging for dripline model A (0.6 L h⁻¹).

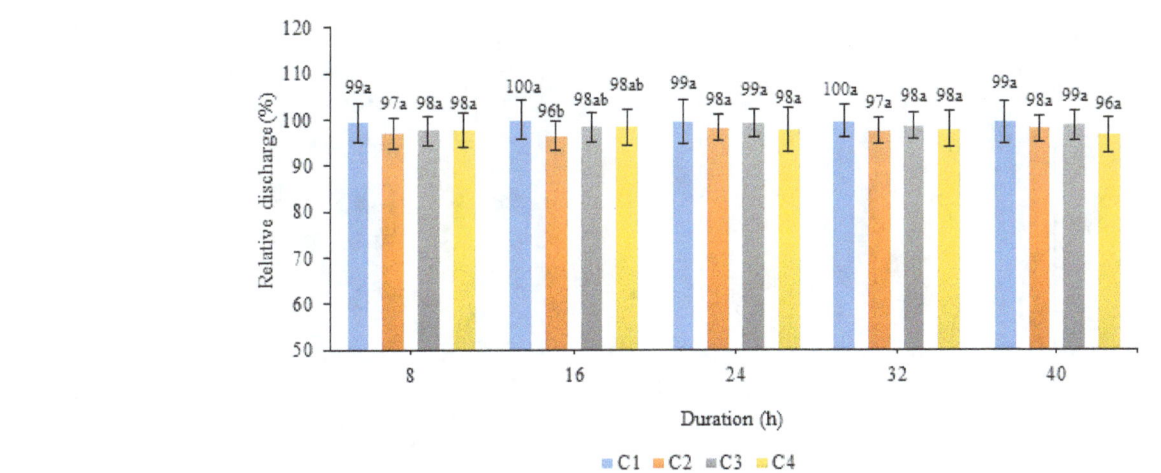

Figure 2: Effect of the concentration of Kt particles in the suspension — C1 (500 mg L⁻¹), C2 (750 mg L⁻¹), C3 (1,000 mg L⁻¹), and C4 (2,000 mg L⁻¹) — on the clogging for dripline model B (1.7 L h⁻¹).

In evaluating the effects of the concentration of Mt on the performance of dripper model A, throughout the test, the concentrations C3 and C4 resulted in the greatest reduction in the flow rate, although there was no significant flow rate difference between them (Figure 3). The concentrations C1 and C2 yielded smaller flow rate reductions, being statistically similar and differing from the concentrations C3 and C4. The largest flow rate reductions (18% and 14%) occurred after 8 h of testing, at the concentrations C3 and C4, respectively. During the test, the flow rate reduction decreased progressively, reaching 9% for both concentrations at the end of the test.

The Mt concentrations caused different flow rate behaviours for dripper model B compared with model A. In general, the flow rate reductions were smaller, with the largest reduction observed for the concentration C3 after 8 h of testing (14%) (Figure 4). After 40 h of testing, there was no flow rate reduction for the concentration C1, and there was a 3% reduction for the concentration C2. This reduction is statistically similar to that (4%) for the concentration C4.

The Mt particles induced more fluctuations on relative flow rates than Kt particles, though the effects of the particle concentration on the relative flow rate of the drippers were statistically similar for the two types of clay. In general, the flow rate reduction tends to be greater at clay concentrations higher than 1,000 mg L⁻¹. For future studies using clay particles, we suggest the use of concentrations higher than that, as lower concentrations did not result in significant effects on the performance of the drippers. In addition, higher concentrations of salt and other pH values could be tested to change the potential of particles aggregation. Finally, exposure times longer than 40 h could influence flow rate fluctuations of drippers.

The concentrations of clay particles that led the greatest flow rate reductions were higher compared with the classification for a severe clogging risk (400 mg L⁻¹) caused by suspended particles [10]. The results of the present study support the classification that considers particles with diameters less than 0.031 mm to be unlikely to cause clogging of drippers [7].

Physical clogging caused by suspended particles depends on two factors: i) the concentration and ii) the diameter of particles. A suitable

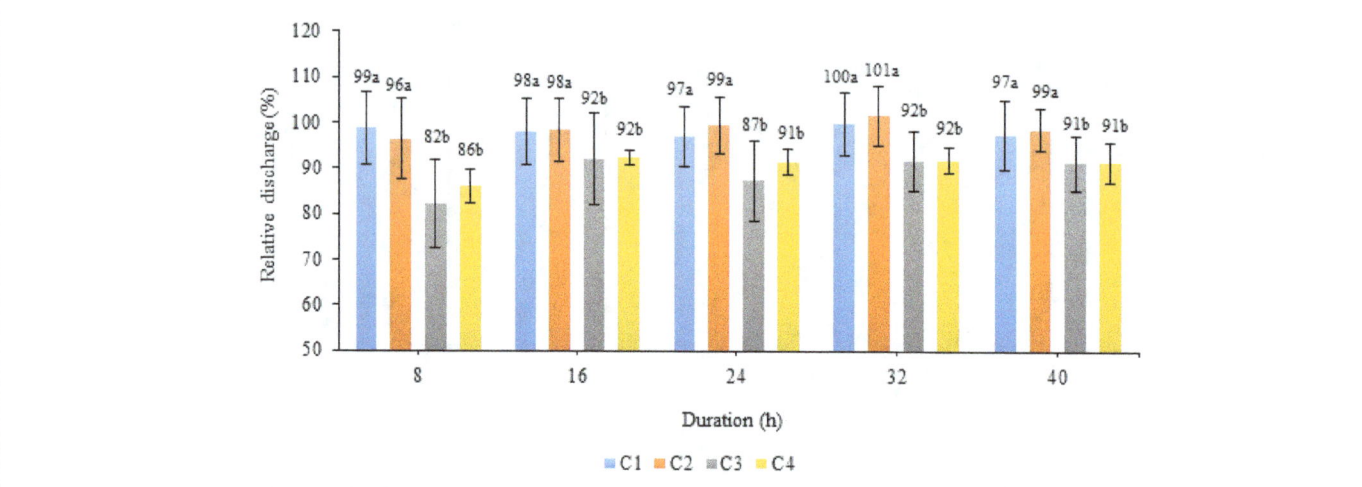

Figure 3: Effect of the concentration of Mt particles in the suspension — C1 (500 mg L⁻¹), C2 (750 mg L⁻¹), C3 (1,000 mg L⁻¹), and C4 (2,000 mg L⁻¹) — on the clogging for dripline model A (0.6 L h⁻¹).

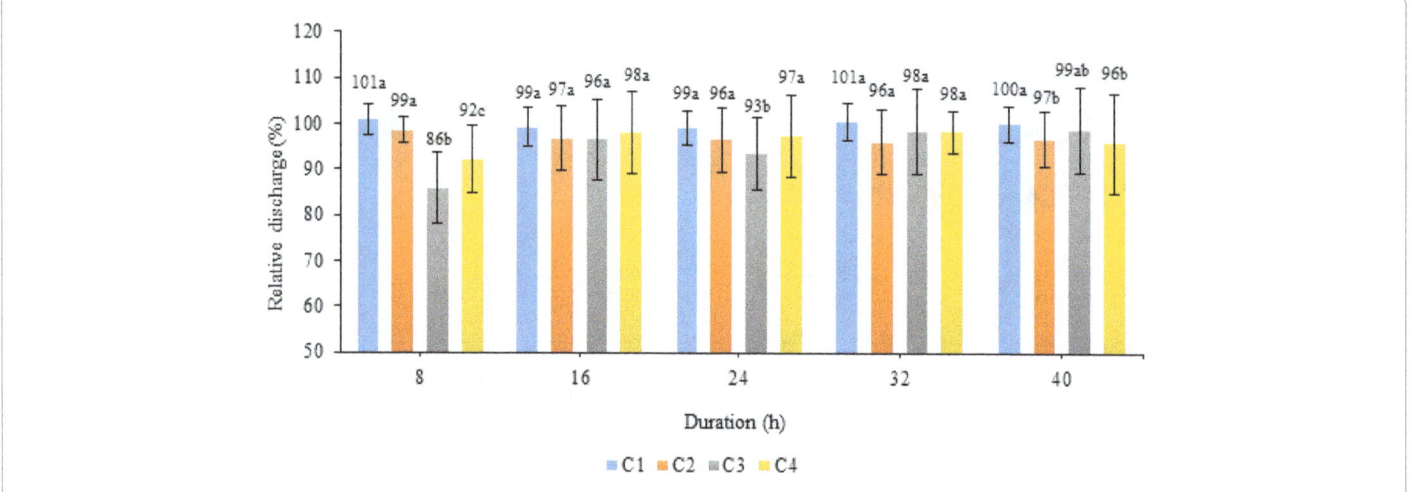

Figure 4: Effect of the concentration of Mt particles in the suspension — C1 (500 mg L⁻¹), C2 (750 mg L−1), C3 (1,000 mg L⁻¹), and C4 (2,000 mg L⁻¹) — on the clogging for dripline model B (1.7 L h⁻¹).

classification for the risk of clogging caused by suspended particles might factor in the effect of the interaction between these two variables on the dripper performance.

The fluctuation of flow rates over time was observed in previous studies for small particles, such as fine sand, silt [7], and clay [9]. The results are attributed to a self-cleaning phenomenon that occurs inside the labyrinths each time irrigation system is switched on. In researches that aim to improve the performance of drippers, self-cleaning is considered an alternative that should be explored in drippers design phase to provide emitters less susceptible to clogging [11]. Additionally, when adding clay particles, the flow rate of non-pressure compensating drippers [9] and pressure-compensating drippers [12] has been reported to increase in some test conditions.

Concerning the different dripper models under study, Model A (0.6 L h⁻¹) was more susceptible to flow variations than model B (1.7 L h⁻¹) (Figures 1-4). The greater sensitivity to clogging of drippers of lower flow rates, compared with higher-flow rate drippers, is mainly attributed to the smaller dimensions of the labyrinths in the low-flow

rate drippers (Table 1), which offer a smaller cross-section for passage and thus results in a greater sensitivity to clogging [13-15].

The deposition of particles that promote drip clogging occurs mainly at the first baffles of the labyrinth [7] near vortices and stagnation regions, where the flow velocity and turbulent kinetic energy are lower than other regions of the labyrinth [9,16].Concerning the type of clay particles, although the Mt caused a greater reduction of the flow rate, at the concentrations C3 and C4, at the end of the first day of testing, both Kt and Mt exhibited a reduction of 9% in the flow rate, for the model a dripper.

For these types of clay to manifest their potential for clogging, it is necessary to change in the physicochemical characteristics of the solution, e.g., increasing the concentration of salts and/or changing the pH value of the solution. In this manner, conditions that favour the aggregation process can be obtained by increasing the size of the aggregates. Larger particles can change the water flow inside the labyrinths, interfering with the dripper performance [9,16].

Regarding to methods to assess the sensitivity of drippers to clogging under controlled conditions, the following remarks should be considered by further related activities. The technical committee ISO/TC23/SC18 is discussing a standard protocol to evaluate the sensitivity of drippers to clogging related to water characteristics in controlled (laboratory) and natural (field) conditions [17]. Considering the effect of electric charges of clay particles and the possibility for these particles to build up aggregates [7], it would be advised using a mixing of different sizes of particles (i.e., clay, silt and sand), as proposed in the framework of the committee ISO/TC23/SC18.

Acknowledgements

The authors are grateful to the São Paulo State Scientific Foundation (FAPESP-Brazil, Project No. 2015/19630-0) for the financial support, the USP-COFECUB program of academic cooperation between French (IRSTEA, Montpellier) and Brazilian (ESALQ/USP) researchers (Project No. 2015-3), and the National Institute of Science and Technology - Irrigation Engineering (INCT-EI).

References

1. Camargo AP, Molle B, Tomas S, Frizzone JA (2013) Assessment of clogging effects on lateral hydraulics: proposing a monitoring and detection protocol. Irrigation Sci 32: 181-191.

2. Zhou B, Li Y, Pei Y, Zhang Z, Jiang Y (2013) Quantitative relationship between biofilms components and emitter clogging under reclaimed water drip irrigation. Irrigation Sci 31: 1251-1263.

3. Zhou B, Li Y, Liu Y, Xu F, Pei Y, et al. (2015) Effect of drip irrigation frequency on emitter clogging using reclaimed water. Irrigation Sci 33: 221-234.

4. Borssoi A, Vilas Boas MA, Reisdorfer M, Hernández RH, Follador FA (2012) Water application uniformity and fertigation in a dripping irrigation set. Engenharia Agrícola (Agriculture engineering) 32: 718-726.

5. Bucks DA, Nakayama FS, Gilbert RG (1979) Trickle irrigation water quality and preventive maintenance. Agr Water Manage 2: 149-162.

6. Hillel D (2003) Introduction to Environmental Soil Physics. Academic Press.

7. Niu W, Liu L, Chen X (2013) Influence of fine particle size and concentration on the clogging of labyrinth emitters. Irrigation Sci 31: 545-555.

8. Al-Muhammad J, Tomas S, Anselmet F (2016) Modeling a weak turbulent flow in a narrow and wavy channel: case of micro-irrigation. Irrigation Sci 34: 361-377.

9. Bounoua S, Tomas S, Labille J, Molle B, Granier J, et al. (2016) Understanding physical clogging in drip irrigation: in situ, in-lab and numerical approaches. Irrigation Sci 34: 327-342.

10. Capra A, Scicolone B (1998) Water Quality and Distribution Uniformity in Drip/Trickle Irrigation Systems. J Agr Eng Res 70: 355-365.

11. Liu HS, Li Y, Liu Y, Yang P, Ren S, et al. (2010) Flow characteristics in energy dissipation units of labyrinth path in the drip irrigation emitters with DPIV technology. J Hydrodyn 22: 137-145.

12. Pinto MF, Molle B, Alves DG, Mouheb NA, Camargo AP, et al. (2017) Flow rate dynamics of pressure-compensating drippers under clogging effect. Revista Brasileira de Engenharia Agrícola e Ambiental (Brazilian Journal of Agriculture and Environmental Engineering) 21: 304-309.

13. Li Y, Yang P, Xu T, Ren S, Lin X, et al. (2008) CFD and digital particle tracking to assess flow characteristics in the labyrinth flow path of a drip irrigation emitter. Irrigation Sci 26: 427-438.

14. Pei Y, Li Y, Liu Y, Zhou B, Shi Z, et al. (2014) Eight emitters clogging characteristics and its suitability under on-site reclaimed water drip irrigation. Irrigation Sci 32: 141-157.

15. Wu D, Li Y, Liu H, Yang P, Sun H, et al. (2013) Simulation of the flow characteristics of a drip irrigation emitter with large eddy methods. Math Comput Model 58: 497-506.

16. Jun Z, Wanhua Z, Yiping T, Zhengying W, Bingheng L, et al. (2007) Numerical investigation of the clogging mechanism in labyrinth channel of the emitter. Int J Numer Meth Eng 70: 1598-1612.

17. ISO/TC 23/SC 18 (2017) ISO/DTR 21540 (under development) - Test methods to evaluate the sensitivity of irrigation emitters to clogging related to water characteristics in controlled (laboratory) and natural (field) conditions.

19

The Effect of Different Irrigation Methods in Biodiesel Production from Sunflower

Karatasiou E*, Papanikolaou C and Makrantonaki MS

Laboratory of Agricultural Hydraulics, Department of Agriculture, Crop Production and Rural Environment, University of Thessaly, Volos, Greece

Abstract

The rational use of irrigation water and the production of energy from renewable sources are among the major concerns of the international scientific community in recent years. Against this background, the Agricultural Hydraulics Laboratory of the University of Thessaly conducted research related to the effect of two different irrigation methods in the production of biodiesel from sunflower crop. Two treatments were organized, in four replications for the growing seasons of the years 2011 and 2012. The treatments were: a) the surface drip irrigation in which the irrigation was scheduled by the Penman-Monteith method and b) the surface drip irrigation in which the irrigation was scheduled by an automatic evaporation pan.

Keywords: Sunflower; Methods of irrigation; Biodiesel; Saving irrigation water

Introduction

The global energy issue which is observed in modern times is growing due to the technological progress and the growth of the world population [1]. As reported by Raum [2], 60% more energy than today will be used in 2030 and according to estimations of experts, oil reserves are sufficient at least until then, although from 2015 onwards in many parts of the world the deposits will begin to run out [3]. Therefore, the scientific interest of the world scientific community about the use and the promotion of renewable energy sources including the energy crops are increasing [4,5]. The biomass that is produced by energy crops is one of the most promising alternatives to fossil fuels, because of its many advantages [6] and mainly because of the effective management of the emissions of greenhouse gases [7,8]. The term "biomass" characterizes each material derived from living or recently deceased, organisms either plant or animal [9]. In Greece, it is possible to cultivate different energy plant species [10], one of which is the sunflower (*Helianthus annuus* L.), which is a plant of high economic importance because of the oilseed.

The sunflower is cultivated worldwide mainly for the edible oil production [11,12] and for other edible compositions because of its high content of oil (about 50 wt%), that includes a high amount of protein, which is up to 50-60% [13]. Industrially it is used for the production of liquid biofuel, the biodiesel [14]. As a plant, it is considered to be relatively resistant to drought due to its deep root system [15], but its greater demands of water are presented from the unfolding of the inflorescences till the anthesis. In Greek conditions, quite satisfactory yields can be achieved by irrigation even if it is deficient [16]. Water saving and biodiesel production, use for irrigation methods applicable on sunflower plant. Furthermore, because of its root system, it uses effectively the nutrients of the soil elements [17], so the plant com is cultivated under low input circumstances.

Water constitutes an essential natural resource for the economic development of each country, based either on the farming or the industrial sector. But now, it is a crucial natural resource in limited availability. The main water user worldwide is agriculture. It uses approximately 80% of the total amount of water. In Central and Northern European countries, irrigation is carried out only during dry summers to improve production, while for the countries of southern Europe irrigation is necessary for the crops [18]. That is why in Greece, agriculture is the largest water consumer, with participation equivalent to the 87.4% of water. Many researchers have studied the dependence of agricultural production on the frequency and quality of irrigation water [19]. Generally it has been found that the existing regime in the irrigation sector is characterized by lack of water or waste of the existing. Based on the above, water saving and effective management, which requires the accurate determination of crop water needs, and are internationally recognized as key priorities [20,21]. The purpose of the effective management of irrigation water is to increase the economic value of a crop, reducing water and energy consumption [22,23]. Nowadays, the global scientific interest focuses on irrigation water saving, through the study of known irrigation scheduling methods using new technologies, as a small percentage saving can achieve huge amounts for the other competitive uses of water. This issue has preoccupied several researchers, including Sakellariou - Makrantonaki et al. [24]. For all the above reasons, it was necessary to study the effect of two irrigation scheduling methods, during the development and the production of the sunflower, as well as the irrigation water saving. The programming methods that were investigated were the widely known method of Penman-Monteith and the automatic evaporation pan. This research aims at maximizing energy benefits, namely the liquid biofuel production from the cultivation of the energy sunflower plant, using modern irrigation systems and also maximizing the irrigation water saving.

Materials and Methods

The experimental procedure for achieving this objective was held at the Farm of University of Thessaly, Greece, during the years 2011 and 2012. A completely randomized design was used and included two treatments, in four replications (Figure 1). The treatments were organized as: a) surface drip irrigation, where the irrigation was scheduled by the Penman – Monteith method (PM) and b) surface drip irrigation where the irrigation was scheduled by an automatic evaporation pan (AUTO (E)). The irrigation dose which was applied

---

*Corresponding author: Karatasiou E, Laboratory of Agricultural Hydraulics, Department of Agriculture, Crop Production and Rural Environment, University of Thessaly, Volos, Greece, E-mail: kareilar@yahoo.gr, msak@uth.gr

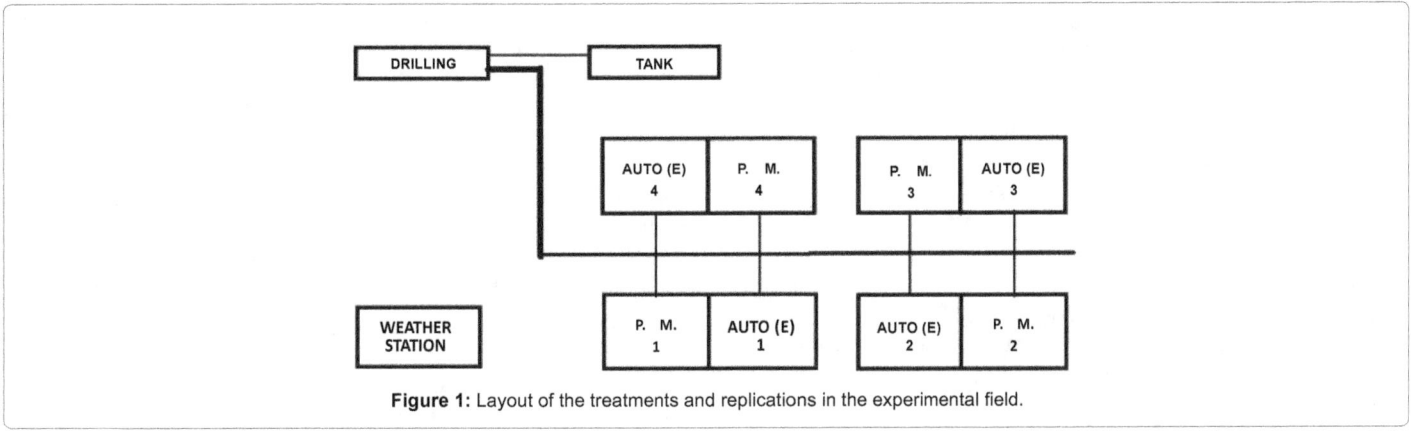

Figure 1: Layout of the treatments and replications in the experimental field.

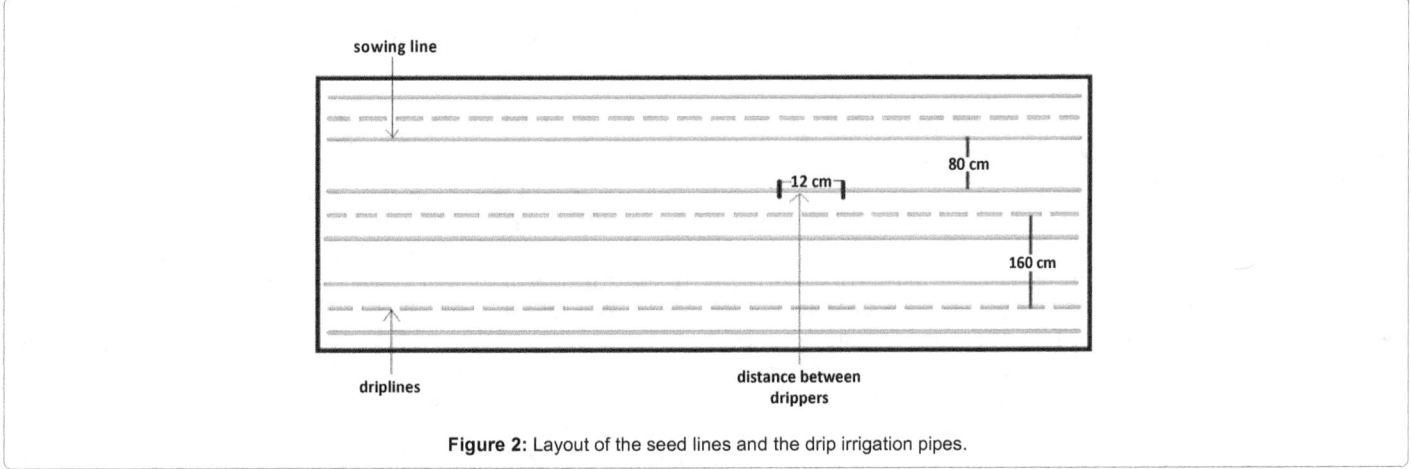

Figure 2: Layout of the seed lines and the drip irrigation pipes.

was equal to 100% of the crop water needs, for both treatments. Furthermore, the cultivation tasks that were used were the same for all the treatments and in accordance with the general cultivation practice in the region.

Soil characteristics

The mechanical composition of the soil of the experimental plot upon analysis was 48% sand, 23% clay and 29% silt, while the soil pH was 7.8 and the organic substance was 0.97% [25]. Before the initial installation of the crop in the experimental field, soil samples were taken, in order to measure in the laboratory the soil parameters which included the field capacity, the permanent welting point and the bulk density.

Cultivation

The area of each experimental unit (plot) was about 56 m^2 and included six seeding lines. The distance between rows was 80 cm, while the distance of the plants on the line was about 12 cm. The drip lines were constructed from 20 mm diameter polyethylene and spaced apart 160 cm (Figure 2). The drippers were self-cleaning and self-regulating, with an equidistance of 80 cm. The value of the emitter flow rate was 2.3 litre/h. The sowing was performed using a crop seeder for linear cultivations with four units, in the first ten days of April for both periods. A Pioneer brand sunflower hybrid was used (PR64A63), in an amount of 0.89 kg/ha for both periods. During the two growing periods pre-emergent and post-emergent herbicides were used. The quantity of the application of STOMP herbicide was 350 cc/ha. During the growing periods no kind of fertilization was carried out in the context of low input.

Measurements

An automatic weather station was used for the collection of meteorological data (MetosCompact, of Pessl Instruments GmbH Company), and was located within 50 m from the experimental plot. It had the ability to record at 12 min the air temperature (°C), the relative air humidity (%), the rainfall (mm), the wind speed at 2 m above ground level ($m*s^{-1}$) and the solar radiation ($W*m^{-2}$). Moreover, it could calculate the reference evapotranspiration with Penman-Monteith method, through an equation. In surface drip irrigation, in which the scheduling of irrigation was based on the method of Penman – Monteith, the meteorological station and a programmer, which defined the beginning and the duration of irrigation, had been used. The start of the irrigation was carried out when 30 mm of evaporation was concentrated. In both treatments, the limit of 30 mm of irrigation dose was set so as to be less than or at most equal to the practical irrigation dose, as estimated by the hydrodynamic parameters of the experimental field soil (the water retention, the permanent wilting point, the specific gravity ground effect) the depth of the root zone of the crop, the daily evapotranspiration reference etc. The interruption of the irrigation was set to be done when a specific number of hours were completed. The number of hours was related to the hourly flow of water drippers, their distance on and between the irrigation lines.

In surface drip irrigation, the irrigation scheduling was performed by the automatic evaporation pan, in which there was a water-level measuring probe (WL_1), and the operation of the irrigation was based on the method of evaporation pan type A. By using this system, the physical presence of the operator to obtain the indication of evaporation, for the beginning and the end of irrigation, is not required. The sensor recorded the change in the value of the electric potential in the perforated pipe within the basin of the evaporation pan, where it was positioned. The change in value of the electric potential was transferred to the Data logger and translated by the equation in mm of water which existed in the pan. In the Data logger a command was set to sum the difference of daily evaporation rates and then give the command in Relay to start the irrigation when the sum mentioned before reached the 30 mm.

By the use of the statistical package SPSS Version 18, data processing was carried out. The statistical analysis was held by the method of Analysis of Variance (ANOVA) (at the 5% significance level) and the classification of averages was done by the application of Duncan's multiple-range test [26].

During the growing periods 2011 and 2012 weekly measurements of crop growth characteristics were carried out from the middle sowing lines of each plot, in order to avoid interactions from neighboring seed lines and experimental plots. Also, indications of the hydrometers were taken before and after for each irrigation. After the appropriate laboratory analysis of the final seed production, the production of sunflower oil was revealed and therefore the energy (biodiesel), which is the main objective of this investigation.

Results

The climate data of the growing periods 2011 and 2012 and of the last twenty five years are presented in Figure 3. The figure shows that the air temperature during the period of the study did not fluctuate much from the values of an average year. Generally, in the last 25 years (including 2011 and 2012), the daily average air temperature ranging from about 20°C in mid-May to 25°C in late June remained constant at about 24–27°C in July and early August and dropped in values between 18 and 23°C from mid-August to mid-September. The total average

rainfall, in June and July over the past 25 years, has been about 44 mm. The rest of the growing period is usually dry with only 96.4 mm of rain falling from mid-July until mid-October. Especially during the years 2011 and 2012, the mean daily air temperature did not differ much from the average values of the past 25 years. However, the mean daily precipitation was higher during May of 2012 than the average values of the last 25 years while during the August of 2011 the precipitation was higher than the average. Under these circumstances and more generally under the climatic conditions in Central Greece, most summer crops, including sunflower, need irrigation to reach acceptable yields.

The amount of water that was applied for irrigation is presented in (Table 1). During the two growing periods, no difference occurred between the treatments at the 5% significance level. Water saving was carried out in the treatment where the irrigation scheduling was based on the automatic evaporation pan (AUTO (E)) (about 3, 5% of the average of two years). In addition, statistically significant difference in the applied amount of water between the growing periods was not presented.

The same table presents the final seed production (kg/ha) for both periods, which is the final product that is marketed by the producer and it is important for this research. There was no difference at the 5% significance level for treatments between the growing periods, while statistically significant difference between treatments was presented during the growing season of 2011 due to the randomness of the sample. The PM treatment showed a slight tendency of superiority as for the final product. That superiority seems to be a result due to the random sampling and less because of the effect of any other factor. Finally, the same table also presents the irrigation water use efficiency (W.U.E.), which is the ratio of the total production to the total irrigation water [26]. No difference appeared in the W.U.E. at the 5% significance level for the treatments between the same growing season and between the two growing periods.

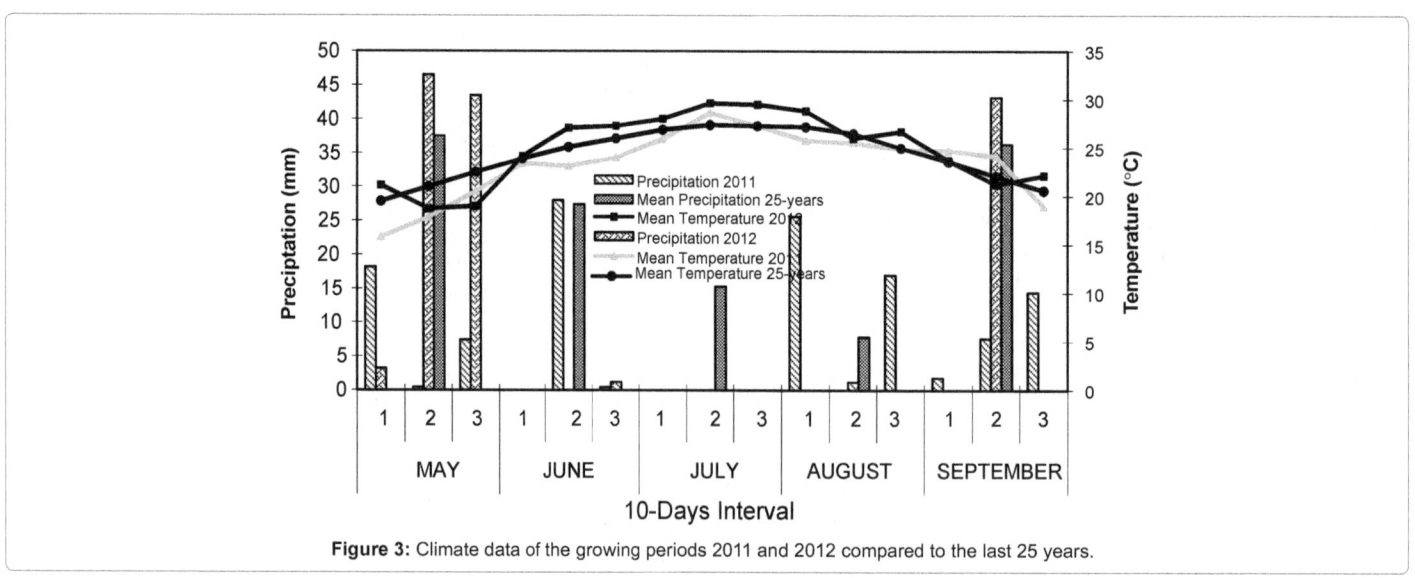

Figure 3: Climate data of the growing periods 2011 and 2012 compared to the last 25 years.

Treatment	Irrigation Water Quantity (m³/0,1ha)		Final Production Seed (kg/0,1 ha)		W.U.E.	(kg/mm H₂O)
	2011	2012	2011	2012	2011	2012
PM	541ᵃ	521ᵃ	399,4ᵃ	395,8ᵃ	0,74ᵃ	0,75ᵃ
AUTO (E)	508ᵃ	517ᵃ	348,4ᵇ	361,6ᵃ	0,69ᵃ	0,69ᵃ

Table 1: Total quantity of irrigation water (m³/0,1ha) applied per treatment and growing period, final seed production (kg/0,1ha) and irrigation water use efficiency (kg/mm H₂O).

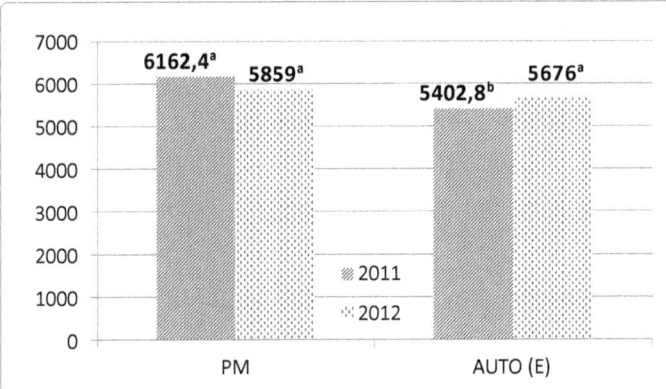

Figure 4: Energy production from sunflower oil (MJ/0,1 ha.) per treatment for the years 2011 and 2012.

Considering the fact that each liter of oil corresponds to 1.13 L of sunflower oil, the calorific value of the oil is 44 MJ/L and the calorific value of the sunflower oil is 33.5 MJ/L, the quantity of sunflower energy extracted is shown in Figure 4. There was a statistically significant difference between the treatments during the growing period of 2011, because of the difference that occurred in the seed production. A statistically significant difference did not appear during the growing period of 2012. No difference was presented between the two growing periods at the 5% significance level.

Conclusion

From the above it is clear that for Greece in the near future, sunflower could be an alternative cultivation due to the advantages it offers. Biodiesel is produced from oil seeds and solid biofuels from crop residues, so the sunflower can be included in the system of "contract farming". As energy plant it could be used in rotation systems in low input agriculture so as to be maximized the economical profit for the farmers.

In this research the automated method of surface drip irrigation using automatic evaporation pan was studied, which showed good results as to the quantity of water consumed, as well as to the production. Its use brought many advantages such as reducing travel to and from the field, optimization of water use and avoidance of the waste of energy. Furthermore, the irrigation programming with evaporation pan class A is still convenient and friendly to the farmers in comparison with the more complicated and expencive Penman – Monteith method as it needs many climatic parameters to give the reference evapotranspiration through an agrometeorological station. It also gave slight irrigation water saving while it was close enough to the accuracy of the Penman – Monteith method.

The results of laboratory analysis of sunflower seeds show that their oil content is not directly correlated with the amount of seed. The smaller amount of irrigation water applied to AUTO (E) during the two growing periods had apparent effect on seed production, but not in the production.

Based on the results obtained from this research, the cultivation of sunflower is suggested for the production of energy using the Penman - Monteith method for more accurate irrigation planning of irrigation when an agrometeorological station is available. The higher accuracy of this method tends to give higher seed production, which is the final product marketed by the producer, without requiring the application of significant additional amounts of irrigation water. It can also be used with automations utilizing the advantages of using new technology in irrigations.

References

1. Sakellariou-Makrantonaki M, Papanikolaou C, Tzimopoulos C (2009) Application of smart irrigation systems in energy plants. Proceeding of 11th conference of Greek Hydrotechnical Association and 7th conference of Greek Scientific Water Management Committee. Greece.

2. Raum E (2008) Fueling the future. Fossil Fuels and Biofuels. Heinemann Library.

3. Colin C (2005) Oil crisis. Multi-Science Publishing. Brentwood.

4. Lemus R, Lal R (2005) Bioenergy crops and carbon sequestration. Crit Rev Plant Sci 24: 1-21.

5. Sakellariou-Makrantonaki M, Nakas C, Dimakas D (2013) Application of biosolids combined with deficit irrigation to energy crop. Proceeding of 8th National conference of Agricultural Engineering. Greece.

6. Bilgic E, Yaman S, Haykiri-Acma H, Kucukbayrak S (2016) Limits of variations on the structure and the fuel characteristics of sunflower seed shell through torrefaction. Journal Fuel Processing Technology 144: 197-202.

7. Ragauskas AJ, Williams CK, Davison BH, Britovsek G, Cairney J, et al. (2006) The path forward for biofuels and biomaterials. Science 311: 484-489.

8. Ageridis G, Christou M (2006) Biofuels and their developmental role to the industry and agriculture. In: two days gala. pp: 1-6.

9. (2007) Biomass Energy Center.

10. Danalatos NG, Archontoulis SV, Ciannoulis KD, Pasxonis K, Tsalikis D, et al. (2008) Cynara, sunflower, sweet and fiber sorghum on farm yields in north, central and south Greece. Proceedings of the International conference on agricultural Engineering. Greece.

11. Haddadi P, Yazdi-Samadi B, Langlade NB, Naghovi MR, Berger M, et al. (2010) Genetic control of protein, oil and fatty acids content under potential drought stress and late sowing conditions in sunflower (Hellianthus annuus). Afr J Biotechnical 9: 67-82.

12. Hu J (2008) Sunflower as a potential biomass crop. Proceeedings of International Symposium on BioEnergy and Biotechnology. Huazhong Agricultural University. China.

13. Salgin K, Doker O, Calmli A (2006) Extraction of sunflower oil with supercritical CO_2: Experiments and modeling. Journal of Supercritical Fluids 38: 326-331.

14. Shehata MS, Razek SMA (2011) Experimental investigation of diesel engine performance and emission characteristics using jojoba/diesel blend and sunflower oil. Fuel 90: 886-897.

15. Stone LR, Goodrum DE, Schlegel AJ, Jaafar MN, Khan AH (2002) Water depletion depth of grain sorghum and sunflower in the central high plains. Agron J 94: 936-943.

16. Steduto P, Hsiao T, Fereres E, Raes D (2012) Yield response to water FAO. Irrigation and Drainage 66: 500.

17. Valchovski I (2002) Influence of heavy rate of nitrogen fertilizer on oil content and fatty acid composition of different varieties and hybrids. Rastenievdni Nauki 39: 338-341.

18. Wriedt G, Velde VM, Aloe A, Bouraoui F (2009b) Estimating irrigation requirements in Europe. J Hydrol 373: 527-544.

19. Doorenbos J, Kassan AH (1979) Yield Response to Water FAO. Irrigation and Drainage 33: 193.

20. Sakellariou-Makrantonaki M, Kalfountzos D, Vyrlas P (2001) Irrigation water saving and yield increase with subsurface drip irrigation. Proceedings of the 7th International Congress of Environmental Science and Technology. Greece.

21. Sakellariou-Makrantonaki M, Papanikolaou C (2008) Water saving by using modern irrigation methods. Proceedings of Ag Eng 2008 International Conference on Agricultural Engineering. Greece.

22. Pandy RK, Maranville JW, Admou A (2000) Deficit irrigation and nitrogen effects on maize in a Sahelian environment. I Grain yield and yield components. Agric Water Manage 46: 1-13.

23. Panda RK, Behera SK, Kashyap PS (2004) Effective management of irrigation water for maize under stressed conditions. Agric Water Manage 66: 181-203.

24. Sakellariou-Makrantonaki M, Giouvanis V (2009) Reuse of wastewater for water saving by irrigating energy plants. Proceedings of 11th International conference on Environmental Science and Technology (CEST). Greece.

25. Mitsios I, Toulios M, Haroulis A, Gatsios F, Floras S (2000) Land study and mapping of the farm of the University of Thessaly in the area of Velestino. Zymel Publications. Athens.

26. Howell TA, Cuenca RH, Solomon KH (1990) Crop yield response. In: Hoffman (ed.) Management of Farm Irrigation Systems. ASAE. pp: 312.

20

Determination of Optimal Irrigation Scheduling for Maize (Zea Mays) at Teppi, Southwest of Ethiopia

Muktar BY and Yigezu TT*

Teppi National Spices Research Center, Teppi, Ethiopia

Abstract

Appropriate irrigation practice is relevant for increased crop productivity and conservation of water resources. No or little concern has been given to the necessity and extension of existing irrigation technologies while the impacts of climate change are visible throughout Ethiopia. A field study was carried out for determining optimal irrigation scheduling for maize production at Teppi, South west Ethiopia for three successive years. The objectives of the study were to evaluate the effect of different irrigation regime (different soil moisture depletion levels) on yield and water use efficiency of hybrid maize (BH-140). The treatments were set based on the recommended soil moisture depletion levels for maize (MAD=0.55). Then five levels of soil moisture depletion were selected for evaluation of optimum irrigation scheduling namely SMD1 (60%), SMD2 (80%), SMD3 (100%), SMD4 (120%), and SMD5 (140% of the recommended value, 0.55). The result indicated that SMD4 has significantly (P<0.05) increased the grain yield and water use efficiency of maize crop on a clay loam textured soil. In addition the total crop water requirement was 535.60 mm. However, the reduced soil moisture depletion level below the recommended values (SMD1 and SMD2) has resulted lower both grain yield and crop water use efficiency. This study also revealed that the appropriate irrigation interval at each crop growth stage should be identified for ease of work to the users.

Keywords: Irrigation scheduling; Water use efficiency; Maize; Teppi

Introduction

Irrigation practice is one of the measures for increasing crop production for Ethiopia whose major economic development is dependent on agricultural production. The country has experienced with severe drought occurrences for the last four decades even though ample amount of water resource from precipitation, surface and sub-surface exist in its periphery.

Ethiopia is one of the largest maize producing countries in Africa. Maize, in Ethiopia, is the main food securing crop that accounted 16.7% in terms of calorie intake, surveyed nationally at 2004/05 [1].

However, the cultivation of maize is mostly dependent on rainfall. Awulachew has explained that the country should double its cereal production to meet the rapidly growing population food demand by 2025 [2]. For this reason, the country is fortunate to cultivate more lands through irrigation especially in its Southwest part where deep fertile soil resources exist.

In the study area, little concern has been given to the necessity and extension of irrigation technologies due to the presence of sufficient rainfall. However, recently, the occurrence of erratic rainfall or impact of climate change drastically reduced crop production. Consequently, traditional irrigation practices are being used for cultivating vegetables in different areas. However, both crop and irrigation water requirement including irrigation scheduling are not known. For better production of medium matured maize crop, Doorenbos has recommended 500 to 800 mm depth of water depending on the climate [3]. In addition, Allen has expressed the soil moisture depletion level for maize should be 0.55 [4]. However, the recommendations are needed to be verified on the operational environment since the crop water requirement is dependent on the type of crop (variety) and climatic condition. For effective use of available water resource, it is relevant to determine the actual crop water need and the right time of water application (irrigation scheduling). Hence, this study was conducted to determine the optimum irrigation scheduling based on the soil moisture depletion levels for hybrid maize (BH-140) at Teppi (Figure 1 & 2). The identified information is important for increased crop production and productivity, improved irrigation water management, and conservation of the environment.

Materials and Methods

General description of the study area

The experiment was conducted at Teppi National Spice Research Center, on station. It is found in Southwest of Ethiopia which is 611 km far from Addis Ababa. It is located at 7.180 N latitude and 35.420 longitudes E with an altitude of 1200 masl. The mean maximum and minimum monthly temperature is 29.850°C to 18.010°C. The area is categorized as hot to warm humid/sub-humid low lands with an annual rainfall of 1563.24 mm. The soil has deep clay loam texture, and 7.3 mm/h intake rate. The source of irrigation water is Shay River which is suitable for irrigation purpose.

Experimental design

The experiment was done for three consecutive years (2013-2015). It was arranged in randomize complete block design with three replications. The treatment was rated for five levels of soil moisture depletion (SMD). The recommended allowable soil moisture depletion for maize is 55% of the total available soil moisture that was used as 100% of SMD. The rates were 60%, 80%, 100%, 120%, and 140% of SMD. The total number of plots was fifteen where the size of each plot was 4 m². Hybrid maize variety (BH-140) was sown at the seed rate of the area (25 kg/ha) and all the recommended practices for the area were applied during the growing season (Figure 3).

Climatic and soil data collection

Climatic data and Reference evapotranspiration: Long-term (20

***Corresponding author:** Yigezu TT, Teppi National Spices Research Center, Teppi, Ethiopia, E-mail: tesfaye.ma@gmail.com

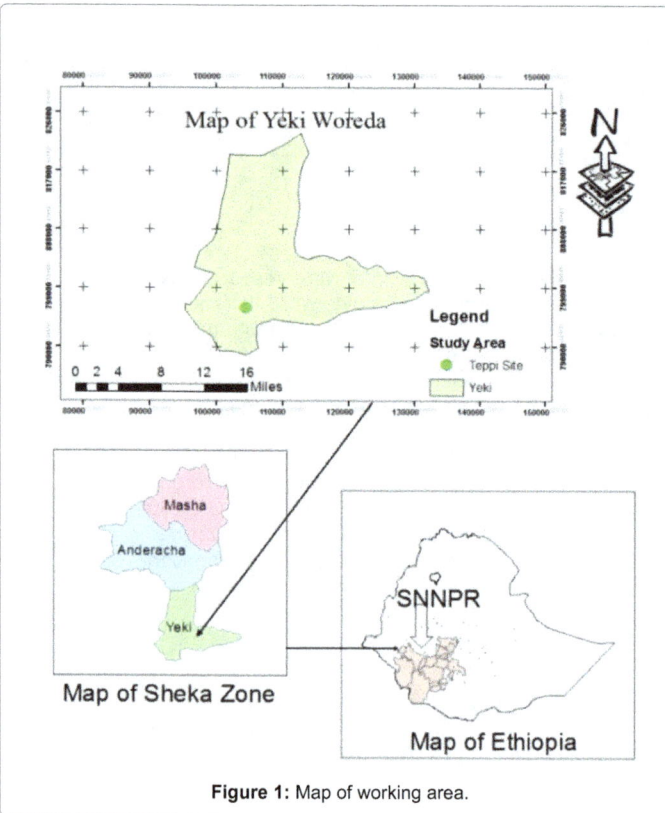

Figure 1: Map of working area.

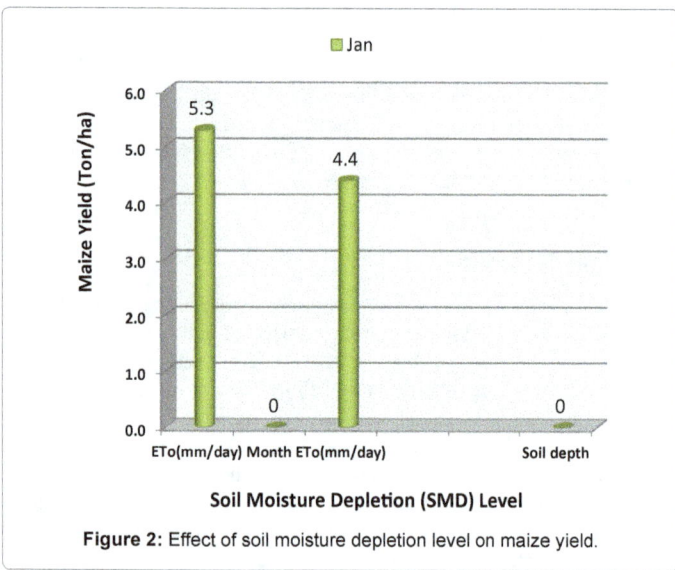

Figure 2: Effect of soil moisture depletion level on maize yield.

years) monthly climatic data of Teppi area was collected from National Meteorological Agency of Ethiopia. The parameters included are rainfall, maximum and minimum temperature, relative humidity, wind speed, and sunshine hours. The monthly reference evapotranspiration of Teppi area was estimated by using FAO CROPWAT 8 program by using long term climatic data (Table 1).

Crop water requirement

For research purpose, the crop water requirement is determined by summing the net depth of water required (dnet) at each irrigation event throughout the crop growing season. The amount of water applied to

the crop rot zone was applied based on the soil moisture depletion level at each growth stage (Table 2). The net irrigation requirement was calculated using the water balance formula.

$$NIR = d_{net} - Pe - GW - \Delta SW \qquad (1)$$

Where,

NIR=Net irrigation requirement, mm

dnet=Net depth of water required, mm

Pe=Effective precipitation, mm

GW=Ground water recharge, mm

ΔSW=Change in soil water content, mm

Water table of the experiment site is deep enough and vertical towards the crop root zone was assumed as negligible. Hence, the ground water recharge is negligible.

The net depth of water required (dnet) was determined by the equation provided by [5].

$$d_{net} = TAW \times Zr \times P \qquad (2)$$

Where,

d_{net}=Net depth of water required (mm)

P=Allowable soil moisture depletion by the crop (0.55).

TAW=Total available soil moisture (mm/m).

$$TAW = 10 \times (\theta_{FC} - \theta_{PWP}) \times Zr \qquad (3)$$

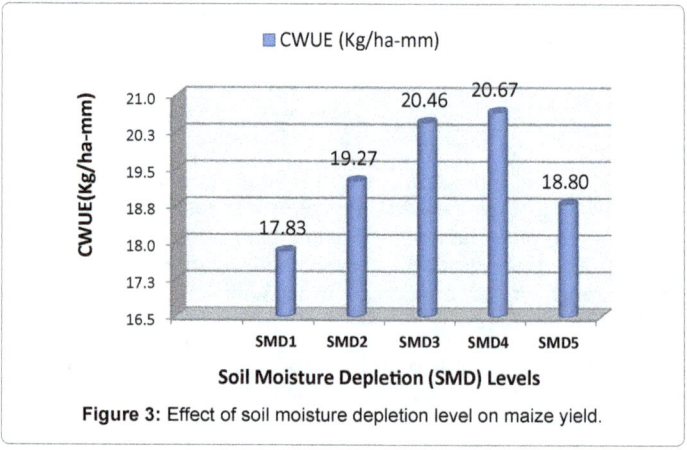

Figure 3: Effect of soil moisture depletion level on maize yield.

Month	Jan	Feb	Mar	April	May	June	Average
ETo (mm/day)	5.3	5.9	5.9	5.3	4.9	4.7	
Month	July	Aug	Sep	Oct	Nov	Dec	5.1
ETo (mm/day)	4.4	4.6	5.1	5.2	5.3	5	

Table 1: Monthly reference evapotranspiration of Teppi, Southwest of Ethiopia.

Soil depth	Bulk density	Field capacity	Wilting point	Available water/ depth
	g/cm³	%	%	mm
0-30cm	1.27	37.1	26.8	130.17
30-60cm	1.27	37.8	25.5	156.21
60-90cm	1.27	38	25.6	157.48
Average	1.27	37.6	26	147.96

Table 2: Soil physical characteristics of study area.

Where,

TAW=Total available soil moisture, mm/m

θ_{FC}=Volume moisture content held at field capacity, %

θ_{PWP}=Volume moisture content held at wilting point, %

Zr=Effective crop root depth, m

Effective rainfall was computed using the National Resource Conservation [6] method and it is described in the following equations.

$$Pe = \frac{\left[P \times \left(125 - 0.2 \times 3 \times P\right)\right]}{125}; \text{ for } P<250/3 \quad (4)$$

$$Pe = \frac{125}{3} + 0.1P; \text{ for } P>250/3 \quad (5)$$

Where,

Pe=Effective precipitation determined in mm/decade.

P=Total precipitation occurred in the crop growing season in the area, in mm/decade.

Gross irrigation requirement (GIR)

GIR is determined using the following formula developed by [5].

$$GIR = \frac{NIR}{(1-LR) \times Ea} \times 100 \quad (6)$$

Where,

GIR=Gross irrigation requirement (mm)

NIR=Net irrigation requirement (mm)

LR=Leaching requirement (fraction)

Ea=Application efficiency (%)

Irrigation scheduling

Irrigation frequency: The number of days between two subsequent irrigations, irrigation frequency, was determined by using equation (7).

$$IF = \frac{Y}{ETc} \quad (7)$$

Where,

IF=Irrigation frequency (days)

d_{net}=Net depth of water required (mm)

ETc=Crop evapotranspiration (mm/day)

The crop evapotranspiration used in irrigation frequency determination was estimated by multiplying crop coefficient with reference crop evapotranspiration [7].

Yield and water use efficiency

The fresh maize grain yield and water use efficiency was selected as dependent variable. CWUE is the quantity of crop yield (Kg/ha) produced per unit depth (mm) of water used [8].

$$CWUE = \frac{Y}{ETc} \quad (8)$$

Where,

CWUE=Crop water use efficiency, kg/ha-mm

Y=Yield of crop, kg/ha

ETc=Crop water requirement, mm

Data collection and analysis

From three consecutive years, data on grain yield and water use efficiency of maize were recorded. The results of yield and water use efficiency were subjected to Analysis of Variance test using the general linear model (GLM) in SAS 9.2 program. The least significant difference (LSD) test at 5% of probability was employed to distinguish among the treatment means.

Results and Discussion

Crop water requirement

The maize, BH-140 variety, was planted on February 26, 2013-2015. As shown detail in Table 3, ten irrigation events with 390.66 mm total irrigation water supplied in the entire crop growing period. The amount of rainfall occurred during cultivation time was very small and the presence of irrigation water could show its importance.

Maize grain yield

The results of pooled mean from the three consecutive years showed that the use of different soil moisture depletion levels were significantly effective (P<0.05) in maize production. As described in Table 4, the mean maize grain yield was gained as 10.4 ton/ha. The maximum yield was obtained when the soil moisture depletion level was reached 120% (SMD4) of the recommended level (55%). However, the yield has declined by 12.7% when the soil moisture depletion level was reduced from the recommended by 40%. The least grain yield was obtained from SMD1 which was practiced with frequent irrigation application or increased number of irrigation event that has the most payments for labors.

Maize water use efficiency

The effect of different management allowable depletion levels were significant (P<0.05) on maize water use efficiency. As described in Table 4, the efficiency of an individual crop to convert irrigation water to maize grain was high in treatment SMD4 which has given 24.3 kg/ha-mm. The minimum crop WUE was 21.01 kg/ha-mm that has showed the least effectiveness of using water for making maize grain. The response of crop water use efficiency had increasing tend when the soil moisture depletion rose from SMD1 to SMD4 but SMD5 was seen with longest irrigation interval and crop stress that has led to reduced water use efficiency.

Conclusion

This study showed that managing the soil moisture content at

Irrigation Event	CWR	Pe	NIR	GIR
	mm	mm	mm	mm
1	7.12	0	7.12	10.17
2	8.9	0	8.9	12.71
3	14.24	3.05	11.19	15.98
4	32.04	6.2	25.84	36.91
5	64.04	5.2	58.84	84.06
6	101.99	4.8	97.19	138.84
7	103.22	0	103.22	147.45
8	103.3	15	88.3	126.14
9	100.77	30.6	70.17	100.24
Total	535.6	64.95	470.65	672.36

Note: CWR=Crop water requirement; Pe=Effective rainfall; NIR=Net irrigation requirement; GIR=Gross irrigation requirement

Table 3: Crop water requirement and irrigation scheduling for (120% MAD).

Treatment	GY	CWUE
	Ton/ha	Kg/ha-mm
SMD1 (60%)	9.567[b]	17.833[b]
SMD2 (80%)	10.32[ab]	19.267[ab]
SMD3 (100%)	10.97[ab]	20.456[ab]
SMD4 (120%)	11.078[a]	20.667[a]
SMD5 (140%)	10.07[ab]	18.800[ab]
Mean	10.4	19.4046
LSD (5%)	1.44	2.7118
CV (%)	14.38	14.47
Note: GY=Grain yield of Maize; CWUE=Crop water use efficiency		

Table 4: Response of maize (BH140) for different irrigation regimes.

different depletion level has influenced the production and water use efficiency of hybrid maize. The total crop water requirement is 535.60 mm. Reducing the soil moisture depletion level by 40% from the recommended fraction (0.55) has significantly reduced both the grain yield and crop water use efficiency. In addition under clay loam soil texture, the use of frequent irrigation is not advisable due to water logging problem. In other hand, by increasing the soil moisture depletion level with 20% over the recommendation or depleting 66% of the total available soil moisture, the grain yield and crop water use efficiency could be improved. Since the area has humid climatic condition and heavy textured soil, longer irrigation interval is advised against water logging. This study also revealed that the appropriate irrigation interval at each crop growth stage should be identified in the area for ease of work to the users.

Acknowledgement

The authors would like to thanks the Ethiopian Institute of Agricultural research for the financial support and Teppi National Spice Research Center (TNSRC) staff members for their holistic support. It is also our gratitude to recognize Mr. Wondiferw Derib for his technical assistance and Mr. Degif G/meskel for his support at field work.

References

1. Berhane G, Paulos Z, Tafere K, Tamru S (2011) Food grain consumption and calorie intake patterns in Ethiopia. Addis Ababa: IFPRI.

2. Awulachew SB, Merrey DJ, Kamara AB, Koppen VB, Penning de Vries F, et al. (2005) Experiences and opportunities for promoting small-scale/micro irrigation and rainwater harvesting for food security in Ethiopia. International Water Management Institute working paper no. 98.

3. Doorenbos J, Kassam AH (1986) Yield response to water.

4. Allen RG, Pereira LS, Raes D, Smith M (1998) Crop evapotranspiration: guidelines for computing crop water requirements. Food and Agricultural Organization (FAO), Rome.

5. Savva AP, Frenken K (2002) Crop water requirement and irrigation scheduling. Irrigation Manual, Harare.

6. United States Department of Agriculture (1997) Natural resources conservation national engineering handbook. Natural Resource Conservation Service. p: 754.

7. Doorenbos J, Pruitt W (1977) Crop water requirements. Irrigation and Drainage, Rome.

8. Tennakoon SB, Milroy SP (2003) Crop water use and water used efficiency on irrigated cotton farms in Australia. Agricultural Water Management 61: 179-194.

21

Fertigation Uniformity under Sprinkler Irrigation: Evaluation and Analysis

Zerihun D[1]*, Sanchez CA[2], Subramanian J[1], Badaruddin M[1] and Bronson KF[3]

[1]*Maricopa Agricultural Center, University of Arizona, Maricopa, USA*
[2]*Departments of Soil, Water and Environmental Science and Maricopa Agricultural Center University of Arizona, Maricopa, USA*
[3]*USDA-ARS Arid-Land Agricultural Research Center, Maricopa, USA*

Abstract

In modern farming systems, fertigation is widely practiced as a cost effective and convenient method for applying soluble fertilizers to crops. Along with efficiency and adequacy, uniformity is an important fertigation performance evaluation criterion. Fertigation uniformity is defined here as a composite parameter consisting of irrigation and fertilizer application uniformity indicators. The field and computational procedures for sprinkler irrigation uniformity evaluation have been the subject of various studies. The objective of the study reported in this paper, however, is the development of an analytical framework for the evaluation and analyses of test-plot scale fertilizer application uniformity under solid-set sprinkler irrigation systems. Irrigation uniformity indices are adapted for use in fertilizer application uniformity evaluation. Fertilizer application rate, given as a function of irrigation depth and fertilizer concentration, is identified as the appropriate variable to express fertilizer application uniformity indices. Pertinent mathematical properties of the uniformity indices along with their practical fertigation management implications are outlined. Carefully designed hypothetical fertigation scenarios were analyzed to examine the significance of the interactive effects, of the local spatial trends of depth and concentration data, on the test-plot scale uniformity of the resultant fertilizer application rate data. The results of the study show that the spatial overlap patterns between depth and concentration data sets are the main determinants of test-plot scale fertilizer application rate uniformity. The study also shows that often the uniformity levels of irrigation and fertilizer concentration data sets cannot be uniquely related to the uniformity of the resultant application rate data. However, some practically useful qualitative relationships between the uniformity of irrigation depth, solute concentration, and application rate data sets are defined. Application of the approach presented here in the evaluation and analysis of fertigation uniformity data sets, measured under sprinkler irrigated conditions, is highlighted.

Keywords: Sprinkler irrigation; Fertigation; Irrigation uniformity; Fertilizer application rate; Application rate uniformity

Notations

c_k=Concentration of nitrogen fertilizer in the *kth* rain gage of the test-plot $[M/L^3]$;

d_k=Irrigation depth in the *kth* rain gage of the test-plot $[L]$;

DU_{lq}=Test-plot scale low-quarter distribution uniformity $[-]$;

k=Rain gage index;

K=The number of rain gages in a test-plot;

UCC=Test-plot scale Christiansen's uniformity coefficient $[-]$;

x_k=Irrigation depth, nitrogen concentration, or nitrogen application rate in the *kth* grid square of the test-plot ($[L]$, $[M/L^3]$, or $[M/L^2]$);

x_{av}=Average depth, concentration, or application rate in a test-plot ($[L]$, $[M/L^3]$, or $[M/L^2]$);

Introduction

In modern farming systems, soluble fertilizers, such as inorganic sources of nitrogen, are commonly applied to crops through fertigation. Compared to conventional fertilizer application methods, fertigation presents a number of potential advantages. It allows a more precise matching of available soil fertilizer content with crop needs through the season [1-3]. In addition, reduced soil compaction, crop damage, and energy and labor costs are also cited as some of the benefits of fertigation [2,4,5]. The additional investment in fertilizer injection and safety equipment are some of the disadvantages of fertigation [1,6]. However, the wide spread use of fertigation with sprinkler systems suggest that the advantages of fertigation far outweigh the disadvantages. Sound system design and management aimed at

maximizing fertigation performance is a key to the realization of these benefits of fertigation. The irrigation method considered in this study is solid-set sprinkler systems. However, the analytical framework described here, for fertigation uniformity evaluation, can be readily adapted to other sprinkler irrigation systems.

An important fertigation performance indicator along with that of efficiency and adequacy is uniformity [7]. In the context of solid-set sprinkler systems, the practical significance of uniformity as a performance criterion stems from the fact that high uniformity is a requirement for the attainment of adequate and efficient fertigation [2,8]. Moreover, uniformity indices are generally considered as indirect indicators of the potential for soil water deficit, deep percolation losses, and nutrient leaching and groundwater pollution from fertigation.

Considering that fertigation is a process that applies both water and fertilizer to crops, it is evident that fertigation uniformity evaluation requires the use of a composite parameter consisting of irrigation and fertilizer application uniformity indicators. Because of its practical significance, and to a certain extent due to the relative simplicity of the required measurement and computational procedure, irrigation uniformity is the performance index that has been most commonly evaluated based on field measurements [8-11]. The factors and physical

*Corresponding author: Zerihun D, Maricopa Agricultural Center, University of Arizona, Maricopa, USA, E-mail: dawit@ag.arizona.edu

mechanisms affecting sprinkler irrigation uniformity and the field and computational methods for evaluating it were examined by various authors [12-21]. The objective of the study reported here, however, is the development of an analytical framework for the evaluation and analyses of test-plot scale fertilizer application uniformity under solid-set sprinkler irrigation systems.

In the study presented here, irrigation uniformity equations are adapted for use in fertilizer application uniformity evaluation. Fertilizer application rate is identified as the appropriate variable for expressing fertilizer application uniformity indices. The mathematical properties of the uniformity equations along with their practical implications are described. Fertilizer application rate uniformity is shown to be a function of the interactive effects of the local spatial trends of the corresponding irrigation depth and solute concentration data sets. The study has also shown that often fertilizer application rate uniformity cannot be uniquely related to the uniformity of irrigation and solute concentration. However, some practically useful qualitative relationships between the uniformity of irrigation depth, solute concentration, and application rate data sets are defined. Application of the approach presented here in the evaluation and analysis of fertigation uniformity data sets, measured under sprinkler irrigated conditions, is highlighted.

Fertigation Uniformity, Pertinent Variables and Spatial Scale

Fertigation uniformity indicators

During fertigation, solute concentration may vary spatially through a sprinkler hydraulic network and temporally during the course of a fertigation event. Hence, fertilizer application uniformity cannot be automatically deduced from irrigation uniformity. The implication is that fertigation uniformity is a composite parameter consisting of irrigation and fertilizer application uniformity indicators. Accordingly, throughout this manuscript irrigation and fertilizer application uniformity indices are treated as two distinct, nonetheless, related and equally important aspects of sprinkler fertigation uniformity.

Variables for fertigation uniformity evaluation

Agricultural inputs for crop production, including irrigation and fertilizers, are typically expressed in terms of application rates: volume or mass of the input per unit area of cropland. Accordingly, sprinkler irrigation uniformity is often defined as a function of irrigation depths. For sprinkler applications, the equivalent variable, to irrigation depth, for expressing fertilizer application uniformity is mass of fertilizer per unit area of field (e.g., gram per square meter), a variable commonly referred to as fertilizer application rate. The mass of fertilizer in irrigation water cannot be measured directly. Hence, fertilizer application rates need to be computed as a function of the directly measureable physical quantities of concentration and irrigation depth.

Spatial scale for uniformity evaluation

The basic field unit for solid-set sprinkler fertigation system uniformity evaluation is a test-plot. Typically, a uniformity evaluation test-plot consists of a rectangular area with dimensions equal to the sprinkler spacing along laterals and the lateral spacing. Rain gages arranged in a grid pattern, with suitably selected spacing, are used to measure sprinkler precipitation depths and fertilizer concentrations over the test-plot. The data collected as such is then used to calculate the test-plot scale fertigation (i.e., irrigation and fertilizer application rate) uniformity indices. Many of the factors affecting uniformity

(including system hydraulics, setting, and maintenance) can be spatially variable, as a result test-plot scale fertigation uniformity may not be representative of field-scale uniformity. A realistic evaluation of field-scale fertigation uniformity may, therefore, require the use of more than one test-plots suitably distributed over the field. In which case, the test-plots can be considered as sampling points of the field-wide variability of irrigation depth and fertilizer application rates. Test-plot scale uniformity indices can then be scaled-up to field-level with an appropriate procedure. A simple approach for deducing field-scale fertigation uniformity from test-plot scale measurements is described in reference [8]. In this paper, however, discussion on fertigation uniformity is limited to test-plot scale evaluations.

Fertigation Uniformity Equations, Properties and Practical Implications

Uniformity can be considered as a measure of the spatial variability inherent in a data set. The practical significance of uniformity as a fertigation performance index is less intuitive than those of the application efficiency and adequacy indices. Nonetheless, in sprinkler applications high uniformity is a necessary condition for adequate and efficient fertigation.

Various indices have been proposed for use as irrigation uniformity metrics [19-23]. Some indices are developed specifically for applications in only certain irrigation methods, e.g., emission uniformity [23] and design uniformity coefficient [22] for trickle irrigation systems. In principle, the uniformity index [20], which uses the coefficient of variation of the irrigation depth data to measure variability, has a broader scope of applicability. Nonetheless, it is not commonly used in practice. Christiansen's coefficient of uniformity [21] and the low-quarter distribution uniformity [19] are currently the most widely used irrigation uniformity indices.

Often variability (uniformity) of a data set is expressed with reference to the average. In this paper two standard indices that are designed to measure different aspects of data variability, with respect to the mean value, are used to quantify fertigation (i.e., irrigation and fertilizer application rate) uniformity: Christiansen's uniformity coefficient, UCC [-], and the low-quarter distribution uniformity, DUlq [-]. Although these indices are customarily used to evaluate irrigation uniformity [9-12], there is no limitation as regards their application to quantifying the spatial variability of any agricultural input applied with irrigation water.

Fertigation uniformity evaluation test-plots are generally rectangular and are further divided into elemental areas of the same shape and dimension (typically squares because of simplicity and symmetry). Each of the elemental areas are associated with a rain gage. Note that the ratio of the catchment area of the rain gage to the elemental area should be sufficiently large for the measured precipitation depth and concentration to be considered a representative average for the elemental area [15,16]. Forms of the UCC and DUlq equations applicable to the conditions described above are presented in the following section.

Christiansen's uniformity coefficient

The equation for test-plot scale Christiansen's uniformity coefficient, UCC [-], of a farm input applied with irrigation water can be given as:

$$UCC = 1.0 - \frac{\sum_{k=1}^{K} |x_k - x_{av}|}{x_{av}} \qquad (1)$$

where k is rain gage index, K is the number of rain gages in a test-plot, x_k is application rate of a farm input (irrigation or fertilizer) computed based on measurements in the kth rain gage ($[L]$ or $[M/L^2]$), and x_{av} is the arithmetic average application rate for the test-plot ($[L]$ or $[M/L^2]$). Note that in order to maintain consistency with the definition used for fertilizer application rate, we chose here the phrase irrigation application rate in reference to the volume of irrigation per unit field area (irrigation depth). Observe that this is different from the customary usage of the phrase in the irrigation literature, where irrigation application rate refers to irrigation depth applied per unit of time.

Low-quarter distribution uniformity

The equation for test-plot scale low-quarter distribution uniformity, DU_{lq} $[-]$, of a farm input applied with irrigation water is given as:

$$DU_{lq} = \frac{x_{lq}}{x_{av}} \tag{2}$$

where x_{lq} is the arithmetic mean of the lowest quarter of the application rates within the test-plot ($[L]$ or $[M/L^2]$). As will be noted in subsequent discussion Equations 1 and 2 are also used to compute nitrogen concentration uniformity, in which case the variables x_k, x_{lq}, and x_{av} in Equations 1 and 2 will have the dimensions of M/L^3.

It can be noted from Equation 2 that distribution uniformity is a measure of the significance of extreme negative deviations from the average application rate. Different forms of distribution uniformity (e.g., distribution uniformity based on the minimum or lower-half of the application rate data) are commonly used, each assigning different levels of stringency to the definition of what constitutes extreme negative deviations from the average. However, the low-quarter distribution uniformity, DU_{lq}, is used here, because it has been widely applied in irrigation uniformity evaluations. The Christiansen's coefficient of uniformity, UCC, on the other hand, can be viewed as an index designed to measure the test-plot scale data variability from the average. Although in this study UCC and DU_{lq} are, the indices, used to evaluate fertigation uniformity, it is important to note that the use of any suitably selected uniformity indices along with, or in place of, these indices is equally valid.

When Equations 1 and 2 are used to quantify fertilizer application uniformity, the variable x_k represents fertilizer application rate, which can be computed as the product of fertilizer concentration in irrigation water, c_k $[M/L^3]$, and irrigation depth, d_k $[L]$:

$$x_k = c_k d_k \tag{3}$$

Properties of the fertigation uniformity equations

Although more general forms of the uniformity equations that are not limited by the shape of the test-plot or the shape and dimensions of the elemental areas constituting the test-plot can be formulated [8], Equations 1 and 2 are the most commonly used forms. The following is a list of the properties of Equations 1-3 and their practical computational implications.

(1) Considering test-plot scale irrigation depth or fertilizer application rate data, UCC and DU_{lq} indices remain unaffected if each element of the data set is multiplied by a constant.

The implication is that the volume of precipitation collected in rain gages, instead of depth, can be used directly to compute irrigation uniformity. Note that this is especially convenient if rain gages graduated in volumetric units are used in fertigation uniformity evaluation.

Likewise, the mass of fertilizer in the rain gages, instead of fertilizer application rates, can be used to calculate application rate uniformity, if the spatial distribution of fertilizer is expressed as such. This property also implies that the uniformity of a data set remains unchanged if the data is normalized with a suitably selected characteristic variable.

(2) If the fertilizer concentration over a test-plot is constant, the fertilizer application rate uniformity will be equal to irrigation uniformity. Observe that this is a corollary to the property stated above.

In such a scenario, the problem of fertigation uniformity evaluation reduces to that of irrigation uniformity evaluation. In practice this scenario can be approximated in a sprinkler system in which the effect of solute transport processes on the spatial distribution of fertilizer concentration is limited and fertilizer concentration at the system inlet is nearly constant throughout the duration of irrigation.

(3) Test-plot scale UCC and DU_{lq} are independent of the spatial distribution of the application rate data points within a test-plot.

This implies that two test-plots with the same number of data points, but different spatial distributions of application rate data, can have the same UCC and DU_{lq} provided the data sets can be shown to be equivalent after having been sorted separately in ascending/descending order. In other words, the uniformity indices associated with a given irrigation depth or fertilizer application rate data set remain unchanged under any possible spatial permutation of the data. Although the computation of irrigation uniformity or fertilizer application uniformity is independent of the spatial distribution of the data points, it should be noted that the computation of fertilizer application rates from depth and concentration data sets, Equation 3, requires a proper accounting of the spatial distribution of the data points within the test-plot.

(4) Test-plot scale fertilizer application rate uniformity is an aggregate index of the interactive effects of the local spatial trends in the irrigation depth and fertilizer concentration data sets.

This property of the uniformity indices is less intuitive than those described above, but it is key to understanding and defining the factors that affect fertilizer application uniformity and has potential fertigation system design and management implications. Hence, in the subsequent section a combination of simplified hypothetical examples and intuitive mathematical reasoning will be used to show its validity.

The Relationships between Irrigation Depth, Fertilizer Concentration and Application Rate Data Sets

Considering that fertilizer application rate is a multiplicative function of irrigation depth and fertilizer concentration, Equation 3, it can be reasoned that the interactive effects of the spatial trends and scale of variability, inherent in the irrigation depth and the concentration data sets, are the main determinants of the uniformity of the resultant application rate data. In other words, depending on the local monotonic property and scale of variability of the depth data in relation to that of the solute concentration data, the three-dimensional response surface representing the resultant application rate data can get vertically stretched (becomes relatively more variable and less uniform) or it can become more compact (less variable and more uniform) compared to the depth and/or the concentration data sets. The spatial trends of measured irrigation depth and fertilizer concentration data sets and related overlap patterns can show significant local variability over a test-plot. Hence analyses of their interactive effects on the variability of the resultant application rate data need to be based on piece-wise (local) spatial behaviors of the (depth and concentration) data sets. Important inferences that stem from the preceding observations are:

(1) In parts of a test-plot where the local spatial trends in the irrigation depth data have the same monotonicity as that of the concentration data, the local spatial variability of the resultant application rate data tends to be larger than the variability inherent in both the depth and concentration data sets.

(2) In any given section of a test-plot the relative contributions, of the depth and concentration data sets, to the local variability of the resultant application rate data are proportional to the scale of variability inherent in the depth and concentration data sets.

(3) In parts of a test-plot where the spatial trends in the depth and concentration data sets have opposite monotonicity, the local spatial variability of the resultant application rate data tends to be smaller than that of the depth and/or concentration data set(s).

Note that the term monotonicity is used here, in relation to the spatial trends of depth and concentration data sets, to refer to the mathematical property of the data sets as increasing or decreasing functions of distance. If, for instance, both data sets are locally increasing or locally decreasing functions of distance in some part of the test-plot, then they are described as having same monotonicity there. On the other hand, if depth data is a locally increasing function of distance, whereas the concentration data is a decreasing function of distance or vice-versa, then the functions are considered to be of opposite monotonicity in that part of the test-plot. Note that in subsequent discussion the monotonic properties of depth and concentration data sets are alternatively referred to as spatial trends of depth and/or concentration data set or spatial overlap patterns between depth and concentration data sets. Furthermore, the term local function behavior should imply that a function exhibits a given mathematical property of interest (e.g., monotonicity) in some subset of its domain (which is the test-plot in the current application). Likewise, global function behavior implies that a property of interest spans the entire test-plot.

Furthermore, it is important to note here that the references to scale differences in the spatial variability of depth, concentration, and application rate data sets consider only comparisons between dimensionless depth, concentration, and application rate data sets.

Evidently, actual (field) uniformity evaluations cannot be designed to produce irrigation and concentration (and hence application rate) data sets, each with, a particular spatial pattern. Thus measured fertigation data sets are not well suited for exploring the effects of specific spatial overlap patterns, of depth and concentration, on fertilizer application rate data. Furthermore, measured fertigation data typically show complex local interactions between depth and concentration and hence they are not readily amenable to simple graphical analyses. Therefore the validity of the inferences summarized above is demonstrated here with four pairs of simplified hypothetical examples presented subsequently.

Each hypothetical scenario is designed to examine the comparative significance of the effects, of test-plot scale uniformities and local spatial overlap patterns of depth and concentration, on the resultant application rate uniformity from a different perspective. The first example (section 4.1) presents a relatively simple scenario in which the irrigation and fertilizer concentration data sets have clearly discernible spatial trend that span the entire test-plot. The second example (section 4.2) is a slightly more nuanced case in which the spatial variability patterns of irrigation and concentration data sets are dominated by

local trends. However, the same local overlap patterns, between depth and concentration data sets, are repeated over the entire test-plot. The third example (section 4.3) is similar to the second, but the spatial variability of the depth and concentration data sets are of significantly different scales. Thus it is primarily designed to explore the interactive effects, of scale of variability and overlap patterns between depth and concentration, on the uniformity of the resultant application rate data. The fourth example (section 4.4) presents a relatively more realistic fertigation scenario in which the variability of both the depth and concentration data sets are dominated by local spatial trends, but also the local spatial overlap patterns are not the same over entire the test plot. Note that in many of the hypothetical fertigation scenarios considered the variability in the irrigation and/or concentration data sets is deliberately exaggerated, the objective is to make their interactive effects on the resultant application rate data readily discernible.

Scenario I: Data sets with clearly discernible dominant spatial trend that spans the entire test-plot

Consider an example in which the spatial distribution of irrigation depth and fertilizer concentration each can be approximated by a plane surface. Such an example can be considered as a simplified representation of a measured data set that has clearly discernible global spatial trend (a trend spanning the entire test-plot) with minor perturbations about a best-fit plane. Furthermore, assume that both the depth and the concentration surfaces are level in a direction parallel to one of the horizontal axes (say along the laterals) and have constant slopes along the mainline. The implication is that each data set (irrigation or concentration) is uniform in a direction parallel to the laterals and is variable only along the mainline. It can then be shown that for such a data the test-plot scale uniformity indices have the same value as those computed for a data set on any transect parallel to the mainline. Evidently, for such a data set the spatial pattern along a given transect (parallel to the mainline) will simply be a replication of the pattern along any one of the other transects. This is advantageous, because with this simplification each of the surfaces representing irrigation depth, fertilizer concentration, and application rate can be reduced to a curve with no concomitant loss of pertinent information. In which case, the application rate curve can be superimposed on a graph depicting the corresponding depth and concentration curves. In order to remove scale effects, arising from dimensions and units of measure, and allow a direct comparison between depth, concentration, and application rate, the respective data sets are nondimensionalized. In Figure 1, each of the data sets were normalized by their maximum values, hence they vary in the range 0.0 to 1.0.

Accordingly, a hypothetical example consisting of irrigation depth and fertilizer concentration data sets along a transect, through a test-plot, in a direction parallel to the mainline is depicted in Figure 1a. Both data sets have negative slopes, hence have the same monotonicity through the test-plot. Note that the equations used to generate the data sets presented in Figure 1a are summarized in Table 1. The irrigation depth data has a relatively narrower range of variation (0.27 to 1.0) compared to the fertilizer concentration data set (0.17 to 1.0). The uniformity of the irrigation depth data set is $UCC=0.688$ and $DU_{lq}=0.543$ and that of the solute concentration data set is $UCC=0.610$ and $DU_{lq}=0.429$ (Table 2).

As can be noted from Figure 1a the resultant application rate data, computed with Equation 3, is a decreasing function of distance (hence has the same monotonicity as the depth and concentration data sets), although it is curvilinear. The application rate data set has a wider range of variation (0.05 to 1.0), and hence appreciably lower

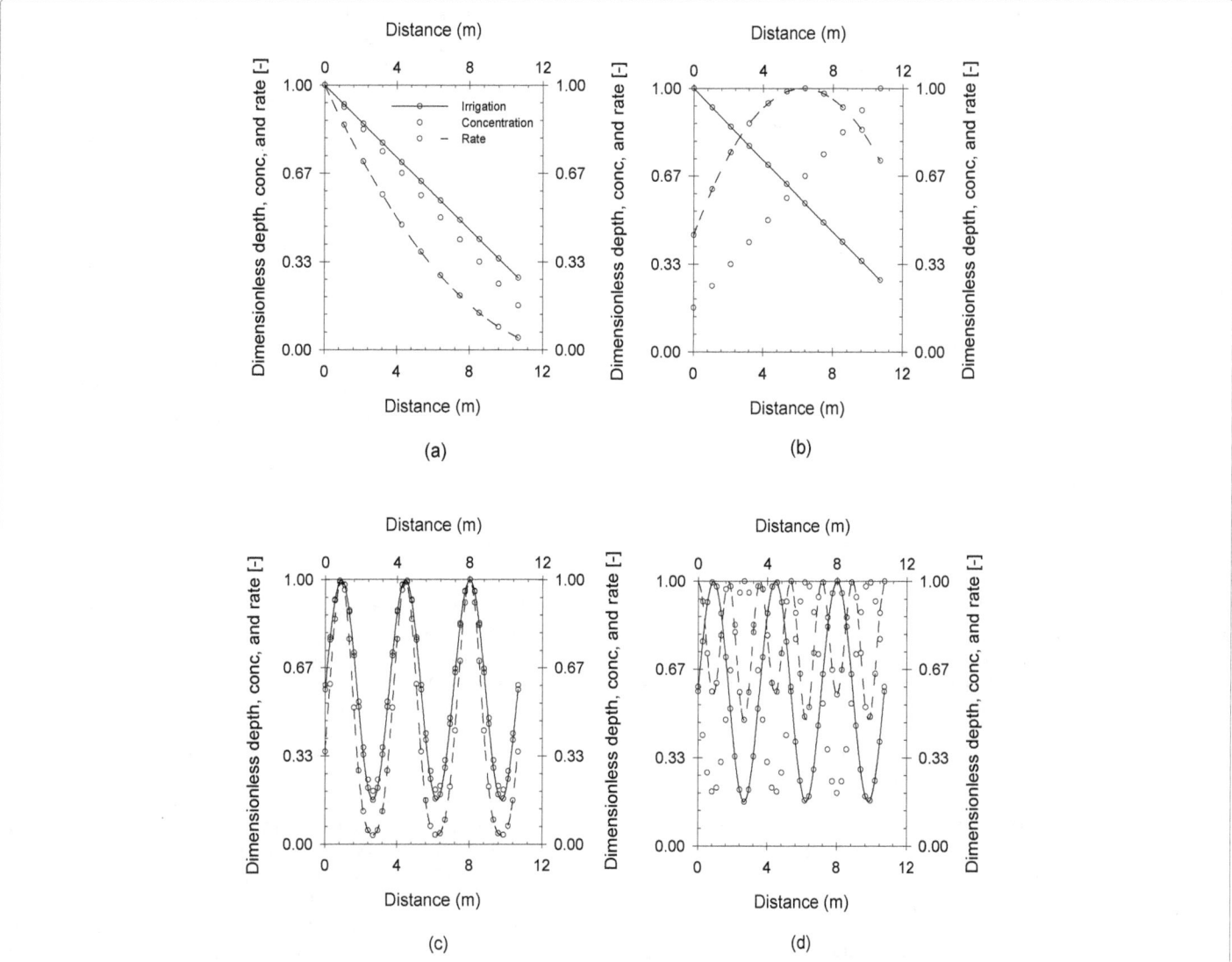

Figure 1: The relationship between spatial trends in irrigation depth, fertilizer concentration, and the spatial variability of the resultant fertilizer application rate. Scenarios where dominant spatial trend spanning the test–plot exists and depth and concentration have: (a) same monotonicity, (b) opposite monotonicity; and Scenarios where local spatial trends dominate and depth and concentration have: (c) same monotonicity, (d) opposite monotonicity.

Data set	Unit	Figures[a]	Equations[b]	Maximum values
Irrigation depth	mm	1a and 1b	$-1.874x + 27.5$	27.5
		1c,1d, and 2b	$17.5 + 12.5\sin(1.77x)$	30.0
		2a, 2c, and 2d	$17.5 + 3.0\sin(1.77x)$	20.5
Solute concentration	mg/L	1a	$-14.058x + 180.0$	180.0
		1b	$14.058x + 30.0$	180.0
		1c and 2a	$105.0 + 70.0\sin(1.77x)$	175.0
		1d	$105.0 + 17.0\sin(1.77x + \pi)$	175.0
		2b	$105.0 + 17.0\sin(1.77x + \pi)$	122.0
		2c	$105.0 + 15.0\sin\left(1.77x + \dfrac{\pi}{2}\right)$	120.0
		2d	$105.0 + 15.0\sin\left(1.77x + \dfrac{2}{3}\pi\right)$	120.0

[a]The spatial spacing between data points in Figures 1a and 1b is 1.067m and in Figures 1c and 1d and 2a-2d is 0.267m, the length of the test-plot along the mainline is 10.67m.
[b]The equations are in dimensional form, they were normalized by dividing them with their respective maximum values and x=distance in meter.

Table 1: Functions used to define irrigation, fertilizer concentration, and application rate curves in Figures 1 and 2.

Scenario	Uniformity indices	Spatial trends of irrigation depth data in relation to fertilizer concentration data					
		Same monotonicity		Rate	Opposite monotonicity		Rate
		Irrigation	Concentration		Irrigation	Concentration	
I	UCC	0.688	0.610	0.384	-	-	-
	DU_{lq}	0.543	0.429	0.209	-	-	-
	UCC	-	-	-	0.688	0.610	0.832
	DU_{lq}	-	-	-	0.543	0.429	0.733
II	UCC	0.557	0.587	0.291	-	-	-
	DU_{lq}	0.358	0.401	0.120	-	-	-
	UCC	-	-	-	0.557	0.587	0.797
	DU_{lq}	-	-	-	0.358	0.401	0.703
III	UCC	0.894	0.587	0.504	-	-	-
	DU_{lq}	0.846	0.401	0.322	-	-	-
	UCC	-	-	-	0.557	0.900	0.633
	DU_{lq}	-	-	-	0.358	0.855	0.434
		Variable monotonicity		Rate	Variable monotonicity		Rate
		Irrigation	Concentration		Irrigation	Concentration	
IV	UCC	0.894	0.908	0.858	-	-	-
	DU_{lq}	0.846	0.869	0.804	-	-	-
	UCC	-	-	-	0.894	0.908	0.897
	DU_{lq}	-	-	-	0.846	0.869	0.860

Table 2: A summary of computed uniformity indices for irrigation depth, fertilizer concentration, and application rate for the hypothetical examples presented in Figures 1 and 2.

uniformity ($UCC=0.384$ and $DU_{lq}=0.209$), compared to both the depth and concentration data sets (Table 2). Note that this result is consistent with the inference stated above in regard to the spatial variability of an application rate data derived from irrigation depth and fertilizer concentration data sets that are of same monotonicity.

Figure 1b presents an example in which both the depth and concentration data sets have a global spatial trend, but in contrast to the preceding example they have opposite monotonicity. The irrigation depth data in Figure 1b is exactly the same as that presented in Figure 1a. The concentration curve, on the other hand, is obtained from that presented in Figure 1a as follows: *(i)* the slope is set equal to -1.0 multiplied by the slope of the concentration curve in Figure 1a and *(ii)* the intercept on the vertical axis is set to the lower limit of the data set (Table 1). As a result, the fertilizer concentration data is an increasing function of distance and has opposite monotonicity with respect to the irrigation depth data (Figure 1b). However, it can be noted that its uniformity remains unchanged as in Figure 1a (Table 2).

In contrast to the preceding example (Figure 1a), the resultant fertilizer application rate curve is not a monotonic increasing function of distance, instead it is increasing in the range 0.0 m to about 6.5 m and is decreasing thereafter (Figure 1b). As a result, the application rate data vary in a much narrower range (0.44 to 1.0) compared to the depth and concentration data sets. The computed fertilizer application rate data also has a considerably higher uniformity ($UCC=0.852$ and $DU_{lq}=0.733$) than the corresponding depth and concentration data sets (Table 2). Note that these results are consistent with the inferences stated above in regard to the spatial uniformity of an application rate data derived from irrigation depth and concentration data sets that are of opposite monotonicity.

A more subtle and interesting point, that follows directly from the above general deductions and is revealed by this example, is that depending on the spatial overlap patterns between depth and concentration substantially different application rate uniformity levels (fertigation scenarios) can be obtained from depth and concentration data sets of given uniformity. This suggests that it is the spatial overlap patterns between depth and concentration data sets, and not their

uniformity levels, that determine the uniformity of the resultant application rate data. A more specific observation of interest here is that a combination of depth and concentration data sets, both with high spatial variability (low uniformity), does not necessarily lead to a low application rate uniformity.

Note that in the current as well as in each of the subsequent examples, different fertigation scenarios are derived from a given pair of depth and concentration data sets through spatial rearrangement of the data points. As can be noted from the properties of the uniformity indices, such a spatial reordering of the data points would leave the uniformity of the irrigation depth and fertilizer concentration data sets unchanged. However, it leads to different spatial overlap patterns between the depth and concentration data sets and hence results in different fertilizer application uniformity levels. It needs to be emphasized that the goal here is only to produce sharply contrasting fertigation scenarios in which the comparative significance of the effects, of test-plot scale uniformities and local spatial overlap patterns of depth and concentration, on application rate uniformity are clearly discernible.

Scenario II: Data sets with dominant local spatial trends in which depth and concentration have either same or opposite monotonicity throughout the test-plot

In this section a second pair of simplified hypothetical examples is presented. These examples are designed to show that the general inferences stated above apply to conditions in which the spatial variability in the irrigation depth and fertilizer concentration data sets are dominated by local trends. Here we consider irrigation depth and fertilizer concentration data sets that follow sinusoidal patterns of variation with distance along the mainline in a direction perpendicular to the mainline and have no gradient in a direction perpendicular to the mainline. It can then be shown that the uniformity indices computed for a data set along any transect through the test-plot in a direction parallel to the mainline is equal to the test-plot scale uniformity indices. The three-dimensional response surfaces of depth, concentration, and application rate each can then be reduced to their equivalent two-dimensional counter parts along a transect, all of which can be presented on a single graph. As discussed above in order to remove scale effects and allow a direct comparison between depth, concentration, and application rate data, the respective data sets are presented in dimensionless graphs.

Accordingly, Figure 1c depicts a scenario in which both irrigation depth and fertilizer concentration are sinusoidal functions of distance (Table 1). The depth and concentration data sets vary in the range 0.17 to 1.0 and 0.2 to 1.0, respectively. It can be noted that in any given segment of the transect, these data sets exhibit comparable scale of variability and are of the same monotonicity. The resultant application rate data acquires the local monotonic properties of the corresponding depth and concentration data sets, but with a larger variability than both data sets (Figure 1c). The range of variation of the application rate data is 0.97, compared to 0.83 and 0.80 for the depth and concentration data sets, respectively. More importantly, the uniformity of fertilizer application rate (UCC=0.291 and DU_{lq}=0.120) is significantly lower than the corresponding irrigation uniformity (UCC=0.557 and DU_{lq}=0.358) and concentration uniformity (UCC=0.587 and DU_{lq}=0.401), Table 2. These results show that the inferences stated above on the spatial variability of the fertilizer application rate data is also applicable to scenarios in which local trends are dominant.

Figure 1d depicts a scenario in which the irrigation and concentration data sets have same uniformity and range of variation as those presented in Figure 1c, but they have opposite monotonicity in any given segment of the test-plot (Table 1). Mathematically this is accomplished by introducing a phase shift of π radian to the concentration curve with respect to the irrigation depth curve, while keeping the depth curve the same as in Figure 1c. Note that this is equivalent to moving the sinusoidal curve for concentration to the left through an angular distance of π radian, which (in the examples presented here) is equivalent to 1.77 m in linear distance. Observe that the concentration data set, in Figure 1d, is a mirror image of the data in Figure 1c about the horizontal line of symmetry. Hence, the transformation of the concentration data (through the introduction of a phase shift) from that given in Figure 1c to that given in Figure 1d in practice involves a simple spatial rearrangement of the same data set. From the property of the uniformity equations it can be noted that such a transformation leaves the uniformity of the data set unchanged at the level of Figure 1c.

From Figure 1d it can be noted that the resultant application rate data set computed with Equation 3 has a much lower range of variation (0.47 to 1.0) compared to the corresponding depth and concentration data sets. A practically more useful point, however, is that the computed application rate data here is significantly more uniform (UCC=0.797 and DU_{lq}=0.703) than the application rate data depicted in Figure 1c (Table 2). Consistent with the inference stated above, these results show that depth and concentration data sets with high local variability can lead to a resultant application rate data with a lower variability (higher uniformity), if they have opposite monotonicity.

Overall the examples presented in Figures 1c and 1d show that given irrigation depth and solute concentration data sets of fixed uniformity and comparable scale of variability, with dominant local spatial trends, substantially different application rate uniformity levels can be obtained depending on whether the depth and concentration data sets are of same or opposite monotonicity.

Scenario III: Data sets with dominant local spatial trends in which the variability of the application rate data is dominated by that of the irrigation or the concentration data set

A third pair of simplified hypothetical examples, summarized in

Figures 2a and 2b, are designed to highlight the fact that the relative contribution of a data set (i.e., depth or concentration) to the variability of the resultant application rate data is proportional to the scale of variability inherent in it. Accordingly, the examples presented here consider combinations of highly uniform irrigation data and highly variable fertilizer concentration data set and vice-versa.

Figure 2a shows an irrigation data set that vary in a relatively narrow range of 0.71 to 1.0 and a concentration data set that spans a much wider range of variation of 0.2 to 1.0. The pertinent sinusoidal functions are given in Table 1. As can be noted from Table 2, with a UCC and DU_{lq} of 0.894 and 0.846, respectively, the irrigation data set can be considered as highly uniform. Comparatively, the uniformity of the concentration data set is significantly lower (UCC=0.587 and DU_{lq}=0.401). The range of variation of the resultant application rate data set is 0.15 to 1.0 and its UCC and DU_{lq} are 0.504 and 0.322, respectively. The range of variation of the resultant application rate data is wider, and its uniformity is lower, than the depth and concentration data sets. This is consistent with the fact that the depth and concentration data sets are of same monotonicity (Figure 2a). In addition, it can be noted that the application rate data closely tracks the highly variable (non-uniform) concentration data set and its uniformity indices are much closer to those of the concentration data set than the irrigation data. This shows that the highly non-uniform concentration data has a dominant effect on the variability of the resultant application rate data. In contrast, the contribution of the highly uniform irrigation data to the variability of the application rate data is marginal. Note that this result is in agreement with the inferences stated above in regard to the significance of scale of variability and monotonicity of the data sets (i.e., depth and concentration) in terms of their effect on the scale and pattern of variability of the resultant application rate data.

The preceding example considers a scenario in which the depth and concentration data sets are of the same monotonicity. However, it can be readily shown that the essential result holds even when the depth and concentration data sets are of opposite monotonicity. The only difference is that with such a case the application rate data would have a slightly higher uniformity than the concentration data set.

Figure 2b depicts a very uniform concentration data set (UCC=0.900 and DU_{lq}=0.855) and a highly variable irrigation data (UCC=0.557 and DU_{lq}=0.358) with opposite montonicity (Table 2). The concentration data set vary in a narrow range of 0.72 to 1.0 compared to the irrigation data, which vary between 0.17 and 1.0. The range of variation of the resultant fertilizer application rate data is 0.23 to 1.0 and its UCC and DU_{lq} are 0.633 and 0.434, respectively. The range of variation of the resultant application rate data and its uniformity indices fall between those of the depth and the concentration data, which is consistent with the fact that the depth and concentration data sets are of opposite monotonicity (Figure 2b). In addition, the resultant application rate curve closely tracks the highly variable irrigation depth data and has a range of variation and uniformity that is comparable to the irrigation data. These results show that the highly non-uniform irrigation data set has a dominant effect on the scale and pattern of variability of the resultant application rate data. By contrast, the contribution of the highly uniform concentration data to the variability of the application rate data is marginal. These observations are consistent with the inferences stated above in regard to the significance of scale of variability and monotonicity of the data sets (i.e., depth and concentration) in terms of their effect on the scale and pattern of variability of the resultant application rate data.

An important practical implication of the results presented in

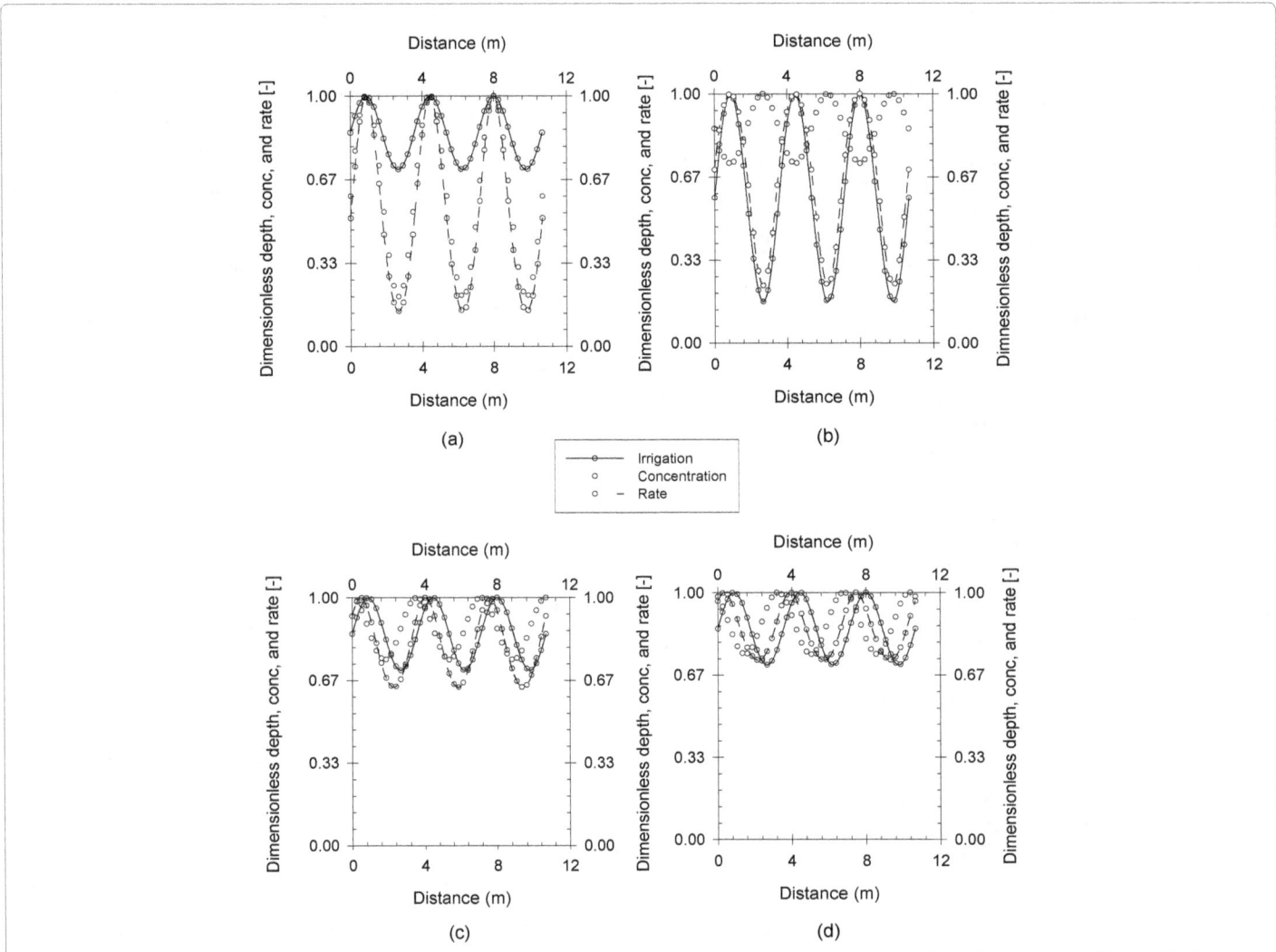

Figure 2: The relationship between spatial trends in irrigation depth, fertilizer concentration, and the spatial variability of the resultant fertilizer application rate. Scenarios where local spatial trends dominate data variability and depth and concentration have: (a) same monotonicity with a combination of high irrigation and low concentration uniformity, (b) opposite monotonicity with a combination of low irrigation and high concentration uniformity, (c) variable spatial overlap patterns over the test-plot with a combination of high irrigation and concentration uniformity, and (d) same as in Figure 2c, but depth and concentration data sets have relatively shorter segments with same monotonicity and longer segments with opposite monotonicity.

Figures 2a and 2b is that a fertigation scenario with high irrigation uniformity and low fertilizer concentration uniformity or vice-versa would likely lead to a low fertilizer application rate uniformity.

Scenario IV: Data sets with dominant local spatial trends in which depth and concentration have same monotonicity in some parts of the test-plot and opposite monotonicity in others

A fourth pair of examples, presented in Figures 2c and 2d, shows a simplified version of a more realistic scenario consisting of variable local spatial overlap patterns. Here we consider irrigation depth and concentration data sets (both following sinusoidal patterns with distance) that have the same monotonicity in some segments of the test-plot and opposite monotonicity in other parts of the test-plot. The goal is to show how the local spatial trends of the resultant application rate curve change as it transitions from a segment of the test-plot where the depth and concentration curves have the same monotonicity to those segments where depth and concentration have opposite monotonicity. In addition, these examples are designed to confirm the fact that it

is the aggregate contribution of these local effects that determine the overall test-plot scale uniformity of the application rate data.

Figure 2c depicts a dimensionless graph of irrigation depth and concentration data sets that follow a sinusoidal pattern superimposed on the resultant application rate curve. The irrigation depth data vary over a slightly wider interval of 0.70 to 1.0 compared to that of the concentration data, which vary in the range 0.75 to 1.0. As can be noted from Table 2, the corresponding irrigation uniformity (UCC=0.894 and DU_{lq}=0.846) and concentration uniformity (UCC of 0.908 and DU_{lq} of 0.869) can be described as very high. In order to form a mix of test-plot segments, where depth and concentration data sets have same monotonicity in some and opposite in others, a phase shift of 0.5π radian is introduced to the concentration data set with respect to the depth data (Table 1).

Figure 2c shows that the resultant application rate curve reaches its peaks, 1.0, and its lowest points, 0.64, in the intervals where the irrigation depth and concentration data sets have opposite monotonicity. On the other hand, in the segments where depth and concentration data sets are of the same monotonicity, the application rate data set not only has the same monotonicity as the depth and concentration data sets, but also has steeper slopes (higher variability) than the curves of both data sets. The implication is that in segments of the test-plot where depth and concentration data sets have opposite monotonicity, the overall effect of the interaction of the local spatial trends is to limit the variability in the resultant application rate data. Conversely, in those segments of the test-plot where depth and concentration data sets have same monotonicity, the overall effect of the interaction of the local spatial trends is to enhance the variability in the resultant application rate data. Note that these results are consistent with the inferences stated above in regard to the effects of the local overlap patterns, of the depth and concentration data sets, on the variability of the resultant fertilizer application rates.

Figure 2d depicts the same irrigation depth data as that presented in Figure 2c. However, in order to show the effect of different local overlap patterns on the variability of the resultant application rate data, here the concentration data is shifted by an angular distance of 0.67π radian instead of the 0.5π radian used in Figure 2c (Table 1). Note that this leaves the uniformity of the concentration data unchanged as in Figure 2c. Evidently, the observations noted in the preceding example (Figure 2c), with respect to the local behavior of the application rate data as affected by the overlap patterns of depth and concentration data sets, hold here as well. However, compared to that of Figure 2c, with the current example the intervals over which the depth and concentration curves have opposite monotonicity are slightly longer and the segments over which they exhibit same monotonicity are slightly shorter. The combined effect of which is to reduce the range of variation of the resultant application rate data to between 0.73 and 1.0 compared to that of 0.64 and 1.0 in the preceding example (Figure 2c). The uniformity of the resultant fertilizer application rate data ($UCC=0.897$ and $DU_{lq}=0.860$) as well is higher than that computed for the preceding example ($UCC=0.858$ and $DU_{lq}=0.804$), Figure 2c and Table 2.

The results presented in Figures 2c and 2d confirm, the preceding observation, that the uniformity indices of the depth and concentration data sets do not determine the specific values of the uniformity indices of the resultant application rate data set. Instead it is the interactive effects of the local spatial trends and scales of variability, inherent in the depth and concentration data sets, that are the main determinants of the test-plot scale uniformity of the resultant application rate data set. In other words, test-plot scale application rate uniformity is an aggregate index of the interplay of these local spatial effects over the test-plot.

Although the spatial overlap patterns of the depth and concentration data sets considered in Figures 2c and 2d are different, for both examples the uniformity of the resultant application rate data sets can be described as high. This is related to the very high uniformity of the corresponding depth and concentration data sets and will be discussed in the next section.

Fertigation scenarios that lead to acceptably high fertilizer application rate uniformity

Considering the irrigation depth and fertilizer concentration functions presented in Figures 2c and 2d, it can be noted that a fairly large number of overlap patterns, and hence different resultant application rate functions with specific uniformity levels, can be derived by simply varying the phase shifts between the irrigation and concentration data sets. Evidently, the minimum fertilizer application rate uniformity, that

can be obtained from these depth and concentration functions, should correspond to an overlap pattern in which the functions are completely in phase, (i.e., they are of exactly the same monotonicity) throughout the test-plot. Accordingly, it can be shown that the minimum resultant fertilizer application rate uniformity indices are $UCC=0.807$ and $DU_{lq}=0.729$. Considering, for instance, a fertilizer application rate uniformity acceptability threshold of $UCC=0.75$ and $DU_{lq}=0.7$ (a criteria that closely parallels the irrigation uniformity thresholds recommended for field crops [11]), this minimum uniformity level can be described as acceptably high. The implication here is that, regardless of the overlap patterns, the uniformity of any application rate data set derived from the depth and concentration data sets, given in Figures 2c and 2d, will remain within the range considered acceptably high.

For comparison let us now consider two additional groups of data sets presented in the preceding sections. Considering depth and concentration data sets presented in Figures 1c and 1d, it can be readily observed that the corresponding minimum fertilizer application rate uniformity is well below the uniformity acceptability threshold given above ($UCC=0.291$ and $DU_{lq}=0.120$). Similarly, for data sets depicted in Figures 2a and 2b, the minimum application rate uniformity is significantly lower than the threshold ($UCC=0.504$ and $DU_{lq}=0.322$). Note that the overlap patterns, between the depth and concentration data sets, that correspond to the respective minimum application rate uniformity levels are the same for all these data sets. Thus, the difference between these data sets lies in the variability (uniformity) of their respective depth and concentration data sets. Considering the data sets presented in Figures 2c and 2d, the uniformity indices for both depth and concentration are very high, exceeding the indicated uniformity acceptability thresholds by a minimum of 15.0%. By comparison, the uniformity indices of the depth and concentration data sets depicted in Figures 1c-1d are both well below the uniformity thresholds (Table 2). For the data sets presented in Figures 2a-2b, however, irrigation uniformity is well above the threshold, but the uniformity of fertilizer concentration is below the threshold by a significant margin and as such it has a dominant effect on the uniformity of the resultant application rate uniformity. A useful observation that stems from the preceding discussion is that high uniformity, of both irrigation depth and fertilizer concentration, is a requisite condition for attaining acceptably high fertilizer application rate uniformity.

Summary of significant results

Evidently the analyses presented in the preceding sections are based on simplified hypothetical examples in which functions of the same form are used to define the spatial variations of both irrigation and concentration. Furthermore, the functions considered have uniform, or locally variable yet repetitive, spatial trends and overlap patterns spanning the test-plot. In addition, the hypothetical data sets considered here generally have higher spatial resolution than typical measured data. Nonetheless, the basic mathematical relationship that determines the effects of the interactions between irrigation and concentration data sets on the spatial patterns of the resultant application rate data is the same for both hypothetical and measured fertigation data. The implication is that results of the preceding analyses, which are based on simplified hypothetical scenarios, have relevance to measured fertigation data sets. Accordingly, interesting inferences with potential applications in the evaluation, design, and management of fertigation systems can be deduced on the relationships between irrigation, fertilizer concentration, and application rate:

(1) The interactive effects of the local spatial trends and scales of

variability, of the irrigation depth and fertilizer concentration data sets, are the main determinants of the uniformity of the resultant application rate data. Thus, test-plot scale application rate uniformity can be considered as an aggregate index of these local effects over the area of the test-plot.

An important practical implication of this observation is that irrigation or fertilizer concentration uniformity alone may not often be adequate to even qualitatively characterize fertilizer application rate uniformity.

(2) The local spatial trends of a fertilizer application rate data are functions of the respective spatial trends of the depth and concentration data sets:

(2a) If irrigation depth and fertilizer concentration data sets have the same local monotonicity in a given section of the test-plot, the resultant application rate data as well will have the same local spatial trend as the depth and concentration data sets;

(2b) If irrigation depth and fertilizer concentration data sets have opposite local monotonicity in a given section of the test-plot and they are of significantly different scale, the resultant application rate data will have the same local spatial trend as either the depth or the concentration data set, whichever has the larger scale of variability; and

(2c) If irrigation depth and fertilizer concentration have opposite local monotonicity in a given section of the test-plot and they are of comparable scale, the local spatial trend of the resultant application rate data may have a larger frequency of spatial variability elements than the underlying depth and concentration data sets and the scale of variability will be less than those of the corresponding depth and concentration data sets (Figure 1d);

(3) The test-plot scale uniformity of a fertilizer application rate data set cannot be predicted based on the uniformity levels of the corresponding irrigation and fertilizer concentration data sets. Note that this inference excludes the unique scenario whereby either depth or concentration or both have perfect uniformity, in which case the local spatial overlap patterns have no effect on the uniformity of the resultant application rate;

(4) The results presented in the preceding sections show that under a special set of conditions a qualitative characterization of the uniformity of the resultant application rate data can be made based on the uniformity of the corresponding depth and concentration data sets, these include:

(4a) Given a fertilizer uniformity acceptability threshold, high uniformity, of both irrigation depth and fertilizer concentration, is a requisite condition for attaining acceptably high fertilizer application rate uniformity.

(4b) A combination of low irrigation and low concentration uniformity may not necessarily lead to low application rate uniformity; and

(4c) A combination of low irrigation uniformity and high concentration uniformity or vice-versa will likely lead to low application rate uniformity.

Applications in Fertigation Uniformity Evaluation

In this section field data sets collected through test-plot scale measurements are presented. The aim is to highlight the practical applications of the results, presented in the preceding sections, in the evaluation and analysis of the uniformity of fertigation data sets.

A concise description of the approach and materials used to measure test-plot scale irrigation depths and concentrations is presented. Computation of the resultant application rate data and the uniformity indices for irrigation, concentration, and application rate with Equations 1-3 is discussed. Finally, the inferences deduced in section 4 are used to explain and analyze the observed relationships between local spatial trends, and test-plot scale uniformities, of measured depth, concentration, and resultant application rate data sets.

Measurements of test-plot scale irrigation depth and fertilizer concentration

As part of a field-scale sprinkler fertigation study a series of fertigation uniformity evaluations were conducted in the Yuma Valley Irrigation Districts of Southwest Arizona in the winter seasons of 2013 and 2014 [8]. Six of the test-plot scale data sets, collected in these evaluations, are presented here. Figure 3 depicts the layout of a uniformity evaluation test-plot used in this study. It covers a rectangular area, circumscribed by four adjacent sprinklers, and measuring 9.14 m × 10.67 m (30.0 ft × 35.0 ft). The test-plot is discretized into 42 grid squares, each measuring 1.524 m × 1.524 m (5 ft × 5 ft). A rain gage is placed in each of the grid squares constituting the test-plot. The rain gages used in these field evaluations were obtained from the Irrigation Training & Research Center, California Polytechnic State University, San Luis Obispo, CA. They have a catchment area of 104.84 cm^2 and are graduated in 5.0 mL increments up to 100.0 mL volume. For measurements ranging between 100.0 mL and 200.0 mL they are graduated in 25.0 mL increments. The maximum measurable depth with these rain gages is about 19.1 mm with an estimated precision ranging between 0.1 mm and 0.5 mm (computed based on assumed volumetric reading errors ranging between ± 1.0 mL to ± 5.0 mL).

Nitrogen fertilizer was applied in the form of ammonium nitrate. The duration of irrigation in each of the field evaluations was three hours. The duration of nitrogen fertilizer application vary from a third of the irrigation application time to the entire irrigation application time. Further details related to the fertigation studies are presented in Zerihun and Sanchez.

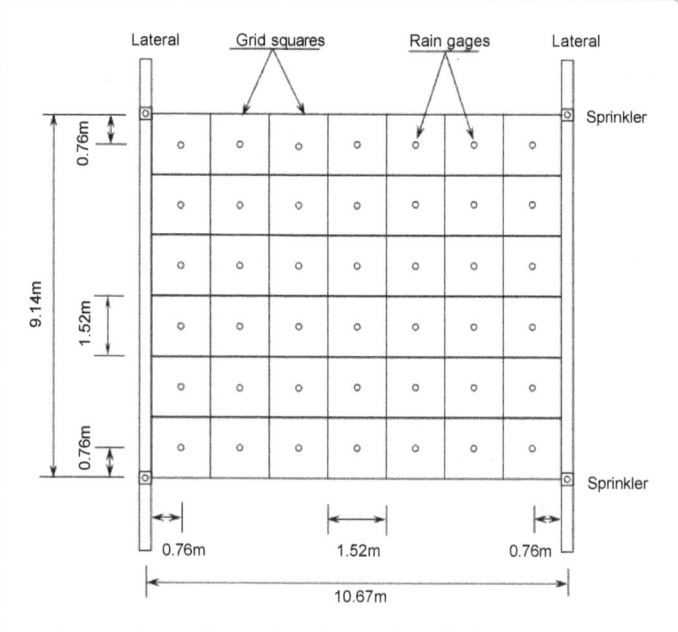

Figure 3: Test-plot layout for field evaluation of fertigation uniformity.

Precipitation depths, collected in the rain gages, were recorded immediately following a fertigation event and are used subsequently to compute test-plot scale irrigation uniformity. Water samples were then collected in appropriately labeled vials from each of the rain gages, which were sealed and frozen shortly after sampling, in order to preserve the integrity of the sample constituents (i.e., mineral nitrogen forms) until laboratory analysis. The total nitrogen concentration here consists of the sum total of elemental nitrogen concentrations present in solution in the form of ammonium- and nitrate-nitrogen. Total ammonium- and nitrate-nitrogen were determined colorimetrically using an Astoria Pacific A2. Ammonium was determined using the indophenol blue method and nitrate was determined using Griess-Ilosovay method after reduction with copperized cadmium [24].

Computation of nitrogen application rates and uniformities

Measured total nitrogen concentrations and irrigation depths were used to compute the resultant application rates with Equation 3. The irrigation depths, nitrogen concentrations, and nitrogen application rates are then used to compute the respective test-plot scale uniformity indices with Equations 1 and 2.

Comparison of the uniformity indices of measured irrigation and nitrogen concentration data sets and the resultant application rate data

Figure 4 presents a comparison of the test-plot scale UCC and DU_{lq} of six field measured irrigation and nitrogen concentration data sets and the corresponding application rate data. For each of the data sets fertilizer application rate uniformity more closely tracks the lower of the irrigation and concentration uniformity levels and often times fall below it (Figures 4a and 4b). Evidently, these results can be explained based on observations made in section 4.3, which states that the data set (i.e., either depth or concentration) with the larger scale of variability (lower uniformity) has a dominant effect on the variability (uniformity) of the resultant application rate data. In addition, the application rate uniformity levels (i.e., the UCC and DU_{lq} values) for data sets 3 and 4 are lower than the concentration uniformity levels by appreciable margins. As can be noted from the discussion in section 4.2, the likely explanation for this is that in a significant fraction of the test-plot area the depth and concentration data for data sets 3 and 4 may have same monotonicity.

Irrigation, nitrogen concentration, and application rate contours

Test-plot scale dimensionless contours of the irrigation, concentration, and application rate data sets for one of the fertigation events, summarized in Figure 4, are presented in this section. The contours are generated with Surfer (Golden Software Inc., Golden, Colorado) using the gridding option of Kriging. The goal is to examine the effects of local spatial trends and overlap patterns, of the measured irrigation and concentration data sets, on the spatial variability of the resultant nitrogen application rate data.

Test-plot scale normalized contours of irrigation depth, nitrogen concentration, and nitrogen application rates for data sets 1 (Figure 4) are shown in Figures 5a-5c. Note that the spatial variability patterns of the nitrogen application rate surface more closely approximate the rather less uniform nitrogen concentration surface than the irrigation depth surface (Figures 5a-5c).

Considering the lower left-hand corner of the test-plot, it can be observed that both irrigation and nitrogen concentration decrease

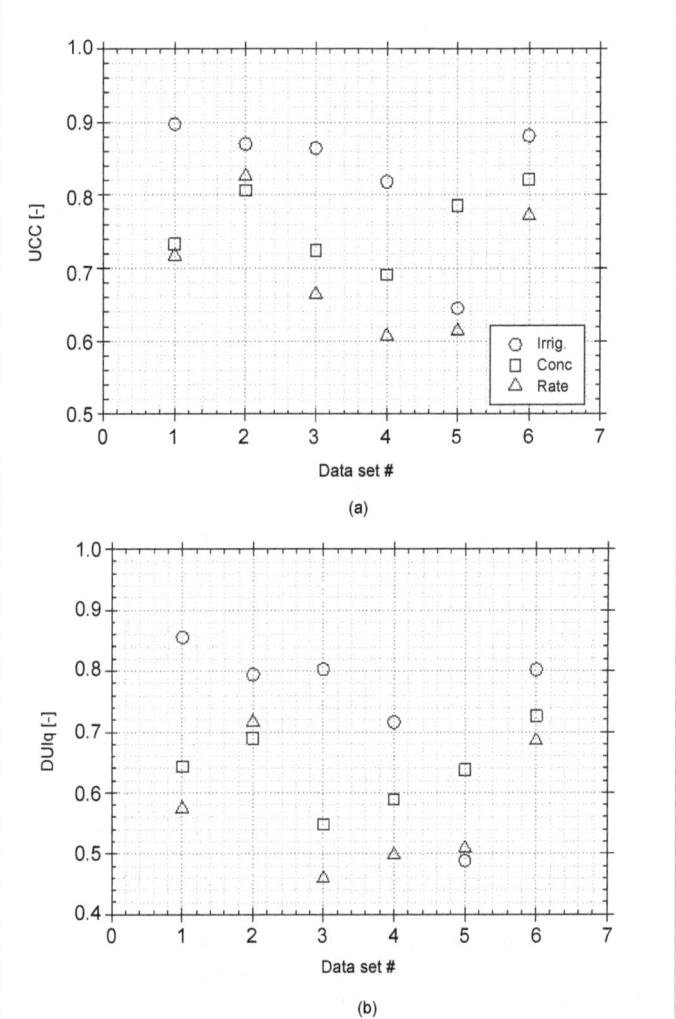

Figure 4: Comparison of irrigation, nitrogen concentration, and nitrogen application rate uniformity for six field measured data sets: (a) UCC and (b) DUlq.

as one moves toward that corner of the test-plot along the direction of the respective arrows (Figures 5a and 5b). Note that the resultant application rate data (Figure 5c) has a similar local spatial trend as the irrigation and nitrogen concentration data sets. Furthermore, a close look at an area of the test-plot that is slightly above the middle section and adjacent to the left edge shows that both the irrigation depth and concentration contours have similar local spatial trends, decreasing in the direction indicted by the arrows (Figures 5a and 5b). Observe that the resultant application rate surface (Figure 5c) as well follows the same general spatial trend as these data sets, but it more closely approximates the relatively more variable concentration surface than that of the irrigation surface. Considering the area slightly above the middle section of the test-plot and adjacent to the right edge, it can be noted that the irrigation and concentration data sets have opposite local spatial trends along the directions indicated by the respective arrows (Figures 5a and 5b). The concentration data set shows a significantly larger local variability than the irrigation data set, as a result the local variability pattern of the application rate data closely approximates that of the concentration data.

Evidently, the local spatial variability and overlap patterns of measured irrigation and concentration data sets are irregular and three

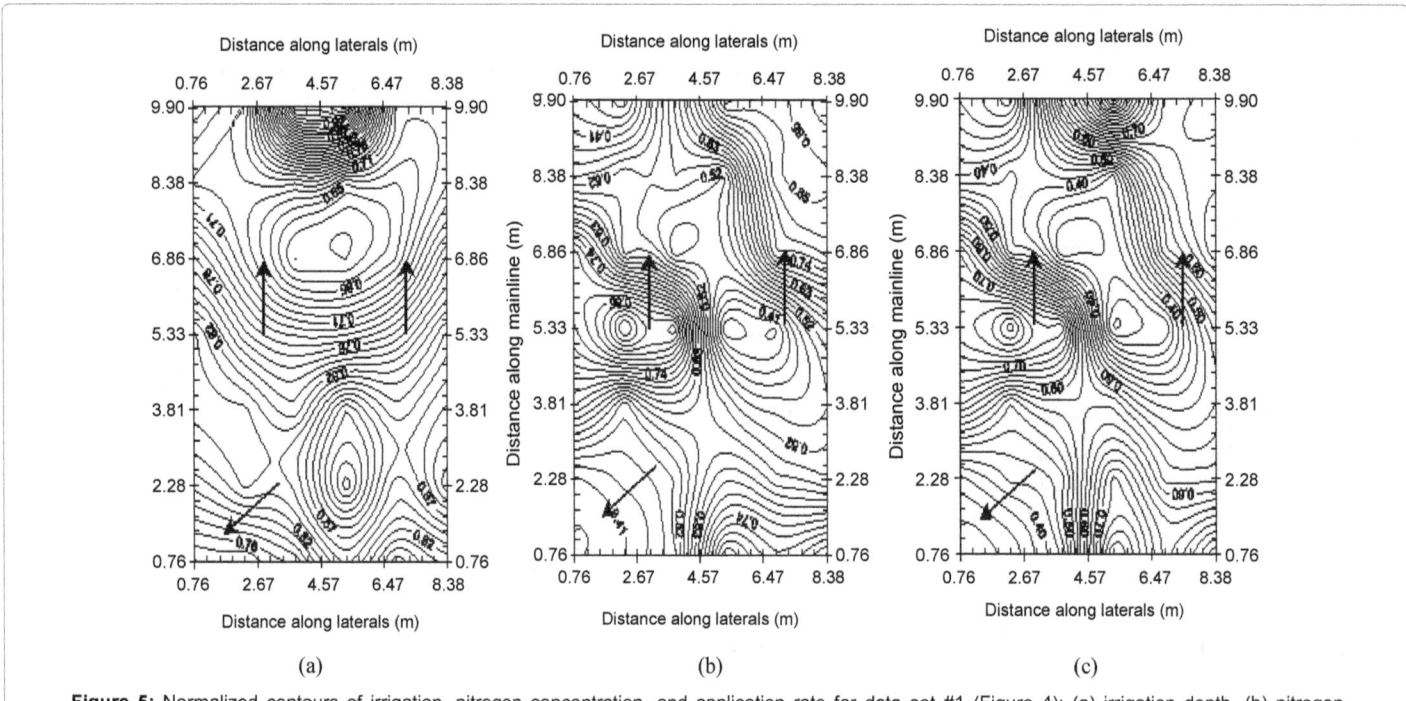

Figure 5: Normalized contours of irrigation, nitrogen concentration, and application rate for data set #1 (Figure 4): (a) irrigation depth, (b) nitrogen concentration, and (c) nitrogen application rate.

dimensional, hence too complex to relate directly to the test-plot scale uniformity of the resultant application rate data. Nonetheless, the results presented above highlight the fact that the local spatial trends and scale of variability of the application rate data (the aggregate effect of which eventually determines the application rate uniformity) are functions of the local overlap patterns and scale of variability of the irrigation and concentration data sets. Note that these observations can be explained by the inferences deduced, in sections 4.1-4.6, based on analyses of hypothetical fertigation scenarios.

Summary and Conclusions

In modern faming systems, fertigation is widely practiced as a convenient and cost effective method for applying soluble fertilizers to crops. Along with efficiency and adequacy, uniformity is an important fertigation system performance criterion. Fertigation uniformity is defined here as a composite parameter consisting of two independent but equally important indices: irrigation and fertilizer application uniformity indicators. The field and computational procedures related to irrigation uniformity evaluation have been studied extensively. Hence, the study reported here focusses on the development of an analytical framework for the evaluation and analyses of fertilizer application uniformity under sprinkler irrigated conditions.

Equations for fertigation uniformity indices are presented. Fertilizer application rate, given as a function of irrigation depth and fertilizer concentration, is identified as the appropriate variable for expressing fertilizer application uniformity indices. Pertinent mathematical properties of the uniformity equations are listed and their practical implications are described. Carefully designed hypothetical examples were analyzed to demonstrate the significance of the effect, of the local spatial overlap patterns between irrigation depth and fertilizer concentration data sets, on the uniformity of the resultant application rate data.

The results of the study show that fertilizer application rate

uniformity is an aggregate index of the interactive effects of the local spatial trends inherent in the irrigation depth and fertilizer concentration data sets. The study also demonstrated that often the uniformity of irrigation and fertilizer concentration cannot be uniquely related to the uniformity of the resultant application rate data set. However, some practically useful qualitative relationships between the uniformity of irrigation depth, solute concentration, and the resultant application rate data sets are presented. Application of the approach presented here in the evaluation and analysis fertigation uniformity data sets, measured under sprinkler irrigated conditions, is highlighted.

References

1. Wright J, Bergsrud F, Rehm G, Rosen C, Malzer G et al., (2013) Nitrogen application with irrigation water: Chemigation.

2. Burt C, O'Conner K, Ruehr T (1998) Fertigation. Irrigation Training and Research Center, California Polytechnic State University.

3. Muirhead WA, Melhuish FM, White RJG (1984) Comparison of Several Nitrogen Fertilizers Applied in Surface Irrigation Systems I. Crop Response Fert Res 6: 97-109.

4. Fares A, Abbas F (2009) Irrigation Systems and Nutrient Sources for Fertigation. College of Tropical Agriculture and Human Resources, University of Hawaii at Manoa.

5. van der GTW, Evans RG, Eisenhauer DE (2007) Chemigation. In Design and operation of farm irrigation systems; pp: 725-753.

6. Threadgill DE (1985) Chemigation via Sprinkler Irrigation: Current Status and Future Development. App Eng Agric 1: 16-23.

7. Zerihun D, Sanchez CA, Farrell-Poe KL, Admsen FJ, Hunsaker DJ (2003) Performance Indices for Surface N Fertigation. J Irrig Drain Eng 1293: 173-183.

8. Zerihun D, Sanchez CA (2014) Evaluation of Sprinkler Fertigation of Vegetables. Report submitted to the Arizona Specialty Crops Council.

9. Heerman DF, Solomon KH (2007) Efficiency and Uniformity. In Design and Operation of Farm Irrigation Systems. Pp: 108-119.

10. Martin DL, Dennis CK, Lyle WM (2007) Design and Operation of Sprinkler Systems. In Design and operation of farm irrigation systems. Pp: 557-631.

11. Keller J, Bliesner R (1990) Sprinkle and Trickle Irrigation. Van Nostrand Reinhold. New York.

12. Burt CM, Clemmens AJ, Strelkoff TS, Solomon KH, Bliesner RD, et al., (1997) Irrigation Performance Measures: Efficiency and Uniformity. J Irrig Drain Eng 123: 423-442.

13. Clemmens AJ, Solomon KH (1997) Estimation of Global Irrigation Distribution Uniformity. J Irrig Drain Eng 123: 454-461.

14. Nderitu SM, Hills DJ (1993) Sprinkler Uniformity as Affected by Riser Characteristics. Trans ASAE 9: 515-521.

15. Fischer GR, Wallender WW (1988) Collector Size and Test Duration Effects on Sprinkler Water Distribution Measurement. Trans ASAE 31: 539-542.

16. Livingston P, Loftis JC, Duke HR (1985) A Wind Tunnel Study of Sprinkler Catch-can Performance. Trans ASAE 28: 1961-1965.

17. Warrick WA (1983) Interrelationships of Irrigation Uniformity Terms. J Irrig Drain Eng 109: 317-332.

18. Kruse EG (1978) Describing Irrigation Efficiency and Uniformity. J. Irrig Drain Div 104: 35-41.

19. Merriam JL, Keller J (1978) Farm Irrigation System Evaluation: A Guide to Management. Utah State University. Utah.

20. Wilcox JC, Swailes GE (1947) Uniformity of Water Distribution by Some Under Tree Orchards Sprinklers. Sci Agric 27: 565-583.

21. Christiansen JR (1942) Irrigation by Sprinkling. Food and Agriculture Organization of the United Nations.

22. Nakayama FS, Bucks DA, Clemmens AJ (1979) Assessing Trickle Emitter Application Uniformity. Trans ASAE 22: 816-821.

23. Keller J, Karmeli D (1975) Trickle Irrigation Design Parameters. Trans ASAE 17: 678-684.

24. Mulvaney RR (1996) Nitrogen-Inorganic Forms. In Methods of Soil Analysis: Part 3–Chemical Methods. Ed. D. L. Sparks Soil Sci Soc Am.

A Developed Criterion for Rationalizing On-Farm Irrigation Water Uses Under Arid Conditions

Arafa YE*

Agricultural Engineering Department, Faculty of Agriculture, Ain Shams University, Cairo, Egypt

Abstract

Exploitation of irrigation water under arid ecosystem conditions becomes the pedagogical problem; therefore, rationalizing irrigation and maximizing water use efficiency based on appropriate developed technologies are the most important aspects in the water and agricultural policies. Therefore, the objectives of this study were to: 1) develop out a criterion to identify irrigation system effectiveness by using a dimensionless analysis; and 2) validate the suggested criterion.

Dimensional analysis outputs revealed that the irrigation efficiency, may be better if it replaced by new developed terminology noted as irrigation effectiveness for calculating the seasonal crop water requirements (SCWR) that represent a ratio of irrigation system performance and irrigated soil characteristics. The developed criterion may be expressed as follows:

$$E_{idc} = \frac{S_{pi}}{I_{pi}} = \frac{\left[\left[\dfrac{\sum (q_i - q_n)^2}{\overline{q}*n*p}\right] * \dfrac{t}{DU} * \dfrac{1}{w}\right]}{\left[\dfrac{M.C_{after} - M.C_{before}}{[F.C - P.W.P]}\right] * \dfrac{d}{T_i} * \ln(k) * [f*n] * coneversion}$$

Moreover, results analysis of the validation process of the application of the developed criterion indicated that SCWR had been improved by applying the developed criterion with about 10.55 − 21.56% comparing with the conventional method that had been recommended by FAO.

Keywords: Micro-Irrigation; Soils; Dimensional analysis; Irrigation Efficiency; Irrigation; Water use efficiency

Introduction

Arid ecosystems conditions could be characterized with dwindling water resources and growing competition for water will reduce water availability for agricultural development processing, while the need to meet growing food demands will require that more food is grown with less water. A more effective water and greater water-unit productivity will be a primary challenge for future development [1,2]. Due to excess or deficient levels of water or nutrients could result in yield reductions, meanwhile, proper design and management processes of micro-irrigation system are essential for successful production. Systems must integrate soil-hydro-physical properties, crop root distribution characteristics, water requirements related to crop growth stage and environmental demand [3-7].

There is no doubt that the average crop yield is a function of the irrigation water application factors (application uniformity; depth of application and the amount of daily evapotranspiration supplied by rainfall), the hydraulic variation of distributors as well as the crop sensitivity to the moisture stress. Application uniformity depends on the manufacture's uniformity of the selected distributors, the hydraulic design and the systems maintenance program [8]. Following early research into the amount of water required for crop production, water use efficiency becomes a widely used agronomic term to express the efficiency of production per unit of water required [9-11]. Supply of water through irrigation involves the caption of water from the source, the transport through the irrigation system (i_{ce}), the distribution on the farm (i_{ef}) and the application to the field and to the crop. Enciso and Al-Jamal mentioned that water management will become an important practice used. Irrigation efficiency (IE) is an important factor into improving water management but so is economic return [12,13].

Patel and Rajput hypothesized that improved yields from subsurface drip systems are most likely due to more water being available to the plants, as compared to surface drip because of less evaporation in subsurface drip system. Burt, Slavil and Zavadil noted that crop evapotranspiration (ET) would be less for a well-watered crop with dry soil and plant surfaces (that is possible with subsurface drip system) than if the crops were to be irrigated with a method that wets the soil and plant surfaces [14,15]. On the other hand, El-Raie and Abdel-Wahab stated that the appropriate selection of the CWU method under diverse micro-climatic regions had to be considered for improving water uses under arid and semi-arid conditions [16]. Kumar, Alazba developed functions that can be used as a guide to yield potential allocation decision related to limited irrigation water [17,18]. However, the transport of on-farm irrigation water to cope the crop-water requirements' is overcoming three levels of management: 1) the on-farm irrigation system managed by an irrigation agency; 2) the farm system managed by a group of farmers and/or an individual farmer and 3) the field system managed by the individual farmers.

Hereby, the objectives of this study were to: 1) develop out a criterion to identify irrigation system effectiveness based on a dimensionless analysis; and 2) validate the suggested criterion under field conditions.

Methodology

Each level of the methodology development is subjected to the technical sophistication of the hydraulic pathway and management of

*****Corresponding author:** Arafa YE, Agricultural Engineering Department, Faculty of Agriculture, Ain Shams University, Cairo, Egypt, E-mail: arafayeh11@gmail.com

the irrigation system, in order to determines to a large extent efficiency of the irrigation systems. However, the first and second levels play a crucial role in the effective management of on-farm irrigation water.

Dimensional and data analysis

Factors affecting the performance analyses of on-farm irrigation systems had been gathered, analyzed and evaluated, in order to observe the dimensionless group. Also, data that represents the soil characteristics for managing the irrigation water had been gathered and evaluated for the same purpose. After then, the observed dimensionless groups had been verified individually and interactional-dynamic; then all groups had been validated under field conditions.

Irrigation performance analyses index (I_{pi}): Irrigation systems that apply more uniformly and in limited amounts to avoid water stress in plants and to prevent excessive drainage is a crucial objective under arid ecosystems conditions. Herby, it can be concluded that the irrigation efficiency of localized irrigation systems is a function of distribution uniformity with which water discharged from the distributor devices.

From a point of view towards pressurize irrigation systems in general and localized irrigation in particularity, wherever, the forced stream through pipelines is the man factors of flow, it can accordingly clarified that both conveyance efficiency and farm efficiency can be negligible, however, their values are approximately to be a comply the maximum values, and their losses can be neglected too. After then, the most important efficiency that affected the performance of those irrigation system is that the application efficiency. Hereby, the factors that may be affecting the optimizing of the localized irrigation systems efficiencies are analyzed below.

Soils performance index (S_{pi}): Application of predicting models for optimizing localized irrigation systems efficiency needs information about hydro-physical properties, especially hydraulic conductivity under saturated and unsaturated conditions. In addition, other parameters such as field capacity and permanent wilting point that indicate soil texture may need to confirm the logical relation between them.

Soil hydraulic resistance: The resistance to vertical flow (R_i) of the i-th soil layer with a saturated thickness d_i and vertical hydraulic conductivity Kv_i is:

$R_i = d_i / Kv_i$

Expressing Kv_i in m/day and d_i in m, the resistance (R_i) is expressed in days. The total resistance (Rt) of the soil profile is:

$Rt = \Sigma\, R_i = \Sigma\, d_i / Kv_i$

where Σ signifies the summation over all layers: i= 1, 2, 3, . . .n The *apparent* vertical hydraulic conductivity (Kv_A) of the soil profile is:

$Kv_A = Dt / Rt$

where Dt is the total thickness of the soil profile: Dt = $\Sigma\, d_i$, with i= 1, 2, 3, . . . n

The resistance plays a role in soil profile where a sequence of layers occurs with varying horizontal permeability so that horizontal flow is found mainly in the layers with high horizontal permeability while the layers with low horizontal permeability transmit the water mainly in a vertical sense. When the horizontal and vertical hydraulic conductivity (Kh_i and Kv_i) of the (i-th) soil layer differ considerably, the layer is said to be anisotropic with respect to hydraulic conductivity. When the *apparent* horizontal and vertical hydraulic conductivity (Kh_A and Kv_A) differ considerably, the soil profile is said to be anisotropic with respect to hydraulic conductivity. When calculating flow to drains though soil profile with the aim to control the water table, the anisotropy is to be taken into account; otherwise the result may be erroneous.

Transmissivity: The Transmissivity is a measure of how much water can be transmitted horizontally, such as to the tile drains. Soil profile may consist of n soil layers. The Transmissivity for horizontal flow T_i of the (i – th), soil layer with a *saturated* thickness d_i and horizontal hydraulic conductivity K_i is:

$T_i = k_i d_i$

Transmissivity is directly proportional to horizontal hydraulic conductivity K_i and thickness. d_i Expressing K_i in m/day and d_i in m, the Transmissivity T_i is found in units m²/day. The total Transmissivity T_t of the soil profile is the signifies the summation over all layers i= 1,2,3,...,n. When the soil layer is entirely below the water table, its saturated thickness corresponds to the thickness of the soil layer itself. When the water table is inside a soil layer, the saturated thickness corresponds to the distance of the water table to the bottom of the layer. As the water table may behave dynamically, this thickness may change from place to place or from time to time, so that the Transmissivity may vary accordingly. However, the estimation of K from grain size could be gotten from Allen-Hazen derived an empirical formula for approximating hydraulic conductivity from grain size analyses:

$K = C(D_{10})^2$

Where: C is the Hazen's empirical coefficient, which takes a value between 0.4 and 10.0, with an average value of 1.0. ρb / ρs ≤ 1 is packing system of soil particles which refers to soil porosity and therefore indicates the efficiency of soil tillage. Small this ratio indicates good aeration in soil and suitable tillage operation. While as this ratio increases this means soil tended to compacted or consolidated and characterized with bad aeration and high penetration resistance needs high power in plowing and through using other soil machines; **n** value in van Genuchten model affects the slope of the Soil Water Characteristics Curve for suctions greater than the air entry suction (y_a). The slope becomes increasing negative as **n** value decreases. The value for **n** is always > 1 generally fluctuated between 1.1 to 1.3 for well-structured clay soil and from 1.4 to 1.8 for medium and light textured soil. The great value of **n** refers to raped and easily water depletion from the soil, so it can be closely related to the period between irrigations. As the soil slope is increases finger flow (Flow of water into macro pores in vertical direction) is decreased while lateral flow tended to increase may resulting in decreasing the efficiency of soil water distribution. However, different hydro-physical characteristics of the Egyptian soils had been gathered and analyzed, as shown in (Table 1).

- **Effect of "n" on hydraulic conductivity:** n value affects the slope of the hydraulic conductivity for suctions greater than the air entry suction (y_a). The slope becomes shallower as n decreases.

- **Effect of "α" on Soil Water Content Curve:** α value (alpha) affects the breakpoint in the curve, commonly referred to as the air entry suction (y_a). The air entry suction increases as α value (alpha) decreases.

- **Effect of "α" on hydraulic conductivity:** α value (alpha) affects the breakpoint in the curve, commonly referred to as the air entry suction (y_a). The break point occurs at higher suctions as α (alpha) value decreases.

Texture Class	N	-- θr -- cm³/cm³		-- θs -- cm³/cm³		-- log(α) -- log(1/cm)		-- log(n) -- log10		-- Ks -- log(cm/day)		-- Ko -- log(cm/day)		-- L --	
Clay	84	0.098	(0.107)	0.459	(0.079)	-1.825	(0.68)	0.098	(0.07)	1.169	(0.92)	0.472	(0.26)	-1.561	(1.39)
Clay loam	140	0.079	(0.076)	0.442	(0.079)	-1.801	(0.69)	0.151	(0.12)	0.913	(1.09)	0.699	(0.23)	-0.763	(0.90)
Loam	242	0.061	(0.073)	0.399	(0.098)	-1.954	(0.73)	0.168	(0.13)	1.081	(0.92)	0.568	(0.21)	-0.371	(0.84)
Loamy Sand	201	0.049	(0.042)	0.390	(0.070)	-1.459	(0.47)	0.242	(0.16)	2.022	(0.64)	1.386	(0.24)	-0.874	(0.59)
Sand	308	0.053	(0.029)	0.375	(0.055)	-1.453	(0.25)	0.502	(0.18)	2.808	(0.59)	1.389	(0.24)	-0.930	(0.49)
Sandy Clay	11	0.117	(0.114)	0.385	(0.046)	-1.476	(0.57)	0.082	(0.06)	1.055	(0.89)	0.637	(0.34)	-3.665	(1.80)
S C L	87	0.063	(0.078)	0.384	(0.061)	-1.676	(0.71)	0.124	(0.12)	1.120	(0.85)	0.841	(0.24)	-1.280	(0.99)
S loam	476	0.039	(0.054)	0.387	(0.085)	-1.574	(0.56)	0.161	(0.11)	1.583	(0.66)	1.190	(0.21)	-0.861	(0.73)
Silt	6	0.050	(0.041)	0.489	(0.078)	-2.182	(0.30)	0.225	(0.13)	1.641	(0.27)	0.524	(0.32)	0.624	(1.57)
Si Clay	28	0.111	(0.119)	0.481	(0.080)	-1.790	(0.64)	0.121	(0.10)	0.983	(0.57)	0.501	(0.27)	-1.287	(1.23)
Si C L	172	0.090	(0.082)	0.482	(0.086)	-2.076	(0.59)	0.182	(0.13)	1.046	(0.76)	0.349	(0.26)	-0.156	(1.23)
Si Loam	330	0.065	(0.073)	0.439	(0.093)	-2.296	(0.57)	0.221	(0.14)	1.261	(0.74)	0.243	(0.26)	0.365	(1.42)

Table 1: Egyptian soil hydro-physical properties under arid conditions.

Theoretical therapy: The concepts of water use efficiency are introduced for the purpose of optimizing localized irrigation efficiency and the technical and agricultural options to increase production with less water elaborated. Efficiency is generally associated with a transformation process in which an input is transformed into an output. Therefore, the overall irrigation efficiency is defined as the fraction of water diverted to the irrigation system, which is ultimately effectively stored in the root zone and utilized effectively by plant in avoidance of plant-water stresses. The effective irrigation efficiency is characterized as a function of both irrigation systems equivalent parameters, soils equivalent parameters and application time.

Validation process:

Field experiments were carried out during two successive growing at a farm located at Longitude 30° 13' 0 E°, latitude 30° 25' 0 N and 25.5 m above MSL, which represents sandy soils conditions of the newly reclaimed soil of the Egypt. The analyses to determine physical and hydro-physical characteristics of the soil site had been conducted according to standard methods and presented in (Table 2). Surface and subsurface drip irrigation systems networks were installed at the experimental site. A split-split plot design was used in this experiment. However, the area was divided into two main, every plot was divided into three sub-plots each (90 × 30 m) for drip irrigation treatments (SD, SSD_{10} and SSD_{20}).

Cultivated crop: Onion crop (*Allium Cepa L.*), Giza 20 for two successive growing seasons (2011-2012) and (2012 -2013), the cultivated area was prepared into leveled basins of (30 × 30 m) for each treatment, and transplanted of onion seeds on Dec. 2011 in the first season and on Dec. 2012 in the second season, meanwhile, harvesting had been taken place on April of each growing season. The sawing was done in row at plant spacing of 14.3 cm between plants and the spacing between plant's rows was varied according to the number of cultivated plant's rows around laterals. All agronomic practices and the rate of applications were applied as recommended by Vegetable Research Institute, ARC, MALR, Egypt. The crop began to show signs of maturity (over 70% dropping of leave head) at 12 and 14 weeks after germination. Harvesting was carried cut about one week after, particularly 10th April 2012 and 13th April 2013. The area of 3 m in long and 1 m in wide in each plot were lifted (without discards), properly labeled and taken to be to laboratory to curve for about two weeks. Therefore, the onion bulbs were separated from the dry matter and weighed.

Calculation methods of the applied amounts of irrigation water: Onion plant-water requirements were calculated and scheduled. However, Reference evapotranspiration of the studied

Soil layer, cm	Particle size distribution, %			Texture class	B. D (gm/cm³)	Moisture content by weight (%)		
	Sand	Silt	Clay			F. C	P.W.P	A.W
0 –20	94.5	3.5	2	Sandy	1.65	8.03	3.33	4.7
20-40	95	3.3	1.7	Sandy	1.56	9.13	3.14	5.99
40-60	95.7	3	1.3	Sandy	1.44	10.07	2.99	7.08

F.C = Field capacity, P. W.P =Permanent Wilting point, A.W= Available water B.D= Bulk density

Table 2: Soil physical properties of the experimental site.

area had been gathered from Central Laboratory of Agricultural Climate (CLAC), Agriculture Research Center (ARC) for the cultivated growing seasons. After then, these gathered data were analyzed and processed according to investigated level of treatments, as described later. Reference evapotranspiration (ETo) was computed using the FAO, modified Penman-Monteith method [19]. ETo data were processed by using CropWat 8.1 model, for all calculation based on FAO method, or were processed to be used for calculation by using the developed criteria method.

$$CWU = kc * ET_0; \quad SCWR_{FAO} = CWU/E_a; \quad \text{and} \quad SCWRE_{idc} = CWU/E_{idc}$$

Treatments:

Actual evapotranspiration treatments: FAO: determination of actual evapotranspiration based on traditional method that had been described by **FAO.**

E_{dic}: determination of actual evapotranspiration based on the developed criterion.

Irrigation systems treatments:

SD: surface drip irrigation system treatments.

SSD_{10}: subsurface drip irrigation system with buried laterals at 10 cm depth treatments.

SSD_{20}: subsurface drip irrigation system with buried laterals at 20 cm depth treatments.

Measurement and calculations:

Soil Water and Salts Distribution Pattern under Deficit Irrigation Treatments: Soil samples were taken periodically during each growing season in order to determine soil moisture distribution patterns under each treatment. Meanwhile, soil samples were taken twice (one before cultivation season and the second after harvesting) in order to determine salt distribution patterns under each treatments, the soil samples were taken around the emitter, at 10 cm depth and 20 cm. Data were exposed to SURFER 7.

Computation of Crop-Water Use (CWU), Seasonal Crop Water Requirements (SCWR): The crop water use, seasonal crop water use and seasonal crop water requirements at each onion plants growing stage were calculated, determined based on the calculation method base.

Results and Discussion

Dimensional analysis outputs of the developed criterion

Irrigation performance parameters index: The observed irrigation performance parameters that affecting the optimizing localized irrigation systems efficiency can be formed as follows:

$$I_{ppi} = \left[\frac{\sum (q_i - q_n)^2}{\overline{q} * n * p} \right] * \frac{t * w}{DU}$$

Where:

q_i	: is the actual discharge of the distributor devices, m³/h.
q_n	: is the nominal discharge of the distributor devices, m³/h.
$\overline{q}$	: is the mean discharge rate of the distributor devices, m³/h.
N	: is a fraction of the clogging risks, fraction.
P	: is a fraction of pressure variation sensitivity (pressure drop fraction).
DU	: is the distribution uniformity, %.
T	: is the operating time of irrigation event, h.
W	: is the effective coverage width of the irrigated soils.

Soils performance parameters index: The observed index comply the following relationships between the soil hydro-physical parameters based on a mathematical and logic analyses. The developed formula maybe summarized as follows:

$$S_{ppi} = \left[\frac{M.C_{after} - M.C_{before}}{[F.C - P.W.P]} \right] * \left[\frac{d}{T_i} * \ln K \right] * [f * n] * coneversion$$

Where:

$M.C_{before}$	: is the soil moisture content before irrigation in equivalent volumetric units, %.
$M.C_{after}$	: is the soil moisture content after irrigation in equivalent volumetric units, %.
F.C	: is the soil field capacity, %.
P.W.P	: is the permanent wilting point, %.
K	: is the soil hydraulic conductivity, cm/h.
T_i	: is the soil Transmissivity, mm²/day
D	: is the applied irrigation water depth at each irrigation event, mm
F	: is the soil porosity, fraction.
N	: is an indices depend on the soil layer texture, which ranged from 1 – 1.6, fraction.

Developed criterion "E_{idc}":

$$E_{idc} = \frac{S_{ppi}}{I_{ppi}}$$

Where:

E_{idc}	: is the effective irrigation developed criterion.
S_{pp}	: is the soil performance parameter.
I_{pp}	: is the irrigation performance parameter.

Validation process outputs of the developed criterion

Soil-moisture and salts distribution patterns: Regarding soil moisture distribution patterns, data illustrated in (Figures 1 and 2) revealed that irrigation water was speeded on a large volume of soil in the treatment of 20 cm subsurface drip irrigation system followed with 10 cm subsurface drip irrigation one. Water distribution pattern under the treatment of 20 cm subsurface drip irrigation system could be represented with a deficit cone. While under surface drip irrigation one a complete cone was formed with a less volume of soil having readily available water (18% of soil water = 55% of Available water) of onion. Generally, subsurface drip irrigation system give a perfect water distribution in the soil based on the high percent of available water and a great amount of stored water localized at active root zone of onion plant. This finding may be due to upward movement of irrigation water from subsurface emitter plus downward one under the effect of gravitational potential. The obtained results indicated also that, using subsurface emitter buried at 20 cm below soil surface could improve water use efficiency of onion by minimizing the evaporative loss and delivering irrigation water directly to the root zone.

For salts distribution patterns, (Figures 3 and 4) indicated that an accumulation process of leached salts directly under emitter, which occupying from 10 up to 25 cm soil depth, while, it reached to 25 up to 35 soil depth horizontally far from emitter with about 10 cm in vertical direction. Due to leaching process which extended up to 10 cm horizontally far from the emitter from each side. Generally, the treatment of 20 cm emitter depth in subsurface irrigation, introduce best salt distribution and give the low value of soluble salt at active root depth of onion seedlings (0-25) beside emitter (10 cm horizontally far from emitter). From data analysis it could be concluded that soil salts are accumulated under emitters as a result of salt transported downward. In the case of 10 cm subsurface emitter, capillary action was more pronounced than in 20 cm case because the weakness of capillary action in this soil (coarse textured soil), which have dominance of macro pores.

Seasonal values of crop water requirements (SCWR): Data illustrated in (Table 3) showed that the general trend of increasing CWU and attributed SCWR from the beginning of cultivation up to the end of bulb formation stage (72 days after sowing seeds), then it decreasing within bulb enlargement and maturity stage. This is normally observation due to the crop water requirements and the change of micro-climate factors and attributed reference evapotranspiration, as well as, changes of either crop coefficient or crop water stress coefficient. In addition, from data analyses it could be noticed that, the highest values of CWU under developed criteria basis had been reduced compared with traditional way of calculation based on FAO under establishment, vegetative growth, bulb fermentation and bulb enlargement and maturity stages respectively. However, the increment percentage of SCWR had been ranged from 10.55 up to 21.21% under developed criterion comparing with the FAO base calculation method.

Conclusion

Rationalizing on-farm water uses has the majority of the agricultural development processes under arid conditions. Hereby, proper management of irrigation water unit by using monitoring and change-detection of dynamic behavior of soil-moisture are essential in this respect. Therefore, the developed criterion maybe consider as an effective index for maximizing water-unit productivity. Generally, observed-data analysis concluded that soil-moisture/or salt distribution

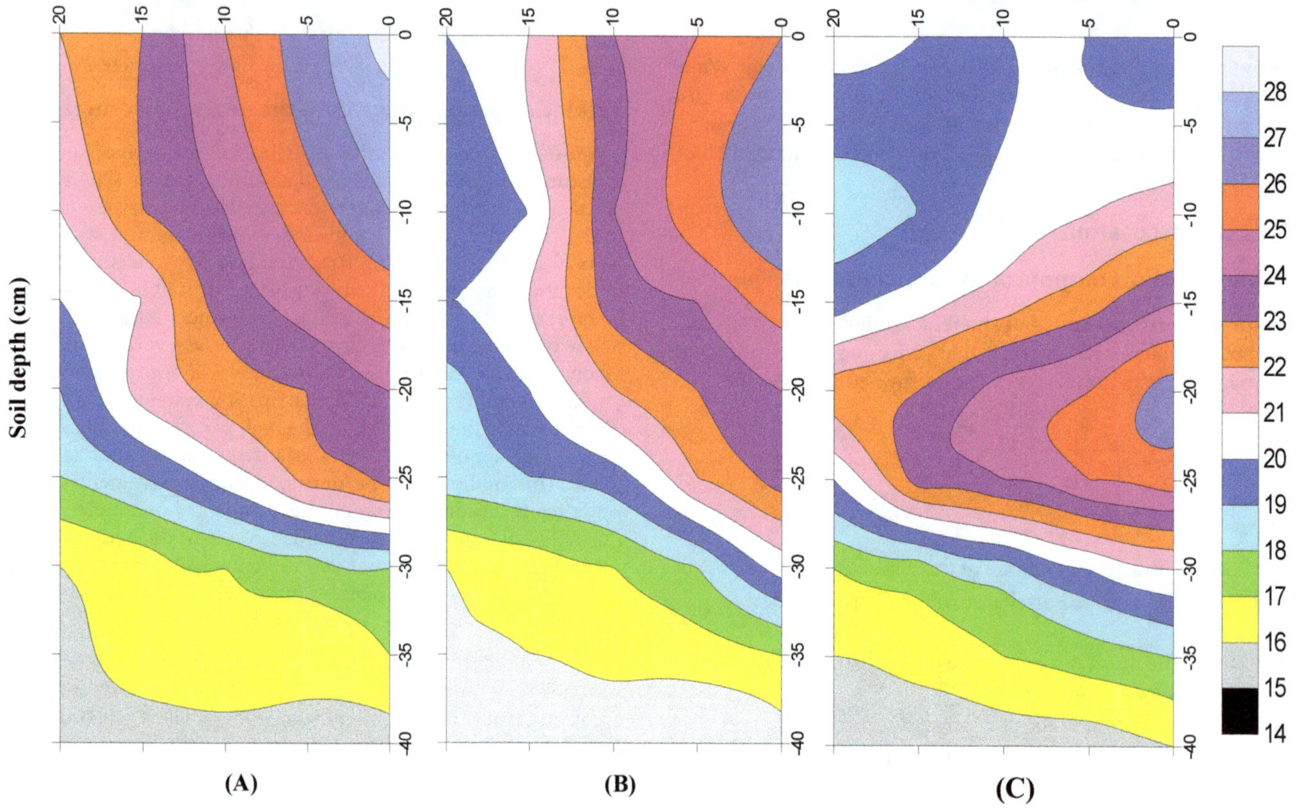

Figure 1: Water distribution around surface (a) and subsurface (b, c) emitter at 10, 20 cm depth based on developed criteria.

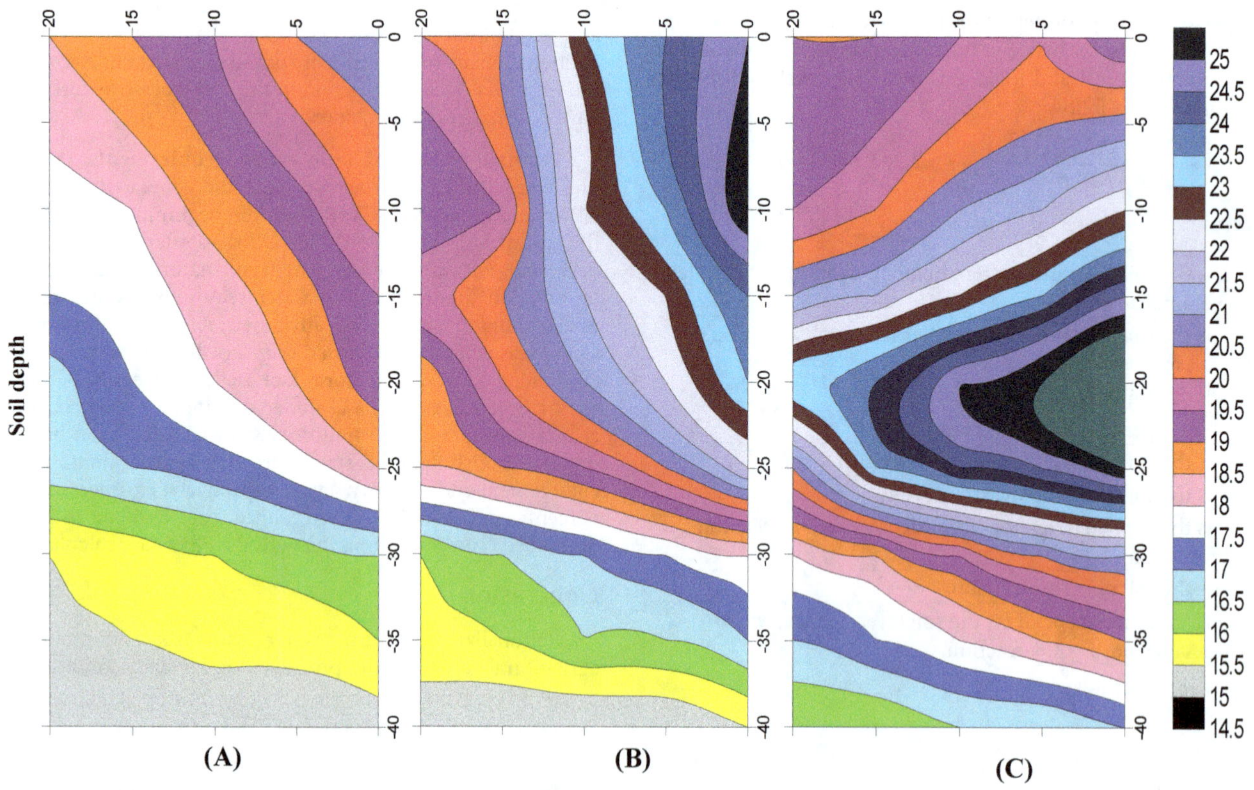

Figure 2: Water distribution around surface (a) and subsurface (b, c) emitter at 10, 20 cm depth based on FAO.

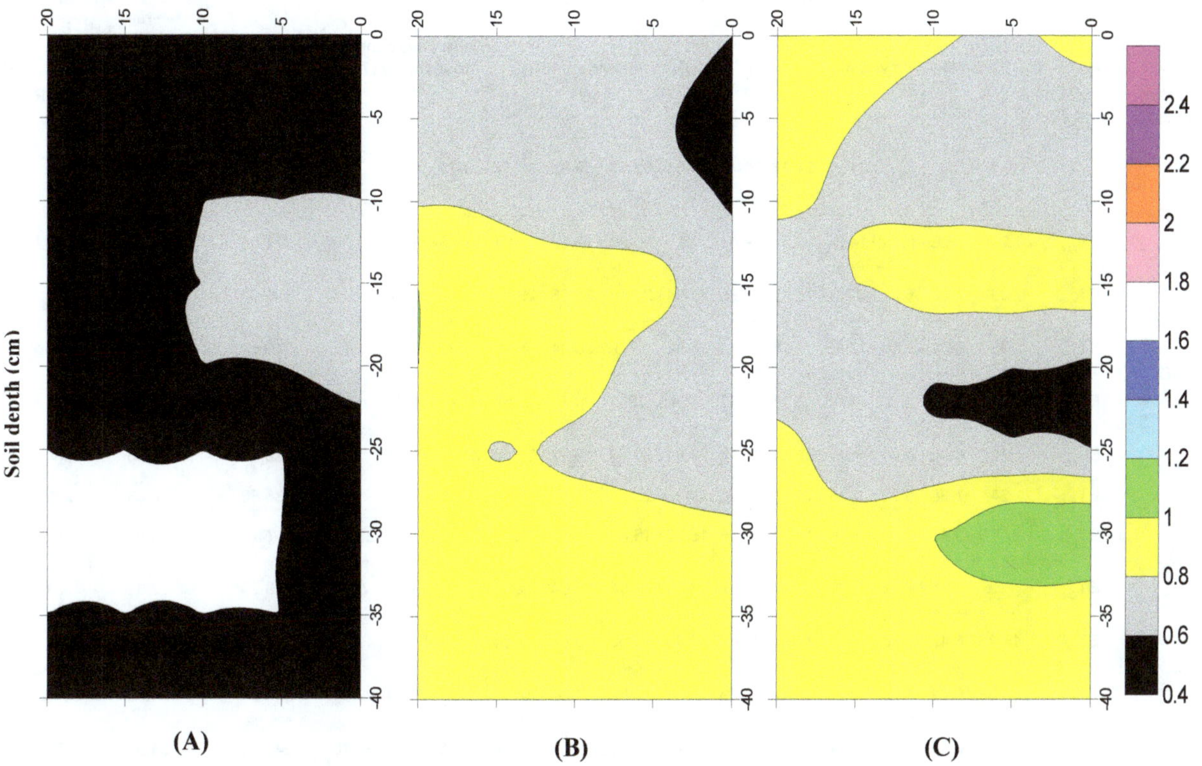

Figure 3: Salt distribution around surface (a) and subsurface (b, c) emitter at 10, 20 cm depth based on a developed criterion.

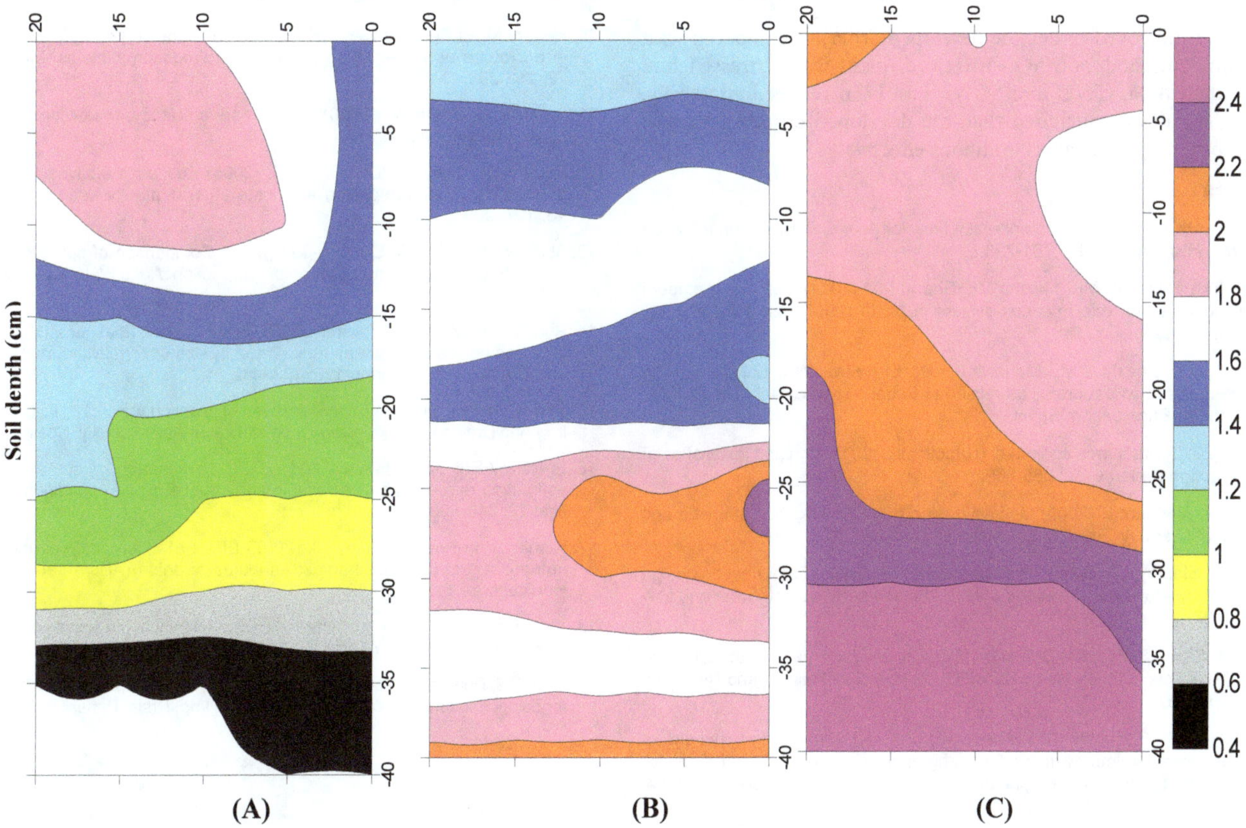

Figure 4: Salt distribution around surface (a) and subsurface (b, c) emitter at 10, 20 cm based on traditional calculation methods of FAO.

Growing season	Drip irrigation system	CWU calculation method	Growing stages / days after sowing seeds														Accumulative CWU, mm/fed	SCWR, m³/fed			Enhancement percentage, %
			Es(0-16)		Vegetative(17-44)				Bulb formation(45-72)				Bulb enlargement to maturity (73-101)				Leaching requirements (10%), mm	SCWR, mm/fed	
			0-9	10-16	17-23	24-30	31-37	38-44	45-51	52-58	59-65	66-72	73-79	80-87	88-101				
2011-2012	SD	ET_{FAO}	29	12.6	10.08	14.84	18.34	19.04	28	30.66	26.32	23.94	21.28	17.76	24.08	324.64	32.46	1499.8	10.61
		ET_{Eidc}	29	7.56	6.58	13.3	17.78	22.96	23.94	25.48	28.56	23.94	26.32	21.12	20.44	290.20	29.02	1340.7	
	SSD10	ET_{FAO}	25	8.54	8.68	12.88	17.08	21.14	26.32	27.3	26.04	22.4	22.82	22.08	28	315.62	31.56	1458.2	11.20
		ET_{Eidc}	25	6.72	6.3	12.46	16.38	20.72	22.4	25.2	26.74	22.54	27.72	20.48	25.2	280.28	28.03	1294.9	
	SSD20	ET_{FAO}	24	8.96	8.68	14.84	21.84	23.66	29.82	30.66	29.12	25.76	22.82	19.68	28	338.64	33.86	1564.5	21.21
		ET_{Eidc}	24	7.28	8.26	9.38	13.86	20.02	24.5	25.06	28.14	23.24	24.78	17.92	19.04	266.83	26.68	1232.7	
2012-2013	SD	ET_{FAO}	21.78	9.42	7.56	9.26	10.04	10.46	15.38	16.86	19.94	24	21.24	17.76	28.42	250	24.96	1152.9	10.56
		ET_{Eidc}	19.6	5.68	4.94	8.28	9.78	12.6	13.14	14.04	21.64	24	26.28	21.16	24.2	223.20	22.32	1031.2	
	SSD10	ET_{FAO}	16.86	6.42	6.46	8.04	9.38	11.66	14.46	15.02	19.66	22.34	22.8	22.14	33.1	245.11	24.51	1132.4	10.55
		ET_{Eidc}	16.86	5.02	4.72	7.8	8.98	11.38	12.32	13.88	20.2	22.56	27.7	20.44	29.84	219.24	21.92	1012.9	
	SSD20	ET_{FAO}	16.2	6.68	6.56	9.24	12.02	13.04	16.42	16.86	22.02	25.82	22.82	19.68	21.16	245.32	24.53	1133.4	16.11
		ET_{Eidc}	16.2	5.42	6.22	5.86	7.6	11.02	13.46	13.76	21.26	23.3	24.84	17.9	22.5	205.80	20.58	950.8	

Table 3: Seasonal Crop water requirements of onion under investigated parameters for the validation process.

patterns had improved under the developed criterion compared with conventional method of FAO. Therefore, plant-water stressed had been avoided. Also, CWU and SCWR had been reduced effectively. Therefore, it can be concluded that, the developed criterion has the integrity for using under arid conditions effectively.

References

1. El-Nemer MK (2014) Adjusted operation time for poor uniformity drip irrigation networks. Misr J Ag Eng 31: 781-798.

2. Khalifa EM, El-Nemer MK, Meleha ME, Sharaf MM (2014) Optimizing amount of applied water for drip irrigation system in North Delta. Misr J Ag Eng 31: 765-780.

3. Arafa YE, El-Shazly AM, Mehawed HS, El-Helew WK (2013) Quality characteristics of wheat grains as related to irrigation systems 3rd Int Conf for Agric and Bio-Engineering. Egypt.

4. ASABE (2012) Standards engineering practices data: Design and installation of microirrigation systems. pp: 865-869.

5. Kamal HA, Vencent B (2005) Optimal design of trickle irrigation submain unit dimension. Misr J Ag Eng 22: 820-839.

6. Clark GA, Stanley CD, Smajstrla AG, Zazueta FS (1999) Microirrigation design considerations for sandy soil vegetable production systems. Int Water and Irrig J 19: 14-27.

7. Burt CM, Styles SW (1994) Drip and Microirrigation for Trees, Vines, and Row Crops (with Special Sections on Buried Drip). Irrigation Training and Research Center. pp: 261.

8. Mehawed HS, El-Shazly AM, Arafa YE (2013) Hydraulic assessment of sprinkler types for improving on-farm irrigation efficiencies, 3rd Int Conf for Agric and Bio-Engineering: Engineering application for sustainable agricultural development. Egypt

9. Hagag AA, Mattar MA (2005) Water economic return of wheat under pivot irrigation system. Misr J Ag Eng 22: 161- 181.

10. Hanafy M, El-Berry AM, Abu-Habsa AR, Bishara BL (2005) The performance of microtube as an emission point in microirrigation systems. Misr J Ag Eng 22: 252-260.

11. Steduto P, Smith M (2000) Water use efficiency, water productivity and biotechnology. FAO. Italy.

12. Enciso JM, Colaizzi PD, Multer WL (2005) Economic analysis of subsurface drip irrigation lateral spacing and installation depth for cotton. Trans ASAE. 48: 197–204.

13. Al-Jamal MS, Ball S, Sammis TW (2001) Comparison of sprinkle, trickle and furrow-irrigation efficiencies for onion production. Agric Water Manage 46: 253-266.

14. Burt CM, Clemmens AJ, Strelkoff TS, Soloman KH, Bliesner RD, et al. (1997) Irrigation performance measures: efficiency and uniformity. Journal of Irrigation and Drainage Engineering 123: 423–442.

15. Slavil I, Zavadil J (1999) Economical use of irrigation water. 17th ICID Int Cong on Irrig And Drainage: Irrigation water conditions of water scarcity. Spain.

16. El-Raie AES, Abdel-Wahed MH (2005) Comparison of some methods for estimating reference evapotranspiration under Egyptian conditions. Misr J Ag Eng 22: 840-860.

17. Kumar S, Imtiyaz M, Kumar A (2007) Effect of differential soil moisture and nutrient regimes on postharvest attributes of onion (Allium cepa L). Scientia Horticulturae 112: 121–129.

18. Alazba AA (2002) Simple mathematical model for water advance determination. Irrigation Science. 21: 75-81.

19. Allen RG, Pereira LS, Raes D, Smith M (1998) Crop evapotranspiration: Guideline for computing crop water requirements. FAO Irrig. Drain. Paper 56: 300.

23

Evaluation of Groundwater Suitability for Drinking and Irrigation Purposes in Toba Tek Singh District, Pakistan

Muhammad Hasan[1]*, Yanjun Shang[1], Gulraiz Akhter[2] and Weijun Jin[1]

[1]*Institute of Geology and Geophysics, University of the Chinese Academy of Sciences, Beijing, China*
[2]*Department of Earth Sciences, Quaid-i-Azam University, Islamabad, Pakistan*

Abstract

Correlating the physicochemical parameters for assessment of the groundwater quality has emerged as a very useful approach for water use. Taking water samples from the Toba Tek Singh District of Pakistan, this study assess the water quality for drinking and irrigation purposes. A sum-total of 72nos. groundwater samples were collected and analyzed for the purpose of different water quality parameters including sodium (Na^+), potassium (K^+), calcium (Ca^{2+}), magnesium (Mg^{2+}), bicarbonate (HCO^-_3), chloride (Cl^-), sulphate (SO_4^{2-}), pH, electrical conductivity (EC), total dissolved solids (TDS) and total hardness (TH). The results obtained were, then, compared with the standard desirable limits of physicochemical parameters prescribed by World Health Organization (WHO), Pakistan Standards and Quality Control Authority (PSQCA) and Pakistan Council of Research in Water Resources (PCRWR) for drinking purposes. In order to classify the groundwater suitability for irrigation purpose, parameters such as sodium adsorption ratio (SAR), percent sodium (PS), permeability index (PI), residual sodium bicarbonate (RSBC), Kelly's ratio (KR), and magnesium adsorption ratio (MAR) were also calculated on the basis of chemical data. After that, the correlation coefficients between different physicochemical parameters were calculated to identify the highly correlated and interrelated parameters for water quality. Different plots like Piper, Durov, Schoeller and Stiff diagrams were drawn to classify the groundwater ability for different purposes. These several classifications show most of groundwater samples falling within the safe limits and thus suitable for drinking and irrigation purposes, except of a few samples with a caution that it may get worse in the future.

Keywords: Physicochemical parameters; Correlation; WHO; Drinking; Irrigation; Pakistan

Introduction

Surface water and ground water are two main sources of water. Surface water includes rivers, canals, streams, fresh water lakes etc., while groundwater is obtained from well and borehole water [1]. Groundwater is under the earth surface originated as infiltration from precipitation, stream flows etc. Under the action of gravity the water moves downwards until it reaches strata to form the groundwater. About 97% of earth's fresh water is groundwater [2]. Groundwater is also an important part of the water cycle and is used to maintain soil moisture, wetlands, stream flow, and is also the main source of drinking and irrigation worldwide. Groundwater contains about 40 to 70% of the world water resources being used for drinking and irrigation purposes [3]. Due to decrease in surface water resources and partially pollution, the groundwater is becoming more important for drinking and irrigation purposes [4,5].

Water quality is very important for the suitability of groundwater for drinking and irrigation purposes. Water has ability to suspend, absorb and dissolve different compounds. By nature water is not pure as it gets contaminants from its surrounding caused by humans, animals and other biological actions [6,7]. Water seeps from the porous soil and gets dissolved salts and gases, metals, organic compounds, nitrates and sulphates [8-10]. Water contains the minerals useful for human nutrition [11]. The groundwater quality is one of the most important parts of water resource studies [12,13]. Scarcity of fresh water is one of most important environmental problems in the world today [14].

The groundwater quality depends on different processes starting from condensation in atmosphere to the water discharge from the well, and is controlled by dissolved salts, material type and disposal system. The water quality is deteriorated by both natural as well as anthropogenic factors [15]. The groundwater quality is controlled by contaminated activities, discharge recharge pattern and nature of the rocks [16]. The different contaminants present in the groundwater above the standard of World Health Organization (WHO) can cause different ailments in humans and are not safe for industrial uses [17,18]. The good quality water prevents disease and improves quality of life [19]. The distribution of trace elements in the groundwater has a large range of chemical composition [20]. This composition depends on aquifer lithology, quality of water recharge and human activities [21]. Univariate statistical analysis is applied to interpret the trace elements. Multivariate method can be used to explain the correlations between different variables [22,23]. This multivariate method is used widely to interpret relationships among different variables to manage the environmental system better [24,9].

The majority of people in developing countries use the water from shallow wells and boreholes which have high contaminations [25]. The pollution of groundwater aquifers causes the wells unfit for consumption [26]. Dramatic increase in world population has resulted in huge consumption of water resources [27]. The population explosion caused by the increase in urbanization, industrialization and development activities has produced the water crises [28]. Low water flow, industrial discharges and municipal effluents may cause the poor water quality [7]. The poor or bad water quality has its effects on life expectancy and the health; it is a constant hazard for soils and crops. Therefore, it is important to check the assessment and the estimation of the quality for ground water whether it is suitable for agricultural, domestic and industrial uses.

The water quality for irrigation and domestic uses is assessed by

*Corresponding author: Hasan M, Institute of Geology and Geophysics, University of the Chinese Academy of Sciences, Beijing, China
E-mail: hasan.mjiinnww@gmail.com

geochemical study [29-31]. Water quality is more important than water quantity in any drinking water supply planning [32,33]. Water quality standards are based on quality control program and treatment process. These standards are helpful for the identification of water quality problems caused by the discharge of waste water from active or abandoned mixing sites, fertilizers and sediments. These standards are very useful to assess the water quality conditions [34,35].

Generally, the water quality assessment is based on physicochemical analysis. World Health Organization (WHO) provides the guidelines for drinking water based on water quality parameters such as sodium (Na^+), potassium (K^+), calcium (Ca^{2+}), magnesium (Mg^{2+}), bicarbonate (HCO_3^-), chloride (Cl^-), sulphate (SO_4^{2-}), pH, electrical conductivity (EC), total dissolved solids (TDS) and total hardness (TH). These parameters should be in permissible limits of concentration for drinking water quality [36,37]. If these parameters cross this limit, it may cause health diseases. Thus, the determination of concentrations of these parameters based on their guideline values is required for the assessment of suitability of drinking water [38].

The guideline for irrigation water quality is provided by the Food and Agriculture Organization (FAO) of the United Nations [39]. Salinity and sodium hazard indicators are useful to assess the suitability of irrigation water [40]. The sodium absorption ratio (SAR) is very effective assessment index for irrigation water [39,41,42].

Groundwater is the main source of drinking and irrigation water in Pakistan. The over burden population, unplanned urbanization, inappropriate dumping of solids and liquid wastes and loose governance have resulted in the deterioration of quality and quantity of groundwater in Pakistan [43]. The present study was carried out in Toba Tek Singh District of Pakistan to assess the groundwater quality for drinking and irrigation purposes. Groundwater is the main source

of drinking and irrigation water in that area. The main goal of this study was to interpret the water quality by determining the variations in the physicochemical parameters, investigating the statistical, correlation and graphical analysis. In order to assess the irrigation water quality in the study area, different indices such as sodium adsorption ratio (SAR), percent sodium (PS), permeability index (PI), residual sodium bicarbonate (RSBC), Kelly's ratio (KR), magnesium adsorption ratio (MAR), total hardness (TH) were calculated from standard equations. It is anticipated that this study would be helpful to determine the water quality in District Toba Tek Singh, and to motivate further studies to this effect.

Study Area

Toba Tek Singh District is situated in Rechna Doab which is lying between Chenab River and Ravi River. It contains the latitude 30°33' to 31°2' N and the longitude 72°08' to 72°48' E (Figure 1). It is an important irrigation District of Southern Punjab in Pakistan. It has hot and dry climate in the summer season from April to October. The summer period is lengthy in the area. And May, June and July are the hottest months. During the summer season, 42°C and 29°C are mean maximum and minimum temperatures. The winter season is comparatively short. December, January and February are the coldest months. The mean maximum and minimum temperatures during the winter season are 29°C and 5°C, respectively. Monsoon is the rainy season in the district from July to September and most of the precipitation is during this season, but the rain in winter is scarce. The annual precipitation in the district is 158 mm.

Toba Tek Singh District is an alluvial plain where sand is the dominant subsurface lithology starting from some feet depth that formed good aquifers in the area. The drainage of the district depends

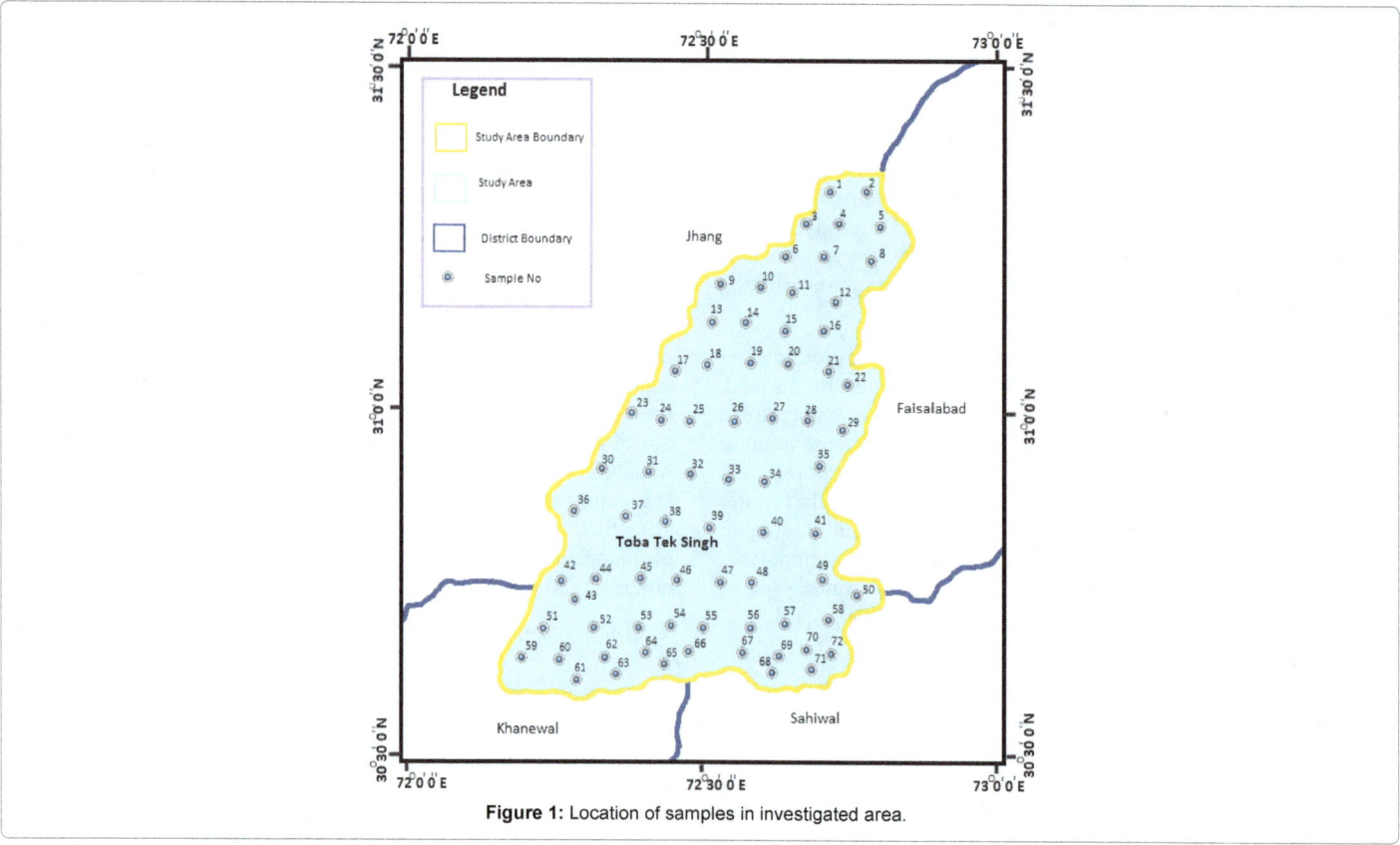

Figure 1: Location of samples in investigated area.

on its rivers and the canal system. Chenab and Ravi supply most of the water through inter river link canals. Ravi River also provides some water during flood season. The ground surface generally slopes in southwest direction. The groundwater is the main source of drinking and irrigation water with the supplement of canal water in the study area. The recharge of groundwater in the area is mostly through the river and canal system. Rainfall also plays a role to recharge the groundwater during monsoon. Through infiltration, the precipitation reaches the water reservoir under the ground. The rainfall in the area is not enough to recharge the ground water for drinking and irrigation purpose, so the rivers are the main source to recharge the groundwater resources. The analysis of data shows that groundwater quality in the study area is generally good or marginally acceptable for drinking and irrigation purposes. However, the decline in groundwater quality is visible and can cause long term sustainability issues, if the proper actions are not taken in time.

Materials and Methods

Sample collection and preservation

A sum-total 72 nos. groundwater samples were collected from different bore-wells at average depth of 50 m to 120 m. Locations of these samples are presented in Figure 1. Sampling collection and preservation were carried out according to the standard procedures [44,45], and then the tested data were interpreted on the basis of physiochemical analysis. Samples of physicochemical parameters were taken into 2L pre-cleaned polythene kegs. In order to avoid from contamination, special care was taken during sample collection, processing and transportation. The bottle was rinsed three times with groundwater filtered through 0.45 mm mixed cellulose ester membrane before the sample collection. Before analysis samples were preserved at approximately 4°C.

Sample analysis

Physicochemical parameters like total hardness, $CaCO_3^-$, Ca^{2+}, Mg^{2+} of collected samples were analyzed as per standard methods [46,47]. The analysis of samples for water quality was carried out for main anions, such as, sulphate, bicarbonate and chloride; the cations such as calcium, sodium, magnesium and potassium; and the physiochemicals such as pH, electrical conductivity and total dissolved solids as per standard procedures. Seventy two water samples were collected and analyzed for 11 physicochemical water quality parameters including sodium (Na^+), potassium (K^+), calcium (Ca^{2+}), magnesium (Mg^{2+}), bicarbonate (HCO_3^-), chloride (Cl^-), sulphate (SO_4^{2-}), pH, electrical conductivity (EC), total dissolved solids (TDS) and total hardness (TH) in the laboratory of Pakistan Council of Research in Water Resource (PCRWR), using Standard Methods for the Examination of Water (Table 1).

In order to verify the accuracy, the international standard referential materials and the synthetic solutions were applied to the samples. The mean values were used for calculations. The recovery was greater than 95% for the standard referential materials. The methods applied for the determination of the other variables, not practiced in the referential materials, were checked by using synthetic aqueous solutions. The data were also checked at two different independent laboratories, and a maximum of ± 5% deviation was observed.

Results and Discussion

Statistical analysis

The classical use of groundwater hydrology is to produce

S. No	Parameters	Analytical Method
1	Bicarbonate	2320, Standard method (1992)
2	Calcium (mg/L)	3500-Ca-D, Standard Method (1992)
3	Chloride (mg/L)	Titration (Silver Nitrate), Standard Method (1992)
4	Conductivity (mS/cm)	E.C meter, Hach-44600-00, USA
5	Hardness (mg/L)	EDTA Titration, Standard Method (1992)
6	Magnesium (mg/L)	2340-C, Standard Method (1992)
7	pH at 25°C	pH Meter, Hanna Instrument Model 8519, Italy
8	Potassium (mg/L)	Flame photometer PFP7, UK
9	Sodium (mg/L)	Flame photometer PFP7, UK
10	Sulfate (mg/L)	SulfaVer4 (Hach-8051) by Spectrophotometer
11	TDS (mg/L)	2540C, Standard method (1992)

Table 1: Methods used for water samples preservation and analysis.

information about water quality. It not only provides information about the environment where the water has circulated, but also helps understanding the suitability of drinking and irrigation water. Drinking and irrigation water quality was assessed on the basis of analytical results. World Health Organization [48], Pakistan Standards and Quality Control Authority [49], and Pakistan Council of Research in Water Resources [50] limits were considered as standard for drinking water quality. The irrigation water quality was assessed on the basis of the guideline provided by the Food and Agriculture Organization (FAO) of the United Nations [39]. Suitability of groundwater for drinking and irrigation was also assessed on the basis of several other classifications. The results of physicochemical analysis for drinking and irrigation water quality collected from different locations in Toba Tek Singh District are given in Tables 2 and 3, respectively. The analytical results of physicochemical parameters for drinking and irrigation water quality were transformed into descriptive statistical parameters, such as minimum, maximum, mean, median and standard deviation for the entire study period (Table 4).

Drinking water quality

Water quality is the physical, chemical and biological characteristics of water [51]. The water quality varies and depends on the variations in geological formations. Different elements present in the groundwater depend on the associated rock bodies and the time it has been in contact with geological material. Groundwater is the only safe and reliable source of drinking water in Toba Tek Singh District. The parameters such as sodium (Na^+), potassium (K^+), calcium (Ca^{2+}), magnesium (Mg^{2+}), bicarbonate (HCO_3^-), chloride (Cl^-), sulphate (SO_4^{2-}), pH, electrical conductivity (EC), total dissolved solids (TDS) and total hardness (TH) are regarded as critical determinants for most development studies of water quality [52]. On the basis of the comparison of groundwater quality data with World Health-Organization [48], Pakistan Standards and Quality Control Authority [49] and Pakistan Council of Research in Water Resources [50], the suitability of drinking water was determined. The number and percentage of the samples which exceeded the permissible limits are given in Table 5. It was observed from Table 5 that majority of the samples were in the safe range, except some of the samples such as pH, electrical conductivity (EC), total dissolved solids (TDS), sulphate (SO_4^{2-}), chloride (Cl^-), magnesium (Mg^{2+}) and calcium (Ca^{2+}). Generally, the groundwater quality in most of the study area is suitable for drinking purposes.

pH: pH is regarded as one of the most commonly used parameters to test soil and water. It indicates the acidic or alkaline potential of water and is calculated on a scale of 1-14. The pH value of 7 represents neutral water; less than 7 indicates acidic and greater than 7 shows basic water. Generally, water with pH range of 6.5~8.5 recommended by WHO and

S. No	Ca²⁺ mg/L	Mg²⁺ mg/L	Na⁺ mg/L	K⁺ mg/L	HCO₃⁻ mg/L	Cl⁻ mg/L	SO₄²⁻ mg/L	TDS mg/L	E.C µS/cm	TH mg/L	pH
1	56	35	82	8.6	195	56	180	525	955	285	7.9
2	32	17	16	2.9	115	14	43	201	366	150	7.2
3	40	24	17	4.2	135	15	77	262	477	200	7.8
4	84	22	78	8.5	175	49	187	528	960	300	7.5
5	46	18	14	3.8	145	13	53	228	415	190	7.6
6	48	17	21	5.9	140	27	47	257	468	190	7.6
7	32	11	17	2.5	85	24	41	194	352	125	8.3
8	42	29	41	4	145	35	98	331	602	210	8.2
9	76	53	121	8.9	280	57	286	795	1325	410	7.7
10	60	36	95	11.6	270	45	202	635	1059	300	7.9
11	72	49	111	10.7	275	69	277	756	1260	380	7.3
12	40	10	7	2.3	110	10	27	166	301	140	7.5
13	30	9	6	2	100	4	28	138	250	110	7.3
14	76	46	32	3.3	95	6	23	138	250	380	7.1
15	74	29	31	8	220	25	106	429	780	305	7.5
16	34	26	28	13	200	4	68	289	526	190	8.1
17	36	9	25	3.4	95	34	37	209	380	125	7.4
18	34	9	7	2.8	100	5	27	152	277	120	7.4
19	42	18	82	4.4	230	24	94	418	760	180	7.7
20	34	9	142	5.2	245	34	43	479	870	120	7.9
21	42	18	7	2.5	100	7	19	141	260	180	6.8
22	34	9	6	5	80	7	22	120	218	120	7.7
23	32	11	5	2.3	110	3	20	140	254	125	7.1
24	24	6	33	4.8	150	56	51	319	580	85	7.2
25	36	6	8	2.7	95	9	20	141	256	115	7.2
26	48	17	6	2.7	95	3	24	134	244	190	7.2
27	32	10	5	2	100	10	22	141	257	120	7.2
28	38	4	10	2.5	120	5	30	172	312	110	6.9
29	24	11	7	2.9	320	190	32	407	740	105	7.4
30	40	10	8	4.1	125	10	29	182	330	140	8.0
31	88	24	62	3	370	14	58	499	908	320	7.6
32	70	45	35	1.7	160	21	85	317	576	360	8.1
33	162	21	10	5.3	335	5	16	396	720	340	7.5
34	60	15	7	2.4	130	12	38	202	367	210	6.9
35	28	7	10	4.2	110	9	43	146	338	110	8.2
36	34	15	12	3.8	110	9	21	147	268	145	7.5
37	32	12	5	2.6	90	9	31	146	265	130	7.5
38	34	16	9	3.5	140	6	26	190	345	150	8.2
39	36	5	8	3	145	10	25	206	375	110	7.8
40	16	17	7	2.8	130	5	30	191	348	110	8.0
41	38	17	9	2.5	100	49	151	424	770	165	8.5
42	40	10	59	5.6	130	13	42	205	373	140	7.8
43	40	24	10	3	100	4	21	143	260	200	8.2
44	28	19	7	3.3	145	36	83	309	562	150	8.2
45	28	10	6	2	180	5	21	226	410	110	8.0
46	20	5	8	3.9	95	2	21	132	236	70	8.2
47	42	26	5	2.1	75	6	23	144	262	210	8.5
48	28	10	6	3.1	300	15	90	453	823	110	7.8
49	26	9	62	4.4	120	9	42	193	350	102	8.2
50	56	78	180	8.4	350	106	418	1248	2080	460	7.7
51	36	12	21	2.8	335	10	142	539	980	140	9.6
52	32	10	7	2.6	70	7	30	136	248	120	9.5
53	42	11	7	3.2	25	3	20	98	174	150	10.0
54	48	26	187	8	275	480	391	1842	3070	225	7.3
55	30	19	5	3	45	1	20	99	155	155	9.2
56	24	5	8	2.1	135	1	32	189	324	80	8.2
57	36	13	6	3.5	75	4	24	123	224	145	8.2
58	20	17	5	2	430	91	121	754	1350	120	7.2
59	38	23	29	4.5	135	4	38	210	381	190	8.2

60	60	12	198	7.8	115	61	72	342	622	200	7.1
61	36	10	32	3.7	105	6	27	153	261	130	6.8
62	30	11	5	2.5	100	3	26	151	274	120	6.9
63	50	28	4	3.6	90	1	23	130	236	140	6.9
64	36	2	5	1.9	75	9	29	134	244	100	7.0
65	32	7	6	2.2	90	6	24	136	247	110	6.5
66	38	12	6	3.2	70	5	27	115	202	145	7.1
67	60	17	42	9.8	85	4	25	135	246	220	6.6
68	20	5	4	3	20	4	25	91	165	70	9.4
69	36	2	5	2.1	85	7	28	135	244	100	6.3
70	32	5	47	4.7	100	10	18	132	234	100	7.2
71	40	15	7	3.6	170	45	40	246	447	160	7.5
72	30	9	22	2.3	285	11	123	582	1058	110	7.4

Table 2: Parameters of groundwater quality (mg/L).

S/N	Ca^{2+} meq/L	Mg^{2+} meq/L	Na^+ meq/L	K^+ meq/L	HCO_3^- meq/L	Cl^- meq/L	SO_4^{2-} meq/L	PS %	PI %	KR -	SAR -	RSBC meq/L	Mg^{2+}/Ca^{2+} -	MAR %
1	2.8	2.92	3.56	0.22	3.20	1.6	3.75	39.79	57.64	0.62	2.10	0.4	1.04	51.04
2	1.6	1.42	0.69	0.07	1.88	0.4	0.89	20.10	55.56	0.23	0.56	0.28	0.89	47.01
3	2	2	0.74	0.11	2.21	0.43	1.60	17.52	46.97	0.18	0.52	0.21	1	50
4	4.2	1.83	3.39	0.22	2.87	1.4	3.90	37.45	53.97	0.56	1.95	-1.33	0.44	30.34
5	2.3	1.5	0.61	0.10	2.38	0.37	1.10	15.74	48.81	0.16	0.44	0.08	0.65	39.47
6	2.4	1.42	0.91	0.15	2.29	0.77	0.98	21.72	51.23	0.24	0.66	-0.11	0.59	37.17
7	1.6	0.92	0.74	0.06	1.39	0.68	0.85	24.1	58.86	0.29	0.66	-0.21	0.57	36.50
8	2.1	2.42	1.78	0.10	2.38	1	2.04	29.37	52.74	0.39	1.18	0.28	1.15	53.53
9	3.8	4.42	5.26	0.23	4.59	1.63	5.96	40.04	54.91	0.64	2.59	0.79	1.16	53.77
10	3	3	4.13	0.30	4.43	1.28	4.21	42.47	61.55	0.69	2.38	1.43	1	50
11	3.6	4.01	4.83	0.27	4.51	1.97	5.77	40.12	55.90	0.63	2.48	0.91	1.11	52.69
12	2	0.83	0.30	0.06	1.80	0.28	0.56	11.28	52.45	0.11	0.25	-0.2	0.41	29.32
13	1.5	0.75	0.26	0.05	1.64	0.11	0.58	12.11	61.38	0.11	0.24	0.14	0.5	33.33
14	3.8	3.83	1.39	0.08	1.56	0.17	0.48	16.15	29.26	0.18	0.71	-2.24	1.01	50.19
15	3.7	2.42	1.35	0.20	3.61	0.71	2.21	20.21	43.51	0.22	0.77	-0.09	0.65	39.54
16	1.7	2.17	1.22	0.33	3.28	0.11	1.42	28.61	59.58	0.31	0.88	1.58	1.27	56.03
17	1.8	0.75	1.09	0.09	1.56	0.97	0.77	31.63	64.26	0.43	0.96	-0.24	0.42	29.41
18	1.7	0.75	0.30	0.07	1.64	0.14	0.56	13.12	57.48	0.12	0.27	-0.06	0.44	30.61
19	2.1	1.5	3.56	0.11	3.77	0.68	1.96	50.48	76.84	0.99	2.65	1.67	0.71	41.66
20	1.7	0.75	6.17	0.13	4.02	0.97	0.89	72	94.84	2.52	5.57	2.32	0.44	30.61
21	2.1	1.5	0.30	0.06	1.64	0.2	0.39	9.09	40.53	0.08	0.22	-0.46	0.71	41.66
22	1.7	0.75	0.26	0.13	1.31	0.2	0.46	13.73	51.83	0.11	0.23	-0.39	0.44	30.61
23	1.6	0.92	0.22	0.06	1.80	0.09	0.42	10	56.99	0.09	0.19	0.2	0.57	36.50
24	1.2	0.5	1.43	0.12	2.46	1.6	1.06	47.69	95.80	0.84	1.55	1.26	0.42	29.41
25	1.8	0.5	0.35	0.07	1.56	0.26	0.42	15.44	60.34	0.15	0.33	-0.24	0.28	21.73
26	2.4	1.42	0.26	0.07	1.56	0.08	0.5	7.95	36.98	0.07	0.19	-0.84	0.59	37.17
27	1.6	0.83	0.22	0.05	1.64	0.28	0.46	10	56.63	0.09	0.20	0.04	0.52	34.16
28	1.9	0.33	0.43	0.06	1.97	0.14	0.62	18.01	68.93	0.19	0.41	0.07	0.17	14.79
29	1.2	0.92	0.30	0.07	5.24	5.43	0.67	14.86	106.9	0.14	0.29	4.04	0.77	43.39
30	2	0.83	0.35	0.10	2.05	0.28	0.60	13.72	56.03	0.12	0.29	0.05	0.41	29.32
31	4.4	2	2.69	0.07	6.06	0.4	1.21	30.13	56.67	0.42	1.50	1.66	0.45	31.25
32	3.5	3.75	1.52	0.04	2.62	0.6	1.77	17.71	35.79	0.21	0.80	-0.88	1.07	51.72
33	8.1	1.75	0.43	0.13	5.49	0.14	0.33	5.38	26.97	0.04	0.19	-2.61	0.22	17.76
34	3	1.25	0.30	0.06	2.13	0.34	0.79	7.81	38.67	0.07	0.20	-0.87	0.42	29.41
35	1.4	0.58	0.43	0.11	1.80	0.26	0.89	21.43	73.51	0.22	0.43	0.4	0.41	29.29
36	1.7	1.25	0.52	0.10	1.80	0.26	0.44	17.37	53.65	0.18	0.43	0.1	0.74	42.37
37	1.6	1	0.22	0.07	1.47	0.26	0.64	10.03	50.87	0.08	0.19	-0.125	0.62	38.46
38	1.7	1.33	0.39	0.09	2.29	0.17	0.54	13.67	55.65	0.13	0.32	0.59	0.78	43.89
39	1.8	0.42	0.35	0.08	2.38	0.28	0.52	16.23	73.65	0.16	0.33	0.58	0.23	18.91
40	0.8	1.42	0.30	0.07	2.13	0.14	0.62	14.28	69.82	0.13	0.28	1.33	1.77	63.96
41	1.9	1.42	0.39	0.06	1.64	1.4	3.14	11.94	45.03	0.12	0.30	-0.26	0.75	42.77
42	2	0.83	2.56	0.14	2.13	0.37	0.87	48.82	74.57	0.90	2.15	0.13	0.41	29.32
43	2	2	0.43	0.08	1.64	0.11	0.44	11.31	38.61	0.11	0.30	-0.36	1	50
44	1.4	1.58	0.30	0.08	2.38	1.03	1.73	11.31	56.18	0.10	0.24	0.98	1.13	53.02

45	1.4	0.83	0.26	0.05	2.95	0.14	0.44	12.20	79.42	0.12	0.25	1.55	0.59	37.21
46	1	0.41	0.35	0.1	1.56	0.06	0.44	24.17	90.75	0.25	0.42	0.56	0.41	29.17
47	2.1	2.17	0.22	0.05	1.23	0.17	0.48	5.95	29.60	0.05	0.15	-0.87	1.03	50.81
48	1.4	0.83	0.26	0.08	4.92	0.43	1.87	13.23	99.52	0.12	0.25	3.52	0.59	37.21
49	1.3	0.75	2.69	0.11	1.97	0.26	0.87	57.73	86.36	1.31	2.66	0.67	0.58	36.58
50	2.8	6.5	7.83	0.21	5.74	3.03	8.71	46.37	59.69	0.84	3.63	2.94	2.32	69.89
51	1.8	1	0.91	0.07	5.49	0.28	2.96	25.92	87.68	0.32	0.77	3.69	0.55	35.71
52	1.6	0.83	0.30	0.07	1.15	0.2	0.62	13.21	50.27	0.12	0.27	-0.45	0.52	34.15
53	2.1	0.91	0.30	0.08	0.41	0.08	0.42	11.20	28.39	0.10	0.24	-1.69	0.43	30.27
54	2.4	2.17	8.13	0.20	4.51	13.71	8.14	64.57	80.74	1.78	5.38	2.11	0.90	47.48
55	1.5	1.58	0.22	0.08	0.74	0.03	0.42	8.87	32.73	0.07	0.18	-0.76	1.05	51.29
56	1.2	0.42	0.35	0.05	2.21	0.03	0.67	19.80	93.23	0.22	0.39	1.01	0.35	25.92
57	1.8	1.08	0.26	0.09	1.23	0.11	0.5	10.83	43.60	0.09	0.22	-0.57	0.6	37.5
58	1	1.42	0.22	0.05	7.05	2.6	2.5	10.04	108.9	0.09	0.20	6.05	1.42	58.67
59	1.9	1.92	1.26	0.11	2.21	0.11	0.79	26.40	54.07	0.33	0.91	0.31	1.01	50.26
60	3	1	8.61	0.2	1.88	1.74	1.5	68.77	79.15	2.15	6.09	-1.12	0.33	25
61	1.8	0.83	1.39	0.09	1.72	0.17	0.56	36.01	67.20	0.53	1.21	-0.08	0.46	31.55
62	1.5	0.92	0.22	0.06	1.64	0.08	0.54	10.37	56.84	0.09	0.20	0.14	0.61	38.01
63	2.5	2.33	0.17	0.09	1.47	0.03	0.48	5.11	27.65	0.03	0.11	-1.03	0.93	48.24
64	1.8	0.17	0.22	0.05	1.23	0.26	0.60	12.05	60.69	0.11	0.22	-0.57	0.09	8.62
65	1.6	0.58	0.26	0.06	1.47	0.17	0.5	12.8	60.34	0.12	0.25	-0.13	0.36	26.60
66	1.9	1	0.26	0.08	1.15	0.14	0.56	10.49	42.16	0.09	0.21	-0.75	0.53	34.48
67	3	1.42	1.83	0.25	1.39	0.11	0.52	32	48.14	0.41	1.23	-1.61	0.47	32.12
68	1	0.42	0.17	0.08	0.33	0.11	0.52	14.97	46.82	0.12	0.20	-0.67	0.42	29.57
69	1.8	0.17	0.22	0.05	1.39	0.2	0.58	12.05	63.88	0.11	0.22	-0.41	0.09	8.62
70	1.6	0.42	2.04	0.12	1.64	0.28	0.37	51.67	81.79	1.01	2.03	0.04	0.26	20.79
71	2	1.25	0.30	0.09	2.79	1.28	0.83	10.71	55.50	0.09	0.23	0.79	0.62	38.46
72	1.5	0.75	0.96	0.06	4.67	0.31	2.56	31.19	97.23	0.43	0.90	3.17	0.5	33.33

Table 3: Parameters of ground water quality (meq/L).

Parameters	Units	Minimum	Maximum	Mean	Median	S.D
pH	-	6.3	10	7.7	7.6	0.71
EC	(μS/cm)	155	3070	529.25	351	465.12
TDS	(mg/L)	91	1842	303	193.5	287.49
Ca^{2+}	(mg/L)	16	162	42.36	36	21.04
Mg^{2+}	(mg/L)	2	78	17.14	12.5	13.09
Na^+	(mg/L)	4	198	31.15	9.5	44.6
K^+	(mg/L)	1.7	13	4.17	3.25	2.5
HCO_3^-	(mg/L)	20	430	151.11	120	87.73
Cl^-	(mg/L)	1	480	27.33	9.5	61.67
Cl^-	(meq/L)	0.03	13.71	0.78	0.27	1.76
SO_4^{2-}	(mg/L)	16	418	66.71	31.5	80.97
TH	(mg/L)	70	460	172.66	142.5	86.13
RSBC	(meq/L)	-2.61	6.05	0.36	0.07	1.44
PS	%	5.11	72	22.79	15.94	16.02
MAR	%	8.62	69.89	37.65	36.87	12.12
SAR	-	0.11	6.09	0.95	0.36	1.27
KR	-	0.03	2.52	0.35	0.16	0.47
PI	%	26.97	108.91	60.15	56.65	19.47
Mg^{2+}/Ca^{2+}	-	0.09	2.32	0.67	0.58	0.38

Table 4: Statistical distribution of physicochemical parameters in the groundwater (n=72).

PSQCA is considered as safe for drinking purpose. PCRWR suggested range of pH is 6.5~9.2. In study area, pH varied from 6.3 to 10, with mean value of 7.7, median of 7.6, and standard deviation of 0.71 (Table 4). It was observed from Table 5 that most of the samples had pH levels within the safe limits of WHO, PSQCA and PCRWR. 8% of samples were not within the permissible limit of WHO and PSQCA, while 6% of samples were out of the safe limit set by PCRWR.

Electrical conductivity (EC): Electrical conductivity measures the water ability to conduct an electric current. It signifies the amount of total dissolved salts and is very useful for assessing the purity of water [53]. It is generally used to estimate the amount of total dissolved solids and minerals. It increases with the reaching of dissolved minerals. In the study area, EC values ranged from 155 to 3070 μS/cm, with mean value of 529.25 μS/cm, median of 351 μS/cm, and standard deviation of 465.12 μS/cm (Table 4). EC standard limit for drinking water is 1500 μS/cm as recommended by WHO and PSQCA. PCRWR safe limit

Parameters	Permissible limits								
	(WHO, 2008)			(PSQCA, 2004)			(PCRWR, 2005)		
	Range	Samples exceeding limit	samples %age	Range	Samples exceeding limit	samples %age	Range	Samples exceeding limit	samples %age
pH	6.5- 8.5	6	8	6.5- 8.5	6	8	6.5-9.2	4	6
EC (µS/cm)	1500	2	3	1500	6	3	2343	1	1
TDS (mg/L)	1000	2	3	1000	2	3	1500	1	1
Ca^{2+} (mg/L)	100	1	1	200	-	-	200	-	-
Mg^{2+} (mg/L)	50	2	3	100	-	-	150	-	-
Cl^- (mg/L)	250	1	1	500	-	-	600	-	-
SO_4^{2-} (mg/L)	200	5	7	400	1	1	400	1	1
Na^+ (mg/L)	200	-	-	200	-	-	200	-	-
K^+ (mg/L)	55	-	-	50	-	-	50	-	-
HCO_3^- (mg/L)	600	-	-	500	-	-	500	-	-
TH (mg/L)	500	-	-	500	-	-	500	-	-

Table 5: The comparison of groundwater parameters with international standards for drinking.

of EC for drinking water is 2343 µS/cm. Table 5 showed that 3% of samples were out of safe limit of WHO and PSQCA. Only 1% samples were not within standard limit of PCRWR. Overall 97% of the samples were safe for drinking water.

Total dissolved solids (TDS): The Total dissolved solids generally indicate the amount of minerals and solids dissolved in water. High values of TDS change the taste, corrosive property and hardness of the water [54-56]. High concentrations of TDS are due to the presence of sulphates, chlorides, bicarbonates, carbonates and calcium [57,58]. The measurement of specific conductivity is the most commonly used method for determining TDS [59]. EC values can be converted to TDS values by multiplying EC by a factor varying with the type of water. This factor ranges from 0.5 to 0.9 [60]. The maximum contaminant limit of TDS for drinking water is 1000 mg/L as given by WHO and PSQCA, and 1500 mg/L by PCRWR standard. In the study area, TDS values varied from 91 to 1841 mg/L with mean, median and standard deviation of 303 mg/L, 193.5 mg/L and 287.49 mg/L, respectively (Table 4). According to Table 5, 3% of the water samples were classified as unacceptable using WHO and PSQCA standard, only 1% samples were not within safe limit of PCRWR. Most of the samples were found within safe limit for drinking water.

The palatability of drinking water studied by panels of taters based on its TDS level was given in Table 6 [52,59]. According to Table 6, 68% of samples belonged to excellent water class and 24% samples were found in good water class as categorized by WHO [59]. According to the water class given by Kumar et al. [52], 97% samples were placed in fresh water category, while only 3% in brackish water class. Hence, based on different classifications of TDS, the water in the study area is good for drinking purpose.

Total hardness (TH): Total hardness depends on calcium and magnesium ions [61]. It was calculated by Ragunath [62] using the formula:

$$TH=(Ca^{2+}+Mg^{2+}) \times 50 \tag{1}$$

TH values in the study area varied from 70 to 460 mg/L with mean values of 172.66 mg/L, median of 142.5 mg/L and standard deviation of 86.13 mg/L (Table 4). The permissible limit of TH recommended by WHO, PSQCA and PCRWR is 500 mg/L as given in Table 5. None of the samples in the study area exceeded this limit. The groundwater for drinking was also classified into four different categories like soft, moderately hard, hard and very hard based on TH in Table 6 [63]. Table 6 shows that 57% of samples belonged to moderately hard category, while 29% samples were regarded as hard, and only 11% samples fell in

TDS (mg/L)	Water class	Number of samples	samples %age
<300	Excellent	49	68
300-600	Good	17	24
600-900	Fair	4	5
900-1200	Poor	-	-
>1200	unacceptable	2	3
0-1000	Fresh	70	97
1001-10000	Brackish	2	3
10001-100000	Salty	-	-
>100000	Brine	-	-

Table 6: Suitability of groundwater for drinking based on the values of TDS.

very hard water category. The results show that the water quality based on TH is over all permissible for drinking purpose in the study area.

Chloride (Cl⁻): Chloride is the major ion associated with Individual Septic Disposal (ISDSS) [64]. It is found in all natural waters with relatively small amounts. It can also be derived from human sources. Chloride can affect the food taste [47]. However, it does not cause any health hazard. In the study area, the range of chloride values was from 1 to 480 mg/L with mean, median and standard deviation of 27.33 mg/L, 9.5 mg/L and 61.67 mg/L, respectively (Table 4). The permissible limit of chloride for drinking water is 250 mg/L set by WHO, 500 mg/L by PSQCA, and 600 mg/L by PCRWR. It was observed in Table 5 that only 1% of the samples exceeded the permissible limit of WHO, while no sample exceeded the safe limit of PSQCA and PCRWR. Hence, the water quality for drinking purpose is permissible based on chloride in the study area.

Sulphate (SO_4^{2-}): Sulphate occurs in groundwater in the form of inorganic sulphate and dissolved gas (H_2S). It is not a harmful substance, although high values of sulphate in groundwater may have laxative consequence. The concentrations of sulphate in the study area varied from 16 to 418 mg/L with mean of 66.71 mg/L, median of 31.5 mg/L and standard deviation of 80.97 mg/L (Table 4). The safe limits of sulphate given by WHO is 200 mg/L, 400 mg/L recommended by PSQCA and PCRWR. Table 5 showed that 7% samples were not in the safe range of WHO, while only 1% samples exceeded the permissible limit of PSQCA and PCRWR. So, overall water quality is good in the study area on the basis of sulphate concentrations.

Bicarbonate (HCO_3^-): The main source of bicarbonate ions in groundwater is the dissolution of carbonate rocks and the carbonate species, and the pH of water is usually from 5 to 7 [65]. It was observed that all dissolved carbonate species convert to H_2CO_3 below pH=6, while the ratio of CO_3 and H_2CO_3 increases above pH=7 [66]. In the study area, the values of bicarbonate ranged from 20 to 430 mg/L with the mean, median and standard deviation of 151.11 mg/L, 120 mg/L and 87.73 mg/L, respectively (Table 4). The permissible limits of bicarbonate for drinking water given by WHO are 600 mg/L; and 500 mg/L set by PSQCA and PCRWR. All the samples were found within the safe limit of WHO, PSQCA and PCRWR. Hence, the drinking water quality is permissible on the basis of bicarbonate values.

Calcium (Ca^{2+}): Calcium is the fifth most common element present in natural waters and it contributes to the water hardness. Calcite, gypsum, aragonite and anhydrite are the main source of calcium in groundwater, especially in sedimentary rocks. Granitic terrain is the natural source of calcium and has large concentration of such elements [67]. In the study area, calcium values varied from 16 to 162 mg/L with mean values of 42.36 mg/L, median values of 36 mg/L and standard deviation of 21.04 mg/L (Table 4). The standard limit of calcium for drinking water set by WHO is 100 mg/L. The permissible limit given by PSQCA and PCRWR is 200 mg/L. Only 1% of samples crossed the limit given by WHO, while all the samples were within the permissible limits set by PSQCA and PCRWR. So, on the basis of calcium concentration, groundwater is safe for drinking purpose in the study area.

Magnesium (Mg^{2+})": Magnesium is regarded as one of the most common elements within the earth's crust. It is found in all natural waters. It contributes to water hardness. Dolomites and mafic minerals in rocks are the main source of magnesium in natural waters. In the study area, magnesium values ranged from 2 to 78 mg/L with mean values of 17.14 mg/L, median values of 12.5 mg/L and standard deviation of 13.09 mg/L (Table 4). The permissible limit of magnesium for drinking purpose is 50 mg/L given by WHO, 100 mg/L by PSQCA and 150 mg/L by PCRWR. Only 3% of the samples were not within the permissible limit of WHO, while all the samples were found within the safe limits of PSQCA and PCRWR. Hence, groundwater is safe for drinking purpose on the basis of magnesium values.

Sodium (Na^+): Sodium is the most important natural mineral. Granitic terrain decomposition increases the concentration of sodium ion [67]. In the study area, sodium values were in the range of 4 to 198 mg/L with mean of 31.15 mg/L, median of 9.5 mg/L and standard deviation of 44.6 mg/L (Table 4). The permissible limit of sodium for drinking purpose is 200 mg/L given by WHO, PSQCA and PCRWR. All the samples were found within the permissible limits of WHO, PSQCA and PCRWR. Hence, groundwater quality is safe for drinking purpose on the basis of sodium values.

Potassium (K^+): Potassium is considered as one of the most important natural minerals. The decomposition of granitic terrain increases potassium ion concentration [67]. In the study area, potassium values varied from 1.7 to 13 mg/L with mean of 4.17 mg/L, median of 3.25 mg/L and standard deviation of 2.5 mg/L (Table 4). The safe limit of potassium for drinking purpose is 55 mg/L set by WHO; and 50 mg/L given by PSQCA and PCRWR. It was observed that all the samples were within the permissible limits of WHO, PSQCA and PCRWR. Hence, groundwater is safe for drinking purpose on the basis of potassium values.

Most of the samples of water quality parameters were found safe for drinking water. Thus, groundwater quality in the study area is suitable for drinking purpose.

Irrigation water quality

The irrigation water quality depends on the constituents of the minerals present in the groundwater [68]. The concentration of dissolved salts, relative proportion of bicarbonate to calcium, magnesium and relative proportion of sodium to calcium are the important chemical constituents, which affect the water quality for irrigation. The major problems of irrigation water quality are salinity and alkalinity. Salts may affect plant growth. The irrigation of food crops has a possible hazard to food consumers if the irrigation water quality is inadequate. The salinity of groundwater for irrigation also depends on the kinds of crops, composition and permeability of soil, the climate of region, the amount of water used the topography of land, the nature of groundwater, as well as the surface water drainage system. In this study, the discussion of irrigation water quality is mainly based on the concentrations of physicochemical parameters (sodium, potassium, calcium, magnesium, bicarbonate, chloride, sulphate, pH, electrical conductivity, total dissolved solids), and other important irrigation water quality parameters, namely sodium adsorption ratio (SAR), percent sodium (PS), permeability index (PI), residual sodium bicarbonate (RSBC), Kelly's ratio (KR), magnesium adsorption ratio (MAR) and residual Mg^{2+}/Ca^{2+} ratio. The calculation of all these parameters was carried out by ionic concentration (meq/L) [30,69-71].

The concentrations of parameters (sodium, potassium, calcium, magnesium, bicarbonate, chloride, sulphate,) for the assessment of irrigation water quality were converted from mg/L to meq/L using the relation:

Unit of parameter in mg/L/equivalent weight of parameter=Unit of parameter in meq/L   (2)

The equivalent weights of these parameters are given in Table 7. The irrigation water quality parameters were calculated in meq/L. The results of different irrigation water parameters are given in Table 3, summarized in Table 4.

Physiochemical parameters: The irrigation water quality was assessed on the basis of water quality parameters such as sodium, potassium, calcium, magnesium, bicarbonate, chloride, sulphate, pH, electrical conductivity and total dissolved solids [39]. It was observed in Table 8 that all parameters, except of some samples of pH, EC and Mg^{2+}, were found within permissible limit. 8% samples of pH exceeded the safe limit, while only 3% samples of EC and Mg^{2+} were found unsafe for irrigation water. Hence, the groundwater quality is overall good for irrigation purpose in the study area based on physicochemical parameters.

Salinity: The salinity affects the crop water availability. It is measured on the basis of EC and TDS [39]. Based on the interpretation of EC and TDS for salinity in Table 9, it was found that 75% of EC samples were safe, while 24% of EC samples had slight to moderate effect of salinity, and only 1% samples had severe salinity effect, whereas 82% of TDS samples had no salinity effect, while 18% TDS samples had slight to moderate salinity effect. So, most of the samples show no salinity effect.

Parameters	Equivalent weight
Sodium	23
Potassium	39
Calcium	20
Magnesium	12
Chloride	35
Bicarbonate	61
Sulphate	48

Table 7: Parameters with their equivalent weights.

Specific ion toxicity: It affects the sensitive crops. It is estimated on the basis of sodium (Na^+) and chloride (Cl^-) [39]. The classification of sodium and chloride for specific ion toxicity as given in Table 9, showed that 86% of sodium samples were safe for surface and sprinkler irrigation, while 14% samples had slight to moderate effect of specific ion toxicity, but none had severe effect. For surface irrigation, 98% samples of chloride had no effect of specific ion toxicity, only 1% samples had slight to moderate effect, while 1% samples had severe effect of specific ion toxicity. For sprinkler irrigation based on chloride, 96% samples were safe while 4% samples of chloride hade slight to moderate effect of specific ion toxicity. Hence, overall samples show safe irrigation water quality on the basis of specific ion toxicity.

Miscellaneous effect: It affects susceptible. It is measured on the basis of bicarbonate (HCO_3^-), and pH [39]. It was found that 21% samples of bicarbonate were safe, while 79% had slight to moderate miscellaneous. 89% samples of pH had no miscellaneous effect (Table 9). No severe miscellaneous effect was observed based on bicarbonate and pH. Hence, irrigation water quality is acceptable in the study area based on miscellaneous effects.

Sodium adsorption ratio (SAR): The sodium adsorption ratio was measured by the following equation [70]:

$$SAR = Na^+ / \sqrt{[(Ca^{2+} + Mg^{2+})/2]} \qquad (3)$$

All ion concentrations were expressed in meq/L. It provides an idea about the adsorption of sodium by soil. It shows the proportion of sodium to calcium and magnesium, which can affect the water availability of crop. The excess of sodium in water reduces the soil permeability [72]. The calculated values of SAR are given in Table 3. SAR varied from 0.11 to 6.09 with mean and median values of 0.95 and 0.36, respectively, the standard deviation is 1.27 (Table 4). According to the classification of SAR [70] in the study area, Table 10 showed that all the samples of SAR had excellent water class and it is acceptable for irrigation in the study area.

Percent sodium (PS): The percent sodium was calculated by the equation given as:

$$PS = [(Na^+ + K^+)/(Ca^{2+} + Mg^{2+} + Na^+ + K^+)] \times 100 \qquad (4)$$

All the ions were expressed in the unit of meq/L. It is very important to study sodium hazard. High percentage of sodium in the groundwater may affect the plant growth and reduce soil permeability [73]. PS was calculated using above equation and its values are given in Table 3. It ranged from 5.11 to 72 with mean, median and standard deviation values of 22.79, 15.94 and 16.02, respectively (Table 4). The classification of PS [74] is given in Table 10. It was observed in Table 10 that 60% samples were in excellent category, 24% samples in good, 12% samples in permissible, and only 4% in doubtful category. Overall

Water Parameters	Usual Range in Irrigation Water	No. of samples exceeding the permissible limit	%age of samples exceeding the permissible limit
pH	6- 8.5	5	8
EC (µS/cm)	0-3000	1	3
TDS (meq/L)	0-2000	-	-
Ca^{2+} (meq/L)	0-20	-	-
Mg^{2+} (meq/L)	0-5	1	3
Cl^- (meq/L)	0-30	-	-
SO_4^{2-} (meq/L)	0-20	-	-
Na^+ (meq/L)	0-40	-	-
K^+ (meq/L)	0-5	-	-
HCO_3^- (meq/L)	0-10	-	-

Table 8: Water parameters for irrigation water quality.

Potential Irrigation Problem			Range of Values	Degree of Restriction on Use	Number of Samples	%age of Samples
Salinity (affects Crop Water Availability	EC (µS/cm)		<700	None	54	75
			700-3000	Slight to moderate	17	24
			>3000	Severe	1	1
	TDS (mg/L)		<450	None	59	82
			450-2000	Slight to moderate	13	18
			>2000	Severe	-	-
Specific Ion Toxicity (affects sensitive crops)	Na^+ (meq/L)	Surface irrigation	<3	None	62	86
			3-9	Slight to moderate	10	14
			>9	Severe	-	-
		Sprinkler irrigation	<3	None	62	86
			>3	Slight to moderate	10	14
	Cl^- (meq/L)	Surface irrigation	<4	None	70	98
			4-10	Slight to moderate	1	1
			>10	Severe	1	1
		Sprinkler irrigation	<3	None	69	96
			>3	Slight to moderate	3	4
Miscellaneous Effects (affects susceptible	HCO_3^- (meq/L)		<1.5	None	15	21
			1.5-8.5	Slight to moderate	57	79
			>8.5	Severe	-	-
	pH		6.5-8.4	None	64	89

Table 9: Guidelines for interpretations of water quality for irrigation.

Parameters	Range	Water Class	Number of Samples	%age of Samples
PS (%)	0-20	Excellent	43	60
	20-40	Good	17	24
	40-60	Permissible	9	12
	60-80	Doubtful	3	4
	>80	Unsuitable	-	-
PI (%)	>75	Safe	15	21
	25-75	Moderate	57	79
	<25	Unsafe	-	-
Residual Mg^{2+}/Ca^{2+} Ratio	<1.5	Safe	70	97
	1.5-3	Moderate	2	3
	>3	Unsafe	-	-
RSBC (meq/L)	<1.25	Good	57	79
	1.25-2.5	Medium	9	13
	>2.5	Bad	6	8
EC (µS/cm)	<250	Excellent	15	21
	250-750	Good	41	57
	750-2250	Permissible	15	21
	2250-4000	Doubtful	1	1
	>4000	Unsuitable	-	-
SAR	<10	Excellent	72	100
	10-18	Good	-	-
	18-26	Fair	-	-
	>26	Poor	-	-
KR	<1	Suitable	67	93
	>1	Unsuitable	5	7
MAR (%)	<50	Fit	58	81
	>50	Unfit	14	19
TDS (mg/L)	<1000	Non saline	69	96
	1000-3000	Slightly saline	3	4
	3000-10000	Moderately saline	-	-
	>10000	Very saline	-	-
Chloride (C^{l-}) (meq/L)	<0.14	Extremely fresh	15	21
	0.14-0.85	Very fresh	41	57
	0.85-4.23	Fresh	14	20
	4.23-8.46	Fresh brackish	1	1
	8.46-28.21	Brackish	1	1
	28.21-282.06	Brackish salt	-	-
	282.06-564.13	Salt	-	-
	>564.13	Hyper saline	-	-

Table 10: Suitability of groundwater for irrigation based on several classifications.

irrigation water quality is suitable on the basis of percent sodium in the study area.

Residual sodium bi-carbonate (RSBC): The residual sodium bicarbonate was calculated using the formula [75]:

$$RSBC= HCO^{3-}-Ca^{2+} \tag{5}$$

RSBC and concentrations of the constituents were measured in meq/L .The concentration of bicarbonate and carbonate affects the groundwater quality for irrigation. High pH of groundwater increases the concentration of bicarbonate. Therefore, such water makes the irrigated land infertile owing to deposition of sodium carbonate [76]. RSBC values given in Table 3 were calculated using above equation. In the study area, RSBC values varied from -2.61 to 6.05 meq/L with mean, median and standard deviation values of 0.36 meq/L, 0.07 meq/L and 1.44 meq/L respectively (Table 4). The classification of RSBC [70] is given in Table 10. 79% of samples fell in good category, 13% samples in medium and only 8% samples in bad category. Hence, overall water samples are considered safe for irrigation water in the study area.

Magnesium adsorption ratio (MAR): The magnesium adsorption ratio was determined using the given relation [62]:

$$MAR=[Mg^{2+}/(Ca^{2+}+Mg^{2+})] \times 100 \tag{6}$$

All the ionic constituents expressed were in meq/L. The concentration of magnesium in groundwater is one of the most important qualitative criteria to determine the irrigation water quality. Generally, the concentrations of calcium and magnesium maintain the equilibrium state in most of the waters. The soil salinity increases with the increase in magnesium concentration in groundwater [73]. MAR values were calculated using above equation (Table 3). In the study area, MAR values ranged from 8.62 to 69.89 with mean, median and standard deviation of 37.65, 36.87 and 12.12 respectively (Table 4). MAR values in the study area are classified in Table 10 [39]. 81% samples were found in fit category while only 19% samples fell in unfit category. Hence, most of water samples are safe for irrigation water in the study area.

Kelly's ratio (KR): Kelly's ratio is calculated using the following formula:

$$KR = Na^+/(Ca^{2+} + Mg^{2+}) \tag{7}$$

All ionic constituents were presented in meq/L. Kelly's ration with values greater than 1 shows excess concentration of sodium; groundwater is suitable for irrigation with Kelly's ratio less than 1 [71]. KR was calculated using above equation and values are given in Table 3. In the study area, KR varied from 0.03 to 2.52 with mean, median and standard deviation of 0.35, 0.16 and 0.47, respectively (Table 4). The classification of KR in the study area is given in Table 10 [71]. It was observed in Table 10 that 93% samples were found suitable while only 7% samples were studied unsuitable for irrigation. Hence, overall groundwater quality is suitable for irrigation purpose.

Permeability index (PI): The permeability index was calculated using the following formula [70].

$$PI = [\{Na^+ + \sqrt{(HCO^{3-})}\}/Ca^{2+} + Mg^{2+} + Na^+)] \times 100 \tag{8}$$

All ions were expressed in meq/L. It is an important parameter to measure the groundwater suitability for irrigation. It is affected by the long term use of agricultural water; total dissolved solids, sodium bicarbonate and soil type are the influencing constituents. PI calculated values are given in Table 3. In the study area, PI varied from 26.97 to 108.91 with mean, median and standard deviation of 60.15, 56.65 and 19.47, respectively (Table 4). The classification of PI is given in Table 10 [70]. About 21% samples were found complete safe while 79% samples were moderately safe and no sample was found unsafe. Hence, overall groundwater quality is safe for irrigation purpose.

Residual Mg^{2+}/Ca^{2+} ratio": The residual ratio was calculated using the following relation [52]:

$$\text{Residual Ratio} = Mg^{2+}/Ca^{2+} \tag{9}$$

All ions were expressed in meq/L. It is very useful to find the suitability of groundwater for irrigation; groundwater can be classified as suitable or unsuitable on the basis of this residual ratio [52]. This ratio was calculated and values are given in Table 3. In the study area, it ranged from 0.09 to 2.32 with mean, median and standard deviation of 0.67, 0.58 and 0.38, respectively (Table 4). According to the classification of residual ratio as given in Table 10 [70], it was observed that 97% of samples fell in safe category, only 3% samples were found moderately safe and no sample was found unsafe. Hence, groundwater quality is suitable for irrigation purpose.

Total dissolved solids: The total dissolved solids were calculated using the relation (Richards, 1954):

$$TDS \text{ (mg/L or ppm)} = EC \text{ (mmhos/cm or dS/m)} \times 640$$

$$ECW \text{ (mmhos/cm or dS/m)} \times 640 = TDS \text{ (mg/L or ppm)} \tag{10}$$

EC and TDS were expressed in μ-mhos/cm and mg/L, respectively. The salts of calcium, magnesium, potassium, sodium in the irrigation groundwater is harmful to crops; and their excess quantities may affect the osmotic activities of the crops and may prevent adequate aeration. To assess the suitability of water for any purpose, TDS should be less than 500 mg/L [76-78]. The ratio of TDS to EC ranges from 550 to 700 ppm for different salt solutions. The most common salt in saline water is sodium chloride which has TDS of 640 ppm at EC of 1dS/m. Mostly, TDS is calculated from EC using this relation or multiplying by other factors. TDS values are given in Table 2. In the study area, it ranged from 91 to 1842 mg/L with mean, median and standard deviation of 303 mg/L, 193.5 mg/L and 287.49 mg/L, respectively (Table 4). TDS values for irrigation purpose were classified in Table 10 [79]. It was observed that 96% samples had no salinity; only 4% samples were

found slightly saline. Hence, groundwater quality based on TDS is safe for irrigation purpose.

Electrical conductivity (EC): EC and TDS have the relation.

EC and TDS were expressed in μ-mhos/cm and mg/L respectively. It is very important parameter to measure the suitability of groundwater for irrigation purpose. The higher the values of EC, the lesser the water available to crops, even the soil is wet because the plants can only transpire the useful water so, the useable water decreases with the increase in EC. It reduces the yield potential of the crops. EC values are given in Table 2. In the study area, it varied from 155 to 3070 (μS/cm) with mean, median and standard deviation of 529.25 μS/cm, 351 μS/cm and 465.12 μS/cm, respectively (Table 4). EC values for irrigation purpose were classified in Table 10 [70]. According to this classification, 21% samples were found excellent, 57% good, 21% permissible and only 1% doubtable. Hence, overall groundwater quality based on EC is good for irrigation purpose.

Chloride (Cl⁻): Chloride is one of the important parameters to assess the groundwater quality for any purpose. The groundwater suitability for irrigation purpose can be determined on the basis of chloride concentrations [80]. Chloride values in meq/L are given in Table 3. In the study area, it ranged from 0.03 to 13.71 meq/L with mean, median and standard deviation of 0.78 meq/L, 0.27 meq/L and 1.76 meq/L, respectively (Table 4). Chloride values for irrigation purpose were classified in Table 10 [80]. It was observed that 21% samples fell in extremely fresh category, 57% samples in very fresh category, 20% samples in fresh category, 1% samples in fresh brackish category and 1% samples in brackish category. On the basis of this classification it is concluded that most of the water is fresh and no saline water. Hence, overall groundwater quality based on chloride is safe for irrigation purpose.

Hence, the groundwater quality on the basis of different irrigation water quality parameters is suitable for irrigation in the study area (Table 11).

Correlation analysis

The correlation for physiochemical parameters was done by using bivariate technique. The correlation coefficients are worked out to find out the relationship between physicochemical parameters of the water samples [81]. The close examination of correlation matrix was helpful because it can determine relations between variables that can explain the overall coherence of the data set and point out the contribution of the individual chemical parameters in numerous control factors, a fact which commonly occurred in hydrochemistry. According to this method the change is measured between two variables or more and the value remains between -1 and 1. R measures the correlation between the variables, and is called the correlation coefficient and its value ranges between -1 and 1. The value of R around zero shows no relationship between the variables [82]. Its value around 1 shows very strong correlation. If the value of R is greater than 0.7, then this is taken as strongly correlated for the geochemical study. If it ranges from 0.5 values to 0.7 values then correlation coefficient is moderately correlated. If the value of R is negative value then it means that the

Total Hardness(mg/L)	Types	No of Samples	Samples %age
<75	Soft	2	3
75-150	Moderately hard	41	57
150-300	Hard	21	29
>300	Very hard	8	11

Table 11: The classification of groundwater for drinking based on hardness.

value of one variable is decreasing with the increase in other variable value [83]. The correlation was carried out for eleven parameters using the linear regressions as represented in Table 12. The ions correlation for the samples of groundwater is as:

Besides very strong correlation (R=0.98) between EC and TDS, strong correlation (R=0.91) also existed between TDS-SO4, EC- SO4 and Mg-TH. pH showed negative correlation with most of the variables. Overall in the study area, EC-TDS, TDS-Na, TDS-HCO$_3$, TDS-Cl, EC-Na, EC-HCO$_3$, EC-Cl, EC-SO$_4$, Ca-TH, Mg-TH, Na-SO$_4$, had strong correlation more than 0.7. The pairs like TDS-Mg, TDS-K, TDS-TH, EC-Mg, EC-TH, Ca-Mg, Mg-Na, Mg-K, Mg-SO$_4$, Na-K, Na-Cl, Na-TH, K-SO$_4$, K-TH, HCO3-HSO4, Cl-SO$_4$, TH-SO$_4$ were moderately correlated with correlation coefficient from 0.5 to 0.7. Other pairs had weak correlation with correlation coefficient less than 5.

Graphical analysis

The ions leach out and dissolve in groundwater during the water circulation in soils and rock bodies. The geochemistry of the

	pH	TDS	EC	Ca^{2+}	Mg^{2+}	Na$^+$	K$^+$	HCO$_3^-$	Cl$^-$	SO$_4^{-2}$	TH
pH	1										
TDS	-0.02	1									
EC	-0.03	0.98	1								
Ca^2	-0.16	0.28	0.28	1							
Mg^{2+}	0.03	0.56	0.54	0.5	1						
Na$^+$	-0.08	0.73	0.72	0.32	0.52	1					
K$^+$	-0.05	0.53	0.48	0.4	0.52	0.65	1				
HCO$_3^-$	-0.06	0.73	0.75	0.36	0.42	0.43	0.33	1			
Cl$^-$	-0.1	0.79	0.81	0.06	0.24	0.56	0.3	0.41	1		
SO$_4^{-2}$	0.02	0.92	0.91	0.27	0.7	0.74	0.58	0.59	0.65	1	
TH	-0.06	0.52	0.51	0.78	0.91	0.54	0.56	0.44	0.2	0.62	1

Table 12: Correlation between physiochemical parameters.

groundwater is influenced by the factors like geological formations, water-rock interaction and relative mobility of ions [84]. The results of groundwater quality parameters in form of tables may be difficult to interpret. The graphical analysis of groundwater parameters is easy to interpret. In order to assess the groundwater suitability, the graphical interpretation of groundwater parameters was worked out by developing Piper, Durov, Schoeller and Stiff diagrams.

Piper diagram: The concentrations of major anions and cations can be plotted in Piper tri linear diagram to understand the geochemical evolution of groundwater [85]. Rock Ware Aq.QA software was used to plot the Piper diagram. Piper diagrams are the combination of anion and cation triangles which lie on the common baseline; diamond shape between them is used to characterize different types of water. Piper divided the water into four types by placing it near four corners of the diamond. Water plotted at the top of the diamond is considered as high with Ca^{+2}+Mg^{+2} and Cl$^-$+SO$_4^{-2}$, which is the area of permanent hardness. The water plot near right side corner is rich in Ca^{+2}+Mg^{+2}; this water region is temporary hardness. The water plot at the lower corner is composed of alkali carbonates (Na$^+$+K$^+$ and HCO$_3^-$+CO$_3^{-2}$). The water near left hand side may be saline water (Na$^+$+k$^+$ and Cl$^-$ +SO$_4^{2-}$).

The groundwater samples were plotted in Piper diagram using Rock Ware Aq.QA software in Figure 2. It was observed in piper diagram that the nature of groundwater present in investigated area is sodium sulphate form. Thus, Piper diagram not only identifies the nature of water samples but also uncovers their relationships among each other. The geologic units along with chemically similar water can be predicted and classified followed by trend of water chemistry analysis along with flow path [86].

Durov diagram: Durov diagram can help to identify the types of water for the assessment of quality of groundwater [87]. This diagram

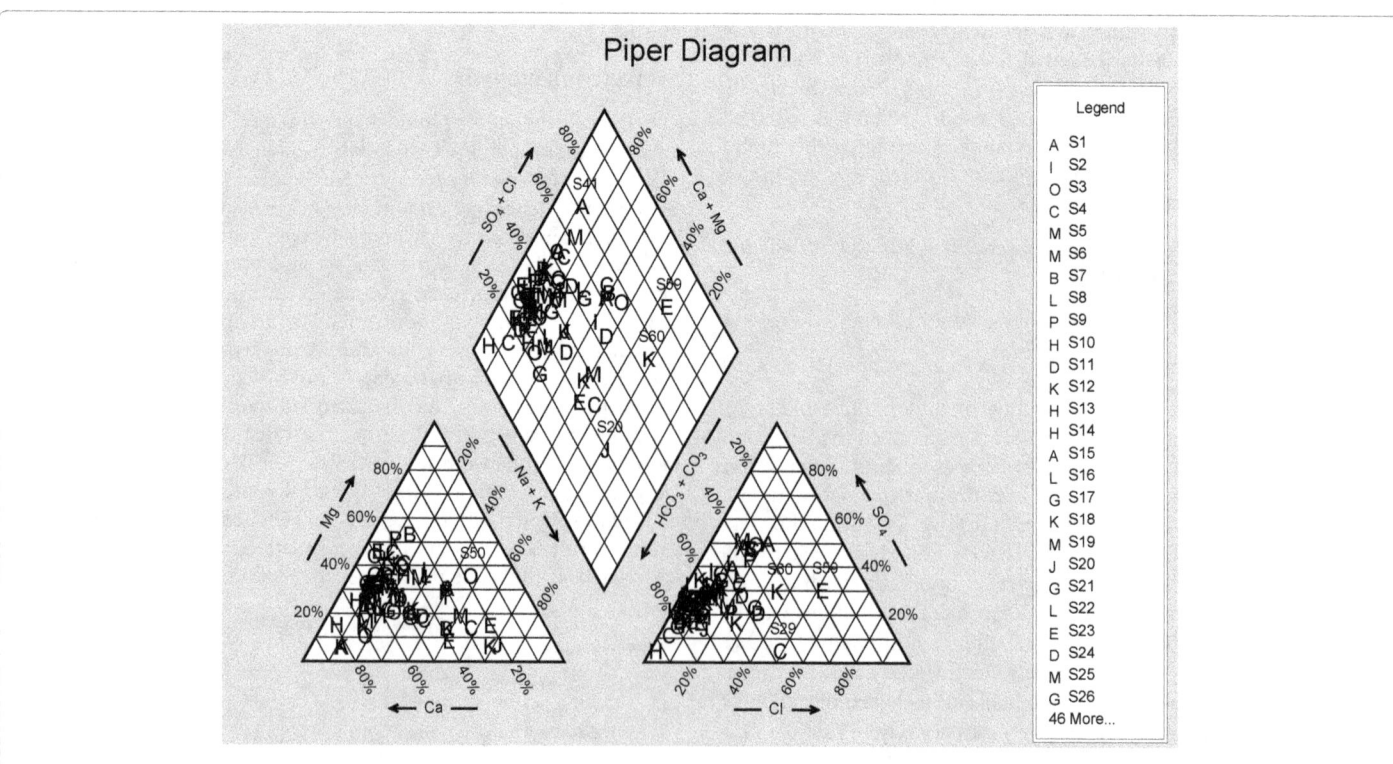

Figure 2: Piper Plot for groundwater parameters in the study area.

is an alternative form of the Piper tri-linear diagram. In Durov diagram, the major cations and anions with percentage of meq/L are set equal to 100% in two base triangles; and the expanded version includes electrical conductivity (μS/cm) and pH data added to the sides of the plot for further comparisons. The data points of the two base triangles are projected in square form which is perpendicular to each triangle. It represents the clustering of data points and the possible geochemical processes which can affect the water quality. This diagram was plotted by using Rock Ware Aq.QA software (Figure 3). The Durov Plot specifies that most of the samples in the study area indicate no dominant anion or cation showing water exhibiting simple dissolution or mixing. Water type of many samples is dominated by Ca^{+2} and HCO_3 ions which show ion exchange process and Na^+ ions indicate probable mixing influences. EC and pH part of the plot shows that overall water quality is suitable for drinking and irrigation purpose in the study area.

Schoeller diagram: The Schoeller diagram is useful for the study of comparative changes in the concentrations and ratios of water quality parameters for different samples [88]. The different water quality parameters were plotted with their concentrations in meq/L as shown in Figure 4. This diagram was generated using Rock Ware Aq.QA software. The results of this diagram show that lines of similar slope with concentrations of different parameters indicate the same source of water. It was observed in this diagram that the most water type of high sodium concentration also has high content of chloride.

Stiff diagram: The geochemistry of groundwater can be studied by means of its major ions [89]. Stiff diagram shows graphical representation of different ions in the groundwater. The average ionic composition analysis of stiff diagram is shown in Figure 5. Stiff diagram was plotted using Rock Ware Aq.QA software [90]. It shows

dominance of Na-Cl, while Ca-HCO_3 and Mg-SO_4 are almost equal in their proportion.

Conclusion

The groundwater quality of Toba Tek Singh District was assessed for its drinking and irrigation suitability. This work has presented the levels of physicochemical parameters like pH, electrical Conductivity (EC), total dissolved solids (TDS), total hardness (TH), calcium (Ca^{2+}) magnesium (Mg^{2+}), sodium (Na^+), potassium (K^+), chloride (Cl^-), bicarbonate (HCO^{-3}) and sulphate (SO_4^{2-}) in the well water samples collected from Toba Tek Singh District. The results obtained from the analysis of physicochemical parameters for drinking purpose show that most of the parameters did not exceed the permissible limit set by the world Health Organization (WHO), Pakistan Standards and Quality Control Authority (PSQCA) and Pakistan Council of Research in Water Resources (PCRWR). The analysis of irrigation water parameters such as sodium adsorption ratio (SAR), percent sodium (PS), permeability index (PI), residual sodium bicarbonate (RSBC), Kelly's ratio (KR), and magnesium adsorption ratio (MAR), show that overall groundwater quality in the study area is good for irrigation. Results obtained from graphical analysis (Piper and Durov diagrams) of groundwater samples show that the groundwater is Na-SO_4 type and most of the groundwater samples are in the phase of mixing, dissolution with few in reverse ion exchange. However, the present status of some of the water samples does not meet the international standard of water quality with respect to some constituents, a condition that is possibly to be worst in future. Thus the results obtained from the present investigation shall be helpful for future management of the reservoir water. The physicochemical characteristics of reservoir water suggested that the water in most of Toba Tek Singh District was no harmful to irrigation and drinking water.

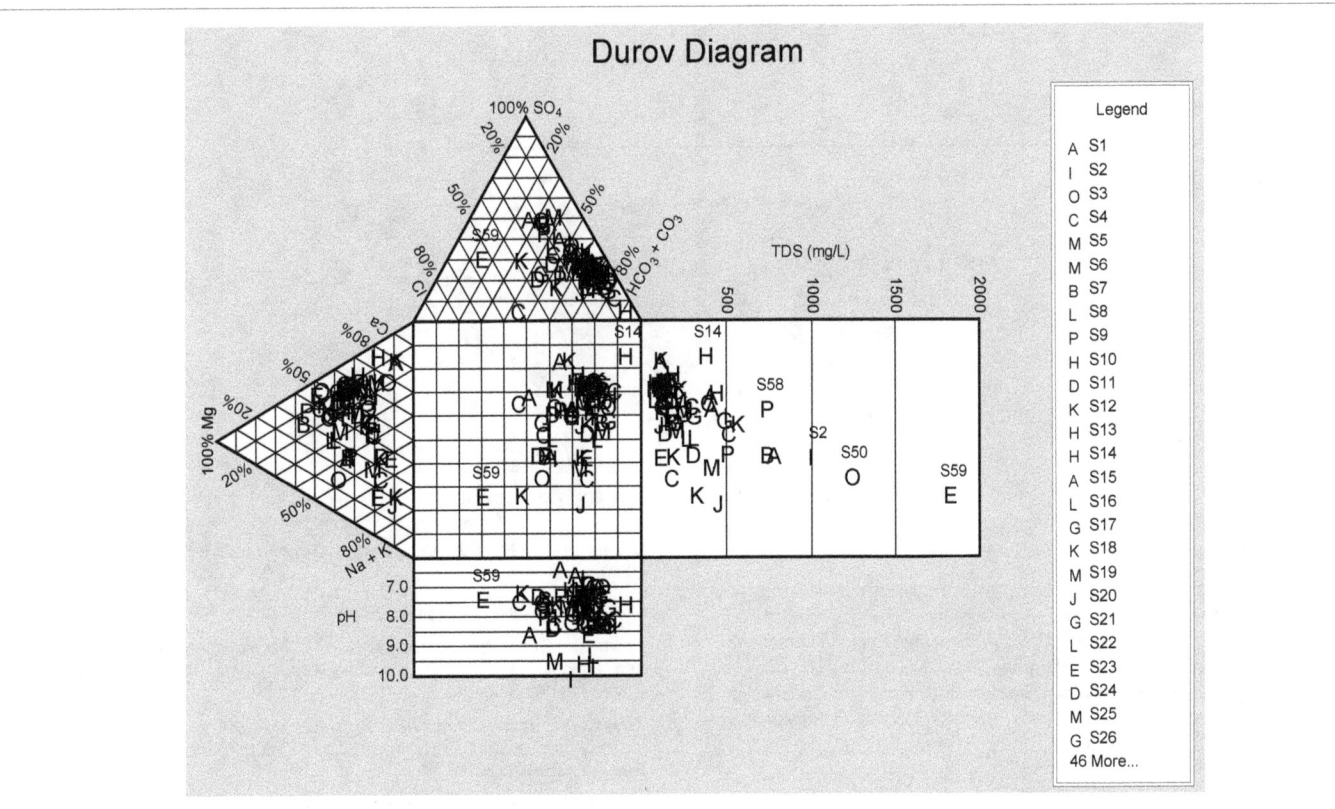

Figure 3: Durov Plot for groundwater parameters in the study area.

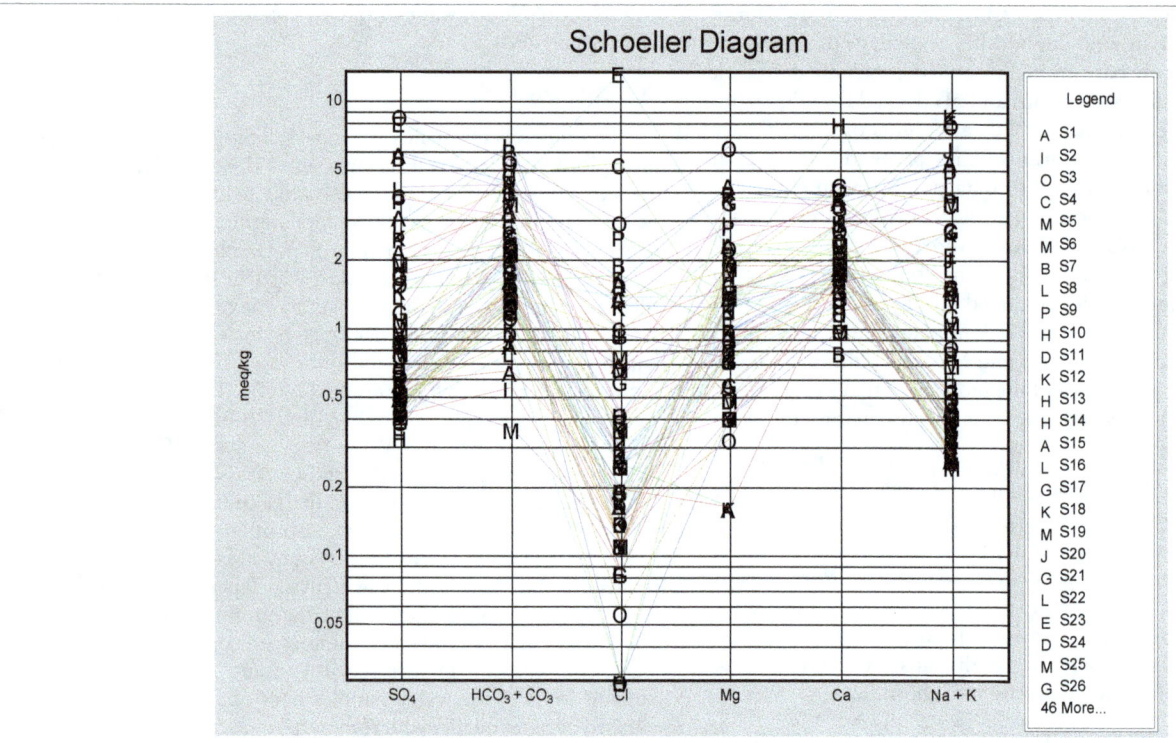

Figure 4: Schoeller Plot for groundwater parameters in the study area.

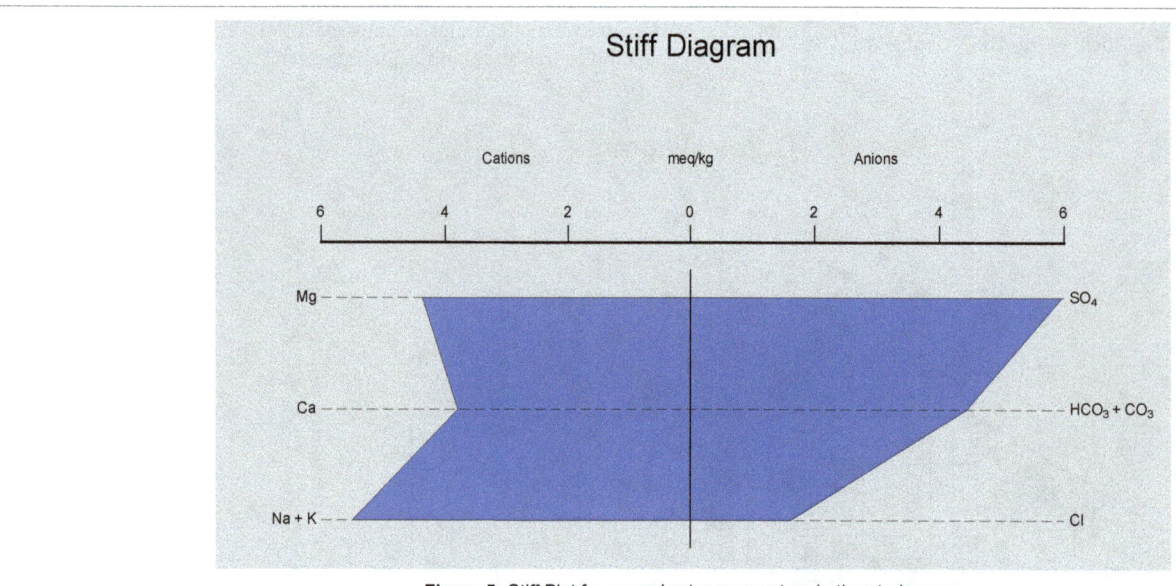

Figure 5: Stiff Plot for groundwater parameters in the study area.

Acknowledgement

This research was sponsored by CAS-TWAS President's Fellowship for International PhD Students; and funded by National Natural Science Foundation of China (Grant No. 41372324) and Xinjiang Uygur Autonomous Region International Science and Technology Cooperation Program (Grant No. 20166017). Authors wish to acknowledge support received from Institute of Geology and Geophysics, Chinese Academy of Sciences, Beijing China; Pakistan Council of Research in Water Resources (PCRWR), Islamabad Pakistan.

References

1. McMurry J, Fay RC (2004) Hydrogen, Oxygen and Water, (4th edn.). In: Hamann KP editor. Mc Murry Fay Chemistry. Pearson Education, pp: 575-599.

2. Delleur JD (1999) The Handbook of Groundwater Engineering. CRC Press, USA.

3. Qiu J (2010) China faces up to groundwater crises. Nature 466: 308.

4. Peiyue L, Quan W, Jianhua W (2011) Groundwater suitability for drinking and agricultural usage in Yinchuan Area, China. International Journal of Environmental Sciences 1: 1241-1249.

5. Tadesse N, Bheemalingeswara K, Berhane A (2009) Groundwater suitability for irrigation: a case study from Debre Kidane Watershed, eastern Tigray, Ethopia. MEJS 1: 36-58.

6. Mendie U (2005) The Nature of Water. In: The Theory and Practice of Clean Water Production for Domestic and Industrial Use. Lacto-Medals Publishers, Lagos, pp: 1-21.

7. Chitmanat C, Traichaiyaporn S (2010) Spatial and temporal variations of physical-chemical water quality and some heavy metals in water, sediments and fish of the Mae Kuang River, Northern Thailand. International Journal of Agriculture and Biology 12: 816-820.

8. Mirecki JE, Parks W S (1994) Leachate geochemistry at a municipal landfill, Memphis, Tennessee. Ground Water 32: 390.

9. Manzoor S, Shah MH, Shaheen N, Khalique A, Jaffar M (2006) Multivariate analysis of trace metals in textile effluents in relation to soil and groundwater. Journal of Hazardous Materials 137: 31-37.

10. Manning JC (1997) Applied principles of hydrology. Upper Saddle River, NJ: Prentice Hall.

11. Versari A, Parpinello GP, Galassi S (2002) Chemometric survey of Italian bottled mineral waters by means of their labelled physico-chemical and chemical composition. Journal of Food Compos Anal 15: 251-264.

12. Ackah M, Agyemang O, Anim AK, Osei J, Bentil NO, et al. (2011) Assessment of groundwater quality for drinking and irrigation: the case study of Teiman-Oyarifa Community, Ga East Municipality, Ghana. Proc Int Acad Ecol Environmental Sci 1: 186-194.

13. Sayyed MRG, Wagh GS (2011) An assessment of groundwater quality for agricultural use: a case study from solid waste disposal site SE of Pune, India. Proc Int Acad Ecol Environ Sci 1: 195-201.

14. Arms K (2008) Environmental Science. Holt, Rinehart and Witson A Harcourt Education Company.

15. Peterson N, Kennedy M (1997) Water quality trends and geological mass balance. John Whiley and Sons, pp: 139-179.

16. Sayyed MRG, Sayadi MH (2011) Variations in the heavy metal accumulations within the surface soils from the Chitgar industrial area of Tehran. Proc Int Acad Ecol Environ Sci 1: 36-46.

17. WHO (2004) Guidelines for Drinking-Water Quality: Training Pack. WHO, Switzerland.

18. Packham RF (1996) Drinking water quality and health. Pollution, causes, effects and control. The Royal Society of Chemistry, UK, pp: 52-65.

19. Dinrifo RR, Babatunde SOE, Bankole RO, Demu QA (2010) Physicochemical properties of rain water collected from some Industries Areas of Lagos State. Nigeria European Journal of Scientific Research 41: 383-390.

20. Aiyesanmi AF, Ipinmoroti KO, Adeeyinwo CE (2004) Baseline geochemical characteristics of groundwater within Okitipupa south-east belt of the bituminous sands field of Nigeria. International journal of environmental studies 61: 49-57.

21. Helena B, Pardo R, Vega M, Barrado E, Fernandez JM et al., (2000) Temporal evolution of groundwater composition in an alluvial aquifer (Pisuerga River, Spain) by principal component analysis. Water research 34: 807-816.

22. Jobson J (2012) Applied multivariate data analysis: volume II: Categorical and Multivariate Methods.

23. Hussain M, Ahmed SM, Abderrahman W (2008) Cluster analysis and quality assessment of logged water at an irrigation project, eastern Saudi Arabia. Journal of Environmental Management 86: 297-307.

24. Chen K, Jiao JJ, Huang J, Huang R (2007) Multivariate statistical evaluation of trace elements in groundwater in a coastal area in Shenzhen, China. Environmental Pollution 147: 771-780.

25. WHO (2011) Guidelines for drinking water quality, 4th edn. WHO press P: 564.

26. Chutia J, Sarma S (2009) Seasonal variation of drinking water quality with respect to fluoride & nitrate in Dhakuakhana sub-division of Lakhimpur District of Assam. Int J Chem Sci 7: 1821-1830.

27. Ho KC, Chow YL, Yau JTS (2003) Chemical and microbiological qualities of the East River (Dongjiang) water, with particular reference to drinking water supply in Hong Kong. Chemosphere 52: 1441-1450.

28. Hujare MS (2008) Seasonal variation of physico-chemical parameters in the perennial tank of Talsande. Maharashtra Ecotoxicol Environ Monit 18: 233-242.

29. Armugan K, Elangovan K (2009) Hydrochemical characteristics and groundwater quality assessment in Tirupur Region, Coimbatore District, Tamil Nadu, India. Environ Geol 58: 1509-1520.

30. Goyal SK, Chaudhary BS, Singh O, Sethi GK, Thakur PK (2010) GIS based spatial distribution mapping and suitability evaluation of groundwater quality for domestic and agricultural purpose in Kaithal district, Haryana state, India. Environmental Earth Sciences 61: 1587-1597.

31. Ketata M, Bouhlila R, Gueddari M (2011) Suitability assessment of shallow and deep groundwaters for drinking and irrigation use in the El Khairat aquifer (Enfidha, Tunisian Sahel). Environ Earth Sci 65: 313-330.

32. Central Public Health and Environmental Engineering Organization (CPHEEO) (1998) A manual on water supply and treatment. Akalank Publication, New Delhi.

33. Blais JF, Tyagi RD, Aucleir JC (1993) Bio-leaching of metals and sewage sludge: effect of temperature. Water Resource. 27: 110-120.

34. Central Ground Water Board (CGWB) (2004) Annual report and other related reports on ground water quality, Central Ground Water Board, New Delhi.

35. Gajendra C, Thamarai P (2008) Study on statistical relationship between ground water quality parameters in Namibiyar river basin, Tamil Nadu, India, Pollution Research 27: 679-683.

36. WHO (1996) Guidelines for drinking water quality. World Health Organization, Geneva 1: 188.

37. Begum A, Krishna HS, Khan I (2009) Analysis of heavy metals in water, sediments and fish samples of Madivala lakes of Bangalore, Karnataka. International Journal of Chem Tech Res 1: 245-249.

38. ISO (1991) Industrial Tyres and Rims-cylindrical and Conical Base Rubber Solid Tyres (Metric Series)-designation, Dimensions and Marketing.

39. Ayers RS, Westcot DW (1985) Water quality for agriculture FAO irrigation and drain. 29: 1-109.

40. Nishanthiny SC, Thushyanthy M, Barathithasan T, Saravanan S (2010) Irrigation water quality based on hydrochemical analysis, Jaffna, Sri Lanka. American-Eurasian Journal of Agricultural &Environmental Science 7: 100-102.

41. Al-Bassam AM, Al-Rumikhani YA (2003) Integrated hydrochemical method of water quality assessment for irrigation in arid areas: application to the Jilh aquifer, Saudi Arabia. Journal of African Earth Sciences 36: 345-356.

42. Richards LA (1954) Diagnosis and Improvement of Saline and Alkali Soils Agric Handbook 60. New Delhi, India, pp: 98-99.

43. World Wide Fund (WWF) (2007) A special report on Pakistan waters at risk: water and health related issues in Pakistan and key recommendations, pp: 1-25.

44. American Water Works Association and Water Environment Federation (1995) Standard Methods for the Examinations of Water and Wastewaters, (19th edn.), American Public Health Association. Washington.

45. Radojevic M, Bashkin VN (1999) Practical environmental analysis. Royal Society of Chemistry.

46. Trivedy RK, Goel PK (1986) Chemical and biochemical methods for water pollution studies. Environmental Publication, Maharashtra.

47. American Public Health Association (2000) Standard methods for the examination of water and wastewater. USA.

48. WHO (2008) Guidelines for drinking-water quality. World Health Organization, Geneva.

49. PSQCA (2004) Pakistan standards specification of bottled drinking water, under compulsory certification marks scheme. Pakistan Standards and Quality Control Authority, Ministry of Science and Technology. Pakistan.

50. Pakistan Council of Research in Water Resources (PCRWR) (2005) Water Quality Status. Pakistan.

51. Diersing, Nancy (2009) Water quality: Frequently asked questions. PDA. NOAA.

52. Kumar M, Kumari K, Ramanathan AL, Saxena R (2007) A comparative evaluation of groundwater suitability for irrigation and drinking purposes in two intensively cultivated districts of Punjab, India. Env Geol 53: 553-574.

53. Dahiya S, Kaur A (1999) Physico chemical characteristics of underground water in rural areas of Tosham subdivisions, Bhiwani district, Haryana. J Environ Poll 6: 281.

54. Balakrishnan P, Saleem A, Mallikarjun N (2011) Groundwater quality mapping using Geographic Information System (GIS): A case study of Gulbarga City,

Karnataka, India. Afr J Environ Sci Technol 5: 1069-1084.

55. Hari Haran A, varshya RC (2002) Evaluation of drinking water quality at Jalaripeta village of Visakhapatnam district Andhra Pradesh. Nat Environ Pollut Technol.

56. Joseph K (2004) A cleaner production approach for minimization of total dissolved solid in reactive dying effluents. Centre for Environmental Studies, Anna University, Chennai.

57. Sohani D, Pande S, Srivastava V (2001) Ground water quality at Tribal Town: Nandurbar (Maha rashtra). Indian J Environ Ecoplan 5: 475-479.

58. Subba RN (1998) Groundwater quality in crystalline terrain of Guntur district andhra Pradesh. Visakha Sci J 2: 51-54.

59. WHO (World Health Organization) (1996a) Guidelines for drinking water quality, (2nd edn.), Health Criteria and Other Supporting Information, Switzerland.

60. Sawyer CN, McCarty PL, Parkin GF (1994) Chemistry for environmental engineering, (4th edn.), McGraw-Hill, Singapore.

61. Tatawat RK, Chandel CPS (2007) Quality of ground water of Jaipur city, Rajasthan (India) and its suitability for domestic and irrigation purposes. Applied Ecology and Environment Research 6: 79-88.

62. Raghunath IM (1987) Groundwater, (2nd edn.), New Delhi, India.

63. Sawyer GN, McMcartly DL (1967) Chemistry of sanitary engineers, (2nd edn.), McGraw Hill, New York, p: 518.

64. Canter IW, Knox RC (1985) Septic Effects on Ground Water Quality Michigan. Lewis Publishing Inc, p: 336.

65. Taylor EW (1958) The Examination of Water and Water Supplies. Soil Science, p: 226.

66. Drever JI (1988) The Geochemistry of Natural Water Englewood Cliffs. Prentice Hall, New Jersey, pp: 383-390.

67. Jameel AA, Sirajudeen J (2006) Risk Assessment of Physico-Chemical Contaminants in Groundwater of Pettavaithalai Area, Tiruchirappalli, Tamil Nadu- India. Environmental Monitoring and Assessment 123: 299-312.

68. Raihan F, Alam JB (2008) Assessment of Groundwater Quality in Sunamganj of Bangladesh. Iranian Journal of Environmental Health Science and Engineering 5: 155-166.

69. Gupta DP, Sunita SJP, Saharan JP (2009) Physiochemical Analysis of Ground Water of Selected Area of Kaithal City (Haryana) India. Researcher 1: 1-5.

70. Richards LD (1964) Notes on water quality in Agriculture, Published as a water science and Engineering. Department of water science and engineering, University of California.

71. Kelly WP (1963) Use of saline irrigation water. Soil Sci 95: 355-391.

72. Biswas SN, Mohabey H, Malik ML (2002) Assessment of the Irrigation Water Quality of River Ganga In Haridwar District. Asian J Chem.

73. Joshi DM, Kumar A, Agrawal N (2009) Assessment of the irrigation water quality of River Ganga in Haridwar District India. Rasayan J Chem 2: 285-292.

74. Wilcox LV (1955) Classification and Use of irrigation Waters. United States Department of Agriculture, p: 969.

75. Gupta SK, Gupta IC (1987) Management of Saline Soils and Water. Oxford and IBH Publication Coy. New Delhi, India, p: 399.

76. Eaton FM (1950) Significance of carbonate in irrigation water. Soil Sciences 69: 123-134.

77. Catroll D (1962) Rain water as a chemical agent of geological process: a view. USGS Water Supply 1533: 18-20.

78. Freeze RA, Cherry JA (1979) Groundwater. Prentice-Hall, Englewood Cliffs.

79. Robinove CJ, Longfort RH, Brooks JW (1958) Saline water resources of North Dakota, US Geol. Surv Water Supply Paper 1428. P: 72.

80. Stuyfzand PJ (1989) Nonpoint source of trace element in potable ground water in Netherland. In: Proceeding of the 18th TWSA Water Working, Testing and Research Institute, KIWA, Nieuwegein, the Netherlands.

81. Usharani K, Umarani K, Ayyasamy PM (2010) Physico-Chemical and Bacteriological Characteristics of Noyyal River and Ground Water Quality of Perur, India. Journal of Applied Science and Environmental Management 14: 29-35.

82. Srivastava SK, Ramanathan AL (2008) Geochemical assessment of ground water quality in vicinity of Bhalswa landfill, Delhi, India, using graphical and multivariate statistical methods. Environ Geology 53: 1509-1528.

83. Giridharan L, Venugopal T, Jayaprakash M (2008) Evaluation of the seasonal variation on the geochemical parameters and quality assessment of the groundwater in the proximity of River Cooum, Chennai, India. Environ Monit Assess 143: 161-178.

84. Yousef AF, Saleem AA, Baraka AM, Aglan O Sh (2009) The impact of geological setting on the groundwater occurrences in some Wadies in Shalatein-Abu Ramad area, SE desert, Egypt. European Water 26: 53-68.

85. Piper AM (1944) A graphic procedure in the geochemical interpretation of water analysis. American Geophysical Union Transactions 25: 914-923.

86. Zhang P (2013) EAS 44600 Groundwater Hydrology.

87. Durov SA (1948) Natural waters sand graphic representation of their composition. Dok, pp: 87-90.

88. Schoeller H, Konoplyantsev AA, Ineson J (1967) Geochemistry of ground water. An international guide for research and practice, UNESCO 15: 1-18.

89. Stiff HA (1951) The interpretation of chemical water analysis by means of patterns. Journal of Petroleum Technology 10: 15-17.

90. Bartram J, Ballance R (1996b) Water quality monitoring: A practical guide to the design and implementation of freshwater quality studies and monitoring programmes. United Nations Environment Programme and the World Health Organization.

24

Gis Based Diagonastic Analyis of Doni Sifa Small Scale Irrigation Scheme: In Upper Awash Ethiopia

Gemechis T[1]*, Quraishi SH[2] and Zeleke T[3]

[1]Soil and Water Engineering Research Case Team, Bako Agricultural Engineering Research Center, Oromia, Ethiopia
[2]School of Natural Resource and Environmental Engineering, Institute of Technology, Haramaya University, Ethiopia
[3]Soil and Water Management Research Case Team, Research and Development, Sugar Corporation, Ethiopia

Abstract

This study attempted to conduct ArcGIS based diagnostic analysis of Irrigation scheme with intention of introduction of wise use of limited natural resources in case of Upper Awash for irrigation purpose. Experimental site selection criteria were that Boset woreda was one of the most chronically food in-secure districts, availability of secondary data and organizational set up of the irrigation projects. As to the output, simple model to calculate weighted overlay analysis of irrigation suitability was conducted and mapped, and from the map 4454 ha area was identified to be in suitable range for irrigation. Rainfall data records of 10 years (2003-2012) and Awash River stream flow of 21 years (1975-1995) records were adjusted for missing data, checked for consistency by developing double mass curve using three adjacent stations, analyzed for 80% stream flow dependability and was found to be 23.78 m³/s after giving allowance of 30% to downstream water use right. Finally, crop pattern of the study area was organized in CROPWAT 8.0 model along with necessary data such as climate, soil, and crop, cropping pattern. Maximum irrigation requirement with irrigation intensity of 100% was found to 137.5 mm/month in May from which net scheme supply design was calculated to be 5.67 m³/s and assuming 80% application efficiency total scheme supply design was found to be 7.1 m³/s .

Keywords: Diagnostic analysis; Weighted overlay analysis; Irrigation suitability; Scheme supply design

Introduction

In Ethiopia, rainfall is becoming more erratic and unreliable from time to time as a result of global climate change and manmade climate changing factors like that of disturbance of ecosystem, environmental degradation. Most rain falls intensively, often as convective storms, with very high rainfall intensity and extreme spatial and temporal variability. These rainfall patterns affect crop and livestock production and contribute to volatility in food prices, which ultimately affects overall economic development [1].

Oromia national region which is the largest states in Ethiopia with respect to population and areal coverage has relatively better natural set up, but suffering from food insecurity as one third of the region is low land prone to drought. Due to the drought and unreliability of rainfall, Fentale, Boset, Dodota-sire, Merti and Jeju were the chronically food in-secured districts in the region [2-3].

According to IWMI, (2007) due to lack of water storage, large spatial and temporal variations in rainfall, there is not enough water for most farmers to produce more than one time per year and also there are frequent crop failures due to dry spells and droughts which has resulted in a chronic food shortage currently facing the country.

Undoubtedly, irrigation plays an important role in food production, self-sufficiency and security, but potential increase in irrigation water and land resource is limited. Despite the higher risks in rainfed agriculture, it is widely accepted that the bulk of the world's food will continue to come from rainfed systems [4].

Therefore accelerated and sustainable development in agriculture sector needs transformation of rainfed agriculture to be irrigated agriculture. Irrigated agriculture with respect to other inputs like fertilizer etc., can secure food security and food self-sufficiency because it minimize the risk of inadequate and uneven distribution of rainfall and also enhance production of superior crops and growing of crops more than once in a year. Irrigable land and water resources are limited by nature. Better use of these limited resources, during planning and design, strong logistics to diagnostic opportunity and limitation by using integrated approach along with advancing technology like GIS in the field of Irrigation Engineering is needed to optimize the return from the project. So far in Ethiopia, such strong logistics, design and monitoring have received little attention by engineers, planners and policymakers.

Therefore, irrigation project planning and development policies need to become more strategic to bring higher returns from agricultural water as precipitation in rainfed areas is characterized by a low annual rainfall, and for unfavorable distribution over the crop's growing season, with high year-to-year fluctuations. Accordingly, by understanding the risk of rivers' stream discharge fluctuations a result of global climate change and inappropriate water management, selection of the best-fit land use and irrigation type is a pre-requisite for wise utilization of scarce physical resource of land and water. If the land and water resources are not wisely developed and properly managed, widespread and severe environmental and ecosystem disturbance is unavoidable. Not only this, but also community like Doni Sifa small scale irrigation suffers from severe drought and his brother famine from year to year even during good year because of late coming, failure in the middle of cropping season or early stopping of rainfall. This calls for a need to conduct detailed to semi-detailed study available potential of irrigable land, river stream and water management system studies and limitations.

Specific objectives

1. To investigate the suitability of irrigation land

---

*Corresponding author: Gemechis T, Soil and Water Engineering Research Case Team, Bako Agricultural Engineering Research Center, Oromia, Ethiopia
E-mail: tesema2010@gmail.com

2. To characterize Awash river stream at or near the study area over cropping seasons

3. To develop optimized scheme design for the land in the acceptable suitability range

Material and Method

Study area

The study area is located in Boset woreda of Oromia Regional state. Boset woreda is found in between 8o25'00"- 8o50'00"N latitude and 39o15', 39o50'E longitude and shares borderlines with Fentale, Adama and Lume districts, Amhara Regional State and Arsi zone. Qombe Gugsa, Sifa Bate and Nura Hasse rural kebeles the specific study area are located in the upper valley of Awash river basin and 33 km North of Sodore town, 52 km from Adama, and 152 km from the capital city of the country, Ethiopia.

Water source and agro-climate of the study area

Awash River is the source of water for Doni Sifa irrigation project. The agro-climatological zone of the anticipated with having mean annual rainfall of 600 mm and with two cropping season namely belg from February to April and meker from June to September. The maximum temperature was observed from 32.4°C to 35°C in the months of February and May while the daily mean minimum temperature was ranging from 10.1°C to 15.6°C in the months of October and July with the mean annual temperature is 22.55°C. The evapotranspiration of the area was estimated as high as 1872 mm per year or 5.10 mm/day.

Materials used in the study

In the processes of land resource investigation, irrigation suitability analysis, Awash River stream analysis and diagnostic analysis of the Doni Sifa irrigation project were summarized in Table 1.

Irrigable land investigation and irrigation feasibility analysis

In delineation of study area, geo-referenced data (easting, northing and elevation) were captured and registered manually following border line of inspected irrigable land within the study area and captured data was converted into appropriate unit from degree; minute and second to only degree in rational form in excel spread sheets. Then appropriate database was established in Arc catalog which was part of AcrGIS.

After appropriate data base was created, the geo-referenced data in spread sheet was imported to Arc catalog where shape file was created. Shape file created on arc catalog was then displayed in Global mapper and there digital elevation model topography map were displayed along with the created shape file.

ArcGIS has edition room or facility to demarcate area of interest but to do so, geospatial analysis extensions should activated. Steps followed to edit were: first, edition tool was selected from tool bar then shape type (line, polyline, polygon etc.) was set according to the need then interconnecting neighborhoods points collected by GPS, displaying

assigned labels in the ArcGIS work area. Then the created shape was masked on DEM, exported and saved in created database where it was used as a template for other analysis like area and slope calculation.

Irrigation feasibility analysis

In irrigation feasibility analysis main factors are water availability and quality, slope, chemical and physical properties of soil, relative location of the water source for small scale irrigation, buffer area were considered and finally come up with the size of the land that can be brought under irrigation. The factors were reclassified and simple weighted overlay analysis was developed with the aid of ArcGIS 10 model which was displayed as map (Figure 1).

Soils and topography

The area was situated on an undulating alluvial plain with open vegetation. In the area surrounding the project site was affected by gully formation up to 10 m deep could be observed and after some depth the soil profile became sandy. The irrigable area lies between a hill to the north and Awash river to the south and southeast. More than 85% of the command area had slope less than 5%. The soil was mostly medium textured ranging from silt loam to sandy loam. The depth of the soils in the area varies from 3 cm to 16 cm.

The assessment of soils for irrigation involves using properties that are permanent in nature that cannot be changed or modified. Such properties include drainage, texture, depth, salinity, and alkalinity [5]. Even though salinity and alkalinity hazards possibly be improved by soil amendments or management practices, they could be considered as limiting factors in evaluating the soils for irrigation [6,7].

This is, just physically investigated, and some common observations like, general depth, color, combinations along with physical and chemical properties of selected sampling pits were examined and the result is summarized in Table 2.

Slope and size of the study area

The other factors in evaluation of land for irrigation purpose are

Material	Source
AcrGIS 10	Free download from ESR websites
CROPWAT 10	Free download from FAO website
Global mapper 11	download from http://www.globalmapper.com
Soil auger, samplers	Bako Agr. Mechanization Research Center
DEM, satellite image, topography sheet	Ethiopian Mapping Agency
GPS, water flow measuring device	Bako Agr. Mechanization Research Center

Table 1: Materials to be used and their sources.

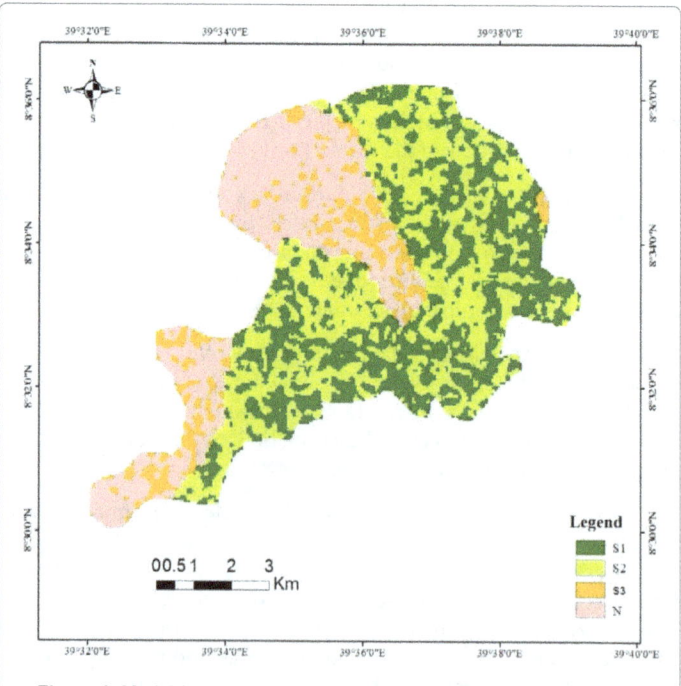

Figure 1: Model for weighted suitability factors analysis and output map.

Pit observed	No. sample	Profile depth (cm)	Location
I	1	0-150	At the beginning of command area
II	2	0-80, 80-170	North part of the command area
III	2	0-100,100-160	Near the boundary of the riverbank
IV	2	0-40, 40-160	At the alluvial part of the riverbank

Table 2: Selected sampling pits.

slope and area. Slope is the gradient of a surface and is commonly expressed as a percent. Slope is important for soil formation and management because of its influence on runoff, drainage, erosion and choice of irrigation types. The slope gradient of the land has great influence on selection of the irrigation methods. According to FAO standard guidelines for the evaluation of slope gradient, slopes which are less than 2%, are very suitable, 2-5% suitable 5-8% marginally suitable for surface irrigation. But slopes, which are greater than 8%, are not generally recommended [8].

Precipitation and stream data analysis

Basically time series data processing was substantially the preliminary step followed in order for the hydrological analysis. In examination, the trend and homogeneity of precipitation and stream data over a season of precipitation in study area, precipitation data for the period of 10 years (2003-2012) were collected from National Metreology Agency of Ethiopia. Collected data can contain errors due to failures of measuring device or the recorder. So, before using the data for specific purpose, the data have to be checked and errors have to be removed. Therefore, simple arithmetic was used to estimate missing information both of rainfall and stream flow data records depending on the variability of the data. The variability of the data was checked by normal ratio method as developed by Paulhus and Kohlar. If the variability of the data is less (Px against Pa, b, c) is less 10% simple arithmetic method can be used to fill the missing data. Therefore, Tibila, Wonji and Below Koka dam stations were used.

$$P_x = \frac{NxP_a}{N_a} + \frac{NxP_b}{N_b} + \frac{NxP_c}{N_c}$$

where Px is precipitation, with missing records at station x and Pa, Pb and Pc were adjacent, Nx, Na, Nb and Nc station's precipitation values for long period of time for x, a, b and c stations. After missing data were filled, consistency of data was checked using double mass curve method and then frequency analysis of data was done

To prepare the stream flow and rainfall data for further application, their consistency was checked using double mass curve analysis. A plot of accumulated discharge/rainfall data at site of interest against the accumulated average at the surrounding stations is generally used to check consistency of stream flow/rainfall data. To check the degree of consistency, the value of coefficient of correlation, Nemec was used [9].

Precipitations data of 21 years (1975-1995) were set in descending order and the ranked and to calculate frequency analysis of stream flow data Weibull plotting formula was used as.

$$P = \frac{m}{N+1} x100$$

The relation between Probability P (%) and return period, T (Years) is:

$$T = \frac{1}{P}$$

Then both for rainfall and stream flow data, design values which

80% of rainfall or stream flow events have a chance of equalizing or exceedance.

Results and Discussion

Irrigable land investigation and slope analysis

In irrigable land investigation, after reconnaissance survey was conducted in the study area, possible irrigable land was delineated through digitization. Then, by clipping the generated shape file by mask to raster data or the so called digital elevation model, the project was enhanced for further analysis. After so doing, an area which was formerly under rainfed agriculture but having potential to be brought under irrigation was calculated along with the slope of which is key determinant in irrigation suitability analysis and method of irrigation to be adopted and the results are summarized in Table 3 and displayed in Figure 2 [10].

A cell or pixel of digital elevation model used was 20 × 20 m² and called count. Therefore to calculate the area of interest, counts are multiplied with its number that gave total area. Literally, 6678 ha in study area was in slope range of 0-8% while majority of the area was in the slope range of 0-5 which makes 83% of the total area. For sake of simplicity both rainfed agriculture land and irrigated agricultural land were included into interest area while irrigation suitability analysis was conducted only in rainfed agriculture as all pits developed and soil samples were taken (Table 2) and were out of irrigated area. A total area of 6678 ha was estimated from both irrigated and rainfed agriculture. As indicated, within the study area, there were two canals while the first canal designed to supply irrigation water to 250 ha out of which 195 ha area was developed for the irrigation but presently only 122 ha had been growing actively. The second canal was designed to supply 210 ha but presently supplying water to less than 40 ha only [11].

Therefore regarding slope of the study area 6,516 ha of the land could be brought under irrigation agriculture excluding areas that is

Value	Count (0.04ha)	Area ha	%
0-2%	72582	2903.28	42.7
2-5%	68583	2743.32	40.4
5-8%	25788	1031.52	15.2
>8%	2936	117.44	1.7

Table 3: Summarized information from slope map of the study area.

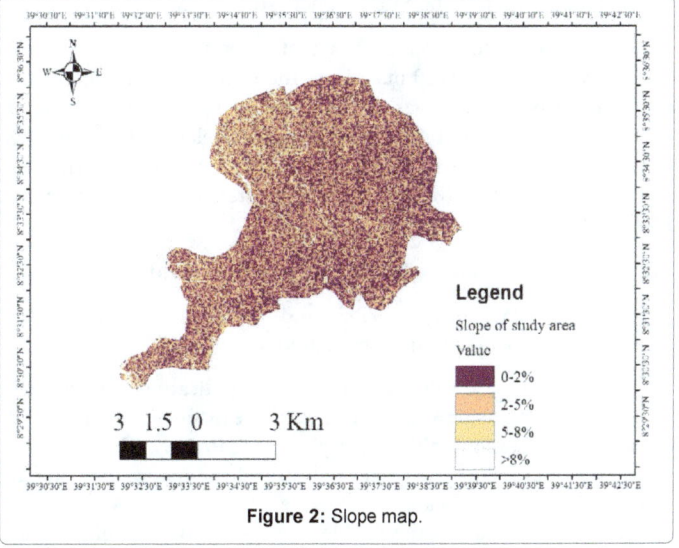

Figure 2: Slope map.

already under irrigation (162 ha); however, elevation and the relative location of water source and its buffer area are also decisive but this problems could be solved through better water abstraction method (gravity or lifted irrigation) and irrigation methods.

Suitability of the study area was categorized as 0-2% which was rated as highly suitable was 42.7% of the total area (2903.28), 2-5% that was rated as moderately suitable with 40.4% of the total area of the study area (2743.32 ha), and 5-8% that was rated as moderately not suitable was 15.2% of the total of the study area (1031.52 ha).

Soil of the study area

As can be seen from soil map, about 80.9% (5402.5ha) of the study area was productive soil of Vitric andosols and Lithosols where as 19.1% (1275.5 ha) of the land was of rocky surface or shallow soil.

Average moisture contents (v/v) at field capacity (FC) and permanent wilting points (PWP) were observed 24.6, 22.5 and 25.51% and 16.8, 17.2 and 19.6%, respectively for the three location of the field. The average moisture content at FC and PWP were 20.7 and 14.25% (w/w). The pH was found in the range of 6.7 to 7.4. The pH value of the soil in the experimental site was almost in suitable range for crop production particularly for common crops production. The optimum soil pH for crop production was considered to be 6.5 to 7.0. The bulk density of the plots was 1.08, 1.0 and 1.25 for the three experimental sites and the electrical conductivity was in the acceptable range [12].

The average calcium content of the soil layer was 59, 72.8 and 56.5% for the three pits one to three respectively. The electrical conductivity of the soil of the study area was 0.04 mmhos/cm on average bases. An average organic carbon of the pits developed within the study area was 1.54, 0.33 and 0.57%. Total nitrogen of the soil layer also found to be 0.01, 0.04 and 0.06 on average Cation exchangeable capacity (Meq/100 g) which were 27.12, 24.58 and 27.6 while average phosphorus were 3.44, 6.38 and 13.18 and exchangeable sodium percentage was 1.48%. According all of the parameters could be rated as acceptable.

Location of water diversion points and buffer area

Elevation of water diversion points and its buffer area is also another important parameter in irrigation suitability analysis as it affects water obstruction. With this respect, the possible water diversion points were assessed and measurement at riverbed was observed by GPS. The study area was divided into four zones depending on the relative elevation of Awash river to the elevation of the whole study area [13].

During surveying, it was noticed that water diversion point for the study area, was about 1260 m.a.s.l and the elevation at command area of interest was in the ranges of 1187-1370 m a.s.l., (Figure 3). About 84.7% (5854 ha) of the study area was found in relative elevation range of (1187-1240 m) which could be brought under surface irrigation method. About 10.1% (824ha) of the study area was found to be above 1260 m above s.l.

Missing data and consistency analysis of rainfall data

In rainfall data analysis, Tibila and Wonji rainfall stations had missing rainfall data records as presented in (Table 4).

Therefore, to use this data for further application, missing data values were filled using simple arithmetic mean since rainfall records of all gauging stations did not vary more than 10%. The rainfall at Nura Hera and Below Koka dam station had no missing rainfall data records (Table 5). Tibila and Wonji were stations with missing rainfall data and filled. Similarly the double mass curve analysis of the Nura

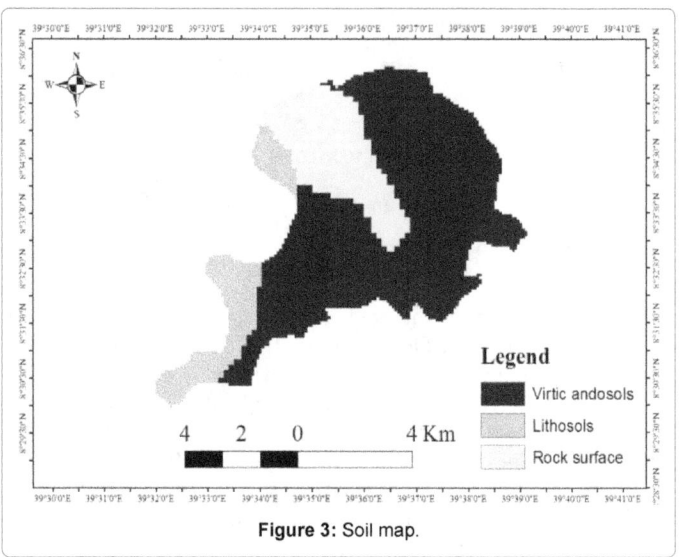

Figure 3: Soil map.

Location	Soil-depth (cm)	pH	Bulk-density (g/cm³)	Ece (ds/m)	FC (%)	PWP (%)	Soil-texture classes
Pit at head	0-20	7.1	1.04	0.66	22.53	14.6	Loam
	20-40	7.2	1.12	0.74	23.49	18.4	Loam
	40-60	7	1.01	0.63	27.78	17.3	Loam
Pit at mid	0-20	6.9	1.1	0.52	22.77	16	silt loam
	20-40	6.8	1.02	0.43	21.39	17.3	silt loam
	40-60	6.7	0.96	0.4	23.37	18.4	silt loam
Pit at tail	0-20	7.4	1.25	0.93	25.51	19.6	Loam

Table 4: Laboratory results of physical properties of soil of the study area.

Hera rainfall stations revealed that the rainfall recorded at the stations are consistent with no significant change of slope on their perspective plots as presented in Figure 4. The correlation factor R2 was very close to 1 which implies that strong and direct relationship exists between adjacent stations (Table 6). This also suggests that the rainfall data recorded at these stations can be used directly for further analysis [14].

Frequency analysis of rainfall data of study area

From frequency analysis, annual dependable rainfall at different probability levels along with their return periods were derived for Nura Hera, tibila, Wonji, and Below Koka Dam stations on average bases. 80% of the probability of exceedance was interpolated on monthly base for those stations and presented in Table 7. The design value was determined from the average of 80% monthly dependable rainfall of 10 years (2003-2012) and found to be 42.5 mm/year.

Irrigation need assessment

Rainfall of 10 years (2003-2012), 80% dependable rainfall was derived and from the 80% dependable rainfall effective rain considered as green water was estimated. Using CROPWAT 8.0 model, net irrigation requirement was generated using climate, soil, crop data and crop pattern as input. The processed rainfall data and crop evapotranspiration data (Table 8) of the study area indicated that the rainfall could not satisfy crop evapotranspiration through the year and supplemental irrigation would be needed.

Missing data, consistency and frequency analysis of stream flow data

In stream flow data analysis, Below Koka Dam, Wonji, Nura Hera

Pit code	Soil parameters							Exchangeable base				
	CaCO3 (%)	E.C, mmhos/cm	O.C, (%)	T.N (%)	C.E.C (Meq/100 gm)	Av. P (ppm)	Esp (%)	Na+	K+	Ca++	Mg++
P1	59	0.04	1.54	0.01	27.12	3.44	1.48	0.27	2.4	9.97	3.46
P2	72.8	0.04	0.33	0.04	24.58	6.38	1.48	0.5	2	12.32	3.33
P3	56.5	0.03	0.57	0.06	27.6	13.18	1.48	0.42	2	10.06	3.13

Table 5: Laboratory results of chemical properties of Soil of the study area.

	Year (months) missed data records
Tibila	2003 (Jan-May), 2010 (Feb-Mar, Jul)
Wonji	2004 (Apr-Jun), 2008 (Aug-Nov)

Table 6: Missing data of rainfall.

Gauging station	Year(months) missed data records
Nura Hera	1975 (Jan-Mar), 1986 (Jan-Feb), 1982 (Jul), 1988 (Feb), 1993 (Aug)
Wonji	1992 (Feb-Mar), 1994 (Jun),
Below Koka Dam	1976 (Mar), 1987 (Jul), 1995 (Aug-Dec)

Table 9: Missing Stream Flow Records.

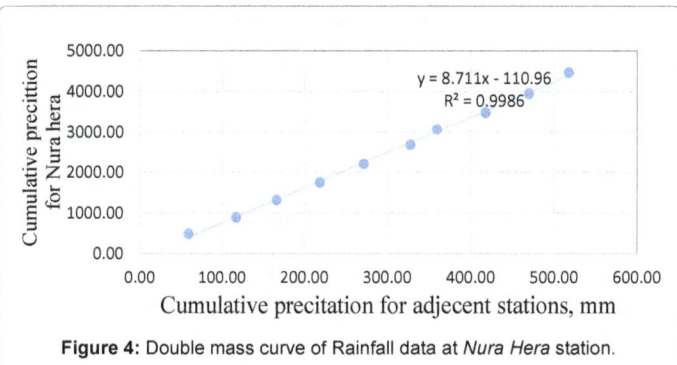

Figure 4: Double mass curve of Rainfall data at *Nura Hera* station.

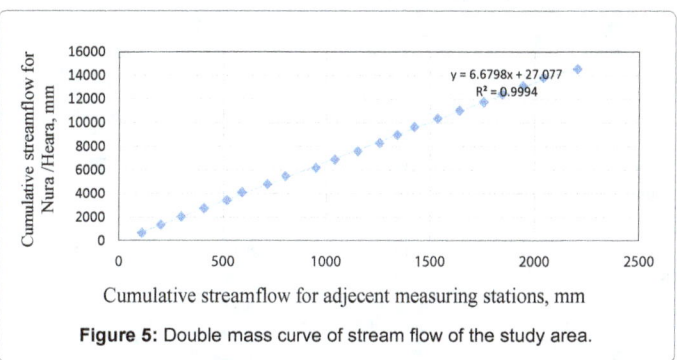

Figure 5: Double mass curve of stream flow of the study area.

Months	Jan	Feb	May	Apr	Ma	Jun	Jul	Aug	Sep	Oct	Nov	Dec
80% RF, mm	4.7	3.1	3.9	3.9	3.9	3.7	3.8	3.8	3.8	3.8	3.9	0.2

Table 7: Summary of monthly frequency analysis of rainfall of stations in the study area.

80% monthly dependable flow, m³/s	Jan	Feb	Mar	Apr	May	Jun	Jul	Aug	Sep	Oct	Nov	Dec
	27.8	28.9	27	29.3	26.6	29.3	41.6	58.7	54.4	33.4	24.7	26

Table 10: Monthly dependable stream flow of Awash river at the study area.

Months	Jan	Feb	May	Apr	Ma	Jun	Jul	Aug	Sep	Oct	Nov	Dec
80% RF	4.7	3.1	3.9	3.9	3.9	3.7	3.8	3.8	3.8	3.8	3.9	0.2
Eff. Rf	4.6	3.1	3.8	3.8	3.8	3.6	3.7	3.7	3.7	3.7	3.8	0.2
Etc,	97.6	47.8	59.4	80.9	137.5	69.5	30.8	60.3	107.6	41.2	77.5	99.1
Deficit	93	44.7	55.6	77.1	133.7	65.9	27.1	56.6	103.9	37.5	73.7	98.9

Note: Dep., dependability, RF, rainfall, Eff. RF, effective rainfall

Table 8: Monthly deficit of rainfall to evapotranspiration in (mm/month).

stream flow gauging stations had missing stream flow data records as presented in Table 9. Therefore to use this data for further application, missing data values were filled.

To check for the consistency of the stream flow data, double mass curve was developed, and its analysis revealed that the stream flow recorded at the stations are consistent with no significant change of slope on their perspective plots as presented in Figure 5. The correlation factor R2 was found close to 1 which implies that strong and direct relationship exists between adjacent stations. This also suggests that like rainfall data analysis that stream flow data of these stations can be used directly for further analysis.

For steam frequency analysis of flow data Weibull plotting formula was used. 80% monthly probability level of flow were found to be 33.97m³/s on average bases and considering downstream water use right of 30%, 23.78 m³/s could be used as design value (Table 10) and Precipitation deficit of the study area was generated by CRPWAT-model (Table 11).

Soil physic-chemical properties and depth

The textures of the soils were classified as loam to silt loam. The

soils were generally described as medium textured with range of bulk density of 0.96 to 1.24 g/cm³. The pH of the soils ranged from 6.75 to 7.39 with sufficient organic matter. The effective depth of soils was in the range of 3 cm to10 cm for existing irrigation. All the selected fields were covered with onion crop. The average lengths of furrows were eight meters with average furrow spacing of 0.6 m. The farmers were applying irrigation water based on their traditional belief (onion gives more yield if watered more). Farmers were applying water regardless of the water requirements of the crop. The average intervals of applying irrigation water for onion on the three fields were 3-4 days.

The average moisture contents (v/v) at field capacity (FC) and permanent wilting points (PWP) were 21.15, 22.0 and 24.2% and 17.17, 18.72 and18.47%, respectively for the three fields located at head, middle and tail of the project (Table 10). The pH was found to be in the range of 6.89 to 7. The pH value of the soil in the experimental site was almost in suitable range for crop production particularly for onion production. The optimum soil pH for crop production is considered to be 6.5 to 7.0 (Prasad and Power, 1997).

Crop water requirement, scheduling and irrigation efficiencies for onion

In calculation of crop water requirement (CWR), irrigation water requirement (IWR) and Scheme supply (SS) of the existing irrigation projects and study area, crop area coverage in percent, soil data (soil class, maximum infiltration rate, and soil water holding capacity) and crop data (growing period, desolation, plant water stress response factor, effective root depth and optionally crop height) and cropping

Prec. deficit	Jan	Feb	Mar	Apr	May	Jun	Jul	Aug	Sep	Oct	Nov	Dec
Onion	102	9	-	-	-	-	-	-	-	-	-	-
Tomato	116	124.1	78.1	-	-	-	-	-	-	-	86.1	110
Pepper	102	8.5	-	-	-	-	-	-	-	-	68.3	89.3
Maize	127	89.4	2.6	-	-	-	-	-	-	-	72.4	108
S/cane	109	141	151.4	155.7	195.7	186.1	108.7	102	133	120.2	31.3	108
Mango	81	96.9	111.1	19.4	175.6	163.4	89.4	91.5	137.3	139.6	39.1	65.1
Orange	55	66.8	60.3	62	98.9	76.8	15.8	29.5	86.5	99.5	112	98.4
Onion	-	-	53.9	104.4	154.9	22.7	-	-	-	-	83	72.8
Tomato	-	-	-	11.7	81.4	114.9	88	91	57.1	-	-	-
Pepper	-	-	-	47.6	142.2	139.6	5.5	-	-	-	-	-
Maize	-	-	-	4.3	127.6	173.6	57	-	-	-	-	-
Onion	-	-	-	-	-	-	8	66.7	135.5	21.8	-	-
Pepper	-	-	-	-	-	-	12.1	17.8	129.5	140.6	-	-
Net scheme irrigation requirement												
mm/day	3	1.7	1.9	2.7	4.5	2.3	1	2	3.6	1.3	2.6	3.2
mm/month	98	48	59.7	81.3	**138.1**	70.1	31.2	60.5	107.7	41.1	77.6	99.3
Irr. A (%)	100	100	98	100	100	100	100	92	92	80	100	100

Table 11: Precipitation deficit of the study area.

Fields	Depth, cm	pH	BD, g/cm³	ECe, dSm	FC, %	PWP, %	Texture
Field H	0-20	7.21	1.06	0.66	22.61	15.82	Loam
	20-40	7.18	1.15	0.74	22.8	18.41	Loam
	40-60	6.89	1.06	0.64	21.05	17.28	Loam
Field M	0-20	6.9	0.98	0.53	21.95	18.73	silt loam
	20-40	6.75	1.01	0.43	21.41	17.97	silt loam
	40-60	6.91	0.96	0.4	22.62	18.73	silt loam
Field T	0-20	7.39	1.24	0.93	23.52	19.57	Loam
	20-40	7.23	1.03	0.67	24.25	18.98	Loam
	40-60	7.21	1.2	0.54	24.81	17.95	Loam

Table 12: Summary of soil data of existing irrigation.

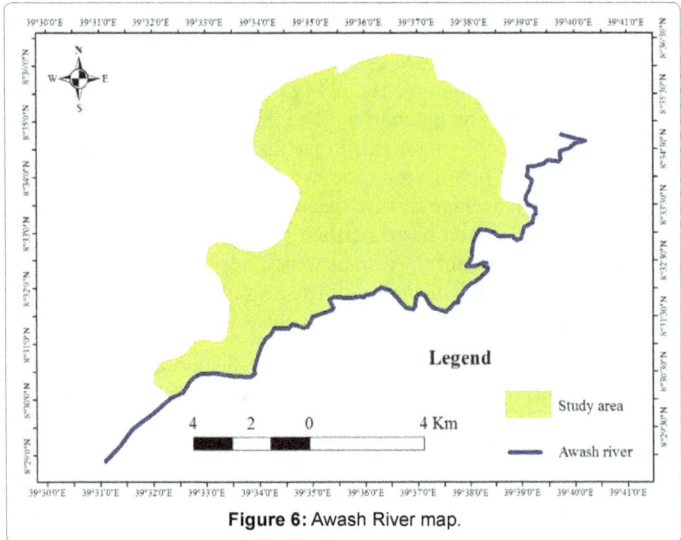

Figure 6: Awash River map.

pattern were used for CROPWAT model. Soil infiltration test was conducted on head and mid location of the irrigation project. Basic infiltration rate was derived from the soil infiltration test data and was found to be 10.9cm/hr. Therefore, the derived value was used as an input for CROPWAT 8.0 model during simulation of crop water requirement, irrigation scheduling and design of scheme supply (Table 12).

Surplus water from steam flow and rainfed farm land

Within the study area, the only source of irrigation water is Awash

river and the map of study area and the Awash river is displayed in Figure 6. The important irrigation suitability factors were used in the simple model developed, calibrated and used in weighted overlay analysis of the project suitability (Figure 7). Therefore, after reclassifying, calibrating the weighted overlay analysis was conducted. The result was displayed in Figure 8. So, the weighted effect the determinant suitability factors are rated S1, S2, S3 and N with respective numerical value of 1087.43, 1234.91, 2292.86 and 2062.79 ha with total irrigable land of 4615 ha. Therefore land that can be brought under irrigation: Irrigable Land=LASR–LUSR–IL where LASR stands for land with acceptable suitability range, LUSR stands for land within unacceptable suitability range and IL, for irrigated land. Therefore irrigable land =6678–2062-162=4,454 ha.

In design of the scheme, data of planting, climate, crop and soil, cropping pattern of that particular area were a key factors. The study area is now under rainfed agriculture. Existing cropping pattern cannot be representative for irrigation need assessment as irrigation water need assessment conducted assuming that the study area was brought under irrigation (Figure 9). It is not doubtful that introduction of the irrigation will change the cropping pattern of the area. Having this fact as an input, cropping pattern at Doni irrigation project was taken as representative for the whole study area.

In the study area, there are generally three cropping seasons, two dry seasons (October-February, March-June) and one wet season (July-September). In assessment of irrigation water requirement, crop type and its dominancy in terms of area coverage was taken into consideration along with all relevant soil and crop data.

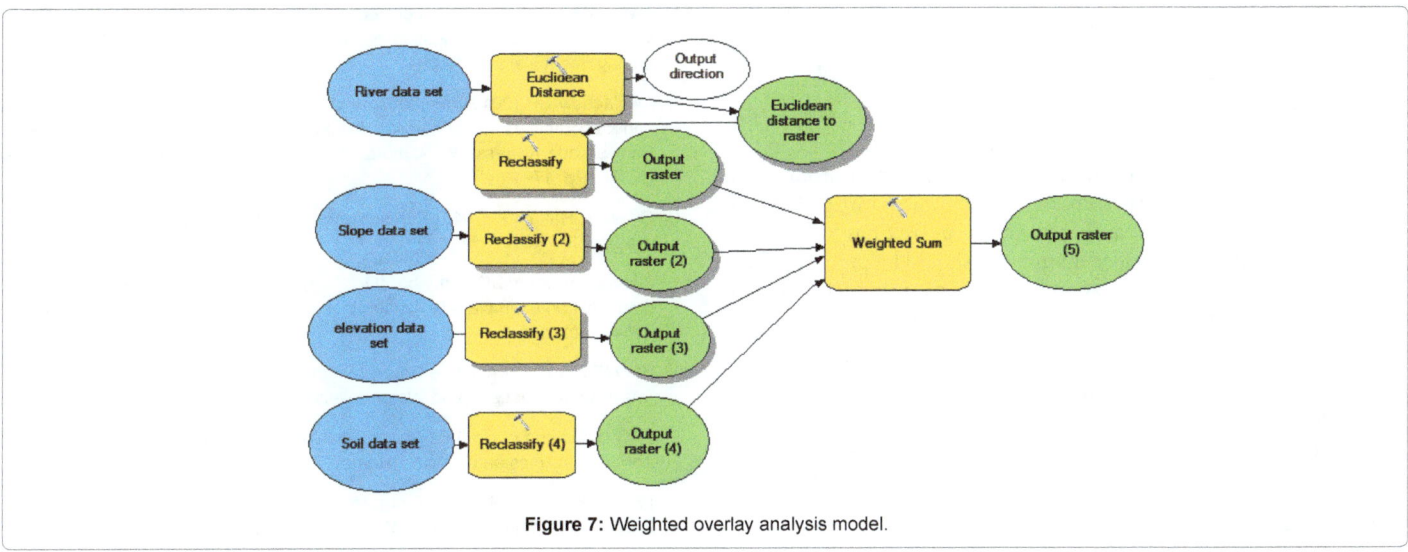

Figure 7: Weighted overlay analysis model.

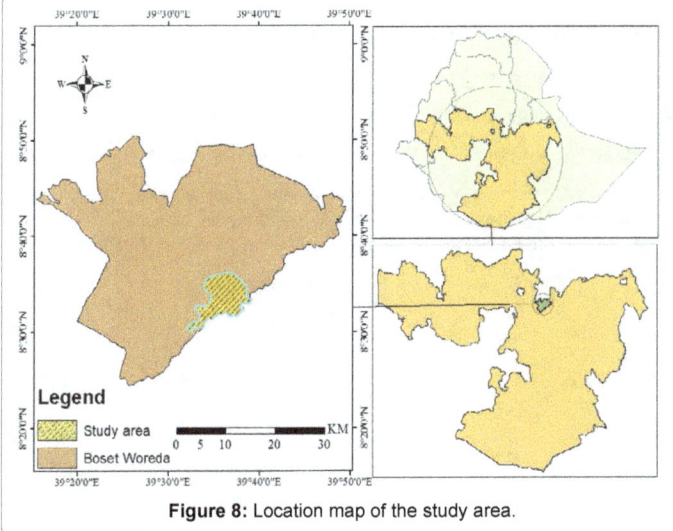

Figure 8: Location map of the study area.

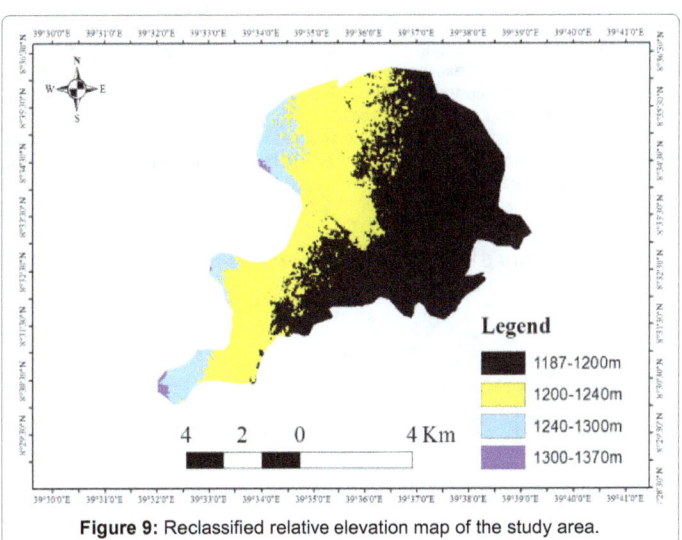

Figure 9: Reclassified relative elevation map of the study area.

Design irrigation scheme

From irrigation land suitability study, 4454 ha was found to be suitable for irrigation in the study area. From characterization of Awash River stream flow through frequency analysis of rainfall and stream flow at 80% probability level was estimated as 42.5 mm/year, 23.78m³/s. From Table 9, with irrigation intensity of 100% irrigation requirement of the study area was found to be 137.5 mm/month in May allowance and 70% for downstream buffer area regime required.

$$Q \text{ (scheme design supply)} = \frac{A \times D}{t}$$

Area (A)=44540000 m², Water depth (D)=0.1375 m Time (t)= month (10 h working time)

=44540000x0.1375/(30×10×3600)=5.6706 m³/s

Water availability is 23.78 m³/s and water need is only 5.6706 m³/s therefore, water supply is in excess of the water need. Therefore, we can that the availability of water from the river flow (23.78 m³/s) is far more than the need of irrigation (5.67 m³/s) of irrigable area (4,454 ha) of the scheme.

Conclusion

In the study area, the finding of this research show that irrigation was absolutely necessary as evapotranspiration was far greater than precipitation even in good year. The investigation on land that could be brought under irrigation was estimated as 4454 ha; however, not necessary for surface irrigation only. Awash River stream analysis near the study area also indicated that water was not the liming factor because the 80% dependable stream of the river was found more than 23.78 m³/s and was far greater than irrigation requirement.

During observation, it was seen that recently working and failed scheme were not design realistically especially drainage land scape was not given due attention.

Recommendations

1. In the light of increasing national economy and environment protection, planning designing and constructing irrigation projects have comprehensive that show available potential and limitation; the irrigation development also has to be future sensitive.

2. Not only investing huge money in construction hydraulic structure; but also, enhancing farmers with basic skill and knowledge through training and follow up is essential.

3. Assigning DA (development agent) and office assistant for the water use association have a paramount importance to the improvement of irrigation projects and used as a mechanism to develop a healthy perception of farmers about irrigation and capacity build to enhance manage the project has to be given due attention. And close monitoring should be practiced than completely left the operation for the farmers. Especially issues like crop water requirements have to be given much emphasis.

4. The last but not is that the result of genuine research works should have some means to be collected, materialized and communicated to the end user. In this respect, concerned bodies (University, Research Institute etc.,) should have coordination and common data base of research to be shared.

Acknowledgements

First of all, I would like to thank the 'Almighty God for his mercy and passionate love on me, and gave me patience, wisdom, and understanding to begin and finish this work successfully. I would gratefully like to recognize Oromia Agricultural Research Institute for awarding me scholarship to pursue my studies. I consider it a profound pleasure to express my deep sense of indebtedness, gratitude and profound thanks to my supervisor, Prof. Shoeb Quraishi, without whom the entire work would not come into existence. I would really like to remember and appreciate the role of my wife Ms. Sonan Fite in my success. Now I realize that source of my strength is your strength and your patience. Finally I would like to acknowledge all authors I used this paper.

References

1. FAO (2005) AQUASTAT-FAO's information system on water and agriculture.

2. Awulachew SB, Yilma AD, Loulseged M, Loiskandl W, Ayana M et al. (2007) Water resources and irrigation development in Ethiopia. International Water Management Institute (IWMI).

3. Awulachew SB (2010) International Water Management Institute Constraints and opportunities for enhancing the system with contributions from: Teklu Erkossa and Regassa E. Namara. International Water Management Institute (IWMI). Pp: 17- 22.

4. Oweis T, Hahum A (2012) Supplemental Irrigation, a highly efficient water use practice. ICARDA.

5. Fasina AS, Awe GO, Aruleba JO (2008) Irrigation suitability evaluation and crop yield an example with Amaranthus cruentus in Southwestern Nigeria. Afr J Plant Sci 2: 61-66.

6. FAO (Food and Agriculture Organization) (1997) Irrigation Potential in Africa: A Basin Approach: Land and Water Bulletin 4. Rome.

7. FAO (Food and Agriculture Organization) (1997) Small-Scale Irrigation for Arid Zones: Principles and Options. Rome.

8. FAO (1999) The future of our land facing the challenge. Guidelines for integrated planning for sustainable management of land resources. Rome.

9. Nemec (1973) Engineering Hydrology. McGraw–Hill publishing company Limited. New Delhi.

10. CSA (Central Statistics Authority) (2009) CSA Statistics. Addis Ababa, Ethiopia.

11. FAO (1976) A framework for land evaluation. FAO Soils Bulletin No. 32. Rome.

12. FAO (2001) Irrigation Water Management: Irrigation Methods. Italy.

13. FAO (1985) Guidelines Land Evaluation for Irrigated Agriculture. Rome. Pp: 290.

14. Staney WC, Bernard Y (1992) Improvement of soil serious for agricultural development: guide line for soil sampling and fertility evaluation. Ministry of natural resource development and environmental protection. Ethiopia.

25

Comparative Performance of Okra (*Abelmoschus esculentus*) Under Subsistence Farming Using Drip and Watering Can Methods of Irrigation

Oshunsanya SO*, Aiyelari EA, Aliku O and Odekanyin RA

Department of Agronomy, University of Ibadan, Nigeria

Abstract

Application of water to crops during the dry season in areas of scare water supply is very important to meet the food demand of the ever-increasing population globally. Thus, modified irrigation technique that poor resource farmers can afford and use easily was evaluated in this paper. A field experiment was conducted at the University of Ibadan Teaching and Research Farm to evaluate the performance of okra under modified bucket drip kit (BDK) and watering-can (WC) methods of irrigation for three growing seasons between 2011 and 2013. The BDK irrigation treatment had a higher mean percentage plant survival (92.9%) than those under the WC method (90.7%). The number of okra leaves, plant height and stem diameter were consistently higher under BDK irrigation than under WC irrigation for the three growing seasons. Harvested number of fresh fruits was only significantly higher under BDK method than WC method by 40.8% and 11.1%, respectively, for the second and third seasons. The plots under BDK irrigation system produced higher fresh fruit weight than the WC method by 0.1, 1.1, 7.4 t ha-1 respectively for the first, second and third seasons. The BDK performed better than WC in terms of okra growth and yields.

Keywords: Soil water; Bucket drip-kit irrigation; Watering-can irrigation; Okra yield

Introduction

Water supply is important for crop growth and production particularly in arid and semi-arid areas. Agriculture utilizes globally about 70% of all the water managed by man, and about 80% of the water used in the developing world [1]. However, the competition among the various sectors-agriculture, communities, industry, nature, etc. becomes stiffer and agriculture is most under pressures for scarce water resources, as the output per unit water is significantly lower than in the other economic sectors. The advent of climate change has brought about considerable uncertainty and curtailment of farming due to frequent drought occurrences and this has had severely negative impacts on the livelihood and food security-especially for rural communities.

In developing countries, rain-fed grain yields are on the average, 1.5 t ha-1 compared with 3.1 t ha-1 for irrigated yields, and increase in production from rain-fed agriculture has mainly originated from land expansion [2]. To identify management options for upgrading rain-fed agriculture, it is essential to assess different types of water stress in food production. Especially important is distinguishing between climate- and human-induced water stresses and between droughts and dry spells. In semi-arid and dry sub-humid agro-ecosystems rainfall variability can generate dry spells (short periods of water stress during critical growth stages) almost every rainy season. However, when rainfall is scarce, supplemental irrigation can increase yields significantly compared with completely rain-fed systems, and in arid regions this increase can be substantial [3].

Based on health impact from wastewater, the World Health Organization (WHO) classified irrigation into three distinct categories: flood and furrow, spray and sprinkler and localized irrigation methods [3]. Flood and furrow irrigation (FI) methods apply water on the surface and pose the highest risks to field workers, especially when protective clothing is not used [4]. Spray and sprinkler are overhead irrigation methods and have the highest potential to transfer pathogens to crop surfaces, as water is applied to edible parts of most crops and because aerosol borne pathogens are carried further. Localized techniques, such as drip-and-trickle irrigation present the lowest risk to farmers and cause minimal pathogens transfer to crop surfaces because water is directly applied to the root [5]. This is one of the best techniques that can be used in applying water to home landscapes, gardens and orchards. However, it is prone to clogging because of the turbidity of polluted water [6]. Martijn and Redwood [7] in their work on assessment of the effects of irrigation methods on crop performance in *Nguruman* irrigation scheme, reported that basin irrigation method had the highest yield (5.6 t ha-1), followed by border strip and furrow, which had 5.0 t ha-1 and 4.2 t ha-1 respectively. Drip irrigation method had the lowest yield of 190 kg ha-1. They stated that the poor performance of drip method was attributed to clogging of the drip nozzles by small particles in the irrigation water.

Recently, cheap bucket drip kit with better potential for use in low income countries are now being developed [8]. This is a controlled, slow application of water to the soil. The water flows under low pressure through plastic pipe or hose laid along each row of plants. Water drips into the soil from tiny holes called orifices which are either precisely formed in the hose wall or in fitting called emitters that are plugged into the hose wall at specified spacing.

Ibragimov et al. [9] conducted an experiment in okra under drip irrigation and reported high yield (4188 kg ha-1) and water use efficiency (8.23 kg ha-1 mm-1). According to the study of, drip irrigation system consistently increased tomato yields (11-80%) due to high water use efficiency [10]. Zotareli [11] reported that the yield obtained for the duration of 15 and 30 minute drip irrigation treatment and basin irrigation was 15.2, 15.1 and 10.8 t ha-1, respectively. The maximum water use efficiency of okra for 15 minute drip irrigation duration was 705.2 kg -1 ha-1 cm-1. The water saving was 60% by adopting drip irrigation compared to basin irrigation.

*Corresponding author: Oshunsanya SO, Department of Agronomy, University of Ibadan, Nigeria, E-mail: soshunsanya@yahoo.com

Okra (*Abelmoschus esculentus* L. Moench) is a fast growing and highly nutritional crop that is commonly grown in the tropics mainly for its pod fruits. It is a good source of calcium, derived from fruits and leaves, 90 mg per 100 g and 70 mg per 100 g respectively. The secondary use is the oil from its seeds, which is about 20% of the seed content. The objective of the study was to compare the growth and yield of okra under drip and watering-can methods of irrigation.

Materials and Methods

Experimental site

The experiment was conducted at the Teaching and Research Farm of the University of Ibadan, Nigeria. Ibadan lies between latitudes 7°25′ to 7°31′ N and longitudes 3°51′ to 3°56′ E. The site has a mean altitude of 180-190 m above sea level. The experiment was a randomized complete block design (RCBD), with bucket drip kit and watering-can methods of irrigation as treatments replicated four times. Growth and yield data were analysed using analysis of variance and simple correlation.

The drip system was laid out to supply water, each having 3 drip lines (laterals), measuring about 35 m. The bucket which served as a reservoir for the drip system was placed 1 m high on a table to allow for flow of water by gravity to the plants through the network of pipe.

Each lateral had 10 slits which allow for supply of drops of water to the plants in rows as presented in Figure 1. Plant spacing of 30 cm within row and 45 cm between rows was used. Soil samples were collected randomly from different spots at 0-30 cm depth bulked together and analyzed before planting to ascertain the base line properties. The bulked sample soil was analysed for physical and chemical parameters such as pH, exchangeable acidity, exchangeable bases, heavy metals, and particle size distribution as described by Vincent et al. [12]. Change in soil moisture was monitored with the aid of tensiometer in the first season. The measurements of okra growth parameters such as number of leaves, plant height, and stem diameter began at 2 weeks after planting to fruition. This was done on a weekly basis. Number of fruits per plant and fresh fruit weight on weekly basis were also determined. Data collected were subjected to analysis of variance (ANOVA) using SPSS 16.0 and where the F-value was found to be significant, the means were separated using LSD.

Results and Discussion

Properties of the soil and irrigation water

The result of the pre-cultivation analysis of the soil of the

experimental plot is presented in Table 1. The soil has an organic carbon content of 17.86 g kg⁻¹ and total nitrogen content 0.85 g kg⁻¹. It is a slightly acidic soil with a pH of 6.20. Available phosphorus, manganese and zinc are in adequate amounts of 7.81 mg kg⁻¹, 98.30 mg kg⁻¹, and 10.23 mg kg⁻¹ respectively with a sandy loam texture. The result of the water analysis is presented in Table 2. This showed that the water used for irrigation belonged to Class 1, excellent, non-saline water because it has EC of < 5 dS/m and TDS of < 500 mg/l [2]. The average soil water retained under BDK irrigated plot was higher than under WC irrigated plot by 10.3%, indicating that soil under drip irrigation absorbed and retained more water for crops to use than soil irrigated with watering can. Figure 5 shows the mean water use efficiency for drip irrigation

Parameter	Content (2011)	Content (2012)	Critical values (FAO 2012)
pH (water)	6.2	6.5	6.5-7.5
Organic Carbon (g kg⁻¹)	17.86	4.0	10-15
N (g kg⁻¹)	0.85	1.0	1.0-1.5
P (mg kg⁻¹)	7.81	12	10-20
Exchangeable Bases (cmol kg⁻¹)			
Ca	1.41	0.9	2.0-5.0
Mg	0.7	0.9	0.3-1.0
Na	0.8	0.6	0.6
K	0.71	0.4	0.15-0.3
Exchangeable Acidity (cmol kg⁻¹)	0.2	0.2	0.1-0.5
Exchangeable micronutrients (mg kg⁻¹)			
Mn	98.3	24.8	20-25
Fe	36.3	26.8	20-25
Cu	1.69	2.0	1.2-2.0
Zn	10.23	10.0	1.0-5.0
Sand (g kg⁻¹)	792	800	850
Clay (g kg⁻¹)	128	130	50
Silt (g kg⁻¹)	80	70	100
Textural class	Sandy loam	Sandy loam	

Table 1: Soil physical and chemical properties of experimental site prior to sowing.

Elements	Quantity	Critical value (FAO, 2012)
Ca (mg/l)	40.9	60.0
Mg (mg/l)	8.7	20.0
K (mg/l)	13.7	20.0
Na (mg/l)	43.0	70.0
Mn (mg/l)	0.2	0.2
Fe (mg/l)	0.1	0.3
Cu (mg/l)	0.003	0.3
Zn (mg/l)	0.2	2.0
Cd (mg/l)	0.01	0.01
Co (mg/l)	0.1	0.05
Cr (mg/l)	0.0	0.1
Pb (mg/l)	0.5	5.0
Ni (mg/l)	0.2	0.2
pH	7.6	6.5-8.4
EC (dS/m)	0.3	5.0
HCO₃⁻ (mg/l)	26.8	60.0
SO₄ (mg/l)	0.74	100.0
Cl (mg/l)	73.6	70.0
SAR	8.6	10.0
TSS (mg/l)	214.6	500.0
TDS (mg/l)	211.2	200-4000

Table 2: Irrigation water analysis.

Figure 1: Components of a drip system showing the reservoir, the supply tube, the header tape and the laterals line beside the okra on the field.

across the three planting seasons to be 0.49 kg/l while that of watering can was 0.32 kg/l.

Seed germination

The higher rate of seed germination was recorded under drip irrigation for the first two growing seasons as presented in Table 3. Although, there was no significant difference in the germination percentage, drip irrigated plots were higher in seed germination percentage than watering can irrigated plots by 2.20% and 4.40% respectively for first and second growing seasons. This was most probably because the amount of water flowing from the drip nozzle was completely utilized by the sown seeds without any waste as a result of direct discharge of water at the planted spot in form of droplets for easy absorption by the soil.

Crop performance

The mean plant height of okra under drip and watering can irrigation for the three growing seasons is presented in Figure 2. In the first season, okra plant height obtained from drip irrigated plots was significantly (P = 0.05) higher than plants irrigated with watering can from 4WAS to 8WAS. On the average, drip irrigated okra plants were higher than watering can irrigated plants by 28.8 cm in the first season. The same trend was followed for the remaining two growing seasons. The result justified the fact that the water supplied was available for okra plants under drip irrigation to utilise maximally since water was discharged directly to the base of the plants at low pressure. Also, drip irrigation probably kept the root zone at the field capacity; therefore there was no shortage of water for the physiological functions of the plant. IITA reported that the heights of okra under the surface and drip were 87 and 90 cm, respectively [13]. They attributed this to availability of water for the physiological functions of the plants under drip irrigation system. The mean stem girth and number of leaves for various weeks were consistently higher under the drip method than watering can method of irrigation for three consecutive growing

Treatment	Soil moisture potential (bars)	% germination		
	Season 1 (2011)	Season 1	Season 2	Season 3
Drip	0.43a	88.90	94.4a	95.55a
Wc	0.35a	86.70a	90a	95.55a

Means with the same alphabet(s) in the same column are not significantly different at P = 5%

Table 3: Effects of bucket drip kit and watering can methods of irrigation on soil moisture and percentage of plant survival.

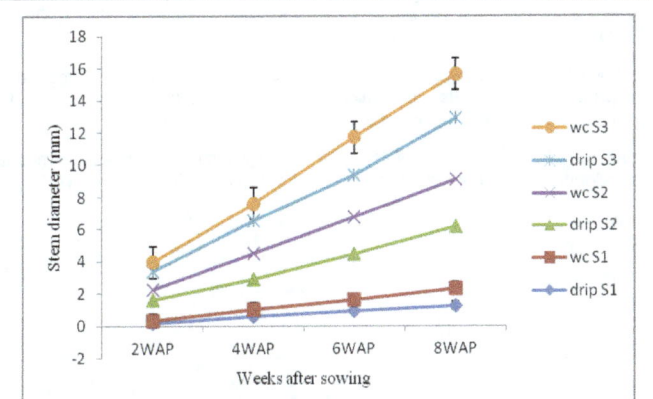

S1: Season 1, S2: Season 2, S3: Season 3, WC: Watering can, Drip: Drip kit
Figure 2: Okra stem diameter as influenced by drip and watering can irrigation methods for three growing seasons.

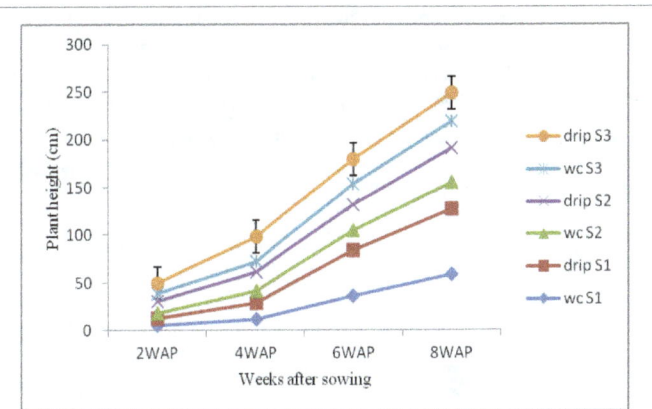

S1: Season 1, S2: Season 2, S3: Season 3, WC: Watering can, Drip: Drip kit
Figure 3: Okra plant height as influenced by drip and watering can irrigation methods for three growing seasons.

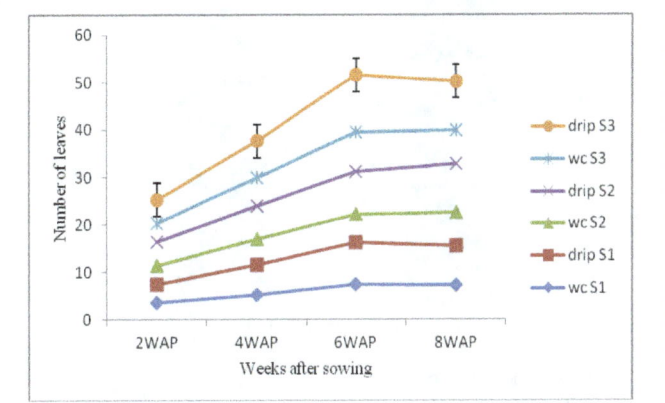

S1: Season 1, S2: Season 2, S3: Season 3, WC: Watering can, Drip: Drip kit
Figure 4: Okra number of leaves as influenced by drip and watering can irrigation methods for three growing seasons.

seasons as shown in Figures 3 and 4 respectively. Averagely, the number of okra leaves obtained from drip irrigated plots was significantly (p = 0.05) higher than those from plots irrigated with watering can method by 14.2%, 28.7% and 27.6% respectively for the first, second and third seasons. This implies that number of leaves available for photosynthesis under drip irrigation were higher than watering can irrigated plants, which could determine to a large extent the assimilates for growth and yield. Similar results were reported by Singh and Rajput [14] that plants grown under drip irrigation had more number of branches and plant heights compared to that of surface irrigated plants. They stated that the above ground matter that included stem diameter had positive significant correlation with yield under drip irrigation (0.99**). It was concluded that the carbon dioxide exchange rates varied considerably under the different methods of irrigation due to difference in irrigation timings and quantity of water applied. On the average basis, the stem diameter of okra plants from drip irrigated plots was consecutively and significantly higher than those of the plants irrigated with watering can by 12.1%, 12.8% and 16.8% respectively for the first, second and third growing seasons of 2011 and 2012. Although, there was no difference between the two methods of irrigation in terms of number of okra fruits in the first season of year 2011, however significant higher number of okra fruits was obtained from drip irrigated plots compared with plots irrigated with watering can in the subsequent seasons (Table 4). In the second and third growing seasons, drip irrigated plots had higher number of fresh okra fruits than plots irrigated with watering can by

Treatment	Number of fruits/ha			Fresh fruit weight (t ha⁻¹)		
	Season 1	Season 2	Season 3	Season 1	Season 2	Season 3
Drip	518,518a	666,666a	888,888a	0.89a	3.03a	12.51a
Wc	518,518a	394,814b	790,369b	0.78a	1.98b	5.1b

Means with the same alphabet(s) are not significantly different. P = 5%

Table 4: Okra yield as influenced by drip and watering can irrigation methods for three growing seasons (2011-2012).

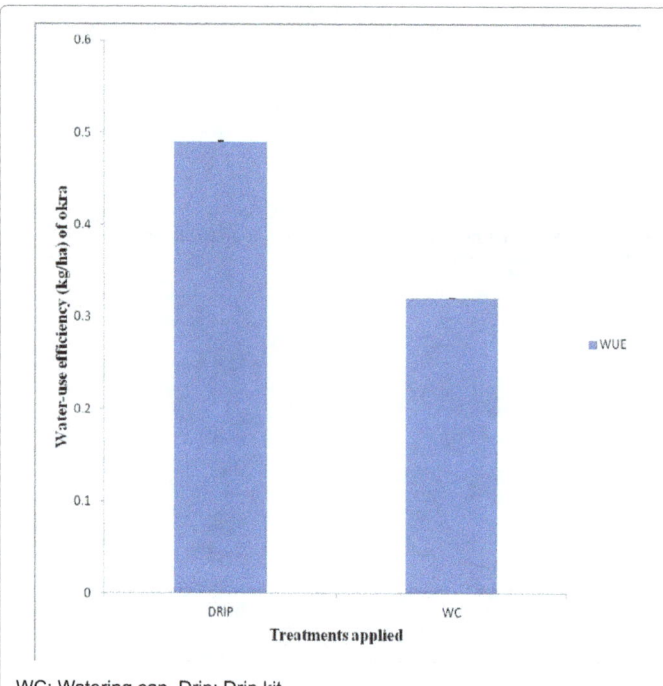

WC: Watering can, Drip: Drip kit

Figure 5: Mean water-use efficiency of okra across three planting seasons.

271, 852 and 98, 519 fresh fruits per hectare, respectively. The total fresh fruit weight obtained from drip was higher than that of watering-can irrigated plots for the three growing seasons as presented in Table 4. Higher significant fresh okra fruit weight was obtained under drip irrigation than watering can method in the second and third season by 1.1 t ha⁻¹ and 7.4 t ha⁻¹, respectively. This is because drip irrigation provides a consistent supply of water to the entire root area on a continuous basis so that "drench and dry-out" stresses are reduced. This is in agreement with the findings of Anthony and Singandhupe [15], who reported that yield increase for okra under BDK irrigation by 20.69% with the water saving of 44.92%. This result also supports Saxena and Gupta [16], who reported that plants receiving drip irrigation had significantly higher yield than furrow and basin for okra.

Conclusion

The findings obtained from this work showed that plant growth parameters and yield were greater with bucket drip kit irrigation as compared to the watering-can method. The difference observed between plant growth parameters under both methods of irrigation indicated the greater water use efficiency by bucket drip kit irrigation over the watering can, as drip irrigation delivers water close to the plant or only to the soil-plant vicinity. The drip system also has the tendency to make large quantity of water available to the plants gradually such that there will be no runoff and deep percolation. The effect is higher percentage germination and plant survival as compared to the watering-can method. Although bucket drip kit irrigation can be

tedious in its installation as compared to the watering can, it saves time, energy, labour and water during the process of water supply to plants after the drip lines have been laid out. The drip kit irrigation is highly affordable for subsistence farming for sustainability of livelihood.

Acknowledgement

The authors appreciate the contributions of Drip Irrigation for Third World Kitchen Gardens headed by Richard D. Chapin (Executive Director, Chapin Living Waters) and William A. Chapin (President, Chapin Watermatics Inc.) for supplying the bucket drip kit materials used for this study at no cost.

References

1. Prinz D (2002) Global and european water challenges in the 21st Century. Regional Conference on Environment-Water, Hungary.

2. Fipps G, Smith E (2012) Irrigation water quality standards and salinity management strategies. Texas A&M Agri Life Extension.

3. Rosegrant MC, Ximing SC, Nakagawa N (2002) The role of rain-fed agriculture in the future of global food production. International Food Policy Research Institute.

4. WHO (2006) Guidelines of the safe use of wastewater excreta and grey water, wastewater use in agriculture.

5. Blumenthal UJ, Deasey A, Ruiz-Palacios G, Mara DD (2000) Guidelines for waste water reuse in agriculture and aquaculture: recommended revisions based on new research evidence. WELL study.

6. Pescod MB (1992) Wastewater treatment and reuse in agriculture. Food and Agriculture Organization of the United Nations, Rome.

7. Martijn E, Redwood M (2005) Wastewater irrigation in developing countries-limitations for farmers to adopt appropriate practices. Irrigation and Drainage 54: 63-70.

8. Muya EM, Gachini GN, Maingi PM, Waruru BK (2007) Assessment of the effects of irrigation methods on crop performance in Nguruman Irrigation Scheme, Kajiado District.

9. Ibragimov N, Evett SR, Esanbekov Y, Kamilov BS, Mirzaev L (2007). Water use efficiency of irrigated cotton in Uzbekistan under drip and furrow irrigation. Agric Water Management 90: 112-120.

10. Rekha KB, Reddy GM, Mahavishnan K (2005) Nitrogen and water use efficiency of okra as influenced by drip fertigation. J Trop Agric 43: 43-46.

11. Zotarelli L, Dukes MD, Scholberg JMS, Mun˜oz-Carpena R, Icerman J (2009) Tomato nitrogen accumulation and fertilizer use efficiency on a sandy soil, as affected by nitrogen rate and irrigation scheduling. Agricultural Water Management 96: 1247-1258.

12. Vincent BO, Michael TM, Stanley MM (2005) Effect of cattle manure application on yield and yield indices of okra (Abelmoschus esculentus L. Moench). J FOOD AGRIC ENVIRON 3: 125-129.

13. IITA (1999) Selected methods for soil and plant analysis.

14. Singh DK, Rajput TBS (2007) Response of lateral placement depths of subsurface drip irrigation on okra (Abelmoschus esculentus). INT J PLANT PROD 1: 73-84.

15. Antony E, Singandhupe RB (2003) Impact of drip and surface irrigation on growth, yield and water use efficiency of capsicum (Capsicum annum L.). Agicultural Water Management 65: 121-132.

16. Saxena NP, Gupta S (2004) Performance evaluation of okra (Abelmoshus esculentus) under drip irrigation system. Asian J Agric Res 4: 139-147.

26

Estimation of Yield Response (Ky) and Validation of CropWat for Tomato under Different Irrigation Regimes

Etissa E[1]*, Dechassa N[2] and Alemayehu Y[2]

[1]*Ethiopian Agriculture Research Organization, Addis Ababa, Ethiopia*
[2]*Haramaya University, Ethiopia*

Abstract

Field experiment was conducted at Melkassa Agricultural Research Center with the objectives to determine the optimal irrigation levels for maximum tomato production and to assess the effect of limited water supply on field grown tomato yield and to estimate 'yield response of tomato to soil water (Ky)' and to validate CropWat irrigation model using the data for tomato cultivation during hot-dry season conditions. Three irrigation scheduling levels such as 1) 100% of crop water requirement (ETc) (Full irrigation) 2) 80% ETc (Full) (= 0.80 ETc) and finally 3) 60% ETc (= 0.60 ETc) were used using drip irrigation replicated three times; the tomato was subjected to various levels of water stresses over whole growth period. Yield data such as marketable, unmarketable and total fruit yield were collected at each harvesting and summed at the end of harvesting. The results of data analysis showed that use of various irrigation depth brought a significant effect (P<0.01) effect on the marketable yield of tomato whereas application of various irrigation depths did not bring significant difference (P<0.05) on unmarketable fruit yield of tomato. Use of various irrigation depths had a significant effect (P<0.05) on the total fruit yield of tomato. The mean separation indicated that the highest fresh fruit yield was obtained from full irrigation and the lowest was obtained from 60% irrigation. Thus, the total fresh fruit yield obtained from fully irrigated tomato plot exceeded the fresh fruit yield obtained from tomato plot irrigated with only 60% of full irrigation water by 62.8%. The results showed that with decrease in the depth of irrigation, there was a decrease in total fruit yield in tomato due to reduced uptake of water. The yield response (Ky) of tomato throughout the crop cycle was calculated and found to be 0.999, indicating that the yield reduction is directly proportional to reduced water use. Then the CropWat irrigation model was validated using field data for tomato cultivation. Accordingly, the efficiency of the model was found to be 94%, indicating that the model is a useful decision support system to help tomato growers.

Keywords: CropWat; EToCal; Irrigation regime; Simulations; Yield response

Introduction

The Central Rift Valley (CRV) area of Ethiopia is amongst the pioneers of market-oriented irrigated vegetable crops production in Ethiopia. Using various water sources for irrigation; vegetable production in this area has nowadays expanded where most growers use hybrid seeds and considerable agricultural inputs.

Agriculture in this area is dominated by traditional small scale irrigation at household level with very small farm size [1,2]. Thus, improving small scale irrigated vegetable production system is expected to improve livelihoods and sustain the environment. Demeke and Haile found that vegetable crops growers that have access to small scale irrigation has an important impact on poverty reduction through high income, and improved wellbeing of farming households.

In all parts of Ethiopia, tomato is produced under furrow irrigation in open field. Based on survey conducted by Etissa et al. [2] among the vegetable grower using furrow irrigation, 16.48% replied that the knowledge source of their irrigation management packages was from experience, while 12.08% replied that the knowledge source was obtained from experience and family and all the remaining replied different sources; the survey indicated that vegetable growers got knowledge and practices from variety of sources showing furrow irrigation is totally is not technical and scientific based. Small holder farmers did not indicate that their irrigation scheduling is supported by improved irrigation technologies in the country.

In addition, because of profitability of vegetable crops production using irrigation on one hand and the current low production and productivity of existing vegetable crops and farm lands in the study area on the other hand, 86.31% of growers responded that they have interests in increasing their irrigable farm land area to expand and intensify vegetable production [1,2]. However, due to the expansion of irrigated areas and uncontrolled irrigation water in the upstream of Central Rift Valley, all the downstream of middle and lower Awash Basin, there is not only limited availability of irrigation water, but also critical water shortage, there is a need for optimal irrigation management and scheduling in order to maximize crop yields under water deficit conditions so that efficient use of water for agriculture is increased.

Among irrigation systems, many losses encounter surface and furrow irrigation, like conveyance loss, surface run off, deep percolation etc... From very limited water sources compared to crop water requirements. It is economically necessary to get even more from the water: this may be done in many cases by adopting efficient irrigation methods through improved efficiency, which can apply the scarce water more accurately; minimizing losses through different ways. The water then can be used much more efficiently for supplemental irrigation for much larger areas, or for longer seasons. The experience from many countries show that farmers who changed from furrow system to drip systems can cut their water use by 30% to 60% and crop yields often increase at the same time [3]. The use of drip irrigation system permits reduction of water losses up to 50%. Hochmuth and Hanlon [4] and can increase the yield per unit of land by up to 100% compared with surface irrigation systems [5].

*Corresponding author: Etissa E, Ethiopian Agriculture Research Organization, Addis Ababa, Ethiopia, E-mail: edossa.etissa@gmail.com

In several places in Ethiopia, there are extensive campaigns of water harvesting, tapping ground water and using appropriate technologies like treadle pump, rope and washer pumps with the realization that in many places existing water resources cannot meet the needs of the expanding population. Hence, it is very crucial to assess effect of irrigation levels for maximum tomato production and to assess the effect of limited water supply on tomato growth and yield. The objectives of this study are to determine the optimal irrigation levels for tomato production and to assess the effect of limited water supply on tomato yield; to estimate 'yield response of tomato to soil water (Ky)' and finally to validate CropWat irrigation model using the data for tomato cultivation for Melkassa during hot- dry season.

Materials and Methods

The experiment was conducted at Melkassa Agricultural Research Centre during the hot- and dry season. There was no rainfall since tomato planting to final harvesting during the experimental period. The detail of materials and methods were published in African Journal of Agricultural Research, by Edossa et al. [6].

Treatment arrangement, experimental materials and procedures

Treatment arrangement: Irrigation scheduling treatments include 1) 100% of crop water requirement (ETc) (Full irrigation), 2) 80% ETc (Full) (=0.80 ETc), 3) 60% ETc (=0.60 ETc). 'Melkasholla' semi-determinate tomato variety was subjected to various levels of irrigation levels (water stresses) over whole growth period. The plots were replicated three times.

Experimental procedures: Melkasholla tomato variety was used for field experiment; it is a multipurpose variety released from Melkassa ARC; it is semi-determinate growth habit. The detail of the procedures was published by Edossa et al. [6]. Tomato seeds were sown in a nursery in a row with the row spacing of 10 cm with very dense spacing within rows. The size of the seedbed was 5 m length and 1 m width. The seed was drilled onto the seedbeds and covered with a soil layer of 1/5 cm. 100 g Urea and 200 g DAP were applied per bed and thoroughly mixed with the soil as recommended by Lemma. Watering was done in the interval of three days throughout the growth period of the seedlings in the nursery for both experiments. Field preparation consisted of ploughing by a mould board plough into the depth of 40–50 cm deep followed by 10 to 15 cm deeper thorough operation of disc harrowing before ridging. Plots with the individual size of 7.0 × 4.5 m, total of 31.5 m², with seven rows, and each row accommodating 15 plants was marked out. The spacing between rows was 100 cm and 30 cm between plants. A total of 61 plants and 44 boarder plants were transplanted. Seedlings were transplanted to the permanent experimental field as recommended by researchers. Pre-plant irrigation was applied, since past rainfall was insufficient to replenish the soil profile [7]. Seedlings were transplanted in field at the usual spacing. A total of 60 experimental plants were planted within each plot and before initiating treatments, plants (seedlings after transplant) were irrigated to nearly field capacity for three weeks in order to improve root development [8].

Irrigation system descriptions:

Low-cost gravitational drip structures and installations: The low-cost gravitational surface drip structure used for the experiment comprised water source tanker at the elevated position, filter, water tank connector, straight connector, connector, control valve, main line, lateral pipe, emitter, wood and nail for tanker stand. Four tankers having the capacity of 2000 litres each were placed at the irrigation

regime at the head of strip plot. The tankers were placed in the field at the height of 1.0 m from above the ground so that water would be at the height necessary to provide the water pressure required operating the system. Once the seedlings were well established for 20 days, the irrigation treatments were commenced.

Each plot consisted of lateral drip lines with 5.5 m length. The emitters on laterals were spaced at 0.3 m corresponding distance of tomato plant spacing within a row in the field. The lateral line was laid out along each tomato row. Each tomato plants were planted under emitter so that they would benefit from the water supplied by the emitters. The field was furrow-irrigated before planting and after transplanting for ten days for crop establishment before imposing drought stress treatments.

Three and half meter distance buffer zone separate each plots or side flows were precluded to avoid lateral run-on and run-off (side flows) from other irrigation treatment plots.

Methods for estimation of soil water

Estimation of daily crop water requirement: The initial soil water content for top soil at the time of transplanting is assumed to be close to field capacity as a result of continuous pre-irrigation. This assumption is dictated by the fact that small vegetable seedlings are extremely very sensitive to moisture stress. Then the proper amount of daily irrigation for a crop is the amount of daily ET taking place minus any daily effective rain fall [9].

Application of daily time step irrigation scheduling: Equal amount of irrigation water were applied to each treatment before the initiation of irrigation treatments (sum of daily ETc). Once the drip system was installed, the drip irrigation was done on the basis of ETo [10] value of the previous day. The amount of irrigation water applied, ETm, was determined from the calculated water requirement for tomato as determined from the crop coefficient (Kc) and the daily reference evapotranspiration (ETo) using the following equation:

ETc=ETo * Kc

Irrigation scheduling was based on a check book of soil water balance budget method (ETc=ETo*Kc) where simple accounting approach is used for estimating how much soil-water remains in the effective root zone based on water inputs and outputs. Irrigation was scheduled when the soil-water content in the effective root zone was near the predetermined allowable depletion volume through keeping track of rainfall, evapotranspiration, and irrigation amounts. Irrigation treatments were applied once a day until the required volume of water was completely gone from the tanker. The total amount of irrigation water applied to each treatment was calculated as the sum of water applied during the crop establishment period and the ETc of the remaining period.

Daily reference ETo: The daily ETo data were calculated with the software programme EToCalc developed by Raes [11] on basis of the FAO Penman Monteith equation from Melkassa Weather Station [10].

Net irrigation (IR_n): It is the amount of irrigation water required to bring the soil moisture level in the effective root zone to field capacity [12].

IR_n=ETc - P_e + LR (mm)

Where,

IR_n=Net irrigation requirement (mm)

ETc=Crop evapotranspiration (mm)

P_{ef}=Effective dependable rainfall (mm)

G_e=Groundwater contribution from water table (mm)

W_b=Water stored in the soil at the beginning of each period (mm)

D=Deep percolation/drainage (mm)

LR=Leaching requirement (mm)

Again if the estimated LR is found be less than 10%, it is ignored from the equation.

As a rule, under drip irrigation conditions of high water tables are rare and as a result groundwater contribution to crop water requirements is normally ignored. Similarly deep percolation was assumed to be zero. If assuming that W_b, G_e and D are zero, then the equation becomes:

IR_n=ETc - P_e + LR (mm)

Again if the estimated LR is found be less than 10%, it is ignored from the equation.

Gross irrigation: Gross irrigation requirement is net irrigation requirement plus losses in water application and other losses [12]. This is expressed in terms of overall efficiencies when calculating gross irrigation requirements from net irrigation requirements:

$$IR_g = \frac{IR_n}{E} + LR$$

Where,

IR_g=Gross irrigation requirements (mm),

IR_n=Net irrigation requirements (mm),

E=Field efficiency of the system (drip system assumed to be 85% [10]

Daily irrigation, the amount of water was adjusted according to existing reference *ET* and *Kc*. The irrigation treatments were differentiated by their two meters arrangement for strip, irrigation events were controlled manually by using valve. The valve was put on and off after calculating net irrigation and adding losses (gross) depending on amount of water to be applied at desired level for each strip separately. Records of daily applied water were kept from the start of treatment application up to the final harvest date for each treatment. The records daily applied water was then summed up for each treatment.

Adjustments for Kc for development and late stage and for partial wetting: The values of Kc of tomato used (0.6, 1.15 and 0.80 respectively, in the initial, mid and late season stages) Allen et al. [9]. During the initial and mid-season stages Kc is constant and equal to the Kc value of the growth stage under consideration (*Anon.*); these growth stage represent 25 days for the initial, 34 days for the development, 20 days for mid and 41 days for the late growing stages totalizing 120 days as recommended by Allen et al. [9].

The daily Kc for developmental and late season stages was adjusted using the formula given by Allen et al. [9]. During the crop development and late season stages, Kc varied linearly between the Kc at the end of the previous stage (Kc prev) and the Kc at the beginning of the next stage (Kc next), which is Kc end in the case of the late season stage. The partial wetting for wetting patterns of the drip emitters was measured from sample drippers and adjusted to 0.3 ratios.

Data collection

All yield data such as marketable fruit yield, unmarketable fruit yield were measured at each harvesting and summed up at end of the experiment and the total fruit yield was obtained by adding all fruit yields.

Estimation and quantifying crop water use

Tomato yield response (*Ky*): Water productivity behaviour of 'Tomato variety '*Melkasholla*' and its yield response to water' (*Ky*) was estimated through the following relationship described by Doorebos et al. [13].

$$(1 - \frac{Y_a}{Y_m}) = Ky(1 - \frac{ET_a}{ET_m})$$

Where,

Y_m=Maximum yield (kg)

Y_a=Actual yield (kg)

ET_m=Maximum evapotranspiration (mm/period)

ET_a=Actual evapotranspiration (mm/period)

All data analyse and methods of testing were very similar to the one described in Chapter 5. Data from this experiment were subjected to analyse of variance as strip plot design using linear). Where ever the treatments were significant means were separated using the LSD test at *P*=0.05 probability significance level.

Validation of CropWat

With the help of the CropWat model, the yield reduction will be determined and compared with the actual measured yield reduction of field experimentation using drip experiment. The yield reductions will be expressed as percentage of the tomato yield obtained under full irrigation [12-14].

Results and Discussions

Fruit yields

Use of various irrigation depths brought a significant (P<0.01) effect on the marketable yield of tomato whereas application of various irrigation depths did not bring significant difference (P<0.05) on unmarketable fruit yield of tomato (Table 1). Use of various irrigation depths had a significant (P<0.05) on the total fruit yield of tomato.

The mean separation indicated that the highest fresh fruit yield was obtained from full irrigation and the lowest was obtained from 60% irrigation water with saving of 40% of irrigation water (Table 2). Thus, the total fresh fruit yield obtained from fully irrigated tomato plot exceeded the fresh fruit yield obtained from tomato plot irrigated with only 60% of full irrigation water by 62.8% [15-18]. The results showed that with decrease in the depth of irrigation, there was a decrease in total fruit yield in tomato due to reduced uptake of water (Table 2). The result of this study corroborate that of Muchovej et al. [19] who reported that high quality and yield of vegetable crops are directly associated with proper water management. Birhanu and Katema also found that the fresh fruit yields of *Melkasholla* variety was reduced under deficit irrigation level. Similar findings were reported by Kirnak et al. [8] where egg plants grown under high water stress had less fruit yield and quality than those in the control treatment. Consistent with the results of this study also found that water stress in

Sources of variations	df	Mean square value		
		Marketable fruit yield	Unmarketable fruit yield	Total fruit yield
Irrigation	2	55159.9**	861.09 NS	4397.91*
Error	4	917.8	339.72	315
Total	44			
CV		22.94	28	8.92
Note NS=Indicates non-significant at P<0.05; *significant at P<0.05 and **significant at P<0.01 probability levels, respectively				

Table 1: Mean square values of vegetative growth yield and yield components parameters of tomato as influenced by integrated nutrient managements and application of various moisture regimes.

Irrigation regimes	Marketable fruit	Unmarketable yield	Total fruit yield
	(t ha⁻¹)	(t ha⁻¹)	(t ha⁻¹)
IR I (100% ETc) (Full irrigation)	63.63 A	18.267	81.902 A
IR II (80% ETc)	33.83 B	22.413	56.250 B
IR III (60 % ETc)	27.82 B	23.062	50.868 C
Mean	41.765	20.813	62.916
LSD (0.05)	9.712	NS	5.689
*=Average of three replications. Means within each column with different letters are significantly different at LSD at P=0.05 level of probability			

Table 2: Mean values of various irrigation regimes on fruit yield of tomato grown under drip irrigated condition.

the container grown eggplants produced a very significant reduction in both dry biomass, they found that eggplant fruit yield was reduced by up to 68% in the water stressed plants compared with unstressed plants. Studento et al. [7] also reported that restricted water supply for tomato can suppress new leaf development, resulting in a shortened yield formation period. Similar findings were reported by that water stress significantly reduced final yield of field-grown sweet pepper. Similar findings were obtained where increasing irrigation increased total tomato fruit yield.

Irrigation positively influenced tomato productivity; the result was attributed to the increase in the number of berries per plant and the fruit average weight as irrigation increased. The authors concluded that the total yield and marketable tomato yields were decreased significantly as the deficit level was increased. The reduction in total yield of tomato with an increased amount of water stress level of this test was consistent with previous work conducted on tomato and other crops such as cotton as reported [20].

Irrigation positively influenced tomato productivity; the result was attributed to the increase in the number of berries per plant and the fruit average weight as irrigation increased. The authors concluded that the total yield and marketable tomato yields were decreased significantly as the deficit level was increased. The reduction in total yield of tomato with an increased amount of water stress level of this test was consistent with previous work conducted on tomato and other crops such as cotton as reported [21].

Water production function of tomato under various irrigation scenarios

The relationship between yield and irrigation water applied was sketched in Figure 1. Based on the relationship tested, about 92% of the variation in fresh fruit yield was brought about by irrigation regime treatments (Figure 1). Thus, as irrigation depth increased, total fruit yield increased linearly.

The relationship between yield and irrigation water supplied could be expressed by a linear relationship very well as: *Fresh tomato fruit yield =28.95x -2811, with R²=0.918*; with a slope of about 28.9:1 in terms of reduced applied water: gross kg yield reduction. Bazza conducted an experiment for sugar beet concluded that more than 90% of the yield variation was coming from the variability in depth of irrigation applications.

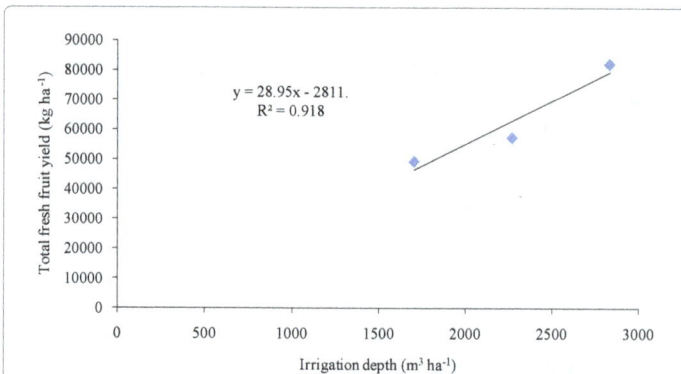

Figure 1: Yield-water relationship (water production function) of drip irrigated tomato *Melkasholla* variety grown under dry and hot season at Melkassa.

Estimation of yield response (Ky)

Relationship between relative yield decrease (1-Ya/Ym) and relative evapotranspiration (1-ETa/ETm) of tomato at Melkassa, yield response to water' (Ky) was determined through the functional relationship described by Doorebos et al. [14]. Thus the yield response (ky) of tomato *Melkaskola* variety at Melkassa was calculated and estimated to be 0.9998 a little bit lower than given by Allen et al. [9] which was 1.05 value (Figure 2).

Although tomato is relatively moderately sensitive crop, and the Ky is estimated to be 1.05 Allen et al. [9] many authors such as Getta, Giardini and Giovanardi found variable value of Ky. The relationship between relative yield decrease (1-Ya/Ym) and relative evapotranspiration (1-ETa/ETm) of tomato at Melkassa was determined through the functional relationship. Thus the yield response (ky) of tomato *Melkaskola* variety throughout the crop cycle at Melkassa was calculated and estimated to be 0.999 indicating the yield reduction in tomato is directly proportional to reduced water use Studento et al. [7] and it is a little bit lower than given by Allen et al. [9] which was 1.05 value (Figure 2).

In this figure, Ky=1 is shown as a reference line, and Ky=1.05 is also shown as a reference line.

Validation of CropWat for tomato

Different levels of irrigation water were applied to tomato crop

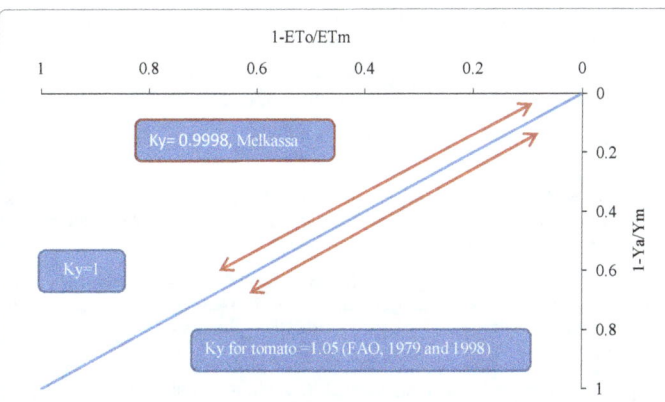

Figure 2: Predicted values and regression line of the functional relationship between relative yield reductions (1-Ya/Ym) and relative evaporation deficits (1-ETa/ETm) of tomato *Melkasholla* variety.

Irrigation treatment	Measured		CropWat
	Yield (kg ha⁻¹)	Yield reduction (%)	Yield reduction (%)
Full ETo	82140	0	0
80% ETo	57300	30.24	19
60% ETo	49300	39.98	34.1

Table 3: Comparisons between yield reductions simulated by CropWat and measured for drip irrigated tomato experiment at Melkassa.

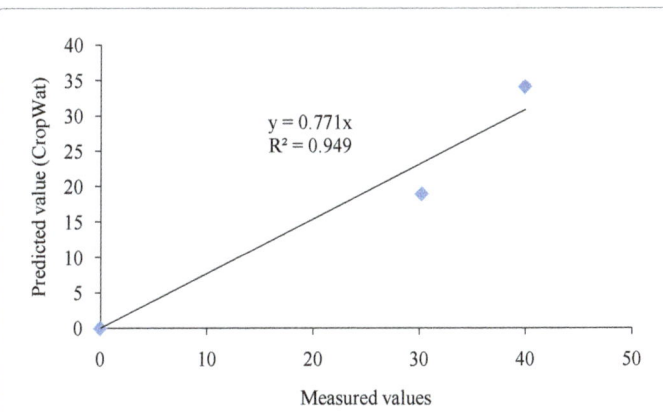

Figure 3: Comparisons of predicted and measured tomato yield reduction under different irrigation regimes at Melkassa.

during the field experiment, inducing water stress throughout the growing season. With the help of the CropWat model the yield reduction was determined and compared with the actual yield reduction of field experimentation [22,23]. Table 3 presents comparison of measured yield reduction with the yield reductions simulated by the CropWat model. The yield reductions were expressed as percentage of the tomato yield obtained under full irrigation.

The CropWat simulation model was combined with 35-year local historical weather data and used as a research tool.

The observed and simulated values for yield are plotted in Figure 3. The model efficiency was calculated and estimated through comparing predicted values to the one-to-one line rather than the best regression line through the origin points. Accordingly, the model efficiency was found to be 94%. This model efficiency was similar to the correlation (r^2) and the r^2 was found to be 95.1% (Figure 3). The measured and simulated tomato total fruit yield showed a good correlation. Furthermore, the simulated results reflected that the impact of stress in

the whole tomato growth cycles was high on fresh fruit yield reduction. The model was confirmed to be a useful decision support system to help farmers to verify the optimal crop management strategy from several points of views.

Summary and Conclusions

An irrigation experiment with drip method was conducted to evaluate and determine the optimal irrigation levels for maximum tomato production, to assess the effect of limited water supply on field grown tomato yield and to estimate 'yield response of tomato to soil water (Ky)' and to validate CropWat irrigation model using the data for tomato cultivation during hot-dry season conditions around Melkassa. Three levels of irrigation regimes with three replications. Among irrigation levels tested, highest yield of 82.14 t ha⁻¹, was recorded from full irrigation treatment (100% ETc) followed by 57.30 t ha⁻¹ from 80% ETc irrigation levels and lowest yield 50.86 t ha⁻¹ from 60% ETc irrigation depth. This indicated that tomato crop should be irrigated at full water requirement to get maximum fruit yield. The relationship between relative yield decrease (1-Ya/Ym) and relative evapotranspiration (1-ETa/ETm) of tomato at Melkassa was determined through the functional relationship and the yield response (*ky*) of tomato *Melkaskola* variety throughout the crop cycle was calculated and estimated to be 0.999 indicating the yield reduction in tomato is directly proportional to reduced water use. This figure is a little bit lower than given by Allen et al. [9] which were 1.05 value. With the help of the CropWat model, the yield reduction simulated by the CropWat was compared with the actual yield reduction of field experimentation. The model efficiency was calculated and estimated through comparing predicted values to the one-to-one line rather than the best regression line through the origin points. Accordingly, the model efficiency was found to be 94%. This model efficiency was similar to the correlation (r^2) and the r^2 was found to be 95.1%. The measured and simulated tomato total fruit yield showed a good correlation. Furthermore, the simulated results reflected that the impact of stress in the whole tomato growth cycles was high on fresh fruit yield reduction. The model was confirmed to be a useful decision support system to help farmers to verify the optimal crop management strategy from several points of views. This further confirm that for rainfed tomato, supplementary irrigation should be switched on during dry spells, and full irrigation should be started on immediately after the rain fall cessation; otherwise much yield loss would occur. This experiment was conducted under drip irrigation conditions whereas all household growers in the study area practice furrow irrigation, thus appropriate irrigation method and irrigation depth estimation should be envisaged in the future for household irrigation water users that maximise yield, improve crop water use.

References

1. Etissa E, Dechassa N, Alamirew T, Alemayehu Y, Dessalegne L (2014) Response of Fruit Quality of Tomato Grown under Varying Inorganic N and P Fertilizer Rates under Furrow Irrigated and Rainfed Production Conditions in the Central Rift Valley of Ethiopia. International Journal of Development and Sustainability 3: 371-387.

2. Etissa E, Dechassa N, Alamirew T, Alemayehu Y, Dessalegne L (2014a) Irrigation Water Management Practices in Small Scale Household Vegetable Crops Production System: The Case of the Central Rift Valley of Ethiopia. Science, Technology and Arts Research Journal 3: 74-83.

3. Sijali IV (2001) Drip Irrigation: Options for Smallholder Farmers in Eastern and Southern Africa. RELMA Technical Hand Book Series 24. Kenya.

4. Hochmuth G, Hanlon EA (2010) Commercial Vegetable Fertilization Principles. The Institute of Food and Agricultural Sciences (IFAS).

5. Cowater International Inc REST (2003) Studies on Water Harvesting Technologies and their Applications in the Region of Tigray. USA.

6. Etissa E, Dechassa N, Alamirew T, Alemayehu Y, Dessalegne L (2014b) Growth and Physiological Response of Tomato to Various Irrigation Regimes and Integrated Nutrient Management Practices. African Journal of Agricultural Research 9: 1484-1489.

7. Studento P, Hsiao TC, Fereres E, Raes D (2012) Crop Yield Response to Water. FAO. Irrigation and Drainage Paper 66. Italy.

8. Kirnak H, Cengiz K, Ismail T, David H (2001) The Influence of Water Deficit on Vegetative Growth, Physiology, Fruit Yield and Quality in Eggplants. Bulg J Plant Physiol 27: 34-46.

9. Allen RG, Pereira LS, Raes D, Smith M (1998) Crop Evapotranspiration (Guidelines for Computing Crop Water Requirements). Irrigation and Drainage Paper No 56. Italy.

10. Muchovej RM, Hanlon EA, McAvoy E, Ozores-Hampton M, Roka FM, et al. (2008) Management of Soil and Water for Vegetable Production in Southwest Florida. Institute of Food and Agricultural Sciences.

11. Doorenbos J, Kassam AH (1979) Yield Response to Water. Irrigation and Drainage Paper 33. Food and Agricultural Organization. Italy.

12. Doorenbos J, Pruitt WO (1977) Guidelines for Predicting Crop Water Requirements. FAO. Irrigation and Drainage Paper 24. Italy.

13. FAO (2009a) ETo Calculator Version 3.1: Land and Water Division. Italy.

14. FAO (2009b) CROPWAT Version 8.0.

15. Hazelton P, Murphy B (2007) Interpreting Soil Test Results. CSIRO Publishing. Australia.

16. Hegde DM, Srinivas K (1989) Studies on Irrigation and Nitrogen Requirement of Tomato. Indian Journal of Agronomy 34: 157-162.

17. Micheal AM (2008) Irrigation Theory and Practice (2nd edn.). Vikas Publishing. India.

18. Fekadu M (2006) Evaluating Small-scale Drip Irrigation System: An Option for Water Harvesting-based Smallholder Farmers' Vegetable Production in Amhara Region. pp: 158.

19. Raes D, Willems P, Gbaguidi F (2006) RAINBOW - A Software Package to Compute Frequency Analyze and Perform Testing of Homogeneity on Hydrometeorological Data Sets. IRRISOFT Software Descriptions and Reviews.

20. Scholberg J, Brian L, McNeal, Jones JW, Boote KJ, et al. (2000) Field-Grown Tomato: Growth and Canopy Characteristics of Field-Grown Tomato. Agron J 92: 152-159.

21. Singh GB, Yadav DV (1992) Integrated Plant Nutrition System in Sugarcane. Fertilizer News 37: 15-22.

22. Spiers JM, Braswell JH (1994) Response of 'Sterling' Muscadine Grape to Calcium, Magnesium, and Nitrogen Fertilization. J Plant Nutr 17: 1739-1750.

23. Bikila T (2007) The Income Contribution of Ziway Lake to Smallholder Farmers and Fishing Households and its Implications for the Sustainability of the Water and Fish Resources. pp: 166.

27

Study on Optimal Irrigation Index of Cotton with Drip Irrigation under Film Mulching based on Pan Evaporation

Xiaojun S[1], Mingsi L[2], Jiyang Z[1], Zugui L[1], Guisen Y[3] and Jingsheng S[1]*

[1]Key Laboratory of Crop Water Use and Regulation, Ministry of Agriculture, Farmland Irrigation Research Institute, Chinese Academy of Agricultural Sciences, Xinxiang, China
[2]Water Conservancy and Architectural Engineering College, Shihezi University, Shihezi, China
[3]Irrigation Experiment Station of Water-conservancy Bureau in Xinjiang Production and Construction Crops, Urumqi, China

Abstract

The objective of this paper is to find a simple and efficient drip irrigation schedule for cotton under film mulching using a homemade evaporation pan with a diameter of 60 cm. A field experiment was conducted to study the responses of cotton growth, seed cotton yield, water consumption and water use efficiency to drip irrigation amounts of 0.5 Ep (pan evaporation), 0.7 Ep and 1.0 Ep at the Irrigation Experiment Station of the Water-conservancy Bureau in Xinjiang Production and Construction Corps in Urumqi, Xinjiang province during the cotton growing seasons of 2007 and 2009. The feasibility of the drip irrigation scheme based on surface water evaporation by using a homemade evaporation pan was analyzed. The results showed that continuous water deficit at the bud and boll stages (irrigation quota equal to 0.5 Ep) produced negative effects on cotton growth, seed cotton yield and the water consumption process during the entire cotton growth period with drip irrigation under mulching compared to the sufficient irrigation treatment (irrigation quota equal to 1.0 Ep); the effect of timely and appropriate water deficit during the bud and boll stages (irrigation quota equal to 0.7 Ep and 0.5 Ep, respectively) on seed cotton yield was not significant; 22.78%~24.88% of the irrigation water was saved, while irrigation water use efficiency was improved by 27.94%~34.85%. It is suggested that the irrigation model with light water deficit during bud emergence (irrigation quota equal to 0.7 Ep), severe water deficit at the late flowering and boll stage (irrigation quota equal to 0.5 Ep), and sufficient water supply at the early flowering and boll stage (irrigation quota equal to1.0 Ep) is a convenient, high-quality, efficient drip irrigation pattern that can be used as a suitable irrigation approach to cotton production with drip irrigation under film mulching in Xinjiang.

Keywords: Drip irrigation under film mulching; Surface water evaporation; Water use efficiency (WUE)

Introduction

The abundant light, heat resources and diversity of the inland desert climate of Xinjiang provide a unique natural ecological environment for high yields and high quality cotton. With the agricultural planting structure adjustment, the Xinjiang cotton district has been China's largest economic district with cotton planting areas and the lint yield ranks the first in China. Cotton production has also been one of the important industries for the local economic development [1,2]. The Xinjiang cotton district is one of the typical irrigated agriculture production areas with little rainfall, and a large evaporation to rainfall ratio. The water needed for cotton growth is mainly obtained from irrigation, in which the water supply resources are not able to meet the water demand. Consequently, with the expansion of irrigated areas and the rapid development of socio-economic, industrial and agricultural production, a more severe water shortage situation confronts the cotton-growing region in Xinjiang. Hence, the efficient use of limited water resources has become one of the important issues facing the cotton production district. Found that water consumption is approximately 450 mm during the entire growth period for cotton high yield in Xinjiang under drip irrigation [3,4]; the average annual precipitation for many years has been less than 200 mm. This has very important implications for the efficient use of limited agriculture water resources, improvement of cotton quality and to achieve the goals of water-saving, high yield, and high efficiency, mainly through scientific irrigation method that can regulate soil water content.

Drip irrigation technology with the advantages of water-saving, high quality and high yield has been promoted and applied to cotton production in Xinjiang, somewhat alleviating the constraints of water supply and demand for cotton production. Drip irrigation under film mulching technology is a combination of the advanced drip irrigation for shallow frequency irrigation and mulching plant cultivation techniques. This is not only prevent deep percolation of irrigation water but also results in a significant reduction in the invalid evaporation between plants and surface soil, simultaneously contributing to increasing of the soil moisture storage, soil temperature and improving the crop hydrothermal environment. The adoption of advanced water-saving irrigation technologies and the development of rational irrigation schemes are required to protect crop yields, save water resources and improve water use efficiency. A number of studies have been conducted on drip tape laying [5,6], emitter discharge rate [7], crop growth [8-10], and yield and quality [11-13] with drip irrigation; these studies propose the corresponding irrigation index [2,14] to further improve water use efficiency of cotton soil under drip irrigation from the beginning of 1990s. However, most of the determinants of these indicators are needed to monitor soil moisture and cotton growth parameters. The measurement and control of these parameter indicators require professional operations and a higher level of theory and practice, which is difficult for ordinary cotton farmers in the Xinjiang cotton production area to master and use.

There is a strong relationship between pan evaporation and ET_0 [15]. A number of studies both at home and abroad investigated

*Corresponding author: Jingsheng S, Key Laboratory of Crop Water Use and Regulation, Ministry of Agriculture, Farmland Irrigation Research Institute, Chinese Academy of Agricultural Sciences, Xinxiang, China 453002
E-mail: jshsun623@aliyun.com

the relationship between pan surface evaporation and crop water consumption under different irrigation conditions [16-20] to develop a scientific irrigation system. Experimental research on A Standard evaporating pan adopted outside China is more common [21-23]. Conversely, the price of the Class A evaporation pan is relatively high and it has specific installation requirements, making it difficult to promote and apply in large areas in developing countries. In this study, field trials during two growing seasons investigated the response of cotton growth, yield and water use efficiency to different irrigation regimes (as different pan coefficients for guiding irrigation). This approach is based on the results of field experiments and will propose a simple and convenient operation and a "fool" water-saving irrigation index for the farmers, which makes the traditional water management model "see the day", "see the ground", and "see crops" for the cotton district in Xinjiang under drip irrigation scientifically achievable.

Materials and Methods

Experimental site

The field experiment was carried out in the Irrigation Experimental Station (43°59′N, 87°23′E) of Water-conservancy Bureau in Xinjiang Production and Construction Crops, Xinjiang, China, during the growth seasons of 2007 and 2009. The station is located in Xinjiang province, 30 km away from Urumqi, which is warm-temperate arid zone with a continental climate. According to statistics of meteorological data for 30 years from the Urumqi National Meteorological Station, average annual sunshine duration is 2864 h; annual accumulated temperatures are 3450°C (≥10°C). Annual precipitation and potential evaporation are 190 mm and 1600 mm, respectively. The soil type of the experimental district is loamy soil, from 0 to 100 cm depth average soil bulk density is 1.51 g·cm^{-3}, the soil weight moisture capacity is 20.7%, water table is consistently 8 m below the ground.

Methods of cotton planting and fertilization

The cotton crops (Xin luzao 9) were planted on April 28, 2007, wholly emerged on May 6, harvested from September 14; and planted on May 2, 2009, wholly emerged on May 14, harvested from September 19. The planting were sowed at low soil moisture and watering at seedling stage. The cotton crops were planted in row spacing 20 cm + 45 cm + 20 cm (one tube controls 4 rows of cotton crop, (Figure 1), with plant spacing 11 cm. Two growing seasons were respectively applied with efficient manure of 150 kg·hm^{-2} (Phosphate, total nutrient content of N, P$_2$O$_5$ and K$_2$O are more than or equal to 60%; the total content of trace elements for Zn, B, Fe, Mn, etc. are more than or equal to 0.5%), and urea 450 kg·hm^{-2} (Total nitrogen content is more than

or equal to 46.4%) during the whole cotton growth. The fertilization through drip irrigation systems was divided into five portion and each time the amount of urea and efficient fertilizer were 90 kg hm^{-2} and 30 kg hm^{-2} (pure N, P$_2$O$_5$, and K$_2$O are 58.7, 7.7 and 5.8 kg hm^{-2}, respectively); while the whole growth period spraying DPC 3 times.

Experimental field watered with drip irrigation under film mulching with an experimental area of approximately 0.32 hm^2, controlled by a branch tube; three plots of each treatment were used as a branch unit (length is 50 m and width is 1.45 m). The valve and water meters were installed at the unit entrance before the installation of the pressure regulating valve and pressure gauges and the insertion of the drip irrigation tube with thin-walled capillary inserted. The emitters are spaced 30 cm apart. The irrigation system pressure was controlled by a regulating valve, while the irrigation of each treatment was controlled by water meters.

Experimental design

The entire cotton growth period can be divided into seedling stage (stage Ⅰ), bud stage (stage Ⅱ), early flowering and boll stage (stage Ⅲ), late flowering and boll stage (stage Ⅳ) and boll opening stage (stage Ⅴ). Dry seeding wet emergence was adopted; 30 mm (2007) and 45 mm (2009) of irrigation water was applied after sowing. According to the local cotton production practice, both seeding and boll opening stages are not irrigated; therefore, four irrigation treatments were designed in the experiment:

Adequate irrigation treatment (T1): Irrigation was applied when soil water content was approximately 50% of field capacity, receiving irrigation water equality that was 100% of the water surface evaporation at the same stage during the entire cotton growing season. The irrigation lower limit is average water content of 0 to 70 cm soil layer.

The 2nd irrigation level (T2): Irrigation time is as the same as T1, and the irrigation quota is 70% of cumulative water evaporation stage.

The 3rd irrigation level (T3): Irrigation time is as the same as T1, and the irrigation quota is 50% of cumulative water evaporation stage.

The 4th irrigation level (T4): Irrigation time is as the same as T1, and the irrigation quota at budding stages is 70% of cumulative water evaporation stage; the irrigation quota at the late flowering and boll stages is 50% of cumulative water evaporation, and the same irrigation water as T1 during the other stages was applied. Three plots were labeled for replications of each treatment, and each plot was randomly arranged.

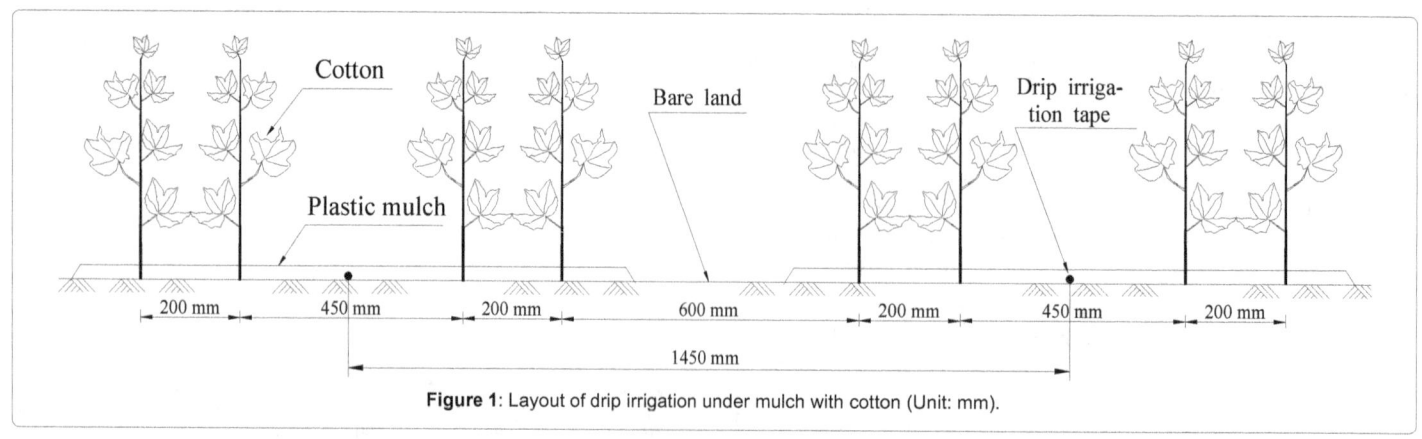

Figure 1: Layout of drip irrigation under mulch with cotton (Unit: mm).

Observation Items and Methods

Soil water measurement:

The soil water contents were taken every 10 cm of the top 0~60 cm depth and 20 cm of the 60~100 cm depth in the planting zone and measured by the oven-drying method. Because field soil moisture of drip irrigation cotton is not a one-dimensional layered distribution and the cultivation characteristics involve the wide and narrow row planting, three sampling points were taken at the wide row (below the drip line under the film mulch), the narrow row (middle of the narrow row under the film mulch) and the bare land outside the mulch. Measurements were taken at an interval of 5~7 days, and extra measurements was taken before and after each irrigation. The average moisture content was calculated according to the reference literature [24].

Morphological index measurement:

Ten tagged plants (five inside and five outside of the row) of every treatment were chosen for the measurement of the morphological index, i.e., plant height, leaf area, the number of bolls, etc. Plant height was measured by steel tapes with an accuracy of 1 mm; length and width of leaves were measured by rulers with an accuracy of 1 mm, and leaf area was determined by multiply of length, width and the coefficient of leaf area. Measurements were taken at an interval of 10 days.

Evaporation of evaporating dish measurement:

Evaporation intensity of free water surface was measured with a simple pan (high 60 cm, diameter 60 cm) made of galvanized iron (0.75 mm of the thick), which was buried in the open space neighboring the experimental field weather station; the open end of the evaporating pan was approximately 20 cm above the ground. A micrometer with an accuracy of 0.02 mm was used to measure the water level of the pan at 8:00 am every day. The water evaporation was calculated by using the difference between the two readings. Replenishment occurred every other day to ensure that there was approximately 5 cm from the water surface to the edge of pan.

Calculation of seed cotton yield and water consumption:

Yield and its components were quantified by measuring the number of bolls, single boll weight, etc. Fifty tagged plants per treatment were randomly chosen at the end of the growing period. For monitoring the actual seed cotton yield of each plot, cotton was harvested twice by hand-picking. All the harvested seed cotton was weighed with electronic weighing scales with an accuracy of 1 g after drying in each plot as the final yield.

Conventional meteorological data:

Meteorological data were gathered at an automatic weather station which was approximately 100 m away from the experimental plots. Meteorological variables measured included air temperature, relative humidity, global radiation, rainfall and wind speed at 2 m above the ground. Changes in reference crop evapotranspiration (ET_0) were calculated with the Penman–Monteith equation.

Experimental Calculation

Crop water consumption according to the water balance equation and water consumption of each treatment was calculated by the formula (1). Because groundwater depths of the experimental plot greater than 15 m, therefore, the effect of groundwater recharge on cotton water consumption was not considered, take $K=0$. Wang showed that when the dripper flow was 2.8 L·h⁻¹ and the irrigation quota was less than 45

mm, deep soil moisture was affected minimally by irrigation [10]. So, there were no deep percolation phenomena; data of soil moisture in the experimental field showed that the irrigation quota is less than 37.5 mm, and irrigation had little effect on soil moisture below the 60 cm soil layer. Therefore, when the irrigation quota is less than 37.5 mm, there is no deep percolation, where $D=0$. Equation (1) can be simplified to the equation (2).

$$ET=P_0+K+M-D+(W_0-W_t) \tag{1}$$

Where: ET is cotton water consumption (mm); P_0 is effective rainfall (mm); K is the amount of groundwater recharge (mm); M is the amount of irrigation (mm); D is the amount of deep percolation (mm); W_t and W_0, respectively, are for the soil reservoir water of the beginning of the period and the end of the period (mm).

$$ET=P_0+M+(W_0-W_t) \tag{2}$$

Crop water use efficiency: Crop water use efficiency means that crops consumed per unit water to gain production and calculated using the following formulas:

$$WUE=Y/ET_a \tag{3}$$

$$IWUE=Y/I \tag{4}$$

Where: WUE and $IWUE$, respectively show water use efficiency and irrigation water use efficiency, kg·m⁻³; Y is seed cotton yield, kg·hm⁻²; ET_a is actual water consumption during the entire growth period, m³·hm⁻²; I is the amount of irrigation for the entire growth period, as irrigation quota, m³·hm⁻².

Results

Meteorological condition:

Changes in reference crop evapotranspiration (ET_0) in two growing seasons were calculated using the Penman–Monteith equation (Recommended by FAO) in (Figure 2). The results from 2007 showed that ET_0 was relatively small at seeding stages (from May 6 to June 20) with an average of 5.61 mm·d⁻¹; with the advance of the cotton growing process, ET_0 gradually increased and reached its maximum at the budding stage (from June 21 to July 11), with an average value of 7.45 mm·d⁻¹; and then it declines gradually, with an average value of 6.24 mm·d⁻¹, 5.97 mm·d⁻¹ and 3.10 mm·d⁻¹, respectively, at the early flowering and boll stage (from July 12 to August 5), late flowering and boll stage (from August 6 to September 1) and boll opening stage (from September 2 to October 20); the average ET_0 was 5.24 mm·d⁻¹ during the entire cotton growth period. The results from 2009 showed that ET_0 still remained small at the seedling stage (from May 2 to June 19), with an average number of 5.46 mm·d⁻¹. The ET_0 gradually increased with the advancement of the cotton growing process, and reached its maximum at the budding stage (from June 20 to July 1), with an average of 6.21 mm·d⁻¹; and then it declines gradually, with an average of 5.90 mm·d⁻¹, 5.18 mm·d⁻¹ and 2.97 mm·d⁻¹, respectively, during the early flowering and boll stage (from July 13 to August 8), late flowering and boll stage (from August 9 to September 3) and boll opening stage (from September 4 to October 20); the average ET_0 was 4.91 mm·d⁻¹ during the entire growth period.

According to the analyzed of surface evaporation, effective rainfall (precipitation was greater than 5 mm) and growth process (Table 1), the alteration process of daily average surface evaporation intensity was approximately the same with ET_0. The data of seeding and boll opening stages were relatively small, but the budding, flowering and boll stages were larger. Precipitation in 2007 and 2009 was approximately 215

Growth stages	Date	Day (d)	Effective rainfall (mm)	Average potential evapotranspiration (mm·d⁻¹)	Evaporation of evaporating pan (mm·d⁻¹)
Seedling	5.6~6.20	46	70	5.61	5.64
Bud stage	6.21~7.11	21	0	7.45	6.06
Early flowering and boll stage	7.12~8.5	25	30	6.24	6.36
Late flowering and boll stage	8.6~9.1	27	45	5.97	5.5
Boll opening	9.2~10.20	49	20	3.1	3.82
Whole stage	5.6~10.20	168	165	5.26	5.24
Seedling	5.2~6.19	49	29	5.46	5.68
Bud stage	6.20~7.12	23	9	6.21	5.61
Early flowering and boll stage	7.13~8.8	27	0	5.9	6.18
Late flowering and boll stage	8.9~9.3	26	8	5.18	4.92
Boll opening	9.4~10.20	47	8	2.97	3.91
Whole stage	5.2~10.20	172	53	4.91	5.15

Table 1: Evapotranspiration and rainfall at each growth stage of cotton with drip irrigation under film mulching.

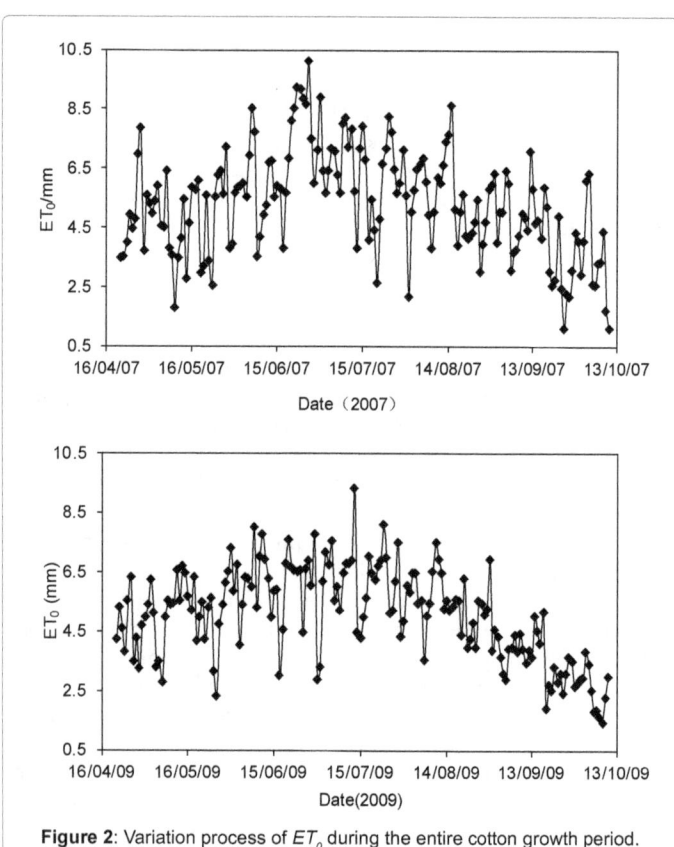

Figure 2: Variation process of ET_0 during the entire cotton growth period.

and 95 mm during the entire growing period, and the effective rainfall (rainfall is more than 5 mm) of the two growing seasons reached 165 and 53 mm, respectively.

Effect of different water treatments on cotton growth:

The cotton plant height is one of the important indexes to measure growth and development of cotton. During different growth stages, especially before topping, the growing rate of plant height is one of the important indexes that reflect the coordination degree between crop vegetative and reproductive growth. An appropriate plant height is suitable for plant type improvement, contributing to a more reasonable distribution of the canopy. The experimental results of two growing seasons (Figure 3) showed that: with drip irrigation under film mulching, as the cotton growing process progressed, plant height gradually increased and the growth rate was faster in both bud and the early flowering and boll stages, and subsequently, the growth rate became very slow. Statistical analysis of the plant height among treatments showed as the following: water treatments are the same during the seeding stages, and the difference of plant height of each treatment was not significantly. However, after entering the bud stages, plant height of T1 was significantly higher than treatment T2 and T3 (P<0.05). The plant height of T4 was significantly lower than T1 in the bud stages (P<0.05), but at the beginning of the early flowering and boll stage (after seeding 81 days in 2007, after seeding 90 days in 2009), plant height of T4 and T1 is roughly equal; the difference was not significant (P<0.05).

Results of leaf area for two growing seasons (Figure 4) showed that: cotton leaf index (*LAI*) increased slowly before bud stages, rapidly growing from the bud stage and reached a maximum peak (after seeding 102 days in 2007, and after seeding 90 days in 2009) at flowering and boll stage, and subsequently, the growth rate gradually decreased. The significant difference analysis showed that: during the seeding stages, the *LAI* of each treatment was not significantly different, but after entering bud stages, *LAI* of T1 was significantly higher than treatment T2 and T3 (P<0.05); *LAI* of T4 was significantly lower than T1 in the bud stages (P<0.05). At the beginning of the early flowering and boll stage (after seeding 90 days in 2009), *LAI* of treatment T4 and T1 was roughly equal and the difference was not significant (P<0.05).

Effect of different treatments on seed cotton yield:

Early summer bolls are mean bolls (diameter larger than 2 cm) that formed before July 15 in the experimental area, while the bolls that formed during July 16 and August 15 are called sweltering summer bolls, where autumn bolls had mean bolls that formed after August 16. Different distributions of bolls in various parts of cotton played diverse roles in the yield due to their formation at different times. The earlier summer bolls formed most favorably in the extension of the effective growing period, and improved the potential of bolls formation per plant [25]. Investigation of early summer bolls revealed that there were more fruit branches, bolls, young bells and early summer bolls for T1, while T3 was the least, but the numbers of bolls of T3 are most. This was due to the continuous small irrigation quota during the bud stage and the flowering and bolls stage and no water compensates in the following stage. Thus, the remaining bolls could not grow and develop normally, and then opened because of the water deficit (in the physical as "drought escape phenomenon"). Sweltering summer bolls are the main component of cotton bolls; their numbers directly determine the seed cotton yield. The results of the investigation showed as the following: the number of sweltering summer bolls for T4 was the highest, followed by T1, while T3 was the lowest. Autumn bolls are outside the space of cotton; their light conditions were best, which have higher boll rate under sufficient water and fertilizer conditions and are also a component of the seed cotton yield. However, Autumn bolls were affected by climate (mainly temperature decline, particularly the end of the frost-free period) at the late growth period, having limits between autumn peach mature and natural boll opening [25]. The investigation found that T1 obtained the most vigorous growth and the highest autumn boll probability, which was due to sufficient water supply, but other treatments were small and showed no significant differences. Results of variance analysis among treatments (Table 2) showed that compared to treatment T1, seed cotton yield of treatments

Year	Treatment	Number of early summer bolls	Number of sweltering summer bolls	Number of autumn bolls	Seed cotton Yield (kg·hm⁻²)
2007	T1	1.40 b	4.80 a	1.30 a	5536.57 a
	T2	1.68 ab	4.20 b	0.75 b	4899.07 b
	T3	2.04 a	2.10 c	0.30 c	2853.30 c
	T4	1.54 b	5.10 a	0.50 bc	5630.19 a
2009	T1	1.18 b	4.08 a	1.06 a	4438.67 a
	T2	1.42 a	3.57 b	0.60 b	3871.04 b
	T3	2.01 b	2.06 c	0.26 b	2895.23 c
	T4	1.23 b	4.34 a	0.39 b	4385.29 a

Table 2: Effects of different water treatments on seed cotton yield.

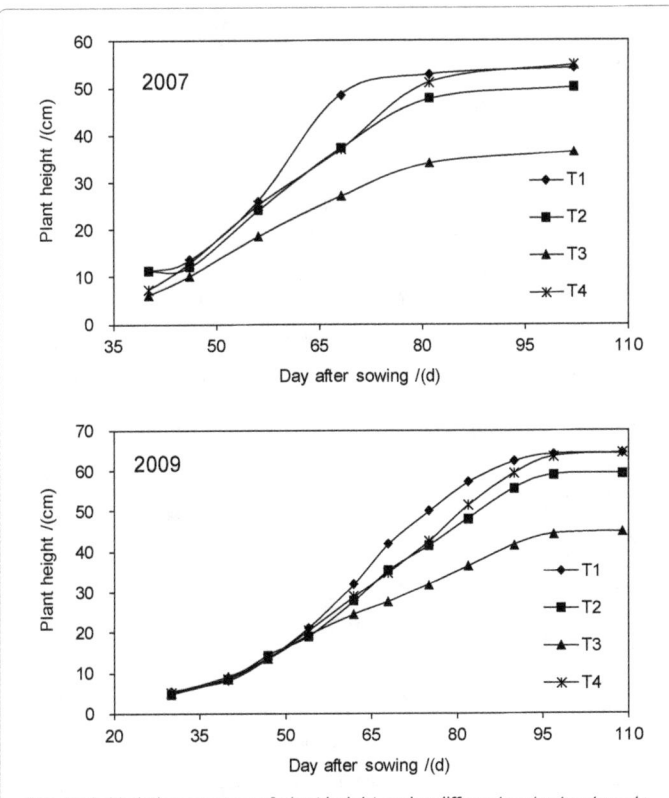

Figure 3: Variation process of plant height under different water treatments.

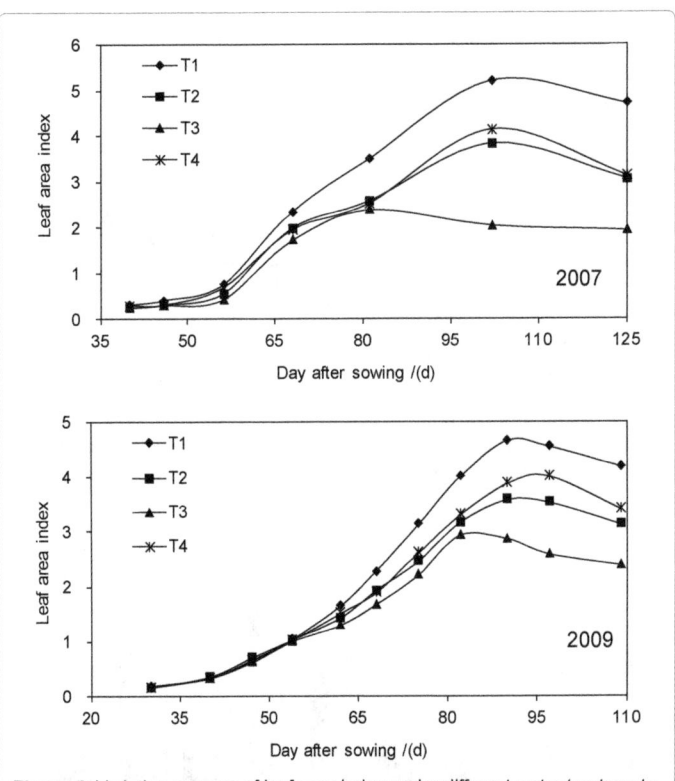

Figure 4: Variation process of leaf area index under different water treatments.

T2 and T3 decreased significantly from 11.51% to 12.79% and from 34.77% to 48.46%, respectively (P<0.05); seed cotton yield of treatment T4 was slightly higher than treatment T1 in year 2007, but slightly lower than treatment T1 in year 2009, and there was no significant difference between two treatments (P<0.05). Comprehensively consideration, treatment T4 was the best irrigation scheme in this experiment.

Effect of different water treatments on cotton water consumption and water use efficiency:

Irrigation of the entire growth period: (Figure 5) showed the irrigation time and irrigation quota for years 2007 and 2009 during the cotton growth period; to ensure emergence, 30 mm irrigation water was applied after sowing. No irrigation was applied during the seedling stage of cotton because of less water consumption and more rainfall (effective rainfall was 70 and 29 mm during 2007 and 2009, respectively). In 2007, irrigation treatments began from the bud stage; the conventional irrigation treatment (T1), the 2nd irrigation level (T2), the 3rd irrigation treatment (T3) and the 4th irrigation level (T4) received the irrigation amount of 420 mm, 303 mm, 225 mm and 315 mm, respectively. Compared to the treatment T1, the treatments T2, T3 and T4 saved irrigation water by 27.85%, 46.43% and 24.88%, respectively. Irrigation treatments in 2009 were as the same as in 2007; T1, T2, T3 and T4 received the irrigation amount of 450 mm, 328.5 mm, 247.5 mm and 347.5 mm, respectively; compared to the treatment T1, the treatments T2, T3, and T4 saved irrigation water by 27.00%, 45.00% and 22.78%, respectively.

Daily average water consumption: The results of water consumption for two growing seasons showed as the following (Figure 6): the seasonal changes in the effect of different water treatments on daily average cotton water consumption were small because of smaller plants and film mulching. Cotton water consumption intensity was small at the seeding stage; access to the nutrition and reproductive

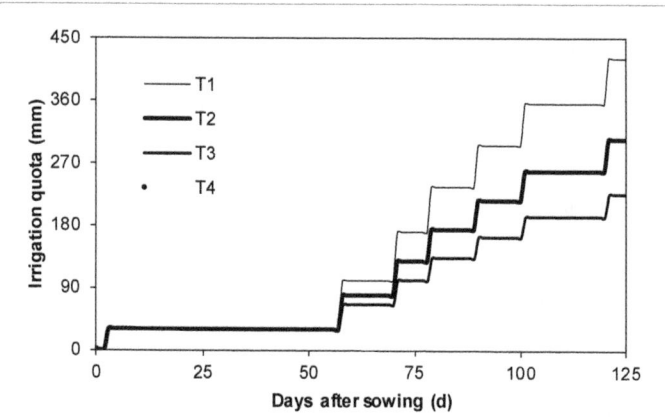

Figure 5: Irrigation treatments of cotton during the whole growth period with drip irrigation under film mulching.

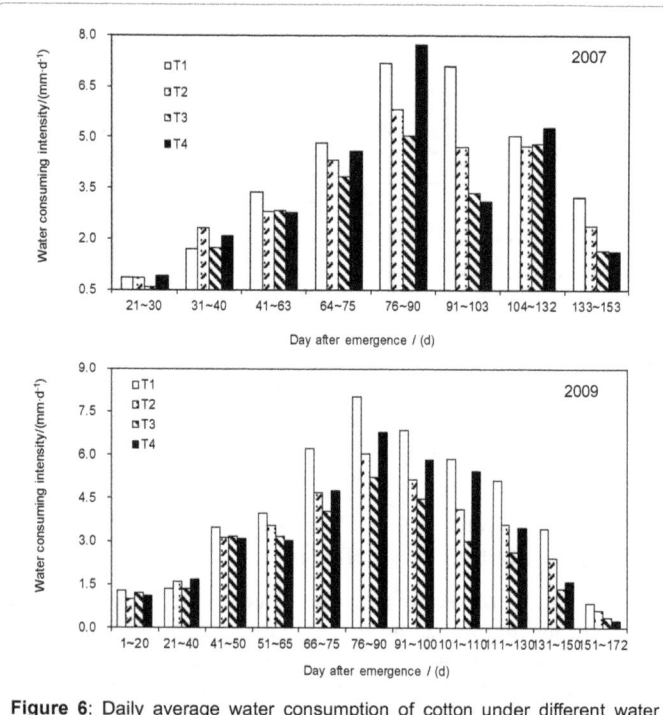

Figure 6: Daily average water consumption of cotton under different water treatments.

growing stages from the bud stage, with the continuous rising temperatures, led to rapidly increasing water consumption. The peak occurred within 75~90 days after sowing. Additionally, daily water consumption decreased gradually with the leaves removed and the temperature decreased from the boll opening stage of cotton. The results from 2007 showed that the water consumption intensity of T2, T3 and T4 appeared to increase within 103~132 days after sowing, due to no irrigation applied from the boll opening stage, and interference caused by the rainfall (the cumulative rainfall exceeded 20 mm). (Figure 5) also showed that water consumption intensity decreased with the decreasing of the irrigation quota.

Effect of different treatments on cotton water use efficiency: Water use efficiency (*WUE*) and irrigation water use efficiency (*IWUE*) of cotton with drip irrigation under film mulching were shown in (Table 3); *WUE* and *IWUE* of Treatment T4 for two growing seasons were the highest, and differences compared to the other treatments

were significant (*P*<0.05). The results showed that *WUE* and *IWUE* of treatment T3 for cotton were both the lowest in 2007 and 2009; and the differences compared to the other treatments were significant (*P*<0.05). In (Table 3), *IWUE* of treatment T4 reached 1.78 kg·m^{-3} (2007) and 1.26 kg·m^{-3} (2009), which was 1.35 and 1.28 times to the conventional irrigation treatment (T1). Seed cotton yield and irrigation water use efficiency was promoted under T4 and it was suggested for applying in production practice.

Cumulative water consumption and accumulative evaporation of cotton and evaporation pan coefficient during the entire growth period

The variation process of cumulative evaporation (*ETp*) and cumulative surface evaporation at the same stages during the entire growth stage in 2007 and 2009 were shown in (Figure 7); the cumulative water consumption was much lower than surface evaporation during the early growth stage, but 70 days after sowing, water consumption

	Treatment	Seed cotton Yield (kg·hm^{-2})	Amount of irrigation (mm)	Water consumption (mm)	WUE (kg·m^{-3})	IWUE (kg·m^{-3})
2007	T1	5536.57 a	420	572.81	0.97 c	1.32 c
	T2	4899.07 b	303	475.3	1.03 b	1.62 b
	T3	2853.30 c	225	414.28	0.69 d	1.27 c
	T4	5630.19 a	315.5	498.34	1.13 a	1.78 a
2009	T1	4438.67 a	450	647.02	0.69 c	0.99 c
	T2	3871.04 b	328.5	500.72	0.77 b	1.18 b
	T3	2895.23 c	247.5	411.92	0.70 b	1.17 b
	T4	4385.29 a	347.5	499.38	0.88 a	1.26 a

Table 3: Water use efficiency and irrigation water use efficiency.

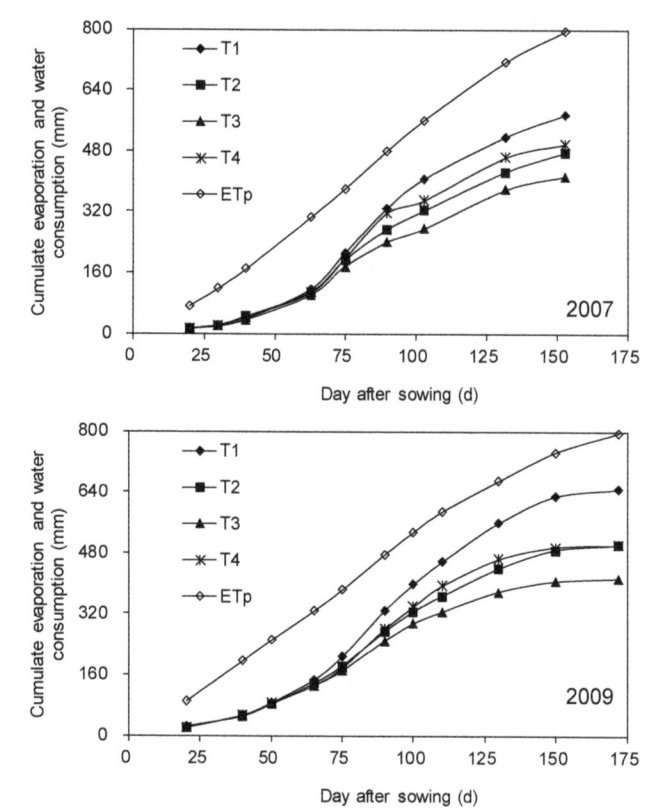

Figure 7: Cumulative water consumption and accumulative evaporation of cotton.

and surface evaporation were approximately the same. The results from 2007 showed as following: cumulative evaporation (*ETp*) was 792.7 mm during the entire growth stage, and cumulative water consumption of cotton during the entire growing periods was 572.81 mm, 475.30 mm, 414.28 mm and 98.34 mm for T1, T2, T3 and T4, respectively. The average pan evaporation coefficient of T1, T2, T3 and T4 was 0.72, 0.60, 0.52 and 0.63, respectively. The results from 2009 showed that cumulative evaporation (*ETp*) was 793.7 mm during the entire growth stage, and cumulative water consumption of cotton during the entire growing period was 647.02 mm, 500.72 mm, 411.92 mm and 499.38 mm for T1, T2, T3 and T4, respectively. The average evaporation pan coefficient of T1, T2, T3 and T4 was 0.82, 0.63, 0.52 and 0.63, respectively. The data analysis also revealed that pan cumulative evaporation has a strong relationship with cotton water consumption, while the correlation coefficient was 0.99 (the number of sample is 11). The evaporation pan coefficient of Treatment T4 in bud stage, the early flowering and boll stage, the late flowering and boll stage was 0.48, 0.87, 0.74 (2007) and 0.52, 0.93, 0.69 (2009), respectively.

Discussion

As the irrigation system changes, plant height also changes. This article investigated the effect of different irrigation treatments on the plant height of cotton with drip irrigation in Xinjiang. The results indicated that compared to the regular treatment, moderate water stress in bud stage, severe water stress in the late flowering and boll stage and sufficient water in the early flowering and boll stage did not significantly reduce the plant height because of the crop water deficit compensation effect, simultaneity, they were all water saving. Conversely, slightly elevated plant height is conducive to high cotton yield. According to the biological characteristics of cotton and the response of crop growth to the water supply mechanism, timely and appropriate moisture stress can reduce crop luxury transpiration. Generally, slightly decline in photosynthetic rate can regulate the ratio of root to crown and coordinate the relationship between vegetative and reproductive growth. The photosynthetic rate has the effect of compensation or overcompensation after water stress to restore the water supply, which is favorable for photosynthetic products to transport to the reproductive organs, thereby enhancing the seed cotton yield [26-28]. In this article, moderate water deficit during bud stage can control vegetative growth and promote reproductive growth; water stress during the late flowering and boll stage is favorable for boll to naturally boll opening. Cotton growth and the development of seed cotton yield is the most sensitive to moisture stress in the early flowering and boll stage; one episode of water stress will cause irreparable negative effects. Timely and appropriate water deficit cannot lower or reduce economic yield, but significantly improves water use efficiency. In our study, with drip irrigation under film mulching, moderate water stress in bud stage, severe water stress in the late flowering and boll stage and sufficient water in the early flowering and boll stage did not lower the yield, on the contrary, it saved irrigation water resources, significantly improved water use efficiency; which was in accord with previous research results [14,29].

This study also analyzed the impact of different water treatments on the water consumption with drip irrigation under film mulching in Xinjiang. The results implied that the allocation of irrigation quota in the entire growth period was different. The stage water consumption and water consumption in the entire growth period of cotton varied and water consumption of cotton in the entire growth period was 411~647 mm. The results were similar to the study of [30]. The performance of the average daily water consumption intensity for the cotton is early

small, middle big, late small. This is because the cotton plant is small during seeding stage, there are fewer leaves, plant transpiration is smaller, average daily water consumption intensity of the cotton fields is lower. At the bud stage, cotton growth began to change from vegetative to reproductive growth, with vigorous growth of cotton plants, leaves and plant transpiration rapidly increased and the average daily water consumption intensity of the cotton fields also significantly increased. The flowering and boll stage is the key period for cotton response to water and fertilizer demand. At reproductive and vegetative growth, response pace is the same where water consumption intensity of the cotton fields at the highest demand, until boll opening stage. However, with lower temperatures and cotton plant aging, water consumption gradually decreases.

Drip irrigation under film mulching had been promoted and applied to a large area in Xinjiang. Since 1998, scholars in China have conducted substantial research on the quality and efficient irrigation system of cotton with drip irrigation, and based on these studies, a variety of high-quality and efficient irrigation indexes have been proposed [3,12]. However, the available indicators applied to the local cotton field moisture management were not very satisfactory, mainly because of the farmers' incomplete irrigation test facilities, and restrictions of the cultural level, which made the physiological and ecological indicators difficult to accurately observe. In addition to these constraints, the soil moisture of cotton fields and environmental indicators that used to determine the irrigation time and irrigation volume was complex. Based on irrigation indicators of water surface evaporation, domestic research mostly concentrated on standard evaporation pan of 20 cm. Study objects were mostly protectorate crops, involving northwest arid zone crops, reporting of drip irrigation under film mulching in Xinjiang is relatively less. In this study, under two growing seasons, it is revealed that: evaporating pan cumulative evaporation has a strong relationship with cotton water consumption; the correlation coefficient was 0.99 (the number of sample is 11). The evaporation pan coefficients of treatment T4 for high yield in bud stage, the early flowering and boll stage, the late flowering and boll stage were 0.48, 0.87, 0.74 (2007) and 0.52, 0.93, 0.69 (2009), respectively. In combination with local climatic conditions and the production practice of cotton, quality and efficient irrigation index can be determined based on simple pan evaporation.

Conclusion

1) Compared to the sufficient water supply treatment of the entire growth period, (T1), water stress significantly inhibits the plant height of the cotton and the growth of the leaf area. However, the plant height and the leaf area index of treatment T4 with moderate water stress in the bud stage, severe water stress in the late flowering and boll stage, sufficient water supply in the early flowering and boll stage, were approximately equal to treatment T1.

2) With the varying irrigation quota, the allocation of irrigation amount in the entire growth period, stage water consumption and water consumption in the entire growth period of cotton were different; with decreasing irrigation quota, the stage water consumption reduced. Compared with treatment T1, water consumption of treatment T4 in the entire growth period decreased by 75~147.64 mm.

3) Compared to treatment T1, the yield of treatments T2 and T3 were reduced by 11.51%~12.79% and 34.77%~48.46%, respectively, and the differences were significant (P <0.05). Seed cotton yield of treatment T4 was slightly higher than treatment T1 in 2007; seed cotton yield of treatment T4 was slightly lower than treatment T1 in 2009, hence the yield differences in treatments T1 and T4 for

the two growing seasons were not significant (P <0.05). Therefore, comprehensively considering the experimental data from two growing seasons, with the desired yield, treatment T4 saved 22.78%~24.88% of irrigation water resources and irrigation water use efficiency was improved by 27.94%~34.85%.

4) The relationship between the cumulative water consumption of high yield fields (T4) and accumulation of 60 cm evaporation pan indicated that: stage cumulative water consumption of high yield fields has a good relationship with accumulation surface evaporation for the same period, and the value of R^2 is more than 0.99. The Kp of the bud stage, the early flowering and boll stage, the late flowering and boll stage for high yield cotton fields with drip irrigation under film mulching were 0.48~0.52, 0.87~0.93 and 0.69~0.74, respectively, which can be beneficial for the evaporation pan evaporation to guide cotton field moisture management. Combining local climatic conditions with the production practice of cotton, quality and efficient irrigation index can be determined based on simple pan evaporation: with dry seeding and irrigation with 30~45 mm emergence water, there is no irrigation in the seedling stage. Irrigation quota is determined in the bud stage, the early flowering and boll stage, the late flowering and boll stage, according to the multiplication between pan cumulative evaporation and K_p (values of the bud stage, the early flowering and boll stage, the late flowering and boll stage are, respectively, taken 0.5, 0.9, 0.7). The irrigation cycle is taken 9~12 days in the bud stage, 5~7 d in the early flowering and boll stage, 7~10 d in the late flowering and boll stage; there is no irrigation in the boll opening stage.

Acknowledgement

We are grateful to the support by the National High-Tech 863 Project of China (2011AA100502), the National High-Tech 863 Project of China (2011AA100509). And we also would like to thank the staffs of the Key Laboratory of Crop Water Use and Regulation, Ministry of Agriculture, P. R. China.

References

1. Xu M, Han XJ, Wang ZS (2005) Study on the application of drip irrigation under plastic film on cotton. Crops 6: 56-58.

2. Wang C, Isoda A, Wang P (2004) Growth and yield performance of some cotton cultivars in Xinjiang, China, an arid area with short growing period. Journal of Agronomy and Crop Science 190: 177-183.

3. Cai HJ, Shao GC, Zhang ZH (2002) Lateral layout of drip irrigation under film mulching for cotton. Transactions of the csae 1: 45-48.

4. Shen XJ, Chen HM, Sun JS (2010) Response of different water deficit on cotton growth and water use efficiency and yield under mulched drip irrigation. Journal of Irrigation and Drainage 1: 40-43.

5. Cai HJ, Shao GC, Zhang ZH (2002) Water demand and irrigation scheduling of drip irrigation for cotton under film mulching. Journal of Hydraulic Engineering 11: 119-123.

6. Li PL, Zhang FC, Jia YG (2009) Coupling effect of water and nitrogen on cotton under different drip irrigation lateral placements. Scientia Agricultural Sinica 5: 1672-1681.

7. Li MS, Kang SZ, Sun HY (2006) Relationships between dripper discharge and soil wetting pattern for drip irrigation. Transactions of the csae 4: 32-35.

8. Li MS, Kang SZ, Yang HM (2007) Effects of plastic film mulch on the soil wetting pattern, water consumption and growth of cotton under drip irrigation. Transactions of the csae 6: 49-54.

9. Sun H, Li MS, Ding H (2009) Experiments on effect of dripper discharge on cotton-root distribution. Transactions of the csae 11: 13-18.

10. Wang YX, Li MS, Lan MJ (2011) Effect of soil wetting pattern on cotton-root distribution and plant growth under film mulching drip irrigation in field. Transactions of the csae 8: 31-38.

11. Zhang ZH, Cai HJ, Yang RY (2005) Relationships between yield, quality and CWSI of cotton under drip irrigation with mulch. Transactions of the csae 6: 26-29.

12. Yan YY, Zhao CY, Sheng Y (2009) Effects of drip irrigation under mulching on cotton root and shoot biomass and yield. Chinese Journal of Applied Ecology 4: 970- 976.

13. Su LT, Abudu S, Hudan T (2011) Effects of under-mulch drip irrigation on soil salinity distribution and cotton yield in an arid region. Acta Pedagogical Sinica 4: 708-714.

14. Shen XJ, Sun JS, Liu ZG (2010) Effects of low irrigation limits on yield and grain quality of winter wheat. Transactions of the csae 12: 58-65.

15. Snyder RL (1992) Equation for evaporation pan to evapotranspiration Conversions. Journal of Irrigation and Drainage Engineering 6: 977-980.

16. Byers PL, Moore JN (1987) Irrigation scheduling for young high bush blueberry plants in Arkansas. Hort Science 1: 52-54.

17. Yang SL, Aydin M, Yano T (2003) Evapotranspiration of orange trees in greenhouse hypsometers. Irrigation Science 21: 145-149.

18. Yuan BZ, Nishiyama S, Kang YH (2003) Effects of different irrigation regimes on the growth and yield of drip irrigated potato. Agricultural Water Management 3: 153-167.

19. Yuan BZ, Sun J, Nishiyama S (2004) Effect of drip irrigation on strawberry growth and yield inside a plastic greenhouse. Bioprocess and Biosystems Engineering 2: 237-245.

20. Liu HJ, Kang YH (2007) Sprinkler irrigation scheduling of winter wheat in the North China Plain using a 20 cm standard pan. Irrigation Science 25: 149-159.

21. Ertek A, Sensoy S, Ibrahim G (2004) Irrigation frequency and amount affect yield components of summer squash. Agricultural Water Management 1: 63-76.

22. Ertek A, Sensoy S, Gedik I (2006) Irrigation scheduling based on pan evaporation values for cucumber (Cucumis sativus L.) grown under field conditions. Agricultural Water Management 81: 159-172.

23. Dogan E, Kirnak H, Berekatoglu K (2008) Water stress imposed on muskmelon (Cucumis Melo L.) with subsurface and surface drip irrigation systems under semi-arid climatic conditions. Irrigation Science 26: 131-138.

24. Shen XJ, Sun JS, Zhang JY (2011) Study on calculation method of soil moisture content under drip irrigation. Journal of Soil and Water Conservation 3: 241-244.

25. Ma FY, Li MC, Yang JR (2002) A study of effect of water deficit of three periods during cotton and thesis on canopy apparent photosynthesis and WUE. Scientia Agricultural Sinica 12: 1467-1472.

26. Liu LD, Li CD, Sun HC (2007) Effect of water stress on carbohydrate metabolism in cotton with varying boll Size. Cotton Science 2: 129-133.

27. Tang W, Luo Z, Wen SM (2007) Comparison of inhibitory effects on leaf photosynthesis in cotton seedlings between drought and salinity stress. Cotton Science 1: 28-32.

28. Ma SC, Zhang XC, Duan AW (2012) Regulated deficit irrigation effect of winter wheat under different fertilization treatments. Transactions of the Chinese Society of Agricultural Engineering 6: 139-143.

29. Shen XJ, Zhang JY, Liu ZG (2012) Effects of different water treatments on yield and water use efficiency of cotton with drip irrigation under film mulch. Agricultural Research in the Arid Areas 2: 118-124.

30. Liu JX, Zhou SQ, Jin LS (2012) Evapotranspiration of a film-mulched cotton field under drip irrigation in North Xinjiang. Arid Zone Research 2: 360-368.

28

Small-scale Irrigation: The Driver for Promoting Agricultural Production and Food Security (The Case of Tigray Regional State, Northern Ethiopia)

Nahusenay Teamer Gebrehiwot[1*], Kassa Amare Mesfin[2] and Jan Nyssen[3]

[1]Department of Cooperative Studies, College of Business and Economics, Mekelle University, Ethiopia
[2]Department of Earth Science, College of Natural and Computational Science, Mekelle University, Ethiopia
[3]Department of Geography, Ghent University, Belgium

Abstract

Small-scale irrigation practices determine the level of food production and, to a great extent, the state of the food security. Considering its importance to the overall growth of the agriculture sector, the Government of Ethiopia gives special emphasis to enhance agricultural production through the promotion of small-scale irrigation schemes. This paper looks at the role of small-scale irrigation in promoting agricultural production and food security. It pays particular attention to the results of small-scale irrigation practices, issues that enhance or impede small-scale irrigation capacities and explore complementary strategies that enhance the development of small-scale irrigation. We use descriptive and regression analysis to examine the relationship between small-scale irrigation and agricultural production. Relying on a survey result and observations, we conclude that, small-scale irrigation schemes could significantly improve agricultural production and the status of food security.

Keywords: Ethiopia; Agricultural production; Food security; Small-scale irrigation; Crop

Introduction

A doubling in global food demand projected for the next 50 years poses huge challenges for the sustainability both of food production and of terrestrial and aquatic ecosystems and the services they provide to society. Developments in agriculture over the last 30 years have brought significant increases in global production, partly as a result of expansion of cropland, partly through changes in technologies over time [1].

Agricultural productivity is very much affected due to variability of rainfall and drought [2]. By supporting the idea of Bhattarai, Warren and Khogali described that the rainfall in Ethiopia has erratic nature and the consequent moisture stress is a major limitation to raising agricultural production [3]. Von Braun and others also argue that farmers in poor areas have suffered from chronic poverty and severe food insecurity being vulnerable to climatic changes and dependent on variable rainfall. This is mainly attributed to a low level of agricultural productivity characterized by persistent rural poverty, and increasing population pressure has often resulted in a vicious circle of poverty and environmental degradation [4].

As many of the low productivity areas have untapped water resources, irrigation development is being suggested as a key strategy to enhance agricultural productivity and to stimulate economic development [2]. To lift and keep millions out of poverty requires that smallholder agriculture be productive and profitable and bring agricultural transformation by which individual farms shift from highly diversified, subsistence-oriented production towards more specialized production oriented towards the market or other systems of exchange. Hussian and Smith in the contemporary literature argued that irrigated farming is recognized as central in increasing land productivity, enhancing food security, earning higher and more stable incomes and increasing prospects for multiple cropping and crop diversification [5,6]. In some places, cereal production is more than doubled between 1995 and 2001 due to the combined effect of expansion of irrigation and the use of high-yielding varieties and fertilizers [7]. It was also claimed that Ethiopia cannot assure food security for its population with rain-fed agriculture alone and without a substantive contribution of irrigation [8]. A review of several empirical studies made by Hussain and Hanjira indicated that there is a strong linkage between irrigation development and poverty reduction through increasing productivity, livelihood diversification as well as creating employment opportunities and income [7].

The history of water harvesting in Ethiopia dated back as early as the pre Axumit period (560 BC). It was a time when rainwater was harvested and stored in ponds for agricultural and water supply purposes. Anthropologists [9] have documented evidences from the remains of ponds that were once used for irrigation during this period. FAO [10] also reported that water harvesting irrigation is an ancient practice in Ethiopia, as an integral part of Ethiopian agriculture, dating back several centuries.

The current Ethiopian government considers water as an entry point for development. It identified as an important policy instrument to stimulate economic growth and rural development in general and in ensuring food security in particul [11]. Irrigation schemes are essential policy options chosen by the government to eradicate poverty and secure food security in Ethiopia and it has a great impact on the livelihood of rural community where agriculture is the bedrock of their life [12]. Research result of Carter and Danert indicated that there are two major classifications of small-scale irrigation, the modern scheme and the traditional scheme. The development of modern small-scale irrigation started since the mid-1980s [13]. They have relatively permanent structure and improved water control system, and are mostly constructed by either the government or non-governmental organizations (NGOs). The traditional one is constructed by the local community.

---

*Corresponding author: Nahusenay Teamer Gebrehiwot, Department of Cooperative Studies, College of Business and Economics, Mekelle University, Ethiopia, E-mail: nahusenay.teamer@mu.edu.et

According to Lakew [14], the types of water harvesting technologies used by farmers are shallow wells, household level structural storage ponds, community ponds and spring water through traditional irrigation. With regard to opportunities and problems of agriculture sector in Ethiopia, Haile [15] described that there is huge potential of agricultural resource; however, the agriculture sector of the country has increasingly been unable to provide adequate food for the rapidly growing population. Sileshi [16] also argued that water resource management in agriculture is a critical contributor to the economic and social development of Ethiopia. However, he also explained that irrigation is not a simple silver bullet: first, it can only work if other components of the agricultural system are also effective (e.g., seeds, extension); second, all the tools in the toolkit will be required to construct a viable solution.

Assessment report of Ministry of Agriculture, Natural Resources Management Directorate [17] indicated that irrigation development, particularly the smallholders has significant importance to raise production and productivity to achieve food self-sufficiency and ensure food security at household level in particular and the country at large. The same report also indicates that the major production constraints impeding development in the irrigation sub-sector among others are:

(i) predominantly primitive nature of the overall existing production systems, (ii) shortage of agricultural inputs and credit systems, (iii) limited access to improved irrigation technologies and inadequate research support, (iv) lack of trained manpower and frequent staff turnover, and (v) unstable institutional set up and inadequate extension services and limited availability of capital.

To see the positive effect of small-scale irrigation on agricultural production and food security, therefore, it is important to understand how small-scale irrigation farmers are doing the irrigation activities and what factors impede them from achieving the optimum level of agricultural outputs. Therefore, this paper tries to assess the results of small-scale irrigation practices in promoting agricultural production and its impact in the improvement of the livelihood of the farmers.

The specific research objectives are formulated as follows:

• Understand the role of small-scale irrigation on agricultural production and food security of rural household farmers.

• Identify issues that enhance or impede small-scale irrigation capacities in the target areas.

• Explore and propose complementary strategies that enhance the development of smallholder irrigation throughout the region.

Materials and Methods

Description of the study area

Tigray regional state is found in the northern part of Ethiopia extending from 12°15' to 14°50'N and from 36°27' to 39°59'E. It covers a little more than 80 000 km^2, most of which are highlands between 1500 and 3900 ma.s.l. As per the projection of the Central Statistical Agency, Federal Democratic Republic of Ethiopia [18], the population of Tigray Regional State is 4,960,003 and from this population about 75.8% is living in rural area.

Tigray region is highly vulnerable to recurrent drought and famine. Rainfall distribution in the region is characterized by high temporal and spatial variability, with annual precipitation ranging from 450 to 980 mm [19]. Moisture deficiency is the most critical problem identified in the study of production systems in the region [20]. Agriculture is the main economic activity that is employed by the majority of farmers in the study areas. Smallholder mixed crop livestock farming is the dominant mode of production in the districts. The system is oriented towards providing subsistence requirements for the farm household.

The target *woredas* are the potential areas whereby small-scale irrigation activities are undertaken by farming communities. In these *woredas*, a lot of conservation activities have been undertaken in the previous years and small-scale irrigation is practiced using micro dams, shallow wells, private ponds, community bond, check dams and diversions. In the target *tabias* farmers produce a mixture of cereals, vegetables, fruits and pulses by using small scale irrigation scheme as well as using rain-fed. Based on the information collected from the five target *tabias*, a total of 2648.4ha was used for irrigation activities in 2013/14 harvesting period. The major crops grown in the area includes teff, barley, wheat, sorghum, maize, faba bean, finger millet, chickpea and others. Different vegetables such as tomato, potato, onion, pepper, lettuce, carrot, garlic and fruits grow in the area using irrigation both at the rainy and dry seasons.

Definition of dependent and independent variables

The dependent variable of this study is agricultural production of the small-scale irrigation scheme in the target *tabias* (tabia is the lowest administrative hierarchy below woreda/district). There are quite variety of factors which can affect agricultural production, both positively and negatively. It is very difficult to enumerate and discuss all the factors that affect agricultural production. However, for the purpose of the research at hand, some of the major factors affecting agricultural production are considered. Based on the economic theory and results of previous empirical studies, each variable is defined with its hypothesis as indicated below.

Sex: This study hypothesized that female headed households are less likely to participate in the small-scale irrigation scheme as compared to males [21].

Age: Older people have relatively greater experience of farming activities and better access to land than younger heads [22].

Household Size: A negative correlation between household size and food security is expected as food requirements increase in relation to the number of persons in a household [23].

Education: Educational attainment could lead to awareness of the possible advantages of modernizing agriculture by means of technological inputs; enable them to read instructions on fertilizer packs and diversification of household incomes which, in turn, would enhance households' food supply [24].

Irrigation farmland size: Under subsistence agriculture, holding size is expected to play a significant role in influencing farm households' food security [24].

Availability of water for irrigation activities: The potential to improve yields depends strongly the practice of irrigation schemes [21].

Application of inputs: Fertilization of farmland can boost agricultural production and influence the food security status of a household.

Application of technologies (motor and treadle pumps): The use of motorized water pump and treadle pump as part of water lifting technology and other water lifting technologies, have increased agricultural production [25].

Farm Income: Net farm income is the single most watched

indicator of farm sector well-being, as it captures and reflects the entirety of economic activity across the range of production processes, input expenses, and marketing conditions that have persisted during a specific time period [26]. In our hypothesis, we assume that irrigation activities enhance agricultural production and thus increase revenue earned by farmer household.

Sampling procedure and data collection

This study was conducted based on the survey undertaken during December 2014 in three potential irrigation *woredas* of Tigray region, Northern Ethiopia. The sample households were selected by utilizing a three-stage stratified sampling procedure as the major method of sampling. During the first stage, three *woredas* were purposely selected, namely Kilte Awlaelo, Ahferom and Tahtay Maichew *woredas* from the eastern and central zone of Tigray regional state. Those *woredas* with better small-scale irrigation interventions were taken into consideration in the selection of *woredas*. Similarly, at the second stage, in consultation with *Woredas* Agriculture and Rural Development offices, five *tabias* (two *tabias* from Kilte Awlaelo *Woreda*, two *tabias* from Ahferom *Woreda* and one *tabia* from Tahtay Maichew *Woreda*) were purposely chosen. The study *tabias* were chosen based on the performance of small-scale irrigation activities, accessibility to the irrigation sites and the presence of a large number of ponds, shallow wells and diversion beneficiaries. At the third stage, household lists that are involved in small-scale irrigation scheme in the selected *tabias* were obtained from *tabia* administration. From the total number of households that are using small-scale irrigations in the selected study areas, 20 beneficiaries were taken from each *tabia* by systematic random sampling.

Based on the multistage sampling process mentioned above, a total of 100 households were selected by systemic random sampling technique. The summary of the number of households selected from each study area is presented in Table 1.

To carry out this study, both primary and secondary data sources were employed. The primary data were collected by employing triangulation method such as key informant interview using semi-structured checklist, focus group discussion, expert interview; semi structured household questionnaire and observation of events in the irrigation scheme. Secondary information that could supplement the primary data were collected from published and unpublished documents obtained from different sources. These included country policy statements, strategies regulations, reports and past case study papers on small-scale irrigation practices.

Issues covered during the data collection were demographic features, socio-economic situation of sampled households, irrigation practices, irrigated land, crops grown, the livelihood impact of irrigation activities, opportunities and challenges of irrigation activities, and areas of attention for the expansion and development of irrigation schemes. Discussions were also held with irrigation experts at the *woredas* Offices of Agriculture and Rural Development. Secondary data

was also collected from respective *woreda* Offices of Agriculture and Rural Development.

Data analysis

The data generated through household questionnaire was analyzed by employing the Statistical Package for Social Science (SPSS version 20) and MINITAB version 14. To analyze the data collected, descriptive methods such as frequency, percentage, average, and standard deviation were used. The Pearson correlation (r) was used to measure the linear association between the dependent and independent variable. It describes the strength of the relationship between the two variables. The coefficient of determination (r^2) was also used to define the percent of the variation in the values of the dependent variable that can be explained by variations in the value of the independent variables. In this regard, r^2 explains the variability in the dependent variable explained by the variability in the independent variable. Multiple regressions was also used to allow additional factors to enter the analysis separately so that the effect of each independent variable can be estimated. It is valuable for quantifying the impact of various simultaneous influences upon a single dependent variable. F-Test was also used to examine whether the two populations have the same variance while the t-test and analysis of variance (ANOVA) was used to compare group means. A t-test was used to examine the mean difference between different continuous variables.

Results and Discussion

Descriptive analysis

Needless to mention that agricultural production is determined by various household attributes. Of these attributes, demographic and socio-economic characteristics are the ones. Hence, this section will discuss household characteristics like sex, age, family size, education status, household access to productive resources, etc. which determine agricultural production of sample respondents. The survey also includes other independent variables that determine agricultural production of the farmers such as cultivated land, available active labour force, availability of water resources such as ponds, diversions, utilization of inputs, trends of agricultural production, and other related activities. The descriptive analysis and the overall relationship of the dependent and independent variables are discussed as follows.

Agricultural production of the sample respondents: The agricultural products produced by the sample respondents in 2013/14 were classified into three categories namely vegetables (onion, tomato, potato and others), fruits (avocado, mango, orange, banana, etc.) and cereal crops (maize, wheat, sorghum, *teff*, etc). Based on the information from the respective *Woredas* Office of Agriculture and Rural Development, the total area covered by small-scale irrigation scheme was about 62.42 ha. From this total amount about 35.77 ha (57.3%) was covered by vegetables with total production of 265,268 kg (7416 kg per ha) and sales volume of Birr 1,209,780.00 (Birr 4.56 per kg). Also, the total area covered by fruits was about 9.4 ha (15.06%) with a total fruit production of 8096 kg (8600 kg per ha) and sales volume of Birr 44,440.00 (Birr 5.49 per kg). The total area covered by crops was about 17.25 ha (27.63%) while the total production was 53310 kg (3090.4 kg per ha). (1USD is equivalent to 20.5Birr as of January 2015) (Figure 1).

Spatial distribution of respondents and agricultural production: In 2013/2014 harvest period, about 27% of the respondents of were producing less than1000 kg, 32% of the respondents were producing above 1000 kg but less than 5000 km while 20% of the respondents

Woredas	Tabias	Frequency
Kilte Awlaelo	Abraha Atsbiha	20
	Mesanu	20
Ahferom	Endamariam	20
	Laelay Megara Tsemri	20
Tahtay Maichew	May Siye	20
	Total	100

Source: survey result, 2015

Table 1: Summary of sample respondents by Woredas and tabias.

Figure 1: Products of small-scale irrigation, Abraha Atsbiha, Kilte Awlaelo.

Total yield in kg in 2013/2014	Tabia Abraha Atsbiha	Tabia Mesanu	Tabia Endamariam	Tabia Laelay Megara Tsemri	Tabia May Siye	Total
No response	0	6	6	9	0	21
Up to 1000 kg	2	2	9	5	9	27
From 1001 kg up to 2000 kg	1	4	4	5	6	20
From 2001 kg up to 5000 kg	2	4	0	1	5	12
From 5001 kg up to 10000 kg	10	2	0	0	0	12
Above 10000 kg	5	2	1	0	0	8
Total	20	20	20	20	20	100

Source: Survey result, 2015

Table 2: Total yield in kg in 2013/2014 harvesting period by tabia.

Total yield in kg in 2013/2014	Sex of the head of the household					
	Male		Female		Total	
	Number	Percent	Number	Percent	Number	Percent
No response	19	21.11	2	20	21	21
Up to 1000 kg	25	27.78	2	20	27	27
From 1001 kg up to 2000 kg	16	17.78	4	40	20	20
From 2001 kg up to 5000 kg	10	11.11	2	20	12	12
From 5001 kg up to 10000 kg	12	13.33	0	0	12	12
Above 10000 kg	8	8.89	0	0	8	8
Total	90	100	10	100	100	100

Source: Survey result, 2015

Table 3: Total yield in kg in 2013/14 by sex.

were producing more than 5000 kg. In this regard, the respondents of Abraha Atsbiha tabia were more productive as compared to other tabias (Table 2).

Sex of the respondents vs. agricultural production: Out of the total sampled household members, 90% respondents constitute male headed households and the remaining 10% respondents are female headed households. The survey result also revealed that about 88% of the respondents are married, 8 % are single and the remaining 4% are either separated or widowed (Table 3).

When we compare the agricultural production by taking sex as a variable, we can see that about 80% female respondents are producing less than 2000 kg while that 66.67 % of men respondents are producing similarly less than 2000 kg. On the other side, no female

was producing more than 5000 kg while 22.22 % of male respondents were producing more than 5000 kg in 2013/14. From the given data Table 2, it is observed that there is higher percentage of production for men headed households than women headed households. This low level of agricultural production of female headed households can be attributed to the low level of input utilization by female respondents. In this regard, about 55.56% of female respondents were utilizing less than 50 kg of inputs in 2013/14 while that of male respondents is 31.51%. On the other side, about 50.68% of male respondents were applying inputs between 51 kg to 100 kg while only about 22.22% of female respondents are applying inputs within this range. Moreover, from the survey result we can also observe that women have less accessibility to irrigation land as compared to men. In this regard, about 44.44% of women respondents are having less than 0.25 ha of irrigation land while that of male is about 15.49%.

When we measure the strength of the linear association between sex and agricultural production through Pearson correlation, it is moderate with p-value of 0.512. Hence the relation between sex and agricultural production is not very much significant and female headed households are likely to participate in the small-scale irrigation scheme like that of males.

Age of the respondents vs. agricultural production: Our hypothesis stated that the higher the age of the household head, the more stable the economy of the farm household; because older people have relatively richer experience of farming activities and better access to land than younger heads. When we see the age structure of the sample respondents, it is observed that 23% of the respondents are within the range of 18 to 35 years of age while 70% of the respondents are representing the age group of 36 to 65, and the remaining 7% is above 65 years of age. The average age of the household members in the study sample was observed to be 46.26 years with the standard deviation of 12.22. When we measure the strength of the linear association between age and agricultural production through Pearson correlation, it is very poor with p-value of 0.059 which is against our hypothesis. Hence the relation between age of the respondents and agricultural production is very poor and this could be attributed as a result of better access to education, input utilization and technologies of the young respondents as compared to older respondents.

Household size vs. agricultural production: Household size was considered as one of the potential variables that could affect agricultural production negatively. About 10 % of the total sample households have a maximum of two family size whereas 53% of the households have the family size that ranges from three to six. The rest 37% of the sample have more than six family sizes. The largest family size of the sample households was found to be 10 and the smallest was 1. On the other hand, the average family size of the sample households is estimated to be 5.63 with the standard deviation of 2.4 and this is similar to the national average of 5 persons (CSA, 1994).

When we measure the strength of the linear association between total household size and agricultural production using Pearson correlation, it is 0.912. Hence, there is very high significant relationship between family size and agricultural production.

Educational status of respondents vs. agricultural production: In our hypothesis, education was considered as one of the factors which influence the agricultural production status and food security of households positively. Educational status of the respondents shows that about 35% of the respondents are illiterate while 6%, 31% and 18% have attended religious learning, primary school first cycle (Grade

1 to 4) and primary school second cycle (Grade 5 to 8) respectively. About 9% of the respondents have joined secondary school (Grade 9-12). Only one respondent is diploma holder in general agriculture. The survey also assessed the number of children attending schools at a household level. Accordingly, about 21% of the respondents do not send their children to schools while 18%, 42% and 16% are sending 1 to 2, 3 to 4 and 5 to 6 children to school respectively. Only 3% of the respondents send above 6 children to schools. When we assess the relationship between education and agricultural production, 15 of the illiterate farmers were producing above 2000 kg while 17 of the educated farmers were producing above 2000 kg (the average agricultural production per household is about 3267 kg with standard deviation of 55.88). When we measure the strength of the linear association between education status of respondents and agricultural production using Pearson correlation, it is weak with p-value of 0.107. This might imply that sample respondents with lower educational attainment were using technological inputs and fertilizer to boost their agricultural yields.

Availability of irrigation land vs. agricultural production: The total irrigation farmland is normally defined as the cultivated own land plus any land operated/owned by other farmers but cultivated with some arrangements such as sharecropping for grain, cash or crop residue. With regard to the irrigated land size among the respondents, about 56% of the respondents are having less than 0.5 ha, 17%, 24% and 3% of the respondents are having irrigation land from 0.51 up to 1 ha, above 1 ha but less than 2 ha and above 2 ha respectively. In this regard, the mean irrigated land size in the study area for 2013/14 harvest period was 0.6242 ha with standard deviation of 0.6102. When we measure the strength of the linear association between the size of the irrigated land and agricultural production through Pearson correlation, it is strong and positive with p-value of 0.000 (significant at 0.01 level, 2-tailed). Therefore, there is very high significant relationship between irrigation farmland and agricultural production, which supports our hypothesis that irrigated lands account for a substantial portion of increased yields.

Availability of water for irrigation activities vs. agricultural production: In our hypothesis we stated that the potential to improve yields depends strongly the practice of irrigation schemes. In this regard, the researchers tried to see availability of water from different corners such as the availability of water sources, investment in ponds, and distance to water points, adequacy of water for irrigation activities frequency of utilization water (access to water) and experience in irrigation activities.

With regard to the ownership of ponds, about 61% of the respondents have their own ponds as a source of water for irrigation activities. The remaining 39% of the respondents do not own ponds but utilize various sources of water such as river, diversions for their irrigation activities. In relation to the adequacy of water for irrigation, about 60% of the respondents also agreed that there was adequate water for irrigation activities. In this regard, a question was raised to respondents in relation to the availability of irrigation water during 2013/14. About 41% of the respondents reply that there was water throughout the year, 9%, 16% and 2% of the respondents confirmed that there was water for 7 to 9 months, for 3 to 6 months and below three months respectively. Based on the information from the respondents, the average availability of water during 2013/14 was 9.058 months with standard deviation of 3.077.

With regard to access to water (frequency of using water per week) indicated in Table 3 above, about 70% of the respondents get access to

Variables	Correlation	Total yield in kg
Source of irrigation water	Pearson Correlation	-.193
	Sig. (2-tailed)	.054
	N	100
Frequency of using irrigation	Pearson Correlation	-.032
	Sig. (2-tailed)	.755
	N	100
Distance to pond in minutes	Pearson Correlation	.416**
	Sig. (2-tailed)	.000
	N	100
Total cost for the construction of pond	Pearson Correlation	.200*
	Sig. (2-tailed)	.046
	N	100
The availability of water ponds	Pearson Correlation	.425**
	Sig. (2-tailed)	.000
	N	100
For how many years you have been using the ponds (experience in using ponds)?	Pearson Correlation	.318**
	Sig. (2-tailed)	.001
	N	100
How long did your pond hold water last year (availability of water during harvesting period)?	Pearson Correlation	.404**
	Sig. (2-tailed)	.000
	N	100
**Correlation is significant at the 0.01 level (2-tailed).		
*Correlation is significant at the 0.05 level (2-tailed).		

Source: survey result, 2015

Table 4: Strengths of linear association of water related variables to agricultural production.

the water once per week, while 19%, 8% and 1% of the respondents get access to irrigation water twice per week, three times per week (once per two day) and on daily basis respectively (Table 4).

The strength of the linear association between water related variables and agricultural production through Pearson correlation was analyzed and its result is found to be as indicated in Table 3.

From the table, we can see that investment in construction of ponds to be significant at 0.05 level (2-tailed) while the linear relationship for the variables distance to pond in minutes, the availability of water ponds, experience in using ponds and availability of water during harvesting period are significant at 0.01 level (2-tailes). Hence, there is very high significant relationship between access to irrigation water and agricultural production.

Application of agricultural inputs vs. agricultural production: Our hypothesis was fertilization of farmland can boost agricultural production and influence the food security status of a household. The application of inputs for agricultural activities was assessed on the basis of whether a household uses fertilizer or not. Accordingly, more than 83% of the respondents were using agricultural inputs for their farming activities during 2013/14 harvesting period. The major types of inputs used in the study area were DAP, UREA and improved seed. When we assess the rate of utilization of inputs, we can see that about 28% of the respondents were utilizing less than 50 kg, 39% of the respondents from 51 kg up to 100 kg, 13% of the respondents from 100 kg to 200 kg and 2% above 200 kg. The average utilization of inputs is 77.72 kg with standard deviation of 70.12 and its average cost is Birr 1056.00 with standard deviation of 1036. When we look up on the strength of the linear association between the fertilizer used and agricultural production through Pearson correlation, it is strong and positive with p-value of 0.000 (significant at 0.01 level, 2-tailed) which is in line with the previous research result of intensive high-yield agriculture is dependent on addition of fertilizers [27] (Figure 2).

Utilization of irrigation technologies vs. agricultural production: In our hypothesis we assume that the use of motorized water pump and treadle pump as part of water lifting technology and other water lifting technologies have increased agricultural productivity of farmers. In our research, about 58% of the respondents own motor pumps while 11% of the respondent rent pumps for irrigation activities and the remaining 31% neither own nor rent pumps. The average price of pump is Birr 8803.00 with standard deviation of 4797 (Figures 3 and 4).

When we measure the strength of the linear association between the use of irrigation technologies and agricultural production through Pearson correlation, it is strong and positive with p-value of 0.004 (significant at 0.01 level, 2-tailed). This result indicates that as household farmers increase their investment in the purchase irrigation technologies, agricultural production increases which leads to better access to food and thereby improved standard of living which is similar to the research results of Nata and Bheemalingeswara [25].

Agricultural production vs. total revenue: Randy [26] identified that net farm income is the most watched indicator of farm sector well-

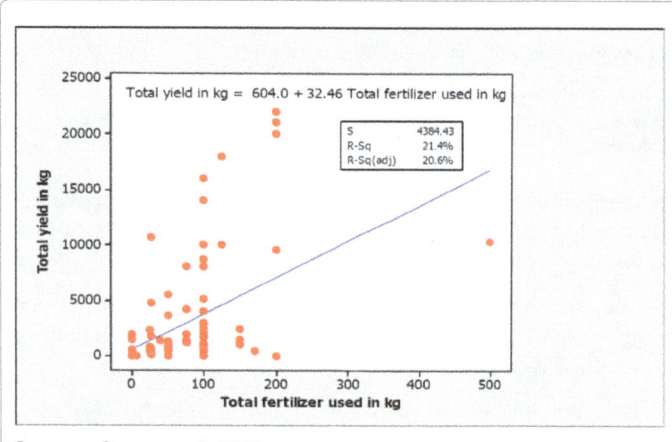

Sources: Survey result, 2015
Figure 2: Relationship of fertilizer used in kg to total yield in kg.

Figure 3: Motorized river diversion, Mesanu.

Figure 4: Treadle pump from ponds, Mesanu.

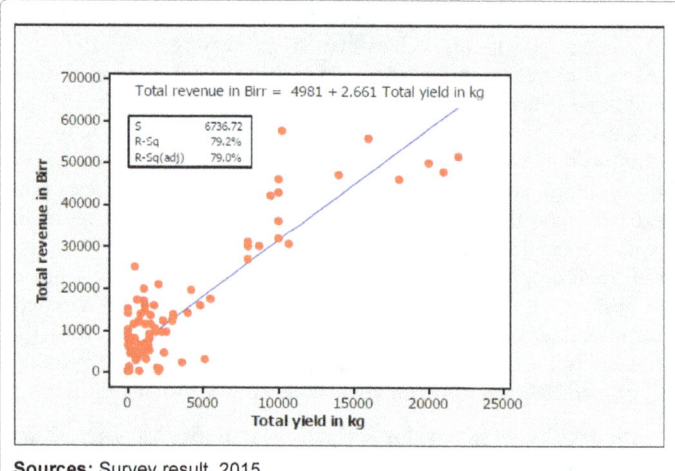

Sources: Survey result, 2015
Figure 5: Relationship of total yield in kg to total revenue in Birr.

being, as it captures and reflects the entirety of economic activity across the range of production processes, input expenses, and marketing conditions that have persisted during a specific time period. During the survey, about 64% of the respondents confirmed that irrigation activities are improving their agricultural production and their income from the sale of agricultural products which leads to improved standard of living of their families. The average income from the sale of agricultural products for the year 2013/14 was Birr 13301.00.

In this regard, the survey result indicated that those farmers who are producing more agricultural products through small-scale irrigation are becoming better-off and they are increasing their income (revenue) from the sale of agricultural products and this leads them to improve their standard of living (Figure 5).

This was further confirmed during the discussion with the respondents. Some of the major results identified in relation to the contribution of small-scale irrigation to agricultural production and thereby food security include increased agricultural production and better food security, getting additional income, access to improved nutritional values (vegetables and fruits), improved feeding habit, improved access of water for drinking livestock development and sanitation, purchase of oxen and cows, pay credit and save money, purchase of household goods, building of houses in towns, cover educational cost for their children and the likes. These all indicate that small-scale irrigation activities are promoting the livelihood of the farmers.

Summary of regression results

The joint effect of a group of independent variables on promoting agricultural production is studied by framing the multiple regression equation of the variable "Y" on the other independent variables. The following model with 6 independent variables was used in the model specification. In classical regression model, each estimate gives the partial effect of a coefficient with the effects of other X variables being controlled.

$$Y = a + b_1x_1 + b_2x_2 + b_3x_3 + b_4x_4 + b_5x_5 + b_6x_6$$

Where: Y=Agricultural production in kg

a = Intercept (constant)

b_1 to b_6= Regression coefficients

Based on the regression analysis, the following results were found. Total yield in quintal = 1.750

0.410 age of the head of household

0.172 schools attended by the head of the household

0.070 Investment (purchase value) of pump

+0.268 total inputs used in kg

+0.090 total irrigated land in ha

+1.030 revenue generated from the sale of agricultural products (Table 5).

From the table of regression analysis (Table 5) we can see that the variables age of the head of the household, schools attended by the head of the household, investment (purchase value) of pump, total inputs used in kg, total irrigated land in ha and revenue generated

from the sale of agricultural products showed a significant impact on agricultural production.

Coefficient of determination result (r^2): The coefficient of determination (r^2) is 91.9%. This denotes that about 91.9% of the total variation of the dependent variable (agricultural production) is explained by the independent variables included in the multiple regressions. Therefore, one must look at other variables beyond the listed independent variables in order to find out the reasons for the increase in agricultural production. Hence, the actual reason for the increase in agricultural production in the farm sector can be the factors such as fertility of soil, ownership of oxen, extension service, mixed cropping, crop rotation, properly application of quality agricultural inputs and others.

The F ratio is also found significant. From the value of "t" statistic corresponding to the regression coefficients, some variables are found to be significant such as age of the head of the household, schools attended by the head of the household, investment (purchase value) of pump, total inputs used in kg, total irrigated land in ha and revenue generated from the sale of agricultural products.

Major problems of small-scale irrigation and possible intervention areas

The major small scale irrigation problems identified during the survey include financial constraints especially for the purchase of motor pumps, shortage of agricultural inputs specially improved seed and pesticides, high cost of irrigation, shortage of water pump technologies, spare parts and gabions, technical problems such as maintenance of motor pumps, insufficient market information and market networks, shortage of ponds and diversion, infrastructure specially road and storage, theft of fruits, diseases and pests such as rust, root ruts, ball worm, blights, powdery mildew, gummosis and water borne diseases, inefficient management of resources such as water, land and labour. As per the discussion with *wereda* irrigation experts the major problems of the irrigation are shortage of motorized pumps, insufficient diversion infrastructure, shortage of improved seed, dependency syndrome on government and on donors (farmers want construction of diversion

Coefficients[a]						
Model		Unstandardized Coefficients		Standardized Coefficients	t	Sig.
		B	Std. Error	Beta		
1	(Constant)	1.750	.441		3.970	.000
	Age of the head of the household	-.410	.127	-.138	-3.229	.002
	Schools attended by the head of the household	-.172	.044	-.166	-3.905	.000
	Investment (purchase value of pump)	-.070	.031	-.108	-2.229	.028
	Total inputs used in kg	.268	.074	.171	3.635	.000
	Total irrigated land in ha	.090	.034	.118	2.606	.011
	Revenue generated from the sale of agricultural products					

a. Dependent Variable: Total yield in kg

ANOVA[a]						
Model		Sum of squares	df	Mean square	F	Sig.
1	Regression	204.549	6	34.092	84.231	.000[b]
	Residual	37.641	93	.405		
	Total	242.190	99			

Model Summary				
Model	R	R Square	Adjusted R Square	Standard error of the estimate
1	.919a	6	0.835	0.636

Source: survey result, 2015

Table 5: Results of multiple regression model.

and water harvesting technologies from the government and donor) rather than introducing such irrigation schemes and technologies, inefficient utilization of resources such as water and land, lack of knowledge and skills in irrigation activities, inappropriate utilization of inputs and the likes.

The respondents were requested to identify priority areas in solving the small-scale irrigation schemes. Accordingly, they highlighted the following issues as a point of attention from the concerned bodies. Improving access to and availability of inputs specially improved seeds and chemicals, improve access to credit and long-term loans, improve access or supply of pumps, spare parts and electric dynamo for water pumps, facilitating training and experience sharing, strengthening/ organizing cooperatives of irrigation, utilize resources efficiently, produce market oriented products/ produce seeds (encouraging multiplying seeds by farmers), improving access to timely market information and market networking, strengthening value chain and supply of processing materials, improve conservation of soil and water, enclosure of irrigation area and reduce grazing.

The respondents were also requested to highlight the role of the government in promoting small-scale irrigation practices throughout the region. Some of the major issues raised by the respondents as the role of government include but not limited to supply of quality and variety of improved seeds for vegetables and fruits, fertilizers, pesticides and chemicals (in some cases subsidizing the supply of inputs), improving access or supply of pumps, spare parts and electric dynamo for water pumps, construction of additional diversions, common well or bore pond and check dams, regular maintenance of water structures, canals and diversions, introducing modern irrigation and water harvesting technologies, technical support including provision of training and extension service experience sharing to beneficiaries on irrigation activities, continuous follow up and supervision, arrangement of transpiration facilities (improve infrastructure facilities), promoting access to credit, promoting integrated farming including moisture conservation and watershed management, supply of construction materials such as cement, gabion and providing timely market information (market networking or linkage).

Conclusions and Recommendations

The study covers the demographic and socio-economic factors that influence small-scale household irrigation in promoting agricultural production and livelihoods. In particular, it targets *woredas* whereby small-scale irrigation activities are undertaken by farming communities. In the target *tabias* farmers produce a mixture of cereals, vegetables, fruits and pulses by using small scale irrigation scheme as well as using rain-fed.

A total of 100 respondents were involved in the survey. From the survey, it was found that the average family size to be 5.59 with the standard deviation of 2.4 and the average irrigation land holding of the respondents is 0.6242 ha with standard deviation of 0.6102. Considering sex of the head of the household as a variable, the survey result revealed that female headed households produce less as compared to men. When we measure the strength of the linear association between different variables and agricultural production through Pearson correlation, variables such as availability of water ponds, distance to water points, availability of water in the harvesting period, utilization of inputs, access to technologies, total irrigated land are playing an important role in promoting agricultural production and there is positive and significant relationship to agricultural production. Similarly from the regression analysis we found that the variables age

of the head of the household, investment (purchase value) of pump, total inputs used in kg, total irrigated land in ha and revenue generated from the sale of agricultural products showed a positive and significant impact on agricultural production and thereby enhancing food security of the rural farmers.

The problems identified during the survey include financial constraints especially for the purchase of motor pumps, shortage of agricultural inputs specially improved seed and pesticides, high cost of irrigation, shortage of water pump technologies, spare parts and gabions, technical problems such as maintenance of motor pumps, insufficient market information and market networks, shortage inadequate ponds and diversion and lack of infrastructure facilities.

Relying on a survey result and observations, we conclude that, small-scale irrigation schemes could significantly improve agricultural production and food security. In order to improve and expand small-scale irrigation activities, it is necessary to solve the above mentioned problems through the involvement and joint effort of all stakeholders including the farming community, government, non-government organizations, private sector and also designing well-structured short-term and long-term plans and development programs to fill the capacity gaps. Moreover, attention should be given for the expansion of irrigation cooperatives so that farmers can join-hand in dealing with irrigation activities based on cooperative principles and values and solve their common problems through members' participation.

References

1. Thrupp LA (1998) Linking agricultural biodiversity and food security: the valuable role of agro biodiversity for sustainable agriculture. Royal institute of international affairs 76: 265-281.

2. Bhattarai M, Sakthivadivel R, Hussein I (2002) Irrigation impacts on income inequality and poverty alleviation: Policy issues and options for improved management of irrigation systems. International Water Management Institute (IWMI).

3. UNSO (1992) Assessment of Desertification and Drought in the Sudano-Sahellank Region, 1985-91. New York.

4. Von Braun J, Fan S, Meinzen-Dick R, Rosegrant MW, Pratt A (2008) International agricultural research for food security, poverty reduction, and the environment. International Food Policy Research Institute (IFPRI) Washington, DC, USA.

5. Hussain I, Marikar F, Thrikawala S (2001) Impact of irrigation infrastructure development on poverty alleviation in Sri Lanka and Pakistan. Journal of Development Studies 21: 29-31.

6. Smith L (2004) Assessment of the Contribution of Irrigation to Poverty Reduction and Sustainable Livelihoods. Int J Water Resource Dev 20: 243-257.

7. Hussain I, Hanjra M (2004) Irrigation and Poverty Alleviation: Review of the empirical evidence. J Irrig Drain Syst 53: 1-15.

8. Van Den Berg M, Ruben R (2006) Small-scale irrigation and income distribution in Ethiopia. J Dev Stud 42: 868-880.

9. Fattovich R (1990) Remarks on the Pre- Axumite Period in Northern Ethiopia. Journal of Ethiopian Studies 23: 1-33.

10. FAO (1997) Agriculture food and nutrition for Africa: A resource book for teachers in agriculture. Food and nutrition division.

11. Fitsum H, Makombe G, Namara R, Seleshi B (2009) Importance of Irrigated Agriculture to the Ethiopian Economy: Capturing the Direct Net Benefits of Irrigation.

12. Habtamu W, Mohammed A, Dessalegn W (2014) How do small-holder farmers organize themselves to manage small-scale irrigation schemes? International Journal of current research 12: 10660-10669.

13. Carter R, Danert K. (2006) Planning for Small-Scale Irrigation Intervention. Farm-Africa working papers No. 4, Farm-Africa, Ethiopia.

14. Lakew D (2004) Concepts of Rainwater Harvesting and Its Role in Food Security: The Ethiopian Experience.

15. Haile T (2008) Impact of irrigation development on poverty reduction in Northern Ethiopia. Department of Food Business and Development. National University of Ireland, Cork.

16. Seleshi B (2010) Irrigation potential in Ethiopia Constraints and opportunities for enhancing the system. Ethiopian Agricultural Portal.

17. Ministry of Agriculture, Natural Resources Management Directorate (2011) Small Scale Irrigation Situation Analysis and Capacity Need Assessment: A Tripartite Cooperation between Germany, Israel and Ethiopia, Addis Ababa, Ethiopia.

18. Central Statistical Agency of Federal Democratic Republic of Ethiopia (2013) Population Projection of Ethiopia for All Regions at Wereda Level from 2014 – 2017.

19. Gabremadhin E, Johnston B (2002) Accelerating Africa's Structural Transformation: Lessons from Asia.

20. Araya A, Addissu G, Daniel, Teka D (2005) Pond water productivity under the present use in Tigray region, Northern Ethiopia.

21. World Bank (1998) World development indicators 1998. Washington, DC, The World Bank.

22. Hofferth (2003) Persistence and Change in the Food Security of Families with Children, 1997-1999. Department of Family Studies, University of Maryland.

23. Paddy F (2003) Gender differentials in land ownership and their impact on household food security: a case study of Masaka district. Master thesis. Uganda.

24. Najafi B (2003) An Overview of Current Land Utilization Systems and Their Contribution to Agricultural Production. Islamic Republic of Iran Asian.

25. Nata T, Bheemalingeswara K (2010) Prospects and Constraints of Household Irrigation Practices, Hayelom Watershed, Tigray, Northern Ethiopia. Department of Earth Science, College of Natural and Computational Sciences, Mekelle University, Ethiopia.

26. Randy S (2014) US farm Income. Congressional Research Service.

27. Pinstrup-Andersen P, Pandya-Lorch R (1996) Food for all in 2020: can the world be fed without damaging the environment? Journal of Environmental Conservation Volume 23: 226-234.

29

Effect of Deficit Irrigation on Water Productivity of Onion (*Allium cepa*L.) under Drip Irrigation

Enchalew B[1]*, Gebre SL[2], Rabo M[2], Hindaye B[2], Kedir M[2], Musa Y[2] and Shafi A[2]

[1]*Melkassa Agricultural Research Center, Ethiopia*
[2]*Department of Natural Resources Management, Jimma University, Ethiopia*

Abstract

Deficit irrigation (DI) improves water productivity and irrigation management practices resulting in water saving by maintaining soil moisture content below optimum level throughout growth season. Field study was carried out on clay loam soil at Melkassa Agricultural Research Center, Ethiopia with the objectives to estimate water productivity of onion and evaluate the effect of water deficit on onion yield and quality using drip irrigation. The experiment contained five DI treatments of 90%, 80%, 70%, 60%, and 50% Crop water use (ETc) and the control (100% ETc) laid out in RCBD design with three replications. Irrigation water was applied at allowable soil moisture depletion (p=0.25) of the total available soil moisture throughout the crops growth stage. Statistical analysis revealed that plant height was not affected by the level of DI while, leaf number, bulb diameter, marketable bulb yield and total bulb yield had shown a highly significant (P<0.01) differences among DI treatments. The highest bulb diameter was observed from a control treatment that was significantly different to all other treatments. The highest total bulb yield of 15,690 kg/ha was observed from a control treatment which was not significantly different with treatment receiving 90% ETc. Highest water productivity of onion bulb yield was observed from treatment receiving 70% ETc and better onion bulb diameter was observed from treatment receiving 100% ETc to 70% ETc. The yield response factor ranged between 0.8 and 1.7. Thus, DI practices should be avoided for Ky values that are less than unity. Considering yield response factors (Ky) is limiting factor, 80% ETc application was a marginal and beyond that yield losses are intolerable. Thus, the practice of DI application up to 20% saved 45 to 108 mm depth of water from the gross onion irrigation water requirement.

Keywords: Deficit irrigation; Drip irrigation; Melkassa; Onion; Water productivity

Introduction

Onion (Allium cepaL.) is the most important, widely grown vegetable crop throughout the world [1]. It is widely cultivated as a source of income by many farmers in many parts of the country. It is also one of the most important vegetable crops in Ethiopia. The crop is widely cultivated as cash crop by small-scale and private large-scale farmers. The country has a great potential to produce onion throughout the year both for local consumption and export. It can be produced throughout the year provided dependable rain and/or irrigation water is available. The majority of onion production is found in the Central Rift Valley (CRV) of Ethiopia; however, rainfall is unreliable and insufficient to support onion production that makes irrigation an indispensable practice.

The crop is shallow rooted and sensitive to water stress. As result the crop is commonly given light and frequent irrigation to avoid water stress [2]. Maximum yield could be obtained with the achievement of the entire crop water requirements. The rift valley area is a semi-arid with limited water resources and increasing demand for water combined with high evapo-transpiration rates limits the production and productivity of the crop. Hence, alternatives need to be explored for effective and efficient use of the existing water resources.

Under conditions of scarce water supply, application of deficit irrigation (DI) could provide greater economic returns than maximizing yields per unit of water. The DI has been considered worldwide as a way of maximizing water use efficiency (WUE) by eliminating irrigation that has little impact on yield [3-5]. With DI, the crop is exposed to a certain level of water stress either during a particular period or throughout the whole growing season [6].

A variety of crops have been found to benefit from DI strategy and many researchers pointed out that yield loss that may result from DI is offset by the benefits of reduced water use [7-10]. The response of

Onion to water deficit has been reported by [11] and [12] that showed DI to increase the water use efficiency of onion.

Drip irrigation is one of the most efficient forms of irrigation technology that will allow to apply light and frequent irrigation. The experience from many countries showed that farmers who switch from surface irrigation to drip systems can cut their water use by 30% to 60% and crop yields often increase at the same time [12]. Therefore, the objectives of this study were to determine water productivity of onion and investigate the effect of DI levels on yield and quality onion bulb under drip irrigation.

Description of the study area

A field experiment was conducted at Melkassa Agricultural Research Center (MARC) of the Ethiopian Institute of Agricultural Research (EIAR) in CRV Ethiopia (8°24' N lat. 39°21' E long. 1550 m.a.s.l.). Central Rift Valley of Ethiopia is a semi-arid environment with mean monthly maximum and minimum temperature of 33°C and 10.8°C, respectively. It is characterized by uni modal low and erratic rainfall pattern with average annual rainfall of 767 mm [13]. The soil is a clay loam type with 35% sand, 28.5 and silt and 36.5% clay. The top 30 cm of the soil at the experimental site has a field capacity of 30.7%, wilting point of 15.8% and bulk density of 1.1 g/cm³ while the total available water was about 49.4.

*Corresponding author: Enchalew B, Melkassa Agricultural Research Center, Ethiopia, E-mail: sintayehulegesse@gmail.com

Materials and Methods

Experimental layout and design

The experiment was laid out in a Randomized Complete Block Design (RCBD) with three replicates. The treatments consisted of five soil moisture deficit levels, *viz.*, 90% ETc (10% deficit); 80% ETc (20% deficit); 70% ETc (30% deficit); 60% ETc (40% deficit) and 50% ETc (50% deficit) and a control treatment of 100% ETc (no deficit) (Table 1). Drip irrigation system was used for applying the required quantity of irrigation water. Each irrigation treatment consisted of three lateral lines of 5 m length. Each lateral line contained emitters spaced at 30 cm interval.

Onion variety Bombay red was raised on nursery bed and transplanted to field plots of 5 m × 2.7 m. Furrows spaced at 60 cm were used and transplanted on both side of a ridge at row and plant spacing of 20 and 10 cm, respectively. To ensure the plant establishment common irrigations was provided to all plots at two days interval before commencement of the differential irrigation. Irrigation water was applied at allowable soil moisture depletion (p=0.25) of the total available soil moisture throughout crops growth stage.

Data collection and analysis

Data collection comprised plant height, leaf number, bulb yield and yield components that include bulb diameter, marketable and unmarketable bulb yield. Water productivity and effect of water stress on crop performance were quantified from WUE and yield response factors (Ky), respectively. Estimation of WUE was carried out as a ratio of total bulb yield to the total water applied [14].

$$WUE\left(kg\ mm^{-1}\right) = \frac{Total\ bulb\ yield\left(kg\right)}{Crop\ Water\ Use\left(mm\right)}$$

Crop water use (ETc) was determined for each treatment for the growing period using the soil-water balance equation [15].

$$ET_c = I + R_e + \Delta S - D + G_e$$

Where: I is irrigation water (mm); Re is effective rainfall (mm) and ΔS is the change in soil water storage for the period (mm). D is drainage below the root zone (mm) and Ge is the groundwater contribution (mm). The contribution of D and Ge were assumed to be negligible. The ΔS were assumed the same at the beginning and at harvest and have no contribution to plant ET.

The yield response factor (Ky) was estimated from the relationship [16].

$$\left(1 - \left(\frac{Y_a}{Y_m}\right)\right) = K_y\left(1 - \left(\frac{ET_a}{ET_m}\right)\right)$$

Where,

Ya is actual harvested yield

Ym is maximum harvested yield

Ky is yield response factor

ETa is actual evapo-transpiration

ETm is maximum evapo-transpiration

The data collected during the experimental period were subjected to statistical analysis using SAS computer program. Whenever treatment effects were significant, Least Significance Differences (LSD) test was used to assess the mean difference among treatments.

Results and Discussion

Statistical analysis has shown a highly significant (P<0.01) difference for leaf number, bulb diameter, total and marketable bulb yield under the different DI treatments. However, no significant difference was observed for plant height and unmarketable bulb yield. The data on Tables 2 and 3 provide plant height, leaf number, bulb diameter, total bulb yield, marketable and unmarketable bulb yield.

Plant height

The DI did not affect much onion plant height. The plant height ranged from 38 and 46 cm. The highest and lowest plant height was observed from treatment receiving 90% and 50% ETc, respectively. However, the control treatment and treatment receiving 80% ETc gave below the average plant height.

The increasing of plant height with adequate soil moisture application is related to water in maintaining the turgid pressure of the plant cells which is the main reason for the growth [17]. In the other side the shortening of plant height under less soil moisture stress may be associated due to the closure of stomata to conserve soil moisture evaporation, this leads to reduce uptake of CO_2 and nutrient. Therefore, photosynthesis and other biochemical reactions are hindred, eventually affecting plant growth [18]. This study outcome is in line with the research that has been done by [19], indicated that soil water supply is directly proportional with plant height growth.

Leaf number

The number of leave per plant ranged from 6 to 8. The highest leaf number was recorded from treatment receiving 90% ETc. This was significantly different to all other treatments at p<0.05 level and had no significant difference with the control at P<0.01 level. Treatment receiving 60% ETc was inferior to treatment receiving 100 to 80% ETc and had no significant difference with treatment receiving 70%

	Soil moisture content/(SMC (mm)								
	Depth (cm)								
Treatment	0-30			30-45			45-60		
	Before irrig.	After irrig.	Net irrigation	Before irrig.	After irrig.	Net irrigation	Before irrig.	After Irrig.	Net irrigation
100% ETc	89.3	102	12.4	129.1	146.1	17	167.3	189	21.6
90% ETc	89.3	101	11.2	129.1	144.4	15.3	167.3	187	19.4
80% ETc	89.3	99	9.9	129.1	142.7	13.6	167.3	185	17.3
70% ETc	89.3	98	8.7	129.1	141	11.9	167.3	182	15.1
60% ETc	89.3	97	7.4	129.1	139.3	10.2	167.3	180	12.9
50% ETc	89.3	96	6.2	129.1	137.6	8.5	167.3	178	10.8
Source: Own data, 2015									

Table 1: Soil moisture content before and after irrigation events and net irrigation requirements.

Treatment	Plant height (cm)	Leaf number
100% ETc	41.2	7.33
90% ETc	46.3	8
80% Etc	41.7	7
70% Etc	43.8	6.67
60% Etc	42.9	6.33
50% Etc	38.2	6.67
Mean	42.3	7
LSD0.05	NS	0.66
0.01	NS	0.94
SE +	2.78	0.21
CV (%)	11.38	5.22
Source: Own data, 2015		

Table 2: Plant height and leaf number as influenced by DI.

Treatment	Bulb diameter (cm)	Bulb yield (kg ha^{-1})	Marketable bulb yield (kg ha^{-1})	Unmarketable bulb yield (kg ha^{-1})	Percent unmarketable yield
100% ETc	6.01	15694.4	10902.8	4791.7	31
90% Etc	5.6	13958.3	9687.5	4270.8	31
80% Etc	5.4	12604.2	7923.6	4680.5	37
70% Etc	5.21	12395.8	8263.9	4131.9	33
60% Etc	4.95	11041.7	6847.2	4194.4	38
50% Etc	4.97	10034.7	6076.4	3958.3	39
Mean	5.4	12621.5	8283.6	4337.9	35
LSD0.05	0.15	1933	2171	NS	
0.01	0.22	2749	3087	NS	
CV (%)	1.6	8.42	14.4	11.85	
SE +	0.05	613.4	688.8	296.8	

Table 3: Bulb diameter, bulb yield, marketable and unmarketable bulb yield as influenced by DI.

and 50% ETc. From the result, it can be observed that up to 20% water deficit was tolerable to obtain at least seven leaf number per plant.

Bulb diameter

The DI water applications affected the size of onion bulb. The highest bulb diameter was recorded from the control treatment and this was significantly different to all other treatments. The least bulb diameter was recorded from treatment receiving 60% ETc and this was not significantly different to treatment receiving 50% ETc. Water deficit up to 20% gave bulb diameter above the mean value of 5.4 cm. This result is in agreement with that of a study conducted by [20], high amount of soil moisture application leads to large photosynthesis area (plant height and large number of leaves), results to large bulb diameter.

Total bulb yield, marketable and unmarketable bulb yield

The total bulb yield was highest for the control treatment and this was not significantly different to treatment receiving 90% ETc. The least bulb yield was recorded from treatment receiving 50% ETc and had no significant difference with treatments receiving 60% ETc at p<0.05 level and with treatment receiving 60% to 80% at p<0.01 level. From this finding, it can be observed that with 10% water deficit to result 11% yield reduction and above 10% water deficit to result total bulb yield below mean value. A study done by [20] also presented similar findings with this result.

Highest marketable bulb yield was recorded from the control treatment and had no significant difference with 90% ETc at p<0.05 level and with 70% to 90% ETc at p<0.01 level. The lowest marketable bulb yield was observed from treatment receiving 50% ETc and had no significant difference with 60% and 80% ETc at p<0.05 level and with 60% to 80% ETc

at p<0.01 level. Here too the finding indicated that with 10% water deficit to result 11% marketable yield reduction and above 10% water deficit to result marketable bulb yield below mean value.

The highest and lowest unmarketable onion bulb yieldswere recorded from the control and 50% ETc treatments, respectively. However, the control and 90% ETc treatment gave the lowest percentage (31%) of unmarketable bulb yield and while 50% ETc treatment gave the highest percentage (39%) of unmarketable bulb yield. From the result, it can be observed that up to 30% water deficit was marginal to obtain the least unmarketable bulb yield. The results presented in this study is inclusive and similar with previous research done by [21], high soil moisture application attributes to vegetation growth and increases plant metabolic activities, which leads to marketable bulb yield increment.

Water use efficiency (WUE)

Table 4 shows the water-use efficiency, applied Water (AW), water saved for onion bulb yield. Treatment receiving 70% ETc resulted in higher WUE and saved 182.5 mm of water (Table 4). The control treatment gave practically similar WUE with 50% and 60% ETc treatments. The lowest WUE was from treatment receiving 90% and 80% ETc. However, 10% water deficit resulted 11% yield reduction and above 10% water deficit resulted bulb yields below mean value for the treatments. The difference in WUE among treatments is very small and considering the yield reduction is a limiting factor, 70% ETc application seems marginal. A study conducted by [22,23] also indicated that WUE is maximum at medium soil moisture level compare to high moisture level treatment.

Yield response factor (Ky)

Observed yield response factors (Ky) for onion bulb production ranged between 0.8 and 1.7, the lowest and highest being for 70% and 90% ETc applications, respectively. The Ky observed was decreasing as irrigation water application decreasing. Treatments receiving 60% and 50% Etc water application showed almost similar yield response factor (Table 5).

The higher Ky values indicate that the crop will have a greater yield loss when the crop water requirements are not met. Generally, the result indicates the sensitivity of the crop to soil moisture deficit. Therefore, DI practices should be avoided for Ky values that are less than unity. This conclusion is in line with a statement given by [24], the decrease in yield is proportionally greater with increase in water deficit. Considering Ky is limiting factor, 80% ETc application was a marginal and beyond that yield losses are intolerable [25]. With DI application up 20%, hence, saved 45 to 108 mm depth of water from the gross IWR of 678 mm depth of water.

Conclusion

Water is scarce resource in Central Rift Valley of Ethiopia and

Treatment	AW (mm)	Bulb yield (Kg ha^{-1})	WUE (kg mm^{-1})	Water saved (mm)	Yield reduction (%)
100%ETc	677.85	15694.4	23.2	0	0
90% ETc	632.74	13958.3	22.1	45.1	11.1
80% ETc	570.07	12604.2	22.1	107.8	19.7
70% ETc	495.4	12395.8	25	182.5	21
60% ETc	469.34	11041.7	23.5	208.5	29.6
50% ETc	429.18	10034.7	23.4	248.7	36.1

Table 4: Applied water, water use efficiency, water saved and percent yield reduction under the control and DI practices.

Treatments	ETc (mm)	Bulb yield (Kg/ha)	Yield response factor (ky)
100% Etc	677.85	15694.4	-
90% Etc	632.74	13958.3	1.7
80% Etc	570.07	12604.2	1.2
70% Etc	495.4	12395.8	0.8
60% ETc	469.34	11041.7	1
50% ETc	429.18	10034.7	1

Table 5: Crop water use and yield response factor under control and DI.

is major limiting factor for crop production. The DI practice under drip irrigation is a suitable and most efficient practice for sustainable production in water scarce area. In this study, DI application of 90% ETc gave highest plant height and leaf number. The maximum bulb diameter, total bulb yield and marketable bulb yield were observed when 100% ETc irrigation water was applied. Drip irrigation with 10% water deficit resulted about 11% yield reduction and above 10% water deficit resulted bulb yields below mean value for the treatments. The highest water productivity of onion was observed at 70% ETc water deficit irrigation application.

The yield response factor of onion ranged between 0.8 and 1.7 and the higher Ky value indicated that onion will have a greater yield loss when the critical water requirements are not met. Thus, DI practices should be avoided for Ky values that are less than unity. Considering Ky is limiting factor, 80% ETc application was a marginal and beyond that yield losses are intolerable. With DI application up 20% saved 45 to 108 mm depth of water from the gross IWR (Irrigation Water Requirement) of 678 mm depth of water.

Recommendation

Water scarcity is the major limiting factor for increased production and productivity. Water is scarce resource in Central Rift Valley of Ethiopia and is major limiting factor for crop production. Onion is one of the major economically important vegetable crops grown in this region. Therefore, DI practice under drip irrigation is a suitable and most efficient practice for sustainable production in water scarce area like Central Rift Valley of Ethiopia.

The maximum bulb diameter, total bulb yield and marketable bulb yield associated with application of 100% ETc irrigation water. However, to obtain the highest water productivity of onion, can be obtained when 70% ETc water deficit irrigation application. If yield response factor (Ky) is considered, 80% ETc application should be a marginal and beyond that yield losses are intolerable. With DI application up to 20%, 45 to 108 mm depth of irrigation water can be saved from the gross irrigation water applied as in the case of no deficit. Deficit irrigation (DI) improves water productivity and irrigation management practices resulting in water saving by maintaining soil moisture content below optimum level throughout growth season.

Acknowledgement

The authors grateful to Melkassa Agricultural Research Center (MARC, Ethiopia) Land and Water Resources Research Process section for providing financial and material support for conducting the research. Our gratitude also extends to Mr. Elias Meskelu, for his genuine advice in technical establishment of the research site.

References

1. Brewster JL (1997) Onions and Garlic. In: Wien HC (ed.) The Physiology of Vegetable Crops. CAB International, UK. pp: 581-619.

2. Doornebos J, Kassam AH (1979) Yield response to water. FAO Irrigation and Drainage, Italy. p: 33.

3. English MJ (1990) Deficit irrigation. I: Analytical framework. J Am Soc Civil Eng 116: 399-412.

4. English M, Raja SN (1996) Perspectives on deficit irrigation. Agric Water Manage 32: 1-14.

5. Kirda C, Moutonnet P, Hera C, Nielsen DR (1999) Crop yield response to deficit irrigation. Kluwer Academic Publishers, Dordrecht.

6. Kirda C (2000) Deficit irrigation scheduling based on plant growth stages showing water stress tolerance. Deficit Irrigation Practices.

7. Karam F, Breidy J, Stephan C, Rouphael Y (2003) Evapo-transpiration, yield and water use efficiency of drip irrigated corn in the bekaa Valley of Lebanon. Agric Water Manage 63: 125-137.

8. Karam F, Masaad R, Sfeir T, Mounzer O, Rouphael Y (2005) Evapo-transpiration and seed yield of field grown soybean under deficit irrigation conditions. Agric Water Manage 75: 226-244.

9. Karam F, Lahoud R, Masaad R, Daccache A, Mounzer O, et al. (2006) Water use and lint yield response of drip irrigated cotton to the length of irrigation season. Agric Water Manage 85: 287-295.

10. Kirnak H, Tas I, Kaya C, Higgs D (2002) Effects of deficit irrigation on growth, yield and fruit quality of eggplant under semi-arid conditions. Austr J Agric Res 53: 1367-1373.

11. Bekele S, Tilahun K (2007) Regulated deficit irrigation scheduling of onion in a semiarid region of Ethiopia. Agric Water Manage 89: 148-152.

12. Nigatu A, Hordofa T, Tesfaye K (2010) Effect of deficit irrigation at different growth stage of onion on bulb production in central rift valley of Ethiopia.

13. Sijali IV (2001) Drip irrigation: Options for smallholder farmers in eastern and southern Africa. World Agroforestry Centre, Kenya.

14. Central Statistics Authority (2011) Atlas of agricultural statistics, Ethiopia.

15. Molden D, Oweis T, Steduto P, BindrabanP, Hanjra MA, et al. (2010) Improving agricultural water productivity: between optimism and caution. Agric Water Manage 97: 528-535.

16. Simonne EH, Dukes MD (2010) Principles and practices of irrigation management for vegetables UF. University of Florida. pp: 17-23.

17. Doorenbos J, Pruitt WO (1977) Crop water requirements. FAO Irrigation and Drainage Paper No. 24, Rome.

18. Vaux HJ, Pruit WO (1983) Crop water production function. In: D Hillel (ed.) Advance irrigation. pp: 16-93.

19. El-Noemani AA, Aboamera AAA, Aboellil OM, Dewedar (2009) Growth, yield, quality and water use efficiency of pea plant as affected by evapotranspiration and sprinkler height. J Agric Res 34: 1445-1466.

20. Al- Moshileh A (2007) Effects of planting date and irrigation water level on onion (Allium cepa L.) production under central Saudi Arabian conditions. J Basic Appl Sci 8: 14-28.

21. Kumar S, Imtiyaz M, Kumar A (2007b) Effect of differential soil moisture and nutrient regimes on postharvest attributes of onion (Allium cepa L.). J Hortic Sci 112: 121-129.

22. Neeraja G, Reddy KM, Reddy IP, Reddy YN (1999) Effect of irrigation and nitrogen on growth, yield and yield attributes of rabi onion (Allium cepa L.) in Andhra Pradesh. Vegetable Science 26: 64-68.

23. Fabeiro C, Martin de Santa Olalla F, de Juan JA (2001) Yield and size of deficit irrigated potatoes. Agric Water Manage 48: 255-266.

24. Sarkar S, Goswami SB, Mallick S, Nanda MK (2008) Different indices to characterize water use pattern of micro-sprinkler irrigated onion (Allium cepa L.). J Agric Water Manage 95: 625-632.

25. Kadayifci A, Tuylu GI, Ucar Y, Cakmak B (2005) Crop water use of onion (Allium cepaL.) in Turkey. Agric Water Manage 72: 59-68.

30

Developed Water Stress and Crop Coefficients of Dripped-Onion Crop under Arid Conditions of Egypt

Arafa YE[1]* and Shalabi KA[2]

[1]Agricultural Engineering Department, Faculty of Agriculture, Ain Shams University, Egypt
[2]On-Farm Irrigation Engineering Department, Agricultural Engineering Research Institute, ARC, Egypt

Abstract

The aims of this study were to develop out crop and water-stress coefficients of onion crop under deficit conditions of arid regions. Therefore, field experiments of fully irrigated and deficit conditions with standard agronomic practices of onion crop had been conducted in two successive growing seasons in 2011-2012 and 2012-2013 in order to develop out the crop coefficients of onion plants at different growing stages.

Results revealed that crop coefficient values at different growing stages under different management parameter considerations indicated that the observed values of kc are ranged above and dawn of the estimated values of FAO. Meanwhile, data analysis revealed that a general trend of increasing CWU and attributed SCWR from the beginning of cultivation up to the end of bulb formation stage (72 days after sowing seeds), then it decreasing within bulb enlargement and maturity stage.

Keywords: Sandy soils; Surface drip; Subsurface drip; Water regimes; Water management

Introduction

Igbadun and Oiganji stated that the kc values of fully irrigated-onion treatments had ranged from 0.36 to 1.15, while those values of the deficit irrigated treatments varied from 0.24 to 1.13 under Samaro district, Nigeria [1]. Meanwhile, Susan et al. [2] speculated that variations of the crop water stress index (CWSI) have been used to characterize plant water stress and schedule irrigations. Usually, this thermal-based stress index has been calculated from measurements taken once daily or over a short period of time, near solar noon or after and in cloud free conditions. A method of integrating the CWSI over a day was developed to avoid the noise that may occur if weather prevents a clear CWSI signal near solar noon. This CWSI and time threshold (CWSI-TT) was the accumulated time that the CWSI was greater than a threshold value (0.45); and it was compared with a time threshold (CWSI-TT) based on a well-watered crop. Meanwhile, Mahmoud revealed that, increasing in soil moisture up to 80% of the field capacity, significantly increased the peas growth characters, a gradual decrease in all yield characters was found as the level of irrigation water decreased, so, the highest level of water supply, i.e., 80% of the field capacity gave the highest values of the above - mentioned characters [3]. The greatest value of water use efficiency was obtained at 60% of the field capacity, while 20% of the field capacity caused the lowest water use efficiency.

Locascio and Smajstrla reported that total marketable yield was almost doubled, while yield of the highest-value extra-large fruit was tripled by irrigation [4]. The four-year average yield with the 10 kPa (10 cb) treatment was largest. In the irrigated treatments, total marketable yield declined linearly as the Natron on garlic cultivated in two interties and drip irrigated in sandy soil mulched with bitumen emulation. They found that 3 days – irrigation period and 4 rows of plants/drip line gave the best results increasing total garlic yield, plant weight, tuper parameters (bulb diameter), number of cloves, air dry weight and over dry weight of garlic cloves. Also, Steduto et al. [5] and Steduro [6] and Robert [7] emphasized that irrigation scheduling simply to know when to irrigate and how much irrigation water to apply, and concluded that an effective irrigation schedule helps to maximize profit while minimizing water and energy use.

Abdel-Mawgoud stated that the highest irrigation treatment enhanced the number of pods significantly compared to lower irrigation levels [8]. Pod quality in terms of length and diameter showed also these positive responses to the increment in irrigation levels. Pod diameter increased with increasing irrigation level to 100% evaporation pan then it was reduced when irrigation level was further increased to 120%. Metin et al. [9] indicated that by increasing irrigation level from 60% and up to 100% of Et_o increased significantly the vegetative growth criteria; i.e., plant height, no. of branches, and pods/plant, leaves area, dry weight of stem and the whole plant dry matter. Moreover, irrigation of bean plants at 80% of the ET_o led to obtaining the highest significant values of pods yield/fed and pod length, compared to either 60 or 100% ET_o. Dennis and Aileen found that irrigation practices of strawberries can consistently produce up to 1.5 kilograms of well flavored marketable fruit per plant each season [10]. With this point of view, Abodi and Shati revealed that irrigation every 5 days was superior in ears numbers per plant, seeds per ears, seed weight and plant yield which reach 138.6 and 159.6 gm/plant for both spring seasons and 154.5 and 174.7 for both fall seasons, respectively [11]. Grain yield reached 8.673, 9.990 and 9.674, 10.938 ton/h. Irrigation depth 12 cm was superior in ears number per plant, seeds per ear, seed weight and plant yield which reached 116.3, 136.8 gm/plant and 142.1, 160.1 gm/plant.

The aims of this study were to develop out crop and water-stress coefficients of onion crop under deficit conditions of arid regions.

Materials and Methods

Field experiments of fully irrigated and deficit conditions with standard agronomic practices of onion crop had been conducted in two successive growing seasons in 2011-2012 and 2012-2013 in order to develop out the crop coefficients of onion plants at different growing stages.

*Corresponding author: Arafa YE, Agricultural Engineering Department, Faculty of Agriculture, Ain Shams University, Cairo, Egypt, E-mail: arafayeh11@gmail.com

Location and description of the experimental field

Experiments were carried out during two successive growing seasons (2011-2012 and 2012-2013) at a farm which located at Longitude 30° 13' E°, latitude 30° 25' N and 25.5 m above MSL. The analyses to determine physical and chemical properties of soil had been conducted according to standard methods and presented in (Tables 1 and 2).

Chemical analysis of irrigation water was carried out by using the standard methods, and presented in (Table 3).

Weather conditions and parameters during the growing season

The minimum and maximum temperature, humidity and wind speed and other meteorological parameters had been gathered and analyzed. However, reference evapotranspiration of the studied area had been gathered from Central Laboratory of Agricultural Climate (CLAC), Agriculture Research Center (ARC) for the cultivated growing seasons. Reference evapotranspiration (ET_o) was computed using the FAO, modified Penman-Monteith method is given.

The cultivated crop

The cultivated area was prepared and transplanted of onion seeds on December of each growing season. The cultivated variety of onion (*Allium Cepa L.*) was Giza 20. The sawing was done in row at plant spacing of 14.3 cm between plants. Meanwhile the spacing between plant's rows was varied according to the number of cultivated plant's rows around laterals. Therefore, the plant population density is varied, and could be summarized as shown in (Table 4). Harvesting took place on April of each successive growing season. The crop began to show signs of maturity (over 70% dropping of leave head) at 12 and 14 weeks after germination. Irrigation was withdrawn that same week and soil

moisture measurement was stopped two weeks after. Harvesting was carried cut about one week after, particularly 10th-13th April. Harvesting was done by lifting onion bulbs the dry matter using a hand hoe. The area of 3 m in long and 1 m in wide in each plot were lifted (without discards), properly labeled and taken to be to laboratory to curve for about two weeks. Therefore, the onion bulbs were separated from the dry matter and weighed. However, all agronomic practices and the rate of applications were as recommended by Vegetable Research Institute, ARC, MALR.

Experimental design and layout

Split-split-split plot design was used in this experiment. The total area of the experiment was (180 × 90 m) and divided into two main plots each was (90 × 90 m) for drip irrigation systems; every plot was divided into three sub-plots each (90 × 90 m) for drip irrigation treatments(SD, and SSD_{10}), and deficit conditions. However, each sub-plot of deficit conditions was divided into three sub-sub plots (90 × 30 m) for irrigation water regime treatments (A, B and C), and every sub-sub-plot divided into three sub-sub-sub plots (30 × 30 m) for plant density treatments (1, 2 and 3), as shown in (Figure 1).

Treatments

Drip irrigation systems

SD: surface drip irrigation system.

SSD_{10}: subsurface drip irrigation system with buried laterals at 10 cm depth.

Deficit conditions

Irrigation water regimes:

A: zero deficit treatments of actual evapotranspiration.

Soil layer, cm	Particle size distribution, %			Texture class	B. D	Moisture content by weight (%)		
	Sand	Silt	Clay		(gm/cm³)	F. C	P.W.P	A.W
0-20	94.5	3.5	2	Sandy	1.65	8.03	3.33	4.7
20-40	95	3.3	1.7	Sandy	1.56	9.13	3.14	5.99
40-60	95.7	3	1.3	Sandy	1.44	10.07	2.99	7.08
F.C=Field capacity W.P=Wilting point A.W=Available water B.D=Bulk density								

Table 1: Soil physical properties of the experimental site.

Soil layer, cm	SAR	pH	E.C, dS/m 25°c	Soluble anions, meq/l				Soluble cations, meq/l			
				CO_3^-	HCO_3^-	Cl^-	SO_4^-	Ca^{++}	Mg^{++}	Na^+	K^+
0-20	1.66	8.23	1.46	0.1	0.93	1.98	9.61	6.23	2.24	3.44	0.51
20-40	1.74	8.11	1.56	0.1	1.15	2.05	9.85	6.45	2.26	3.76	0.58
40-60	1.84	7.97	1.63	0.1	1.33	2.11	10.16	6.65	2.29	3.91	0.65
SAR=Sodium adsorption ratio			EC=Electric conductivity, ds/m								

Table 2: Soil chemical characteristics of the experimental site.

pH	E.C, dS/m at 25°C	Soluble anions, meq/l				Soluble cations, meq/l				SAR	RSC	ESP	%Ca/Na
		CO_3^-	HCO_3^-	Cl^-	SO_4^-	Ca^{++}	Mg^{++}	Na^+	K^+				
7.14	1.18	0.1	4.7	10.6	8.15	1.8	2.8	18.4	0.55	12.1	0.2	78.1	9.8
RSC=Residual sodium carbonate		ESP=Exchangeable sodium percentage											

Table 3: Chemical analysis of irrigation water at the studied site.

No. of rows	Row spacing, cm	No. of plants per m²	No. of plants per each plot
4	25	28	25200
6	16.7	46	41400
8	12.5	56	50400

Table 4: Plant population density at each treatment plot.

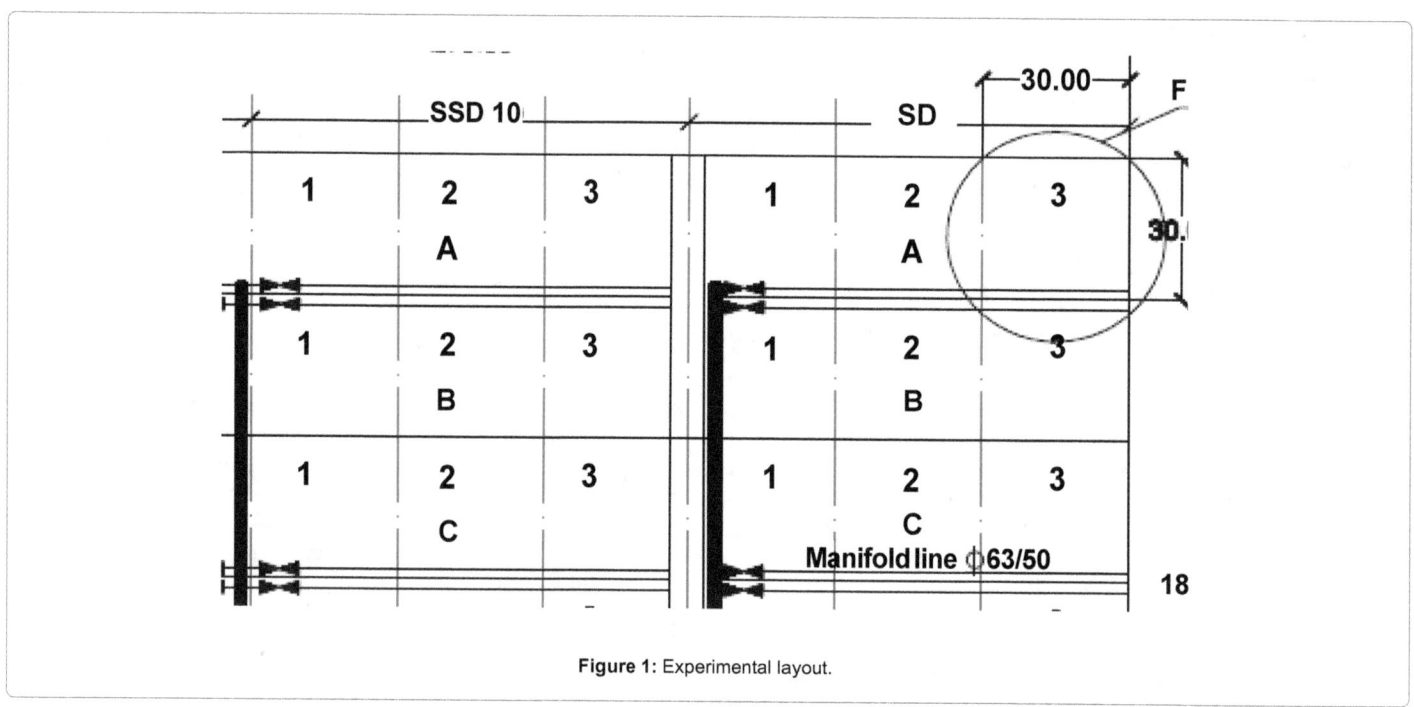

Figure 1: Experimental layout.

Growing season	Drip irrigation system	Index	Onion Growth stages/days after planting											
			Establishment (0-16)	Vegetative (17-44)				Bulb formation (45-72)				Bulb enlargement to maturity (73-101)		
			0-16	17-23	24-30	31-37	38-44	45-51	52-58	59-65	66-72	73-79	80-87	88-101
	Growth stages duration, day		16	7	7	7	7	7	7	7	7	7	7	14
2011-2012	ET_0, mm/day		2	2.1	2.4	2.4	2.4	2.55	2.6	2.8	3.25	3.25	3.45	3.45
	SD	Kc	0.68	0.69	0.76	0.75	0.78	1.04	1.14	1.06	1.05	0.93	0.68	0.47
		Ks	1.12	0.76	0.74	0.81	0.82	0.81	0.80	0.94	0.98	0.97	1.05	1.21
		CWU, mm/time of investigation	24.36	7.74	9.46	10.26	10.68	15.06	16.5	19.52	23.5	20.58	17.2	27.54
	SSD_{10}	Kc	0.5	0.59	0.66	0.7	0.87	1.02	1.06	1.09	1.02	0.93	0.79	0.51
		Ks	1.41	0.77	0.76	0.82	0.83	0.83	0.81	0.96	1.00	0.95	1.03	1.19
		CWU, mm/time of investigation	22.5	6.68	8.38	9.7	12.06	15.08	15.68	20.5	23.3	20.14	19.56	29.24
2012-2013	ET_0, mm/day		1.8	2	2.2	2.2	2.2	2.8	3.4	3.5	5.6	5.6	5.6	5.6
	SD	Kc	0.68	0.68	0.67	0.66	0.78	0.91	1.02	1	1.02	0.83	0.69	0.63
		Ks	1.00	1.00	1.00	1.00	1.00	1.00	1.00	1.00	1.00	1.00	1.00	1.00
		CWU, mm/time of investigation	19.584	9.52	10.318	10.164	12.012	17.836	24.276	24.5	39.984	32.536	27.048	49.392
	SSD_{10}	Kc	0.51	0.56	0.54	0.68	0.98	1.06	1.16	1.09	1.04	0.95	0.76	0.42
		Ks	1.25	0.54	0.54	0.48	0.46	0.42	0.28	0.41	0.40	0.37	0.43	0.38
		CCWU, mm/time of investigation	18.43	4.2	4.46	5.02	7.01	8.74	7.64	11	16.32	13.7	12.7	12.5
Average values of the two growing seasons	SD	Kc	0.68	0.685	0.715	0.705	0.78	0.975	1.08	1.03	1.035	0.88	0.685	0.55
		Ks	1.060	0.882	0.870	0.907	0.908	0.906	0.898	0.970	0.992	0.986	1.024	1.107
		CWU, mm/time of investigation	21.40	5.97	6.96	7.64	8.85	11.90	12.07	15.26	19.91	17.14	14.95	20.02
	SSD_{10}	Kc	0.505	0.575	0.6	0.69	0.925	1.04	1.11	1.09	1.03	0.94	0.775	0.465
		Ks	1.33	0.65	0.65	0.65	0.64	0.62	0.54	0.69	0.70	0.66	0.73	0.78
		CWU, mm/time of investigation	20.465	5.44	6.42	7.36	9.535	11.91	11.66	15.75	19.81	16.92	16.13	20.87

Table 5: Crop coefficient and crop water stress coeficient of onion under fully irrigated conditions, Nubaria district, Egypt.

B: 30% deficit treatments of actual evapotranspiration.

C: 60% deficit treatments of actual evapotranspiration.

Plant population densities:

A: 4 rows of plants cultivated around the lateral line.

B: 6 rows of plants cultivated around the lateral line.

C: 8 rows of plants cultivated around the lateral line.

Growing season	Drip irrigation system	Irrigation regime	plant population /m²	Index	Establishment (0-16) 0-16	Vegetative 17-23	24-30	31-37	38-44	Bulb formation 45-51	52-58	59-65	66-72	Bulb enlargement 73-79	80-87	88-101
Average values of kc according to FAO (zaytoun, 2007)					*0.5*	*0.75*				*1.05*				*0.85*		
2011-2012		ET₀, mm/day			*2*	*2.1*	*2.4*	*2.4*	*2.4*	*2.55*	*2.6*	*2.8*	*3.25*	*3.25*	*3.45*	*3.45*
	SD	Average kc of 100% of Et₀			*0.68*	*0.685*	*0.715*	*0.705*	*0.78*	*0.975*	*1.08*	*1.03*	*1.035*	*0.88*	*0.685*	*0.55*
		I_{70}	28	kc	0.64	0.67	0.66	0.66	0.78	0.91	1.02	1.00	1.02	0.83	0.69	0.63
				ks	0.99	0.89	0.64	0.94	0.98	0.82	1.06	1.10	0.79	0.89	0.89	0.41
			56	kc	0.64	0.67	0.66	0.66	0.78	0.91	1.02	1	1.02	0.83	0.69	0.63
				ks	0.99	0.89	0.64	0.94	0.98	0.82	1.06	1.10	0.79	0.89	0.89	0.41
		I_{40}	28	kc	0.47	0.49	0.46	0.5	0.54	0.68	0.94	0.91	0.88	0.64	0.52	0.54
				ks	0.77	0.70	0.52	0.71	0.81	0.63	0.65	0.69	0.53	0.66	0.68	0.27
			56	kc	0.47	0.49	0.46	0.5	0.54	0.68	0.94	0.91	0.88	0.64	0.52	0.54
				ks	0.77	0.70	0.52	0.71	0.81	0.63	0.65	0.69	0.53	0.66	0.68	0.27
	SSD₁₀	Average kc of 100% of Et₀			*0.505*	*0.575*	*0.6*	*0.69*	*0.925*	*1.04*	*1.11*	*1.09*	*1.03*	*0.94*	*0.775*	*0.465*
		I_{70}	28	kc	0.5	0.59	0.66	0.7	0.87	1.02	1.06	1.09	1.02	0.93	0.79	0.51
				ks	0.95	0.76	0.48	0.55	0.48	0.40	0.56	0.55	0.60	0.79	0.78	0.50
			56	kc	0.5	0.59	0.66	0.7	0.87	1.02	1.06	1.09	1.02	0.93	0.79	0.51
				ks	0.95	0.76	0.48	0.55	0.48	0.40	0.56	0.55	0.60	0.79	0.78	0.50
		I_{40}	28	kc	0.5	0.49	0.61	0.75	0.76	0.98	1.02	1.02	1.1	0.88	0.71	0.51
				ks	1.09	0.83	0.59	0.72	0.94	0.85	0.99	0.96	0.73	0.78	0.93	0.63
			56	kc	0.5	0.49	0.61	0.75	0.76	0.98	1.02	1.02	1.1	0.88	0.71	0.51
				ks	0.63	0.37	0.25	0.40	0.51	0.47	0.48	0.50	0.44	0.45	0.65	0.33
2012-2013		ET₀, mm/day			*1.8*	*2*	*2.2*	*2.2*	*2.2*	*2.8*	*3.4*	*3.5*	*5.6*	*5.6*	*5.6*	*5.6*
	SD	Average kc of 100% of Et₀			*0.68*	*0.685*	*0.715*	*0.705*	*0.78*	*0.975*	*1.08*	*1.03*	*1.035*	*0.88*	*0.685*	*0.55*
		I_{70}	28	kc	0.64	0.67	0.66	0.66	0.78	0.91	1.02	1.00	1.02	0.83	0.69	0.63
				ks	1.06	0.80	0.52	0.60	0.52	0.37	0.43	0.44	0.35	0.46	0.48	0.31
			56	kc	0.64	0.67	0.66	0.66	0.78	0.91	1.02	1	1.02	0.83	0.69	0.63
				ks	0.83	0.70	0.52	0.64	0.59	0.41	0.44	0.48	0.35	0.52	0.55	0.25
		I_{40}	28	kc	0.47	0.49	0.46	0.5	0.54	0.68	0.94	0.91	0.88	0.64	0.52	0.54
				ks	0.64	0.55	0.43	0.48	0.48	0.31	0.27	0.30	0.23	0.38	0.42	0.17
			56	kc	0.47	0.49	0.46	0.5	0.54	0.68	0.94	0.91	0.88	0.64	0.52	0.54
				ks	0.74	0.39	0.36	0.65	0.79	0.62	0.40	0.45	0.32	0.36	0.54	0.19
	SSD₁₀	Average kc of 100% of Et₀			*0.505*	*0.575*	*0.6*	*0.69*	*0.925*	*1.04*	*1.11*	*1.09*	*1.03*	*0.94*	*0.775*	*0.465*
		I_{70}	28	kc	0.5	0.59	0.66	0.7	0.87	1.02	1.06	1.09	1.02	0.93	0.79	0.51
				ks	0.82	0.54	0.44	0.52	0.49	0.41	0.40	0.39	0.34	0.43	0.52	0.39
			56	kc	0.5	0.59	0.66	0.7	0.87	1.02	1.06	1.09	1.02	0.93	0.79	0.51
				ks	0.82	0.54	0.44	0.52	0.49	0.41	0.40	0.39	0.34	0.43	0.52	0.39
		I_{40}	28	kc	0.5	0.49	0.61	0.75	0.76	0.98	1.02	1.02	1.1	0.88	0.71	0.51
				ks	0.82	0.66	0.48	0.49	0.56	0.42	0.42	0.42	0.32	0.45	0.57	0.39
			56	kc	0.5	0.49	0.61	0.75	0.76	0.98	1.02	1.02	1.1	0.88	0.71	0.51
				ks	0.47	0.37	0.28	0.28	0.32	0.24	0.24	0.24	0.18	0.26	0.33	0.22

Table 6: Developed crop coefficient and crop water stress of onion under deficit conditions, Nubaria district, Egypt.

Measurement and Calculations

Computation of crop coefficient under deficit irrigation treatments (k_c)

The crop coefficient (K_C) for the field the irrigated treatments were determined on weekly basis, according to FAO publication and other references under the same conditions [12]. Meanwhile, the crop coefficient of the deficit irrigation treatments (referred to as $K_{c_{deficit}}$) was computed as a ratio of the average daily crop water use of deficit irrigated treatments to the average daily ET_o for the week.

$$Kc_{deficit} = \frac{CWU_{deficit}}{ET_o}$$

Computation of water stress coefficient (K_s)

The water stress coefficient (K_s) integrate the crop and soil factors that make the actual crop water use of the deficit irrigated conditions differ from crop use under fully irrigated conditions. The relationship was exposed by Allen. The values of K_c, $Kc_{deficit}$, and K_s for the four growth stages of onion crop were computed by finding the average of the weekly coefficients values for the growth stages. The water stress coefficients can be classified on the basis of its impact on seasonal consumptive use as (Critical: $0.1 < K_s \leq 0.5$, Severe: $0.5 < K_s \leq 0.75$, Moderate: $0.75 < K_s \leq 0.9$, Minor: $0.9 < K_s \leq 0.99$).

Computation of onion crop water use (CWU) and seasonal crop water requirements (SCWR)

Onion plant-water requirements were calculated and scheduled

Growing season	Drip irrigation system	Irrigation water regime, day	Establishment (0-16) 16	Vegetative 17-23 (7)	24-30 (7)	31-37 (7)	38-44 (7)	Bulb formation 45-51 (7)	52-58 (7)	59-65 (7)	66-72 (7)	Bulb enlargement to maturity 73-79 (7)	80-87 (7)	88-101 (14)	Accumulative CWU, mm/ growth seasn	Leaching requirements, 10%	Irrigation requirements, mm/ed	SCWR, m³/fed
2011-2012	ET_0, mm/day		2	2.1	2.4	2.4	2.4	2.55	2.6	2.8	3.25	3.25	3.45	3.45				
	SD	I_{100}	24.36	7.74	9.46	10.26	10.68	15.06	16.5	19.52	23.5	20.58	17.2	27.54	202.40	20.24	247.38	1038.99
	SD	I_{70}	17.05	5.42	6.62	7.18	7.48	10.54	11.55	13.66	16.45	14.41	12.04	19.28	141.68	14.17	173.16	727.29
	SD	I_{40}	9.744	3.096	3.784	4.104	4.272	6.024	6.6	7.808	9.4	8.232	6.88	11.016	80.96	8.10	98.95	415.59
	SSD_{10}	I_{100}	22.5	6.68	8.38	9.7	12.06	15.08	15.68	20.5	23.3	20.14	19.56	29.24	202.82	20.28	247.89	1041.14
	SSD_{10}	I_{70}	15.75	4.68	5.87	6.79	8.44	10.56	10.98	14.35	16.31	14.10	13.69	20.47	141.97	14.20	173.52	728.80
	SSD_{10}	I_{40}	9	2.672	3.352	3.88	4.824	6.032	6.272	8.2	9.32	8.056	7.824	11.696	81.13	8.11	99.16	416.46
2012-2013	ET_0, mm/day		1.8	2	2.2	2.2	2.2	2.8	3.4	3.5	5.6	5.6	5.6	5.6				
	SD	I_{100}	19.584	9.52	10.318	10.164	12.012	17.83	24.276	24.5	39.984	32.536	27.048	49.392	277.16	27.72	338.76	1422.78
	SD	I_{70}	13.71	6.66	7.22	7.11	8.41	12.48	16.99	17.15	27.99	22.78	18.93	34.57	194.01	19.40	237.13	995.94
	SD	I_{40}	13.7088	6.664	7.2226	7.1148	8.4084	12.481	16.9932	17.15	27.9888	22.7752	18.9336	34.5744	194.01	19.40	237.13	995.94
	SSD_{10}	I_{100}	18.43	4.2	4.46	5.02	7.01	8.74	7.64	11	16.32	13.7	12.7	12.5	121.72	12.17	148.77	624.83
	SSD_{10}	I_{70}	12.90	2.94	3.12	3.51	4.91	6.12	5.35	7.70	11.42	9.59	8.89	8.75	85.20	8.52	104.14	437.38
	SSD_{10}	I_{40}	7.372	1.68	1.784	2.008	2.804	3.496	3.056	4.4	6.528	5.48	5.08	5	48.69	4.87	59.51	249.93
Average values of the two growing seasons	ET_0, mm/day		1.9	2.05	2.3	2.3	2.3	2.675	3	3.15	4.425	4.425	4.525	4.525				
	SD	I_{100}	21.97	8.63	9.89	10.21	11.35	16.45	20.39	22.01	31.74	26.56	22.12	38.47	239.78	23.98	293.07	1230.88
	SD	I_{70}	15.38	6.04	6.92	7.15	7.94	11.51	14.27	15.41	22.22	18.59	15.49	26.93	167.85	16.78	205.15	861.62
	SD	I_{40}	11.73	4.88	5.50	5.61	6.34	9.25	11.80	12.48	18.69	15.50	12.91	22.80	137.49	13.75	168.04	705.77
	SSD_{10}	I_{100}	20.465	5.44	6.42	7.36	9.535	11.91	11.66	15.75	19.81	16.92	16.13	20.87	162.27	16.23	198.33	832.99
	SSD_{10}	I_{70}	14.3255	3.808	4.494	5.152	6.6745	8.337	8.162	11.025	13.867	11.844	11.291	14.609	113.59	11.36	138.83	583.09
	SSD_{10}	I_{40}	8.186	2.176	2.568	2.944	3.814	4.764	4.664	6.3	7.924	6.768	6.452	8.348	64.91	6.49	79.33	333.19

Table 7: Crop water use (CWU) and seasonal crop water requirements (SCWR) of onion under deficit irrigation conditions, Nubaria district, Egypt.

according to investigated level of treatments. ET_o data were processed by using CropWat 8.1 model, for all calculation. The irrigation water requirements at each irrigation events under different onion plants growing stage were computed.

Results and Discussion

Effect of drip irrigation systems on water stress coefficient

Data are illustrated in (Tables 5 and 6) indicated that the highest significant effects were due to the plant population density and drip irrigation systems.

Regarding the k_s response to the investigated drip irrigation systems. Data revealed that, the lowest k_s values at bulb formation stage were 0.57 and 0.59 under SSD_{10} and SD respectively. Based on the k_s classification it could be concluded that the water stress coefficients were minor except at vegetative growth stage of onion plants, wherever, it had moderate effect. The seasonal k_s values of I_{70} and I_{40} had been ranged from moderate to severe under SD compared with SSD_{10}. However, K_s values were noticed to have similar trends for agronomic management parameters and drip irrigation systems SD and SSD considerations that imply the impacts of deficit irrigation condition were consistent irrespective of cropping methodologies and onion crop variety. The abovementioned results are agreed with that had been observed by Moustafa.

Crop coefficient as response to deficit irrigation conditions

Results revealed that there is no significant effect on the onion crop coefficients due to plant population density; however, the highly significant effect was due to the irrigation water regimes treatments. This observation is in agreements with other related works. On the other hand, data indicated that the specific amounts of the applied irrigation water and its distribution patters within soil profile had a significant effect of the crop coefficient within specific growing stages. Moreover, data analysis of crop coefficient values at different growing stages under different management parameter considerations indicated that the observed values of k_c are ranged above and dawn of the estimated values of FAO. This may be indicated that several effects had to be taken into consideration regarding the improving of on-farm irrigation water uses under different domestic macro-climatic. With this point of view, data analysis indicated that the highest k_c values were obtained under I_{100} followed by I_{70} and I_{40} respectively. These data are in agreement with Igbadun and Oiganji [1].

Crop water use (CWU) and seasonal cop water requirements (SCWR) of onion

Data illustrated in (Table 7) showed that the average daily (CWU) and seasonal crop water requirements (SCWR) within the two successive growing seasons, under deficit irrigation conditions. A general trend of increasing CWU and attributed SCWR from the beginning of cultivation up to the end of bulb formation stage (72 days after sowing seeds), then it decreasing within bulb enlargement and maturity stage. This is normally observation due to the crop water requirements and the change of micro-climate factors and attributed reference evapotranspiration, as well as, changes of either crop coefficient or crop water stress coefficient. In addition, from data analyses it could be noticed that, the highest values of CWU reduced under establishment, vegetative growth, bulb fermentation and bulb enlargement and maturity stages respectively. These observations are in agreement with Igbadun and Oiganji [1] and Fereres and David [13].

Conclusion

Finally study concludes that crop coefficient values at different growing stages under different management parameter considerations indicated that the observed values of Kc are ranged above and dawn of the estimated values of FAO. Moreover, data analysis revealed that a general trend of increasing CWU and attributed SCWR from the beginning of cultivation up to the end of bulb formation stage (72 days after sowing seeds), then it decreasing within bulb enlargement and maturity stage.

References

1. Igbadun HE, Oiganji E (2012) Crop coefficients and yield response factors of onion (Allium Cepa. L) under deficit irrigation and mulch practices in Samaru. J of Ag Research 7: 5137-5152.

2. Susan AO, Steven RE, Paul DC, Terry AH (2012) A crop water stress index and time threshold for automatic irrigation scheduling of grain sorghum. Agricultural Water Management 107: 122-132.

3. Mahmoud (2000) Effect of irrigation on growth and yield of pea (Pisum sativum L) Horticulture Science (Vegetable Crops). Department of Horticulture. Minufiya University.

4. Locascio SJ, Smajstrla AG (2013) Tensiometer-controlled, drip-irrigation scheduling of tomato. American Society of Agricultural Engineers 12: 315-319.

5. Steduto P, Hsiao TC, Fereres E (2007) On the conservative behaviour of biomass water productivity. Irrig Sci 25: 189-207.

6. Steduto P (2003) Biomass water-productivity. comparing the growth-engines of crop models. FAO Expert Consultation on Crop Water Productivity Under Deficient Water Supply. Italy.

7. Robert E, Cassel DK, Sneed RE (1996) Soil, water, and crop characteristics important to irrigation sceduling. North Carolina Cooperative Extension Service.

8. Metin S, Yazar SA, Canbolat M, Eker S, Felike G (2005) Effect of drip irrigation management on yield and quality of field grown green beans. Agric Water Management 71: 243-255.

9. Dennis Ph, Aileen R (2008) Irrigation and fertilizer guidelines for strawberries on sandy soils. Department of Agriculture and Food Government of Westren Australia.

10. Abodi HMK, Shati RK (2010) Role of irrigation depth and frequency in growth and yield of maize. The Iraqi Journal of Agricultural Sciences 41: 40-47.

11. Zayton AM (2007) Effect of soil-water stress on onion yield and quality in sandy soil. Misr J of Ag Eng 24: 141-160.

12. Fereres ES, David G (2002) Suitability of stem diameter variations and water potential as indicators for irrigation scheduling of almond trees. Journal of Horticultural Science & Biotechnology 78: 139-144.

13. Abdel-Maywood AMR (2006) Growth, yield and quality of green bean (Phaseolus vulgaris L.) in response to irrigation and compost applications. J of Applied sciences Research 2: 443-450.

PERMISSIONS

LIST OF CONTRIBUTORS

HE Igbadun
Department of Agricultural Engineering, Ahmadu Bello University, P.M.B. 1044, Zaria, Kaduna State, Nigeria

BA Salim
Department of Agricultural Engineering and Land Planning, Sokoine University of Agriculture, P.O. Box 3003, Chuo Kikuu, Morogoro, Tanzania

Greshishchev V, Onikura N and Iyooka H
Fishery Research Laboratory, Kyushu University, Tsuyazaki, Fukutsu, Fukuoka, Japan

Pradip Adhikari
Postdoctoral Fellow, Dept. of Soil Environment and Atmospheric Sciences, Univ. of Missouri, Columbia, Missouri, USA

Manoj K. Shukla and John G. Mexal
Dept. of Plant and Environmental Sciences, New Mexico State University, New Mexico 88003-8003, USA

David Daniel
Dept. of Economics and International Business, New Mexico State University, Las Cruces, New Mexico 88003-8003, USA

Ifabiyi JO and Komolafe SE
Department of Agricultural Extension and Rural Development, University of Ilorin, Ilorin, Nigeria

Etissa E and Alemayehu Y
Ethiopian Institute of Agricultural Research, Addis Ababa, Etiopia

Dechassa N
College of Agriculture and Environmental Sciences, Haramaya University, Etiopia

Xin Liu, Abdolmajid Mohammadian and Julio Angel Infante Sedano
Dept. of Civil Engineering, University Of Ottawa, Ottawa, Canada

Hatiye SD, Hari Prasad KS, Ojha CSP and Kaushika GS
Department of Civil Engineering, Indian Institute of Technology, Roorkee, India

Adeloye AJ
Institute for Infrastructure and Environment, Heriot-Watt University, Edinburgh, UK

Tewedros Fikre Zenebe and Yasir Mohamed
Faculty of Civil Engineering and Applied Geosciences, Stevinweg 1, 2600 GA Delft, The Netherlands

Haile AM
UNESCO-IHE and Spate irrigation Network, P.O. Box 3015, 2601 DA Delft, the Netherlands

Mohammad AZ
Department of Basic Science, Faculty of Marine science, Chabahar Maritime University, Iran

Mohan S and Ramsundram N
Professor, Environmental and Water Resources Engineering Division, Department of Civil Engineering, Indian Institute of Technology Madras, Chennai, India

A.H. AbdelGadir
Biosystems Engineering Department, Auburn University, Research Fellow, 200 Corley Bldg, Auburn, AL 36849-5417

M. Dougherty
Biosystems Engineering Department, Auburn University, Associate Professor, 200 Corley Bldg, Auburn, AL 36849-5417

J.P. Fulton
Biosystems Engineering Department, Auburn University, Associate Professor, 200 Corley Bldg, Auburn University, AL 36849-5417

L.M. Curtis
Biosystems Engineering Department, Auburn University, Emeritus Professor, 200 Corley Bldg, Auburn, AL 36849-5417

T. W. Tyson
Biosystems Engineering Department, Auburn University, Professor, 200 Corley Bldg., Auburn, AL 36849-5417

H.D. Harkins
Tennessee Valley Research & Extension Center, Associate Director, P.O. Box 159, Belle Mina, AL 35615

B.E. Norris
Tennessee Valley Research & Extension Center, Director, P.O. Box 159, Belle Mina, AL 35615

Nithya KB and Shivapur AV
Water and Land Management, Visvesvaraya Technological University, Belagavi, India

Akhter M, Ali M, Haider Z, and Saleem U
Rice Research Institute, Kala Shah Kaku, Lahore, Pakistan

Mahmood A
Ayub Agricultural Research Institute, Faislabad, Pakistan

Mike Rowan and Karen M. Mancl
Department of Food, Agriculture and Biological Engineering, Ohio State University, 590 Woody Hayes Dr., Columbus, OH 43210, USA

Olli H. Tuovinen
Department of Microbiology, Ohio State University, 484 West 12th Avenue, Columbus, OH 43210, USA

Kassahun Tadesse Birhanu, Tena Alamirew, Megerssa Dinka Olumana, Semu Ayalew and Dagnachew Aklog
Natural Resources Management, Debre Markos University, Ethiopia

K N Tiwari, Mukesh Kumar, Santosh D T, Vikas Kumar Singh, M K Maji and A K Karan
Precision Farming Development Centre, Agricultural & Food Engineering Department, Indian Institute of Technology, Kharagpur- 721 302(W.B), India

Engel B and Bralts V
Department of Agricultural and Biological Engineering, Purdue University, USA

Radwan S and Rashad M
Department of Agricultural Engineering, Suez Canal University, Egypt

Hashem A
Department of Agricultural and Biological Engineering, Purdue University, USA
Department of Agricultural Engineering, Suez Canal University, Egypt

Oliveira FC, Lavanholi R, Camargo AP and Frizzone JA
Department of Biosystems Engineering, Luiz de Queiroz College of Agriculture, University of São Paulo, Piracicaba, São Paulo, Brazil

Ait-Mouheb N, Tomas S and Molle B
National Research Institute of Science and Technology for Environment and Agriculture, UMR G-EAU Montpellier, France

Karatasiou E, Papanikolaou C and Makrantonaki MS
Laboratory of Agricultural Hydraulics, Department of Agriculture, Crop Production and Rural Environment, University of Thessaly, Volos, Greece

Muktar BY and Yigezu TT
Teppi National Spices Research Center, Teppi, Ethiopia

Zerihun D, Subramanian J and Badaruddin M
Maricopa Agricultural Center, University of Arizona, Maricopa, USA

Sanchez CA
Departments of Soil, Water and Environmental Science and Maricopa Agricultural Center University of Arizona, Maricopa, USA

Bronson KF
USDA-ARS Arid-Land Agricultural Research Center, Maricopa, USA

Arafa YE
Agricultural Engineering Department, Faculty of Agriculture, Ain Shams University, Cairo, Egypt

Muhammad Hasan, Yanjun Shang and Weijun Jin
Institute of Geology and Geophysics, University of the Chinese Academy of Sciences, Beijing, China

Gulraiz Akhter
Department of Earth Sciences, Quaid-i-Azam University, Islamabad, Pakistan

Gemechis T
Soil and Water Engineering Research Case Team, Bako Agricultural Engineering Research Center, Oromia, Ethiopia

Quraishi SH
School of Natural Resource and Environmental Engineering, Institute of Technology, Haramaya University, Ethiopia

Zeleke T
Soil and Water Management Research Case Team, Research and Development, Sugar Corporation, Ethiopia

Oshunsanya SO, Aiyelari EA, Aliku O and Odekanyin RA
Department of Agronomy, University of Ibadan, Nigeria

Etissa E
Ethiopian Agriculture Research Organization, Addis Ababa, Ethiopia

Dechassa N and Alemayehu Y
Haramaya University, Ethiopia

Xiaojun S, Jiyang Z, Zugui L and Jingsheng S
Key Laboratory of Crop Water Use and Regulation, Ministry of Agriculture, Farmland Irrigation Research Institute, Chinese Academy of Agricultural Sciences, Xinxiang, China

Mingsi L
Water Conservancy and Architectural Engineering College, Shihezi University, Shihezi, China

Guisen Y
Irrigation Experiment Station of Water-conservancy Bureau in Xinjiang Production and Construction Crops, Urumqi, China

Nahusenay Teamer Gebrehiwot
Department of Cooperative Studies, College of Business and Economics, Mekelle University, Ethiopia

Kassa Amare Mesfin
Department of Earth Science, College of Natural and Computational Science, Mekelle University, Ethiopia

Jan Nyssen
Department of Geography, Ghent University, Belgium

Enchalew B
Melkassa Agricultural Research Center, Ethiopia

Gebre SL, Rabo M, Hindaye B, Kedir M, Musa Y and Shafi A
Department of Natural Resources Management, Jimma University, Ethiopia

Arafa YE
Agricultural Engineering Department, Faculty of Agriculture, Ain Shams University, Egypt

Shalabi KA
On-Farm Irrigation Engineering Department, Agricultural Engineering Research Institute, ARC, Egypt

Index

www.ingramcontent.com/pod-product-compliance
Lightning Source LLC
Chambersburg PA
CBHW080413190526
45161CB00003B/231